African Ethnobotany in the Americas

Robert Voeks • John Rashford
Editors

African Ethnobotany in the Americas

Springer

Editors
Robert Voeks
Department of Geography
California State University
Fullerton, CA, USA

John Rashford
Department of Sociology and Anthropology
College of Charleston
Charleston, SC, USA

ISBN 978-1-4614-0835-2 ISBN 978-1-4614-0836-9 (eBook)
DOI 10.1007/978-1-4614-0836-9
Springer New York Heidelberg Dordrecht London

Library of Congress Control Number: 2012944256

Printed on acid-free paper

Springer is part of Springer Science+Business Media (www.springer.com)

Contents

Author Biographies

Stanley B. Alpern served with the US Navy in World War II. He received a Harvard A.B. in 1947 and a Columbia M.A. in 1950. He spent 6 years as a copyreader with the *New York Herald Tribune* and then served with the US Information Agency for 22 years. Among his foreign posts were Abidjan, Côte d'Ivoire, and Ouagadougou, Burkina Faso. For the last three decades, he has been researching and writing about precolonial West African history. His scholarly articles have included "The European introduction of crops into West Africa in precolonial times," *History in Africa*, 19 (1992), 13–43, and "Exotic plants of Western Africa: where they came from and when," *HA*, 35 (2008), 63–102. He has also written *Amazons of Black Sparta: the women warriors of Dahomey* (London & New York, 1998), new ed. 2011, and *Guide to original sources for precolonial West Africa* (Madison, 2006). He is a consultant to the Oxford English Dictionary for words deriving from West Africa, and he lives on the French Riviera.

Tinde van Andel received her Ph.D. degree in 2000 from Utrecht University and now works as a postdoc at the National Herbarium of the Netherlands (NCB Naturalis, Leiden University). She has published several scientific papers on traditional plant use in Guyana and Suriname and two illustrated field guides: Non-timber forest products of the North-West district of Guyana (Tropenbos, 2000) and Medicinal and ritual plants of Suriname (KIT Publishers, 2011). Her current research focuses on similarities between plant use in West Africa and the Caribbean.

Dorothea Bedigian received her Ph.D. in Agronomy at the University of Illinois. She investigates traditional sesame cultivars and the natural history of wild species of *Sesamum*, as well as peoples who use them, in Africa and Asia. Her approach is interdisciplinary, blending agronomy with anthropology, archaeology, botany, chemistry, genetics, geography, history, and linguistic analysis. Her current projects are a taxonomic revision of the genus *Sesamum* and an in-depth review of sesame and subsistence in Africa, historically, to the present time.

Judith Carney is professor of Geography at UCLA. She teaches courses on African ecology and development, Africa and the African diaspora, and comparative food and environmental systems. She is the author of more than 60 research articles and two award-winning books: *Black Rice: The African Origins of Rice Cultivation in the Americas* (Harvard University Press, 2001) and with Richard Rosomoff, *In the Shadow of Slavery: Africa's Botanical Legacy in the Atlantic World* (University of California Press, 2009). *Black Rice* was the cowinner of the Melville Herskovits Book Award (2002) from the African Studies Association and the James M. Blaut Publication Award (2003) from the Association of American Geographers. *In the Shadow of Slavery* is the cowinner of the Frederick Douglass Book Prize (2010). Professor Carney is a recipient of research fellowships from the John Simon Guggenheim Memorial Foundation, the American Council of Learned Societies, the Wenner-Gren Foundation, and the Rockefeller Foundation.

Maria Fadiman received her Ph.D. from the University of Texas at Austin and is currently an associate professor in the department of Geosciences at Florida Atlantic University. Her research focus is on how people use plants in the developing world in relation to conservation and sustainability.

Cari Goetcheus is an associate professor of Historic Preservation at Clemson University. She received her Master's in Historic Preservation from the University of Georgia and her Bachelors of Landscape Architecture from Utah State University. Her research explores the intersection of planning, preservation, and landscape architecture, with a focus on African American communities and their landscapes.

Brian Grabbatin is a doctoral candidate in the Department of Geography at the University of Kentucky. His research focuses on the relationship between property regimes and natural resource management. He is currently working on a dissertation that explores racialized landscapes and landownership in the South Carolina Lowcountry.

Sara Groenendijk contributed to this research in Suriname as a student of the M.Sc. Natural Resources Management (Utrecht University, the Netherlands) in 2006–2007. This research triggered her interest in the field of international development cooperation. Today, she uses her interdisciplinary background in her work, advising governments of developing countries on their environmental policy and practice.

Angela Halfacre is an associate professor with the Department of Political Science and Department of Earth and Environmental Sciences at Furman University. She earned her Ph.D. in Political Science from the University of Florida. Her research examines public perceptions of sustainability issues and conservation approaches. In examining community governance and natural resource decision-making and management, she incorporates an environmental justice framework.

Bruce Hoffman has spent much of the past 20 years working as a field botanist and ethnobotanist in the Guiana Shield region of northeastern South America. He received his Ph.D. in Botany from the University of Hawai'i-Manoa in 2009. He is currently working as honorary research staff at the Netherlands National Herbarium – Leiden University. His research is focused upon cross-cultural cognitive ethnobotany, the biology and use of non-wood forest products, and advances in field identification of tropical plants.

Patrick T. Hurley is an assistant professor of Environmental Studies at Ursinus College. He earned his Ph.D. in Environmental Science, Studies and Policy from the University of Oregon. His research focuses on human-environment interactions, drawing on insights from political ecology and conservation science. In examining non-timber forest products in cities and peri-urban areas, Hurley specifically examines the ways that urbanization transforms the use and management of local ecological systems.

Erica S. Moret has a Ph.D. in Geography from Jesus College, University of Oxford. Her research entailed a multi-scaled analysis of US economic sanctions on Cuba in the post-Soviet era, combining human geography, political economy, geopolitics, and environmental science.

Sonia Peter is Head of the Departments of Chemistry and Environmental Science at the Barbados Community College, Barbados, West Indies. She received her B.S. in Chemistry from the University of the West Indies, Cave Hill Campus, and after a long stint of teaching graduated with her Ph.D. in Natural Products Chemistry in 1994. Her areas of interest include organic chemistry and structure elucidation of natural products from medicinal plants.

Kobeke Van de Putte spent three months in the Saramaccan village Brownsweg (Suriname) for her Master's thesis, where she conducted an ethnobotanical household survey. She has contributed to several scientific papers on traditional plant use in Suriname, and a field guide of medicinal and ritual plants of Suriname (KIT Publishers, 2011). She is currently working at the mycology research group of Ghent University, Belgium, focusing on the taxonomy and systematics of the worldwide species complex of the fishy milkcaps (*Lactifluus volemus* sensu lato).

John Rashford received his Ph.D. in Anthropology from the City University of New York and is currently professor of Anthropology at the College of Charleston in Charleston, South Carolina. His research focuses on the ethnobotany of the Caribbean.

Dale Rosengarten has been studying the tradition of African American coiled basketry since 1984. As guest curator for McKissick Museum at the University of South Carolina, she developed the exhibition and catalog *Row Upon Row: Sea Grass Baskets of the South Carolina Lowcountry* (1986), which enjoyed a 20-year run as a traveling show. Her doctoral dissertation (Harvard University, 1997) placed the Lowcountry basket in a global setting and led to a partnership with the Museum for African Art in New York. With cocurator Enid Schildkrout, she produced the exhibit and book *Grass Roots: African Origins of an American Art* (2008). Pursuing her other field of research—southern Jewish history and culture—Rosengarten founded the Jewish Heritage Collection at the College of Charleston Library. In collaboration with McKissick Museum, she curated the exhibition *A Portion of the People: Three Hundred Years of Southern Jewish Life* (2002) and, with her husband, Theodore Rosengarten, coedited the accompanying volume.

Sofie Ruysschaert is a plant ecologist and ethnobotanist affiliated with the Laboratory of Tropical and Subtropical Agriculture and Ethnobotany at Ghent University, Belgium. She conducted fieldwork in Suriname between 2004 and 2006. In the near future, she hopes to publish her Ph.D. thesis on the use of non-timber forest products by Arawak Indians and Saramaccan Maroons. Currently, she works as a biologist for the Flemish government, but continues to be involved in ethnobotanical field projects.

James Sera received a Ph.D. in Anthropology at the University of Wisconsin. He now resides in Los Angeles, California, where he lectures at California State University, Los Angeles. His research interests include Brazil, African diaspora, race and color, Capoeira, religion, social hierarchy, politics, ethnomusicology, and ethnobotany. He is a Contra-Mestre of Capoeira Angola for the Associação de Capoeira Angola Corpo e Movimento and director of the Brazilian Cultural Foundation, a nonprofit organization based in Los Angeles.

Robert Voeks received his Ph.D. in Geography from the University of California, Berkeley. He is currently a professor of Geography at California State University, Fullerton, and Editor-in-Chief of the journal *Economic Botany*. He is author of numerous publications on tropical ethnobotany, biogeography, and the African diaspora, including *Sacred Leaves of Candomblé: African Magic, Medicine, and Religion in Brazil* (Univ. of Texas Press, 1997).

Contributors

Stanley B. Alpern, A.B., M.A. Villefranche-sur-Mer, France

Tinde van Andel, Ph.D. Netherlands Center for Biodiversity Naturalis (Section National Herbarium of the Netherlands), Leiden University, Leiden, the Netherlands

Dorothea Bedigian, Ph.D. Research Associate, Missouri Bot Gard, St. Louis, MO, USA

Judith Carney, Ph.D. Department of Geography, University of California, Los Angeles, CA, USA

Maria Fadiman, Ph.D. Department of Geosciences, Florida Atlantic University, Boca Raton, FL, USA

Cari Goetcheus, M.H.P., B.L.A. College of Environment and Design, University of Georgia, Athens, GA, USA

Brian Grabbatin, Ph.D. Department of Geography, University of Kentucky, Lexington, KY, USA

Sara Groenendijk, M.Sc. Social Sciences Group, Wageningen University, Wageningen, the Netherlands

Angela Halfacre, Ph.D., M.A. Department of Political Science and Sustainability Science, Furman University, Greenville, SC, USA

Bruce Hoffman, Ph.D. National Herbarium of the Netherlands, Naturalis Biodiversity Center, RA, Leiden, the Netherlands

Patrick T. Hurley, Ph.D., M.S. Environmental Studies Program, Collegeville, PA, USA

Erica S. Moret, Ph.D., M.Sc., B.Sc. Jesus College, Oxford University Centre for the Environment, University of Oxford, Oxford, UK

Sonia Peter, B.Sc., Ph.D. Department of Chemistry and Environmental Science, Division of Natural Sciences, Barbados Community College, Bridgetown, Barbados

Kobeke Van de Putte, M.Sc. Department of Biology, Research Group Mycology (Spermatophytes), Ghent University, Ghent, Belgium

John Rashford, M.A., Ph.D. Department of Sociology and Anthropology, College of Charleston, Charleston, SC, USA

Dale Rosengarten, Ph.D. Marlene and Nathan Addlestone Library, Special Collections, College of Charleston, Charleston, SC, USA

Sofie Ruysschaert, M.Sc. Laboratory of Tropical and Subtropical Agriculture and Ethnobotany, Ghent University, Ghent, Belgium

James Sera, Ph.D. Department of Anthropology, California State University-Los Angeles, USA

Robert Voeks, Ph.D. Department of Geography, California State University–Fullerton, Fullerton, CA, USA

Chapter 1
Introduction

Robert Voeks and John Rashford

This collection explores the ethnobotany of the African diaspora. Drawing on papers presented at the 50th Annual Society for Economic Botany Meeting Symposium "African Ethnobotany in the Americas," held in 2009 in Charleston, South Carolina, as well as contributions by other leading scholars on the subject, it explores the complex relationship between plant use and meaning, historical and current, among African-descended populations in North America, the Caribbean, and South America. A number of sophisticated studies have been carried out that support Herskovits' pioneering research on the African heritage of the Americas, but with several notable exceptions the reader will encounter in this volume, they have largely overlooked the importance of plants in African cultures and in the African diaspora. This collection is offered as a small contribution to filling this gap. It presents various aspects of the botanical legacy of Africans who played a significant part in the making of the modern world (Thornton 1992).

Rooted in European colonial expansion, the field of economic botany, the forebear of currently ethnobotany, has a rich historical and scientific tradition. Indeed, the objectives of the colonial enterprise were as much about the quest for "green gold" as it was for precious metals and sources of labor. Newly encountered useful species—cinchona, guiacum, vanilla, jalapa, and myriad others—as well as novel discoveries regarding the biogeography and utility of long sought-after medicinal spices and flavorings, cinnamon, black pepper, nutmeg, and turmeric, to list a few, enriched the purses of private entrepreneurs and the coffers of colonial powers (Barrera 2002; Schiebinger 2004;

R. Voeks, Ph.D. (✉)
Department of Geography, California State University—Fullerton,
800 N. State College Blvd, Fullerton, CA 92834, USA
e-mail: rvoeks@fullerton.edu

J. Rashford, Ph.D., M.A.
Department of Sociology and Anthropology, College of Charleston,
66 George St., Charleston, SC 29455, USA
e-mail: RashfordJ@cofc.edu

R. Voeks and J. Rashford (eds.), *African Ethnobotany in the Americas*,
DOI 10.1007/978-1-4614-0836-9_1, © Springer Science+Business Media New York 2013

Spary 2000). Other species were introduced to Europe and spread worldwide, including new crop plants such as potatoes, maize, manioc, bananas, and rice, dramatically altering the demography and subsistence patterns of peoples in Asia, Africa, Oceania, and the Americas (Crosby 1972, 2004; Mann 2011; Wolf 1983). In so many ways, the world we live in today, biologically, economically, and culturally, is a living legacy of plant discoveries and transfers initiated with the Columbian landfall. At the same time, the cultural and environmental cost of botanical exploration and exploitation was steep and continues to be so today. Plantation agriculture destroyed viable traditional food production systems, causing massive Amerindian genocide and encouraging the kidnapping and enslavement of millions of Africans. Exotic useful species and weeds often turned invasive, extirpating native plants and causing untold hardships on local cultures (Pfeiffer and Voeks 2008). And colonial scientists, physicians, and missionaries, operating under the racist illusion that indigenous "brutes are botanists by instinct" (Long 1774, p. 381), shamelessly expropriated intellectual property and genetic diversity (cf. Gramiccia 1988; Ly-Tio-Fane 1958).

As the scholarly offspring of economic botany, ethnobotany has a briefer and less exploitive pedigree. Dominated in its infancy by descriptive work and lengthy inventories of useful plants, ethnobotanical research focuses increasingly on problem solving and hypothesis testing by means of qualitative and increasingly quantitative methods (Voeks 2011). It seeks to understand how culturally relevant floras are cognitively categorized, ranked, named, and assigned meaning (Berlin 1992). It investigates the complex strategies employed by traditional societies to manage plant taxa, communities, and landscapes. It explores the degree to which local ecological knowledge promotes or undermines resource conservation and contributes to the solution of global challenges, such as community health, nutrition, and cultural heritage (Turner 2011; Vandebroek et al. 2011). It investigates the economic value and environmental sustainability to local communities of non-timber forest products (Ticktin 2004), as well as the strategies through which individual ecological knowledge and practices encourage resilience to change—modernization, climate change, and many others (Kassam 2010). Most importantly, contemporary ethnobotanical research is grounded increasingly in respect for all cultures, embracing the principles of prior informed consent, benefit sharing, and general mindfulness (ISE 2006).

Because the primary objective of ethnobotanical research has long been focused on unearthing new species and their novel uses, the lion's share of inquiry has been carried out with indigenous people, the perceived custodians of useful plant knowledge. Whether stated or implied, received wisdom suggests that mature ethnobotanical knowledge is space and time contingent, that is, it is the outcome of long-term occupation and gradual cognitive familiarity with the floristic environment. Ethnobotanical narratives frequently allude to the antiquity and continuity of knowledge profiles among indigenous people, and native people are often romantically idealized as repositories of ancient plant wisdom, handed down unchanged as sacred oral text from generation to generation. The primary role of ethnobotanists, in this view, is to salvage this calcified forest wisdom, for the good of the community and of western science in general, before it succumbs to the forces of globalization and cultural erosion. Conversely, the botanical knowledge retained by diaspora communities, whose territorial occupancy is often measured in mere decades or centuries, has until recently been viewed as shallow, degenerative, and unworthy of academic research. Immigrants, it is assumed, simply have not acquired enough knowledge about the local flora to be of much scientific interest. (but see Alexiades 2009; Pieroni and Vandebroek 2008).

But this conceptual interpretation of the nature-society interface implies a degree of temporal and spatial emplacement of human populations and useful plant species that is strikingly inconsistent with the record. It hinges on the notion that indigenous societies have been geographically stable throughout much of their history, which is clearly wrong in almost every instance, and follows, according to Alexiades (2009, p. 25), from the untested assumption that environmental knowledge is the outcome of a "long-term empiricist enterprise." It also assumes that migrants necessarily come to inhabit floristic provinces that are wholly (or nearly so) alien and unknown to them. This biogeographical view of useful plants, that they are prisoners of their geographical place of origin by topography and climate, fails to appreciate the dynamism of the world's semidomesticated and useful wild species, particularly in the post-Columbian context (Crosby 1972; Pfeiffer and Voeks 2008).

Nowhere are these untested and unsupported biocultural assumptions more strikingly apparent than the limited ethnobiological inquiry conducted with African-descended populations in the Americas. Over the course of three and a half centuries, 11 million or more Africans were uprooted from their homeland, transported under appalling conditions across the Atlantic, and distributed as chattel slaves throughout the New World and elsewhere (Eltis and Richardson 2008). Confronted with the challenges of living and working under conditions of racialized economic exploitation, Africans at first glance seem poorly situated to have introduced their rich agricultural and ethnobotanical traditions or to have assimilated the indigenous names and properties of these alien forests and fields and croplands. The native peoples, languages, and religions were different; the crops and cropping systems were different; and, most importantly in this context, the extant floras were different. Two thousand miles of ocean and 100 million years of biological diversification separated Africans from their known landscape. Moreover, given racist ideologies rationalizing racially based economic exploitation, Africa has long been stereotyped as an intellectual backwater, lacking in the innovative powers and creative energies evidenced by their tropical Asian and American counterparts. Accordingly, considering both the monumental barriers to ethnobotanical diffusion and assimilation, and the likely preconceptions of researchers, it is not surprising that the ethnobotany of African diaspora communities has largely been overlooked in favor of indigenous groups.

Objectives and Overview

Ethnobotany is a profoundly interdisciplinary subject, and accordingly, contributors to this volume are drawn from a variety of disciplines, including agronomy, anthropology, botany, chemistry, environmental science, geography, history, and political science, and employ a range of methodologies, including ethnobotany and ethnography, ecological sampling, archival research, phytochemistry, and geospatial techniques. Voucher specimens were collected and stored in local and international herbaria, except in the case of religiously or legally protected species. Each contribution was peer-reviewed by the editors and by at least one additional specialist (the efforts of which the editors gratefully acknowledge). Note also that in contrast to the use of "African" in the restricted sense of the first generation of Africans brought to the New World, the term will be used here in the widest sense, allowing us to keep in focus the continent and its diverse peoples and their descendant populations in the Americas.

Although each contributor brings his or her own paradigms and points of view to the discussion, one of the consistent threads that run through these works is that Africans and their descendants have not been considered serious agents of ethnobotanical innovation and agricultural introduction to the evolving American biocultural landscape. Some 11 million Africans reached the shores of the Americas, for several centuries dwarfing the demographic contribution of Iberians and other Europeans. And yet judging by the paucity of scholarly attention, the effects of these millions of forced immigrants in terms of crop introduction and management techniques, even in the case of cultigens of African origin, were minuscule in comparison to the often ill-informed attempts of temperate zone migrants to introduce their own crops and medicinals. Because Africans were brought from one moist tropical or subtropical landscape to another and were obviously well versed in the exigencies of tropical cultivation and floristic foraging, it seems paradoxical that they should have been considered such ineffective transporters of agricultural and ethnobotanical skills compared to Europeans. This carries over as well into the cognitive realm of naming, classification, exploitation, and management of wild and semidomesticated New World species. Even taking into consideration the time contingency argument, it is curious how infrequently ethnobotanical research is focused on the African diaspora. It seems instructive in this respect to remember that colonial men of science and the cloth in the seventeenth and eighteenth centuries, counted among them the Swedish botanist Daniel Rolander, the Scottish physician Sir Hans Sloane, the French pastor Jean-Baptiste Labat, and many others, held no such preconceptions regarding the botanical knowledge and skills of Africans of the plantation period and their descendants. They were, it seems from their journals, equally eager to plumb the botanical expertise of Amerindians and Africans in the Americas (Labat 1724; Rolander 2008; Sloane 1707/1725). These essays, it is hoped, will help to expand our interest in and understanding of African ethnobotany in the Americas.

The first part of this volume, *Crops and Cultivators*, addresses the long-standing and often contentious issue of African contributions to crop plants and agricultural systems in the Americas. Judith Carney begins by reviewing more than two dozen plants that arrived in plantation societies during the transatlantic slave trade. She explores the agency of Africans in pioneering cultivation of familiar dietary and medicinal plants in their dooryard gardens, and how many of these taxa were initially disdained by plantation owners as "slave food." This wall of culinary segregation gradually disintegrated as signature ingredients of the African diaspora—okra, greens, plantains, black-eyed peas, pigeon peas, sesame, and others—infiltrated the cuisine of slaveholders. Africa's botanical legacy, Carney argues, is built upon this unacknowledged foundation.

Stanley Alpern's contribution focuses on a much contested issue—the question of African agency in the introduction of rice and rice cultivation culture to South Carolina. In defense of the "black rice hypothesis," Alpern marshals a prodigious quantity of primary material on pre-colonial West Africa to demonstrate that African rice (*Oryza glaberrima*) and later Asian rice (*O. sativa*) were widely cultivated staples all the way from Senegal to Ghana. He outlines the evidence for the transshipments of *O. glaberrima* with slaves to the Americas and authoritatively presents the case for African agency and expertise in the development of South Carolina Lowcountry's eighteenth-century rice boom. The achievements of whites and blacks in the Lowcountry, he concludes, were a unique synthesis owing as much to the Africans as to the Europeans.[1]

With reference to a wealth of historical material, Dorothea Bedigian explores the transfer of sesame (*Sesamum indicum*) from Africa to the New World. Linking the crop with the transatlantic slave trade and the geography of colonization, she examines surviving traces pertaining to sesame, from African languages and customs into modern African-descended cultures. She notes, for example, that the preference for mucilaginous foods was widespread throughout Africa, and that this slave tradition was eventually recognized and adopted by masters in the American colonies. Drawing on eyewitness reports, including correspondence of Thomas Jefferson, she analyzes the dynamic nature of contributions by African-born slaves in the dissemination and use of sesame in the Americas.

The next part, *Handicrafts and Crafters*, explores continuity and change in the use of botanical products to create artisanal handicrafts. Dale Rosengarten begins by tracing the tradition of basketmaking in South Carolina's Lowcountry. Having arrived with Africans at least by the late seventeenth century, the skill of fashioning coiled baskets was adapted over time to the local grasses, bulrushes, and palms. As rice plantations spread up and down the Atlantic coast, the diffusion of African-American winnowing baskets followed suit. By the twentieth century, with the demand for agricultural "work" baskets in sharp decline, basket makers focused on producing the sweet grass "show" baskets that have become emblematic of Lowcountry African-American heritage.

Patrick Hurley and colleagues take this topic in a different direction with their analysis of changing patterns of access to sweet grass (*Muhlenbergia sericea*), the principal species used to fashion coiled baskets today. Using ethnography and geographical information systems, they explore the process of displacement of traditional collection habitats by suburban expansion. They discover that collecting has expanded from a primarily local activity to a regional one, and that in spite of

[1] Van Andel's (2010) recent discovery *O. glaberrima* being cultivated by Afro-Surinamese maroon residents further bolsters the "black rice hypothesis" (van Andel 2010).

dwindling supplies, networks of exchange based on friendship, kinship, and entrepreneurship continue to connect basket makers with gatherers and resource supplies. Their work points to the ways that greater attention to African-American cultural traditions might reveal new insights into local ecologies and landscape change.

Moving from the Carolinas to the Esmeraldas region of Ecuador, Maria Fadiman examines the ethnobotany of basketmaking among Afro-Ecuadorian forest dwellers. Focusing on the *piquigua* hemi-epiphyte (*Heteropsis ecuadorensis*), she details gendered roles associated with weaving techniques and ultimate products, as well as the apparently sustainable means by which the aerial roots are collected and managed. Although the current market for *piquigua* is limited, the cultural, nonmarket value to Afro-Ecuadorians is considerable.

James Sera and Robert Voeks investigate the history and ethnobotany of Brazil's *berimbau de barriga*, the iconic musical bow of Afro-Brazilian capoeira. Although *berimbau* is a Bantu term, the precise provenance of the instrument's African prototype is unclear. Employed early on the streets of Brazilian cities to attract customers to street merchants, the *berimbau* did not become associated with capoeira until perhaps the late nineteenth century. The primary botanical material for the staff (*verga*) is derived from biriba (*Eschweilera ovata*), which is harvested principally from the highly threatened Atlantic Rainforest. How the competing demands of *berimbau* crafters and conservationists will be resolved remains to be seen.

The knowledge and use of plants for religious and healing rites holds a prominent position in the ethnobotany of African-descended communities. Given the often degenerate nature of colonial white medicine, African healers were often sought out by Europeans as well as Africans. At the same time, Africans were widely feared by the plantocracy for their perceived mastery of the occult arts, leading in many cases to total bans on the use of magico-medical herbs by the enslaved. The section *Medicinal and Spiritual Ethnofloras* begins with Erica Moret's exploration of the role that African-derived ethnobotanical knowledge has played in Cuba's colonial and postcolonial history. She compares the medicinal plant knowledge in sugar zones, where African influence was considerable, with that of tobacco zones, where Mediterranean influences held sway. She finds that ethnobotanical specialists who reside in selected sugar regions have significantly greater knowledge of West African plant use than that of Mediterranean species, underscoring both the contribution of African knowledge and skills to the Cuban pharmacopoeia and the importance of geography and migrant history in assessing these features.

Recognizing the profound spiritual significance of plants to practitioners of Afro-Suriname's *Winti* religion, Tinde van Andel and colleagues ask the intriguing question—what makes a plant magical? Through a comprehensive survey of over 400 magical species employed by Maroons, they discover that most *Winti* magical species are native to Suriname and are members of several specific plant families. In addition to remarkable growth form, shape, scent, color, and other biological and ecological factors, species that once aided escaped Africans in their flight to freedom have attained magical status among Afro-Surinamese. Unlike the Afro-Brazilian, Afro-Cuban, and Barbadian pharmacopoeias (see Chaps. 9, 11, and 14 by Moret, Peter, and Voeks, this volume), these researchers discovered a high proportion of trees and primary forest species in their spiritual ethnoflora, likely reflecting the unique degree of isolation experienced by Surinamese Maroons.

Culturally derived etiologies for disease and healing play a pivotal role in many African-descended communities, as Sonia Peter discovers in her census of medicinal plants and healing traditions in Barbados. Ninety-three species were identified as useful for a variety of medical problems, most often employing leaves. The most common formulation was of "cooling teas," the regular administration of which is perceived to be beneficial to fortification of the body against common maladies, such as cold and flu, as well as chronic diseases. Qualitative and quantitative estimation of polyphenol validated selected species as significant sources of these agents, which also function as antioxidants. The Barbados population of African or mixed descent, culturally steeped in ethnomedical traditions, continues to benefit from recognition of the value of cooling teas to human health and perhaps longevity.

One of the most significant contributions of the study of diaspora plant knowledge and skill is the degree to which these reflect on questions of ethnobotanical agency, contingency, continuity, and innovation. Are immigrants active in the transport of useful species? Do newcomers substitute similar botanical taxa for elements that could not be transferred? Are traditional biocultural relations sufficiently pliable to allow for borrowing and new species trial and error? These topics are explored in the final section *Ethnobotanical Continuity and Change*. In his essay on Brazil's ritual iroko tree, John Rashford explores the question of what species of fig (*Ficus*) constitute the identity of Candomblé's cosmic tree. In West Africa, iroko is represented by *Milicia excelsa* (Moraceae), which failed to make the Atlantic crossing. Contrary to the widespread notion that Brazil's iroko is represented by a single species such as *Ficus religiosa*, an exotic species, he discovers that several native *Ficus* species serve as Candomblé's iroko, and that the process of species adoption is a subtly nuanced blend of ritual, negotiation among *terreiro* (shrine) members, and serendipity of seed and seedling colonization.

The question of whether African forced immigrants maintain significant ethnobotanical knowledge and skill is addressed by Bruce Hoffman in Suriname. Employing a cross-cultural comparison of Afro-Surinamese Maroons and the indigenous Carib-speaking Trio, he uses plot sampling methods to determine that indigenous people have more names and uses for species than the descendants of Africans, especially of medicinals in old-growth forest. But Maroon knowledge is robust as well, especially in disturbed forests, and with reference to timber and handicraft uses. The basis of indigenous ethnobotanical knowledge appears to be a product of long-term experience and a subsistence economy, whereas the foundation of Maroon knowledge is reflected more in commercial interests and taboos associated with old-growth forest exploitation. Ethnobotanical knowledge in this case is time and space contingent but also very much a reflection of differing sociocultural histories.

In his essay on Candomblé food and medicinal plants, Robert Voeks explores the question of ethnobotanical continuity and mobility among Brazil's African diaspora. Employing a historical ecology approach, he argues that many of the useful species employed by Africans inhabited both shores of the Atlantic generations before the peak of the slave trade. This transatlantic floristic homogenization allowed Afro-Brazilians to continue using Old World species in Brazil and to rediscover New World natives that had long ago arrived and been adopted into use in West Africa. Afro-Brazilians developed a rich and culturally relevant relationship with plants, one dominated numerically by successional, domesticated, exotic, and herbaceous species. Whether this preference for the flora of anthropogenic habitats is a characteristic of diaspora ethnobotany awaits further investigation.

The editors extend their sincerest appreciation to the many hundreds of African-descended field collaborators whose knowledge of botanical nature represents the foundation of this volume. We also thank all of the contributors to this volume, whose decision to explore unorthodox ethnobotanies has expanded the scope of the enterprise significantly, as well as Kelly Donovan, who produced several of the figures, and Leaa Short, who proofread many of the manuscripts and produced the Index.

References

Aleixades M (2009) Mobility and migration in indigenous Amazonia: contemporary ethnoecology perspectives—an introduction. In: Alexiades M (ed) Mobility and migration in indigenous Amazonia: contemporary ethnoecology perspectives. Berghahn, New York, pp 1–43

Barrera A (2002) Local herbs, global medicines: commerce, knowledge, and commodities. In: Smith P, Findlen P (eds) Merchants and marvels: commerce, science, and art in early modern Europe. Routledge, New York, pp 163–181

Berlin B (1992) Ethnobiological classification: principles of categorization of plants and animals in traditional societies. Princeton University Press, Princeton

Crosby AW (1972) The Columbian exchange. Biological and cultural consequences of 1492. Greenwood Press, Westport

Crosby AW (2004) Ecological imperialism: the biological expansion of Europe, 900–1900. Cambridge University Press, Cambridge

Eltis D, Richardson D (2008) A new assessment of the transatlantic slave trade. In: Eltis D, Richardson D (eds) Extending the frontiers: essays on the new transatlantic slave trade database. Yale University Press, New Haven, pp 1–60

Gramiccia G (1988) The life of Charles Ledger (1818–1905): alpacas and quinine. Macmillan, London

International Society of Ethnobiology (2006) International Society of Ethnobiology Code of Ethics (with 2008 additions). http://ethnobiology.net/code-of-ethics/

Kassam KS (2010) Pluralism, resilience, and the ecology of survival: case studies from the Pamir Mountains of Afghanistan. Ecology and Society 15(2):8. http://www.ecologyandsociety.org/vol15/iss2/art8/

Labat JB (1724) Nouveau voyage aux isles de l'Amérique: contenant l'histoire naturelle de ces pays, l'origine, les moeurs, la religion & le gouvernement des habitants anciens & modernes. Alberts R, van Duren J, Gosse P, Husson P, Johnson T, Le Vier C, Paris

Long E (1774) The History of Jamaica. Vol 2. T. Lownudes, London

Ly-Tio-Fane M (1958) Maruitius and the spice trade: the odyssey of Pierre Poivre. Esclapon Limited, Port Louis

Mann C (2011) 1493: uncovering the new world Columbus created. Knopf, New York

Pfeiffer J, Voeks R (2008) Biological invasions and biocultural diversity: linking ecological and cultural systems. Environ Conserv 35(4):281–293

Pieroni A, Vandebroek I (eds) (2008) Traveling cultures and plants: the ethnobiology and ethnopharmacy of human migrations. Berghahn, New York

Rolander D (2008) Daniel Rolander's journal (1754–1756). The Suriname Journal: composed during an exotic journey (trans: from Latin by Dobreff J, Dahlman C, Morgan D, Tipton J). In: Hansen L (ed) The Linnaeus Apostles: global science & adventure, vol 3, book 3. IK Foundation & Company, London, pp 1215–1576

Schiebinger L (2004) Plants and empire: colonial bioprospecting in the Atlantic world. Harvard University Press, Cambridge, MA

Sloane H (1707/1725) A voyage to the islands Madera, Barbados, Nieves, S. Christophers and Jamaica with the natural history of the herbs and trees, four-footed beasts, fishes, birds, insects, reptiles, &c. Printed by the author, London

Spary EC (2000) Utopia's garden: French natural history from old regime to revolution. University of Chicago Press, Chicago

Thornton J (1992) Africa and Africans in the making of the Atlantic world, 1400–1800. Cambridge University Press, Cambridge

Ticktin T (2004) The ecological implications of harvesting non-timber forest products. J Appl Ecol 41:11–21

Turner NJ (2011) Ethnobotany: the study of people-plant relations. In: Anderson E, Pearsall D, Hunn E, Turner N (eds) Ethnobiology. Wiley, Hoboken, pp 133–148

van Andel T (2010) African rice (*Oryza glaberrima* Steud.): lost crop of the enslaved Africans discovered in Suriname. Econ Bot 64(1):1–10

Vandebroek I, Reyes-Garcia V, Albuquerque U, Bussmann R, Pieroni A (2011) Local ecological knowledge: who cares? J Ethnobiol Ethnomed 7:35. http://www.ethnobiomed.com/content/7/1/35

Voeks RA (2011) Ethnobotany, Chapter 26. In: Millington A, Blumler M, MacDonald G, Schickhoff U (eds) The Sage handbook of biogeography. Sage Publications, London, pp 493–509

Wolf E (1983) Europe and the people without history. University of California Press, Berkeley

Part I
Crops and Cultivators

Chapter 2
Seeds of Memory: Botanical Legacies of the African Diaspora

Judith Carney

Abstract The decades following 1492 launched an era of European overseas expansion, which led to an unprecedented intercontinental exchange of plant and animal species. Literature on the Columbian Exchange emphasizes the New World and Asian crops that revolutionized the food systems of Africa but ignores the role of African crops in the New World tropics. This chapter draws attention to the neglected African components of the Columbian Exchange. The movement of African plant and food animals across the Atlantic Ocean in the initial period of plantation development depended on the transatlantic slave trade for their dispersal. Plants and animals arrived on slave ships together with African captives for whom the species were traditional dietary staples, medicinals, and food animals. A proper appreciation of African contributions to New World agricultural systems requires a new perspective on plantation societies, one that shifts standard research from the export commodities that slaves grew to the plants they cultivated for their own needs. This in turn draws attention to the significance of African species as a vital logistical support of the transatlantic slave trade and to the agency of enslaved Africans in pioneering cultivation of familiar dietary plants in their dooryard gardens and food fields.

Keywords Columbian Exchange • African diaspora • Slavery • Subsistence • Food animals • Culinary signatures

Seeds of Memory: Botanical Legacies of the African Diaspora

The decades following 1492 launched an era of European overseas expansion, which led to an unprecedented intercontinental exchange of plant and animal species. Historian Alfred W. Crosby famously called this process the *Columbian Exchange*.

J. Carney, Ph.D. (✉)
Department of Geography, University of California Los Angeles, 1255 Bunche Hall, Los Angeles, CA 90095-1524, USA
e-mail: carney@geog.ucla.edu

R. Voeks and J. Rashford (eds.), *African Ethnobotany in the Americas*,
DOI 10.1007/978-1-4614-0836-9_2, © Springer Science+Business Media New York 2013

In his second book, *Ecological Imperialism*, Crosby drew attention to the significance of ordinary people for the transfer of species across the globe. These were European emigrants to new lands—people who were not operating as administrators, scientists, or representatives of colonial institutions such as botanical gardens, scientific societies, and museums. With the plant and animal species that accompanied them, European settlers transformed the environments of Australia, New Zealand, and South Africa into, in Crosby's words, *Neo-Europes* (Crosby 1972, 1986).

However, the Columbian Exchange literature ignores another important intercontinental species transfer that over the same period of time also occurred as a consequence of immigration. In this instance, the migration was not voluntary but forced, and the agents of dissemination involved not just Europeans but also enslaved Africans. Reference is to the African plants and food animals that proved instrumental for European colonization of the New World tropics. The movement of these biota across the Atlantic Ocean to tropical America in the first century of plantation development depended significantly on the transatlantic slave trade for their dispersal. Plants and animals arrived on slave ships together with African captives for whom they were traditional dietary staples, medicinals, and food animals. Although the Columbian Exchange celebrates the role of New World and Asian crops in revolutionizing food systems of Africa, there is little attention given to the impact of African species in lowland tropical America. Of interest is the contrasting significance of these African species for slaveholders and the enslaved.

This chapter examines the largely unheralded African components of the Columbian Exchange. The discussion refers to more than two dozen plants, which arrived in plantation societies during the transatlantic slave trade (Table 2.1). Most are of African origin. Others, domesticated originally in Asia, reached the continent in prehistory, when Africans adopted them into existing agricultural systems. Each of these plants is documented in the historical record of crops grown in New World plantation societies.

A proper appreciation of African contributions to New World agricultural systems requires a new perspective on plantation societies, one that shifts standard research from the export commodities that slaves grew to the plants they cultivated for their own needs. This in turn draws attention to three additional and interrelated concerns: (1) the significance of African species as a vital logistical support of the transatlantic slave trade, (2) their role in colonization of the New World tropics, and (3) the agency of enslaved Africans in pioneering cultivation of familiar dietary and medicinal plants in their dooryard gardens.

By way of introduction, it is important to understand the growing intersection of European and African experiences that began right at the start of the so-called Age of Discovery. Prior to the colonization of the Americas, there were enslaved Africans in the Iberian Peninsula. They arrived during the Muslim occupation through the trans-Saharan trade in slaves. Documents record their presence in Spain during the fourteenth century. With maritime expansion in the early fifteenth century, the Portuguese diverted the slave trade to direct importation from places their caravels reached along the African coast. By 1448, about a 1,000 slaves had been carried back by sea to Portugal and its colonies in the Atlantic islands (Madeira, Azores).

Table 2.1 African plants mentioned in historical records of plantation societies of tropical America

Common names	Species names
Cereals	
Millet	*Pennisetum glaucum*
Sorghum	*Sorghum bicolor*
Rice	*Oryza* spp.
Tubers	
Yams	*Dioscorea cayenensis*
Plantain/banana	*Musa* spp.
Taro/eddo	*Colocasia esculenta*
Legumes	
Black-eyed pea/cowpea/calavance	*Vigna unguiculata*
Pigeon/Angola/Congo pea/*guandul*	*Cajanus cajan*
Bambara groundnut/*Voandzeia*	*Vigna subterranea*
Lablab/hyacinth/bonavist bean	*Lablab purpureus*
Oil plants and fruits	
Castor bean	*Ricinus communis*
Oil palm	*Elaeis guineensis*
Watermelon	*Citrullus lanatus*
Muskmelon	*Cucumis melo*
Ackee	*Blighia sapida*
Beverages	
Roselle/bissap	*Hibiscus sabdariffa*
Kola nut	*Cola* spp.
Vegetables and spices	
Okra	*Hibiscus esculentus*
Egyptian spinach/jute mallow	*Corchorus olitorius*
Guinea pepper	*Xylopia aethiopica*
Guinea squash	*Solanum aethiopicum*
Forage grasses	
Guinea grass	*Panicum maximum*
Angola/Pará grass	*Brachiaria mutica*
Bermuda grass	*Cynodon dactylon*
Molasses grass	*Melinus minutiflora*
Jaraguá grass	*Hyparrhenia rufa*

Source: Carney and Rosomoff (2009)

By the middle of the sixteenth century, African slaves formed one-tenth of Lisbon's population of 100,000. Seville's slave population exceeded 6,000—one out of every 14 urban dwellers (Fig. 2.1) (Pike 1967; Saunders 1982).

The first African slaves introduced to the New World originated in Seville, where many had been born. In 1510, the Spanish monarchs shipped 200 African slaves to Hispaniola (they were baptized in the Christian faith just prior to their departure). African slaves and freedmen also formed part of the conquistador armies in the New World. One example is Juan Garrido, who was born in West Africa about 1480,

Fig. 2.1 Image by anonymous artist. *Chafariz d'el Rey*, c. 1570–1580 (Reproduced with permission of Collection Joe Berardo, Lisbon, Portugal)

enslaved by the Portuguese as a teenager, and taken to Hispaniola in 1503. He was part of the contingent that landed in Veracruz, Mexico with Cortés and his army in 1519. The anonymous figure depicted in Fig. 2.2 may have been Garrido or another African whose destiny similarly placed him as a participant in the Spanish conquest of the Aztec Empire (Restall 2000).

Food Grown in Africa as a Vital Support of the Atlantic Slave Trade

Africans participated fully in the Neolithic Revolution that led to plant and animal domestication in different parts of the world beginning some 10,000 years ago. African contributions to global food supplies include nine cereals, half a dozen root crops, five oil-producing plants, several forage crops and as many vegetables, three fruit and nut crops, coffee, and the bottleneck gourd (Table 2.2). Most of these food-staples are tropical species and not widely known to Western consumers. These contributions to world food supplies are often overlooked because some of the continent's staples are incorrectly assigned an Asian origin. But the African continent harbors several indigenous food crops, including rice (*Oryza glaberrima*) and eggplant (NRC 1996, 2006; Carney 2001).

Fig. 2.2 Image entitled *El Encuentro de Cortes y Moctezuma* appears in Fray Diego de Duran, *Historia de las Indias de Nueva España y Islas de Tierra Firme*, Plancha 58, c. 1579–1581 (Reproduced with courtesy of La Biblioteca Nacional, Madrid, Spain)

Table 2.2 Food crops of African origin

Savanna	
Adansonia digitata L.	Baobab
Brachiaria deflexa (Schumach.) C.E. Hubb. Ex Robyns	Guinea millet
Ceratotheca sesamoides Endl.	Leaves and seeds
Citrullus lanatus (Thunb.) Matsum. & Nakai	Watermelon
Corchorus olitorius L.	Jute mallow/bush okra
Cucumis melo L.	Muskmelon
Digitaria decumbens Stent	Pangola grass
Digitaria exilis (Kippist) Stapf	Fonio/"hungry rice"
Digitaria iburua Stapf	Black fonio
Hibiscus cannabinus L.	Kenaf
Hibiscus sabdariffa L.	Sorrel/hibiscus/roselle/bissap/vinagreira/cuxá
Lagenaria siceraria (Molina) Standl.	Bottle gourd
Oryza glaberrima Steud.	African rice
Parkia biglobosa (Jacq.) R. Br. Ex G. Don	Locust bean
Pennisetum glaucum (L.) R. Br.	Bulrush or pearl millet
Polygala butyracea Heckel	Black beniseed
Sesamum alatum Thonn.	Sesame: leaves
Sesamum radiatum Schumach. & Thonn.	Beniseed
Solanum aethiopicum L.	African eggplant/garden
Solanum incanum L.	Bitter tomato
Solanum macrocarpon L.	Nightshade
Sorghum bicolor (Linn.) Moench	Sorghum/guinea corn

(continued)

Table 2.2 (continued)

Vitellaria paradoxa C.F. Gaertn.	Karité or shea butter tree
Vigna subterranea (L.) Verdc.	Bambara groundnut/Voandzeia
Xylopia aethiopica (Dunal) A. Rich	Guinea pepper
West African Savanna-Woodland Ecotone	
Aframomum melegueta K. Schum.	Melegueta pepper
Amaranthus spp.	Vegetable amaranth/African Spinach/ bledo/callaloo
Blighia sapida K.D. Koenig	Ackee/akee apple
Cajanus cajan (L.) Millsp.	Pigeon pea/Congo pea/Angola pea/ guandul
Coffea robusta Linden	Coffee (*robusta*)
Cola acuminata (P. Beauv.) Schott & Endl.	Kola nut
Cola nitida (Vent.) Schott & Endl.	Kola nut
Cucumeropsis edulis (Hook. f.) Cogn.	Egusi
Dioscorea bulbifera L.	Air potato yam
Dioscorea cayenensis Lam	Yellow guinea yam
Dioscorea dumetorum (Kunth) Pax	Three-leaved or bitter yam
Dioscorea rotundata Poir.	White guinea yam
Elaeis guineensis Jacq.	Oil palm
Gossypium herbaceum L.	Cotton
Hibiscus esculentus L.	Okra, gumbo
Kerstingiella geocarpa Harms	Kersting's groundnut/Hausa groundnut/ geocarpa bean
Momordica charantia L.	African cucumber/bitter melon
Piper guineense Schumach & Thonn.	Piper seed
Plectranthus esculentus N.E. Br.	Dazo/finger potato
Solenostemon rotundifolius (Poir.) J.K. Morton	Hausa potato/piasa
Sphenostylis stenocarpa (Hochst. ex A. Rich.) Harms	African yam bean
Tamarindus indica L.	Tamarind
Telfairia occidentalis Hook. f.	Fluted pumpkin
Vigna unguiculata (L.) Walp.	Cowpea/black-eyed pea
Ethiopia/East African Highlands	
Avena abyssinica Hochst.	Ethiopian oats
Catha edulis Forssk.	Chat
Coccinia abyssinica (W. & A.) Cogn	Anchote
Coffea arabica L.	Coffee (*arabica*)
Eleusine coracana (Linn.) Gaertn.	Finger millet
Ensete ventricosum (Welw.) Cheesman	Enset
Eragrostis tef (Zucc.) Trotter	Tef
Guizotia abyssinica (L.f.) Cass.	Niger seed, noog
Lablab purpureus Sweet	Lablab/hyacinth bean/bonavist/Egyptian bean
Panicum maximum Jacq.	Guinea grass
Pennisetum clandestinum Hochst. Ex Chiov.	Kikuyu grass
Ricinus communis L.	Castor bean

Source: Harlan (1975), 71–72 and MacNeish (1992), 298–318

Africans also contributed in important ways to the development of several Asian crops that reached the African continent centuries before the beginning of the transatlantic slave trade (McNeill 2000; Alpern 2008). These include the root crops: taro, the Asian yam, ginger, plantain, and the banana. The banana and its cousin, the cooking banana or plantain, arrived in Africa between 2,000 and 3,000 years ago. African experimentation with these related plants led to the development of new cultivars and the emergence of secondary centers of domestication on the African continent (de Langhe 1995; Mbida et al. 2000; Kleiman 2003). In emphasizing the geographical origin of specific plants such as the banana and plantain, rather than the continent where European navigators first encountered them, Columbian Exchange scholars unwittingly depreciate African botanical contributions to global plant transfers. Such views inadvertently perpetuate the misperception of Africa as a continent with few botanical resources of its own, one that has always depended on food introduced from elsewhere for the survival of its peoples.

During the Atlantic slave trade, Africa, in fact, routinely produced surplus food. We know this from ship manifests as well as the logs and drawings of ship captains, who depended on African-grown food to facilitate their commerce in human beings. While slave ships carried some food stores from Europe, captains relied in no small part on African food surpluses to provision their human cargoes across the Middle Passage. Food purchased in Guinea's ports included introduced Amerindian crops—notably maize, bitter manioc, and peanuts emphasized in the Columbian Exchange literature—as well as indigenous African foodstaples (such as millet, rice, black-eyed pea, and melegueta pepper). Captains of slave ships purchased foodstaples in bulk, frequently revealing a distinct preference for traditional African dietary staples because they commonly believed that mortality rates across the Middle Passage improved when captives were given food to which they were accustomed. Slavers purchased provisions for the transatlantic crossing from African merchant middlemen, supplies stocked by European forts along the Guinea Coast, and in local markets (Carney and Rosomoff 2009).

The region from Senegal to Liberia, known during the slave trade as the Upper Guinea Coast, provided indigenous cereals such as millet (*Pennisetum glaucum*), sorghum (*Sorghum bicolor*), and African rice (*O. glaberrima*) in addition to the Bambara groundnut (*Vigna subterranean*) and melegueta pepper (*Aframomum melegueta*). Traditional dietary staples purchased to the south in Lower Guinea included African yams (*Dioscorea cayenensis*, *D. rotundata*), the oil palm (*Elaeis guineensis*), black-eyed pea (*Vigna unguiculata*), and the pigeon pea (*Cajanus cajan*). The Asian root crops Africans had adopted in prehistory also figured in food purchases, notably the Asian yam (*Dioscorea alata*), taro (*Colocasia esculenta*), and plantains (*Musa* spp.). The African kola nut (*Cola nitida*, *C. acuminata*) was also a frequent passenger on slave ships. Long a traditional African medicinal, kola was prized for its ability to improve the taste of stored drinking water. Slave captains quickly borrowed the African practice: by placing kola nuts in shipboard water casks, stagnant water could be refreshed and made palatable again during the long transatlantic voyage (Hair et al. 1992, I: 188).

African Animal Introductions and Forage Grasses

Slave ships also carried live animals to the Americas as fresh meat for their crews. The first to make its appearance in the historical record of plantation societies is the guinea fowl (*Numida meleagris*). A Jesuit priest claimed that the guinea fowl arrived on ships that carried the first boatloads of African slaves to Hispaniola. The African "chicken" was a well-established food animal in Pernambuco, Brazil, when German naturalist Georg Marcgraf described and drew it in 1640. Marcgraf used the existing Portuguese name for the guinea fowl, *galinha d'Angola*, and averred its African provenance (Marcgrave 1942, 192; Donkin 1991, 97). The guinea fowl formed a significant component of the small animal stock that some plantation slaves were able to keep. They raised them in the yard around their dwellings and occasionally sold them to their masters. Today, it remains an important poultry species reared by African Americans in the southern United States. In Brazil, the guinea fowl is also raised for food and is featured in the liturgical practices of the Africa-derived religion, *candomblé* (Pessoa de Barros 1993).

European vessels deliberately transported African livestock to tropical America since the animals were better adapted to the climate than their European counterparts. The African "hair" sheep did not have the woolly coat that made the lowland tropics inhospitable for European breeds. But it did satisfy colonists' demands for animals suited to New World tropical environments. The hair sheep was introduced as a meat animal in the early settlement period of Brazil, Barbados, and Jamaica. Marcgraf noted its arrival in Brazil via ships from western Africa. He also recorded the seventeenth-century Portuguese names for the hair sheep: *carneiro de Guiné* and *carneiro d'Angola* (Marcgrave 1942, 234). Writing about Jamaica later that century, Hans Sloane—founder of the British Museum—indicated that the island's sheep were a breed that came from Africa (Sloane 2001, I: 254). Richard Ligon, who resided in Barbados during the 1640s, identified two locations along the African coast where the hair sheep breed was transported to the English colony: "[They] are brought from *Guinny* and *Binny*, and those have hair growing on them, instead of wool; and liker goats than sheep" (Ligon 1970, 59).

African cattle also comprised some of the livestock introduced to the Americas. By the seventeenth century, Europeans had developed a flourishing trade in live animals, meat, and hides with livestock holding societies of the Upper Guinea Coast. African cattle were sold to slave ships (Gamble and Hair 1999). Jean Barbot's drawing made during a French slaving expedition to West Africa in the late seventeenth century shows Africans transporting cattle (likely the dwarf indigenous *n'dama* breed) in their canoes to waiting slave ships (Fig. 2.3) (Barbot 1752, 99). The image, with its implication of African cattle transfers to the New World, corroborates the findings of recent genetic studies. This research reveals that African cattle were indeed brought to tropical America and influenced the development of New World Creole livestock populations. DNA analysis shows genetic introgressions by African cattle in two ways—one via the Iberian Peninsula, the other directly from West

Fig. 2.3 Barbot (1752, vol. V, Plate E, 99)

Africa. In the first, African cattle entered the Iberian livestock population in two distinct historical periods—during the Bronze Age when they were introduced across the Straits of Gibraltar and during the Moorish occupation of the Iberian Peninsula. These introductions contributed genes from African progenitors to Iberian cattle populations. However, some New World breeds (*Guadeloupe Creole*, other types found in Brazil) were directly influenced by the DNA of cattle imported from West Africa during the transatlantic slave trade. Suited to the high temperature and humidity of lowland tropical America, these cattle were introduced via established Portuguese and French slave routes from West Africa (Cymbron et al. 1999; Magee et al. 2002; Lirón et al. 2006; Carney and Rosomoff 2009; Ginja et al. 2009).

Transport animals, such as donkeys (native to Africa) and camels, were also brought to the Americas in the early colonization period. The Portuguese named one donkey breed they transported, *assinigoes*, likely after the livestock-herding Berber Azenegues, who lived north of the Senegal River and with whom they traded at Arguim Island. Richard Ligon discussed this breed in Barbados and sketched it in his seventeenth-century map of the island (Ligon 1970).

The Spanish initiated the practice of using camels in sixteenth-century Peru as pack animals to carry heavy loads from the mountains and across the coastal desert. These camels did not arrive in the New World directly from Arabia but came from West Africa and the Atlantic islands offshore Senegambia. The English followed Spanish precedent, relying upon camels as transport animals in their sugar colony of Barbados in the initial decades of plantation development. Camels are pictured, with their enslaved African tender, on Ligon's map, which dates to 1647 (Fig. 2.4)

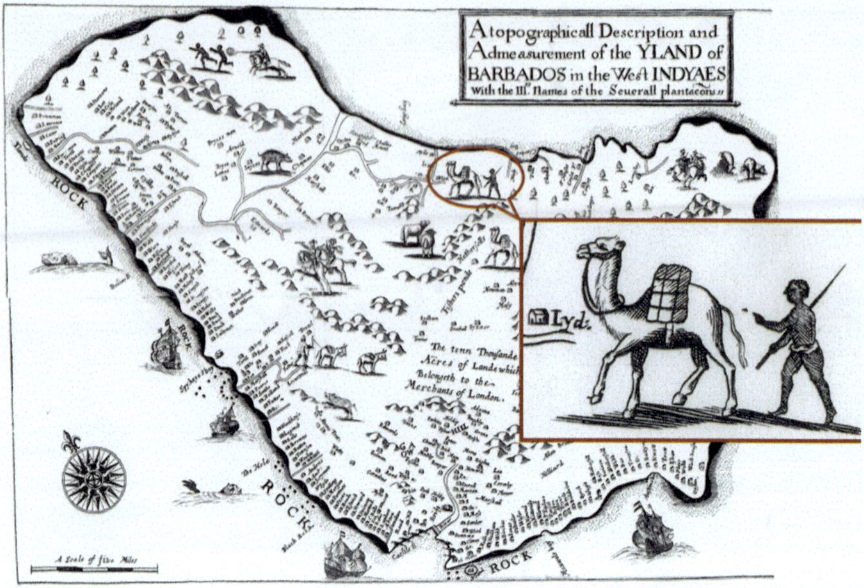

Fig. 2.4 Detail of Ligon's map of Barbados (Ligon 1970 [c. 1647], image at front of book)

(Ligon 1970). Camels had formed an important component of livestock herds in the semiarid region between Morocco and the Senegal River by the third century A.D. When the Spanish completed their conquest of the Canary Islands and its indigenous Guanche people in the fifteenth century, they raided the nearby African mainland (today's Morocco and Spanish Sahara, just 100 km away) for camels, cattle, sheep, goats, and slaves (Rumeo de Armas 1956, 115). Indeed, the mainland Berbers were known as skilled herdsmen. The camel's presence in Peru came to an end in 1615, when runaway slaves killed the last remaining specimen for food. But the use of camels continued in seventeenth-century Barbados and Saint Domingue (Crosby 1972; Mercer 1973; McClellan 1992).

African livestock were accompanied across the Atlantic by the indigenous grasses that were their natural fodder. Five African forage species arrived in this way (Table 2.1). The significance of Africa for their introduction is recognized in the names some of them still bear in colonial languages: *Guinea grass*, *Angola grass*. African pasture grasses likely dispersed on the hoofs of cattle, sheep, goats, and other introduced food animals to tropical America. Repeated introductions ensured the rapid dispersion of these grasses. Angola and Guinea grass were especially appreciated for their nutritive qualities and suitability for tropical cultivation (Parsons 1970). They were sown, cut, and transported to feed stabled dairy cows and horses in Brazil, Cuba, and many other parts of the New World tropics (Chase 1944; Debret 1954; Watts 1987).

The African Castor Bean: Contrasting Significance for Slaveholders and Slaves

The transatlantic slave trade also propelled diffusion to the Americas of several African plant species appreciated for their medicinal properties. These included tamarind, gum Arabic, bitter melon (*Momordica charantia*), melegueta pepper, kola nut, and castor bean plant. The castor plant was established at an early date in Brazil and introduced to Santo Domingo from Africa by 1509. It was present with the founding of the Puritan plantation economy on Providence Island the next century. Documents from the seventeenth century also indicate its presence in the Leeward Islands, Barbados, and Martinique (Piso 1957; Kupperman 1993; Watts 2000). Geographer David Watts called it a major agent of environmental change in the early settlement history of the West Indies (c. 1624–1645) (Watts 1987, 169). Albert Eckhout—a painter accompanying the scientific expedition to Dutch-occupied Brazil (1630–1654)—included the castor plant in his 1641 painting, *Mameluke with a Basket of Flowers* (Fig. 2.5) (Buvelot 2004, 57). Willem Piso, the expedition's naturalist, sketched and described a plant he considered native to tropical America (Fig. 2.6) (Piso 1957, 385–386).

The castor bean plant exemplifies a species that especially facilitated the transatlantic commerce in human beings. Slave merchants on both sides of the Atlantic

Fig. 2.5 Mameluke with a basket of flowers (Buvelot 2004, 57)

Fig. 2.6 Piso (1957[1645], 385)

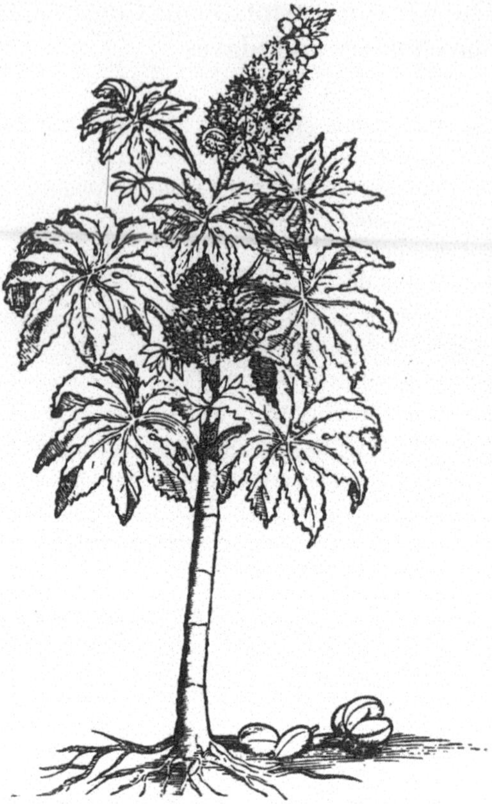

NHAMBU GUAÇÚ OU RÍCINO AMERICANO

learned to appreciate the medicinal properties of the castor plant. As a remedy for many afflictions suffered by captives kept in confined quarters, Jean Barbot in the 1670s noted its cultivation in the garden of the commandant of the slave depot on Gorée Island, Senegal (Barbot 1752, 31). Slave traders learned that the plant was a powerful purgative and could be used to treat skin ailments and head lice. But it was also important to the enslaved. Johann Moritz Rugendas makes this contrasting association explicit in two paintings he made in Brazil during the 1820s. *Novos Negros* reveals the significance of the castor plant to the slave trader. Rugendas illustrates newly landed Africans awaiting sale in a slave depot. One section of the painting shows a disconsolate man with shaved head (Fig. 2.7, left). Beside him is the castor leaf. By positioning the plant next to the African, Rugendas suggests its medicinal role in readying the man's enslaved body for sale, even if we do not know whether the plant was employed as a purgative, a skin treatment, or a delousing agent (Slenes 2002; Corrêa do Lago 2001, 189). In the second Rugendas painting, *Dwelling of Slaves*, the castor plant is once more featured with slaves but this time

Fig. 2.7 (**a**) *Novos Negros* by Johann Moritz Rugendas (Reprinted in Corrêa do Lago 2001, 189). (**b**) *Habitação de negros* (Rugendas 1954, Plates 4/5, following p. 205). (**c**) Image taken by Judith Carney, October, 2005, Minas Gerais, Brazil

in a wholly different context (Fig. 2.7, center) (Rugendas 1954, 205). It is encountered as a crop in a slave dooryard garden (the castor plant is pictured to the right of the hut behind the water carrier and child). In this Rugendas painting, slaves are deliberately cultivating a multipurpose plant whose properties formed part of their shared ethnobotanical heritage (Carney and Rosomoff 2009).

Enslaved Africans in the Americas were well aware of the diverse uses of the castor plant, which is native to sub-Saharan Africa. They employed it to treat skin disorders, ophthalmic infections, venereal diseases, joint pain, gastrointestinal problems, and helminthic infestations. One observer of Jamaica's plantation society described slaves using castor oil to draw out the guinea worm from afflicted legs (Grimé 1979, 173). The distinctive appearance of the castor plant, its ubiquity in tropical plantation societies, and broad curative properties attracted the commentary of many European observers (Grimé 1979; Kimber 1988; Kupperman 1993; Watts 2000). One of them was Samuel Hazard, who traveled to Cuba in 1871, when slavery was still legal. Hazard noted that the castor bean "grows in great quantities all over these mountains, and is prepared by the superannuated negro women, who select the beans and clean them ready for extracting the oil." He marveled at the ability of one elderly slave woman to detect the best beans for expressing oil, even though she was blind. Figure 2.8 is the image he sketched of her (Hazard 1971, 469).

The castor plant was among several African botanical introductions that also became plantation commodities. In eighteenth-century Saint Domingue, plantations

Fig. 2.8 Woman sorting castor beans (Samuel Hazard 1971, 469)

grew it on vast acreages for lamp oil (McClellan 1992). In an era when candles were the only other source of illumination, the castor plant offered an important alternative. It is still planted for this purpose in some remote hamlets of Brazil that were founded by fugitive slaves (*quilombos*). In Minas Gerais, one elderly *quilombo* leader grows the castor plant in his kitchen garden for lamp oil to commemorate his ancestors (Fig. 2.7, right). João Ribeiro recounts the boyhood stories his great grandparents told him of the horrors of the Middle Passage crossing from Angola.[1] He lights the castor-oil lamp when offering devotional prayers to the family's protective saint. In the state of Maranhão, medicinal castor is frequently encountered in *quilombos*. In Bahia, as Robert Voeks observes in *Sacred Leaves of Candomblé*, it is an important plant in the religion's practice (Voeks 1997, 78–79; Pessoa de Barros and Napoleão 1998).

European commercial objectives do not alone offer a satisfactory explanation for the establishment of indigenous African plants in tropical America. Captains of slave ships certainly recognized the immediate value of stocking African dietary preferences as victuals for the Atlantic crossing and as medicines to maximize the survival of the human "commodities" they forcibly migrated to plantation societies. For slave-ship captains, the utility of African crops ended when their victims were disembarked and sold (Fig. 2.9). For the landed African captives, that utility was

[1] Nearly 40% of the slaves traded to the Americas went to Brazil, which exceeded more than three million Africans between the sixteenth and nineteenth centuries. Brazil was the last country in the Western Hemisphere to abolish slavery, which was decreed in 1888.

Fig. 2.9 Sea captains carousing in Surinam by John Greenwood, c. 1752–1758 (Reproduced with permission of the Saint Louis Art Museum)

never lost; it was recast and transformed in the plantation and mining societies of the New World. The familiar foods and medicines that accompanied them across the Middle Passage could now forestall hunger and treat ailments. Here a new narrative emerges, one that engages the role of enslaved Africans in instigating the cultivation of familiar plants in new lands.

Slave Agency in Instigating the Cultivation of African Foodstaples

Over 350 years of the transatlantic slave trade, it took an estimated 30,000 slave voyages to carry the 11 million Africans documented to have landed in the Americas (Eltis et al. 1999). In fact, until the first decade of the nineteenth century, more Africans arrived in the Americas than Europeans. The entire success of a slaving voyage, which might last several weeks or months, depended vitally on an adequate food supply to keep the captives alive en route. Although Amerindian maize and bitter manioc grown in Africa supplied some of the food demand, the provisions slave ships loaded in Africa also included African cereals and root crops. Significantly, despite the subsistence demands of the transatlantic slave trade, and the predations of the trade itself on the ablest segments of the population, societies of western Africa still managed to produce food in such quantity as to provision tens of thousands of slave voyages.

In the early colonial period, plantation owners encountered many new plants growing in the food plots of their slaves. Europeans referred to some species by geographical descriptors or toponyms that indicated their African provenance. Many of these dietary staples are still known in the Portuguese, Spanish, French, and

English languages by the place name "guinea," the name slave traders generally applied to the African continent. In English, we have guinea corn (sorghum), guinea sorrel (*Hibiscus sabdariffa*)—the plant that gives Red Zinger tea its color—guinea squash (*Solanum aethiopicum*), guinea melon (*Cucumis melo*), guinea pepper (*Xylopia aethiopica*), guinea grass (*Panicum maximum*), guinea yam (*Dioscorea cayenensis*), and even guinea fowl.

Other introduced African crops were named after specific African regions or slave ports, where surplus food was usually available and easily purchased for the Middle Passage. For example, the African pigeon pea was called Congo or Angola pea in English, *pois d'Angole* in French. One type of cooking banana or plantain is still known in Brazil as *banana* de *São Tomé*. In the former plantation areas of eastern Cuba, along Colombia's Caribbean coast, and in El Salvador, bananas are called *guineos*, after the region where Europeans first encountered the Old World tropical species.

Such toponyms draw attention to the importance of Africa as the source of food species critical to provisioning slave ships bound for the Americas. In this respect, slave vessels served as an inadvertent conduit for the introduction of African foodstaples to the Western Hemisphere. Even though not every slave ship stocked adequate food supplies, and most were consumed en route, leftover victuals were not an infrequent consequence of repeated Atlantic crossings.

Many of the plant introductions to tropical America are known in colonial languages by their African vernacular names. This draws attention to enslaved Africans who initiated their cultivation and to the sites in plantation societies where they established them. Plantation owners and European naturalists in the Americas encountered many novel plants in slave dooryard gardens and food plots. For the crops that had no existing words in European languages, they borrowed the names by which they were known to the slaves who grew and prepared them. Slave agency in the New World establishment of several introductions is suggested by the African loan words that were adopted by the colonial languages of former plantation societies. *Banana*, for example, is a word of African origin. So are yams and okra, and other plants grown in tropical America, such as *guandu*, *guandul*, *wando* (Portuguese, Spanish, and Dutch for the pigeon pea); *dendê*, *abbay* (Portuguese, Jamaican English for African palm oil); *quiabo*, *quingombó* (Portuguese, Spanish for okra); *bissy* and *eddo* (for kola nut and taro, respectively, in the English Caribbean); *pindal*, *goober* and *benne* (in South Carolina and the English Caribbean for the peanut and sesame, respectively) (Schneider 1991; Cassidy and LePage 2002; Carney and Rosomoff 2009).[2]

Many dish preparations of tropical America also carry African names. These include gumbo and the Caribbean one-pot stews known as *callaloo* and regional dishes that feature important dietary staples of Africa such as *mangú*, *mofongo* made

---

[2] A plant of South American origin, the peanut had not made it as far as mainland North America and parts of the Caribbean in pre-Columbian times. Established in Africa in the early sixteenth century, the peanut arrived in English plantation colonies as leftover provisions on slave ships. Slaveholders in these areas adopted the African names for a foodstaple with which slaves were quite familiar.

with plantains (Dominican Republic, Puerto Rico). A regional favorite of Maranhão, Brazil, is *arroz de cuxá* (rice with sorrel—the loan word *cuxá* deriving from West Africa's rice-growing Mandinka, who still cultivate and make several food preparations with sorrel, *Hibiscus sabdariffa*, which evolved among the local rice plantations whose slave populations included people from Senegambia and Guinea-Bissau). Until recently, Maranhão led Brazil in rice production by state, mostly by its mixed-race smallholders. In this former rice-growing area, *quilombo* descendants of runaway slaves narrate a history of rice beginnings. They attribute the crop's introduction to an enslaved woman who placed some rice grains in her hair as she disembarked the slave ship (Carney and Acevedo Marin 1999; Carney and Acevedo 2004; Carney 2004, 2005). Slave-ship captains often filled their stores with less-expensive unhusked grain from African rice-producing societies. Significantly, unmilled grain left from a slave voyage could have served as seed for planting.

In claiming an African woman initiated the cultivation of rice, the *quilombo* oral history offers a contrasting narrative to the Columbian Exchange, which acknowledges only the role of Europeans in disseminating and establishing plants on other continents (Carney 2001). The Maroon rice narrative substitutes the usual agents of global seed dispersal celebrated in Western accounts—European navigators, colonists, and men of science—with an enslaved African woman whose deliberate effort to sequester rice grains in her hair led to the establishment of an African dietary preference in tropical and subtropical America.

The *quilombo* rice account is also of significance because it identifies a slave ship as the conveyance for transatlantic seed transfers. The cereal grains, medicinal plants, and root crops occasionally remaining aboard slave voyages provided enslaved Africans opportunities to access familiar crops and to quietly cultivate them in their dooryard gardens and subsistence plots. In this way, slaves reinstated many African botanical species that were novelties to the plantation owners who held them in bondage.

While the slave ship served as the vessel by which African plants arrived in the Americas, the initiative for pioneering African foodstaples rested largely with slaves. This is because slaveholders did not readily adopt novel crops that they discovered in the food fields of their slaves. Just as the English in seventeenth-century Ireland scorned the New World potato as "Irish food," many African introductions were initially disdained by plantation owners as "slave food." The social and racial prejudice that divided slaveholders from those they enslaved also initially kept separate the foods they ate. But this wall of culinary segregation gradually disintegrated over time as signature ingredients of the African diaspora stealthily made their way into white kitchens and dining rooms (Wilson 1964). African foodstaples—among them, okra, greens, plantains, black-eyed peas, pigeon peas, sesame—infiltrated the cuisine of slaveholders through the dishes and confections their enslaved female cooks prepared for them. Today, we recognize such culinary signatures in the US South's *Hoppin' John* (made from rice and beans), *benne* (from sesame), hush puppies, and other deep-fried dishes known as fritters; the Caribbean pepper pot or *callaloo*, with greens as a prominent component as well as other one-pot stews featuring okra, such as the Louisiana *gumbo*; and *arroz de cuxá* (rice cooked with sorrel leaves), which is a regional favorite of Maranhão, Brazil.

Conclusion

The African crops carried to tropical and subtropical America during the slave trade owe their establishment to the first generations of Africans enslaved on New World plantations. Slaves grew these to ward off hunger, diversify their diet, reinstate customary food preferences, and to treat illness. The migrations of African plants in the period of plantation slavery are thus ineluctably tied to the institution and processes of the transatlantic trade in human beings. Slave ships carried Africa's botanical heritage, which gave uprooted Africans opportunities to establish them anew. In their dooryard gardens and food fields, slaveholders discovered them and at times even exploited their commercial potential (Carney and Rosomoff 2009).

Ownership of human beings imparted to slaveholders the right to appropriate the practices and cultivation methods that slaves used to secure their daily sustenance. Property rights gave plantation owners the power to claim that knowledge as their own and transmute it over time as proof of their presumed ingenuity. Slavery signifies not only an appropriation of the body and its labor but also the knowledge and ideas held by enslaved human beings. Significantly, it enabled the slaveholder to trade occasional favorable treatment for the knowledge and skills of the enslaved person's mind. This knowledge was crucial to the colonization of New World tropical lowlands.

A critical feature of human migration the world over is the preservation of traditional dietary preferences across space and the dislocations of geography. That the migration of Africans was compelled through extremes of violence and cruelty does not diminish this universal desire or preclude the possibility of achieving it. African staples enabled slaves at times to reinstate some food traditions of specific cultural heritages and to combine ingredients in new ways with Amerindian and European foods. In this way, slaves discretely modified the monotony of any food regimen slaveholders might impose. Africans and their descendants profoundly shaped the culinary traditions of slave societies, combining in new ways the foods of three continents in their struggle to secure daily sustenance. Through the dishes enslaved cooks prepared for their masters, African foods surreptitiously made their way onto planters' tables. In this way, the African plant introductions encouraged the distinctive—and today much celebrated—regional foodways that eventually developed across plantation societies. Africa's botanical legacy in the Americas is built upon this unacknowledged foundation.

Acknowledgments Earlier versions of this chapter were presented at the Michigan State University (2005), the University of Wisconsin (2006), the annual meetings of the Brazilian Studies Association (2008), and the *Society for Economic Botany* in May 2009. I wish to thank these audiences for raising provocative questions that guided this chapter's evolution. I am grateful to Antoinette WinklerPrins for bringing my attention to Fig. 2.5, her father, who kindly sent the Eckhout exhibition book from the Netherlands, and to Chase Langford, the UCLA Geography Department cartographer, for scanning and enhancing that image. I wish to thank Michael O'Grady for assistance with fieldwork in Pará (1996); Rosa Acevedo, my collaborator on *quilombo* fieldwork in Maranhão and Pará (1996, 1997, 2002); Leonard Abrams, Seu Benedito, Dona Maria, and Ivã with research on *quilombos* near Itapecurú, Maranhão in 2002. Additional gratitude is extended to Jacque Chase, my collaborator in the Minas Gerais *quilombo* research in 2005.

References

Alpern SB (2008) Exotic plants of Western Africa: where they came from and when. Hist Afr 35:63–102

Barbot J (1752) A description of the coasts of North and South Guinea; and of Ethiopia inferior, vulgarly Angola; being a new and accurate account of the Western Maritime Countries of Africa. In: Churchill A (ed) A collection of voyages and travels, some now first printed from original manuscripts, others now first published in English, 8 vols, vol V. Printed from Messieurs Churchill for T. Osborne, London, p 99

Buvelot Q (2004) Albert Eckhout: a Dutch artist in Brazil. Waanders Publishers, The Hague

Carney J (2001) Black rice: the African origins of rice cultivation in the Americas. Harvard University Press, Cambridge, MA

Carney J (2004) 'With grains in her hair': rice history and memory in colonial Brazil. Slavery Abolit 25(1):1–27

Carney J (2005) Rice and memory in the age of enslavement: Atlantic passages to Suriname. Slavery Abolit 26(3):325–347

Carney J, Acevedo R (2004) Saberes agrícolas dos escravos africanos no Novo Mundo. Ciência hoje 35(205):26–33

Carney J, Acevedo Marin R (1999) Aportes dos escravos na história do cultivo do arroz africano nas Américas. Estudos sociedade e agricultura 12:113–133

Carney J, Rosomoff R (2009) In the shadow of slavery: Africa's botanical legacy in the Atlantic World. University of California Press, Berkeley

Cassidy FG, Le Page RB (2002) Dictionary of Jamaican English. University of the West Indies Press, Kingston

Chase A (1944) Grasses of Brazil and Venezuela. Agric Am 4:123–126

Corrêa do Lago P (2001) Iconografia Brasileira. Itáu Cultural, São Paulo

Crosby AW (1972) The Columbian exchange: biological and cultural consequences of 1492. Greenwood Press, Westport

Crosby AW (1986) Ecological imperialism: the biological expansion of Europe, 900–1900. Cambridge University Press, New York

Cymbron T, Loftus RT, Malheiro MI, Bradley DG (1999) Mitochondrial sequence variation suggests an African influence in Portuguese cattle. Proc R Soc Lond B 266:597–603

de Langhe E (1995) Banana and plantain: the earliest fruit crops? Annual report, INIBAP, Montpellier

Debret JB (1954) Viagem pitoresca e histórica ao Brasil: 1834–1839, 2 vols. Livraria Martins, São Paulo

Donkin RA (1991) Meleagrides: an historical and ethnogeographical study of the guinea fowl. Ethnographica, London

Eltis D, Behrendt S, Richardson D, Klein H (1999) The trans-Atlantic slave trade: a database on CD-ROM. Cambridge University Press, Cambridge

Gamble DP, Hair PEH (eds) (1999) The discovery of River Gambra (1623) by Richard Jobson. Hakluyt, London

Ginja C, Penedo MCT, Melucci L, Quiroz J, Marínez López OR, Revidatti MA, Martínez-Martínez A, Delgado JV, Gama LT (2009) Origins and genetic diversity of New World Creole cattle: inferences from mitochondrial and Y chromosome polymorphisms. Anim Genet 41:128–141

Grimé W (1979) Ethno-botany of the Black Americans. Reference Publications, Algonac

Hair PEH, Jones A, Law R (1992) Barbot on Guinea: the writings of Jean Barbot on West Africa, 1678–1712, 2 vols. Hakluyt, London

Harlan JR (1975) Crops and man. Crop Society of America, Madison

Hazard S (1971) Cuba with pen and pencil. Hartford Publishing Co., Hartford

Kimber CT (1988) Martinique revisited: the changing plant geographies of a West Indian island. Texas A & M University Press, College Station

Kleiman K (2003) The pygmies were our compass: Bantu and Batwa in the history of West Central Africa, early times to c. 1900 C.E. Heinemann, Portsmouth

Kupperman KO (1993) Providence Island, 1630–1641: the other Puritan colony. Cambridge University Press, New York

Ligon R (1970 [1647]) A true & exact history of the island of Barbadoes. Frank Cass & Co., London

Lirón JP, Bravi CM, Mirol PM, Peral-García P, Giovambattista G (2006) African matrilineages in American Creole cattle: evidence of two independent continental sources. Anim Genet 37:379–382

MacNeish R (1992) The origins of agriculture and settled life. University of Oklahoma Press, Norman

Magee DA, Meghen C, Harrison S, Troy CS, Cymbron T, Gaillard C, Morrow A, Maillard JC, Bradley DG (2002) A partial ancestry for the creole cattle populations of the Caribbean. J Hered 93:429–432

Marcgrave J (Georg) (1942 [1648]) História natural do Brasil. Imprensa Oficial do Estado, São Paulo

Mbida CM, Van Neer W, Doutrelepont H, Vrydaghs L (2000) Evidence for banana cultivation and animal husbandry during the first millennium BC in the forest of southern Cameroon. J Archaeol Sci 27:151–162

McClellan JE III (1992) Colonialism and science: Saint Domingue in the old regime. The Johns Hopkins Press, Baltimore

McNeill JR (2000) Biological exchange and biological invasion in world history. Paper presented at the 19th international congress of the historical sciences, Oslo, 6–13 August

Mercer J (1973) Canary Islands: Fuerteventura. Stackpole Books, Harrisburg

National Research Council (NRC) (1996) Lost crops of Africa, vol 1, Grains. National Academy Press, Washington, DC

National Research Council (NRC) (2006) Lost crops of Africa, vol 2, Vegetables. National Academy Press, Washington, DC

Parsons JJ (1970) The "Africanization" of the New World tropical grasslands. Tübinger Geographische Studien 34:141–153

Pessoa de Barros JF (1993) A galinha d'angola: Iniciação e identidade na cultura Afro-Brasileira. Universidade Federal Fluminense, Rio de Janeiro

Pessoa de Barros JF, Napoleão E (1998) Ewé Òrìsà: Uso litúrgico e terapêutico dos vegetais nas casas de candomblé Jêje-Nago. Betrand Brasil, Rio de Janeiro

Pike R (1967) Sevillian society in the sixteenth century: slaves and freedmen. Hisp Am Hist Rev 43(3):344–359

Piso G (Willem) (1957 [1648]) História natural e médica da India Ocidental. Instituto Nacional do Livro, Rio de Janeiro

Restall M (2000) Black conquistadors: armed Africans in early Spanish America. Americas 57(2):171–205

Rugendas JM (1954) Viagem pitoresca através do Brasil. Livraria Martine Editôra, São Paulo

Rumeu de Armas A (1956) España en el Africa Atlántica. Consejo Superior de Investigaciónes Científicas, Madrid

Saunders ACdeCM (1982) A social history of black slaves and freedmen in Portugal, 1441–1555. Cambridge University Press, Cambridge

Schneider JT (1991) Dictionary of African borrowings in Brazilian Portuguese. Helmut Buske Verlog, Hamburg

Slenes R (2002) African Abrahams, Lucretias and men of sorrows: allegory and allusion in the Brazilian anti-slavery lithographs (1827–1835) of Johan Moritz Rugendas. Slavery Abolit 23(2):147–169

Sloane H (2001) Page 254, vol I. Quoted in Edwards B (2001) [1793] History, civil and commercial, of the British colonies in the West Indies, 5 vols. Adamant Media, Chestnut Hill

Voeks R (1997) Sacred leaves of candomblé. University of Texas Press, Austin

Watts D (1987) The West Indies: patterns of development, culture, and environmental change since 1492. Cambridge University Press, Cambridge

Watts D (2000) Early Hispanic new world agriculture, 1492 to 1509. In: Shepherd V, Beckles HMcD (eds) Caribbean slavery in the Atlantic World: a student reader. Marcus Wiener, Princeton, pp 136–152

Wilson MT (1964) Peaceful integration: the owner's adoption of his slaves' food. J Negro Hist 49(2):116–127

Chapter 3
Did Enslaved Africans Spark South Carolina's Eighteenth-Century Rice Boom?

Stanley B. Alpern

Abstract Beginning in the mid-1970s, students of the rice-growing boom that made South Carolina rich in the eighteenth century began focusing on the role of the enslaved Africans who grew the crop. They decided that the role consisted of much more than sheer brawn, that the slaves brought centuries of rice-cultivation experience in West Africa to the task, and indeed a whole cultural complex. The main advocates of this new historical formulation were Peter H. Wood, Daniel Littlefield, and particularly Judith Carney. Their viewpoint seemed generally accepted until, in 2005, it was sharply criticized by three eminent scholars, David Eltis and David Richardson, historians of the Atlantic slave trade, and Philip Morgan, a historian of American slavery. They contended that the importance of the slaves' contribution to the rice bonanza had been greatly exaggerated. This paper, in turn, challenges their position on many points, large and small. And it concludes that the achievement of white masters and black chattel in the Lowcountry was a unique synthesis owing at least as much to the slaves as to the planters.

Keywords African rice growing • *Oryza glaberrima* • *Oryza sativa* • Atlantic slave trade • Columbian Exchange • Lowcountry rice growing • African/American parallels • Gullah • Division of labor

Introduction

Africanists and students of the African diaspora are deeply indebted to David Eltis and David Richardson for their leading roles in a prodigious research project: the Trans-Atlantic Slave Trade Database, published on a CD-ROM in 1999 and updated online beginning in 2008. At last count, the project had assembled information about nearly

S.B. Alpern, A.B., M.A. (✉)
1575 Avenue Olivula, Quartier Saint Michel, 06230 Villefranche-sur-Mer, France
e-mail: stan.alpern@wanadoo.fr

R. Voeks and J. Rashford (eds.), *African Ethnobotany in the Americas*,
DOI 10.1007/978-1-4614-0836-9_3,

35,000 slaving voyages that brought 10,700,000. Africans to the New World. It was estimated that about 82% of all slaves shipped from Africa were already accounted for in the database (Eltis and Richardson 2008: 9, 43, www.slavevoyages.org/tast/database/index.faces).

It therefore came as a surprise, to me at least, to learn that Eltis and Richardson, along with Philip Morgan, a noted historian of American slavery, had mounted a strong challenge to research indicating that slaves with rice-growing experience in Africa had played a pivotal technological part in the rice-growing-and-exporting boom that made South Carolina the richest area in North America for a good while in the eighteenth century. Their critique first appeared in 2005 in a paper presented at a University of Georgia conference (Eltis, Morgan, and Richardson [hereafter EMR] 2005). In revised form, the paper was published in the *American Historical Review* in December 2007 under the title "Agency and Diaspora in Atlantic History: Reassessing the African Contribution to Rice Cultivation in the Americas."

Eltis, Morgan, and Richardson (sometimes called just "the authors" in this paper) carried their challenge to what they labeled the "black rice hypothesis" (after a book title) quite far, reaching the (for me) startling conclusion that rice cultivation in the Lowcountry – i.e., the marshy coasts and Sea Islands of South Carolina and Georgia – "would have followed a similar trajectory if the slave trade had drawn exclusively on Angola," where rice was not grown. They attributed the targeted hypothesis to "[a] desire to celebrate an African accomplishment in the New World" and contended that such interpretations "are drawn by tracking migrants from one side of the Atlantic to the other, and tracing cultural connections between the societies they left behind and the societies to which they subsequently contributed. Historians engaged in such an exercise," they observed rather caustically, "are unlikely to conclude that such links are unimportant" (EMR 2007: 1357). One might with equal justice (or injustice) contend that historians focused on the slave trade, like Eltis and Richardson, are unlikely to conclude that such cultural connections are important.

The authors concentrated on three books that, "in ever stronger form," made the case that "the major export crop of eighteenth-century South Carolina and Georgia – rice – [was] predominantly a creation of Africans" (EMR 2007: 1332). These were Peter H. Wood's *Black Majority: Negroes in Colonial South Carolina from 1670 Through the Stono Rebellion* (1975), Daniel C. Littlefield's *Rice and Slaves: Ethnicity and the Slave Trade in Colonial South Carolina* (1981), and Judith A. Carney's *Black Rice: The African Origins of Rice Cultivation in the Americas* (2001a).

The three critics found that "[t]he basic argument" of the so-called black-rice school "rests on three core elements," namely, that rice growing was indigenous to Africa and of long standing there, that in the Americas non-slave labor could never be induced to grow rice for export, and that "putative parallels have emerged between rice cultivation in Africa and…the Americas" (Eltis et al. 2007: 1333). (The original 2005 paper spoke of "fascinating" parallels. Presumably, this was changed to "putative" to reduce their weight).

Oryza glaberrima

There is no major dispute over the first two elements, but researching the first is instructive. Rice, specifically *Oryza glaberrima* Steud., was domesticated in West Africa independently and surely without prior knowledge of the Asian domesticate, *O. sativa* L. This probably happened in the first millennium BCE, most likely in what is known as the inland (or central) delta region of the Niger River in the Malian Sahel. The oldest known *O. glaberrima* was found in 1981 in the earliest levels of excavations at the site of Jenne-Jeno, an extinct settlement founded around 250 BCE (McIntosh and McIntosh 1981: 15–16; Shaw et al. 1993: 629, 631; McIntosh 1995: 352, 377; personal communication from her, 2009). Rice remains perhaps dating to the eighth century BCE have been found at Dia, northwest of Jenne-Jeno, but it may have been wild (Murray 2004: 1–3).

In the first millennium CE, West Africans are thought to have carried *O. glaberrima* to the Atlantic coast, first to the far west in Senegambia and later all along the coast from Senegal's Saloum Delta south and east to the Axim area on the southwestern coast of modern Ghana, a stretch of c.1,450 miles, not counting coastal indentations and protuberances. That is roughly the distance from Boston to Miami.

The French botanist Roland Portères, who brought African rice to much wider attention among savants – as knowledgeable an Africanist as George Peter Murdock was unaware of *O. glaberrima* until his classic 1959 work, *Africa: Its Peoples and Their Culture History*, was in press (see his footnote on p. 68) – speculated that sometime between 1500 and 800 BCE, rice-growing pioneers were the builders of megalithic circles found in the Gambia region. He put forward this widely echoed theory in 1950 and maintained it to his death in 1974 (Portères 1950: 490–491, 1976: 444–445). But fresher research indicates that the circles date to the first millennium CE or later, and no link has been found to rice cultivation (Thilmans et al. 1980: 153; Ozanne 1966: 11, 12; Fagan 1969: 150–151; McIntosh and McIntosh 1993: 104–105; Lawson 2001: 32; SK McIntosh, 2009, personal communication).

It would appear, therefore, that rice growing reached and spread along the Atlantic coast in the first millennium CE, but we have no written evidence until European explorers burst upon the scene in the fifteenth century. Sometime between 1446 and 1451, a Portuguese caravel captain, Estêvão Afonso, went ashore in what was probably the Gambia River and reported many rice fields in marshland (Zurara 1960: 246). The Venetian Alvise da Ca' da Mosto confirmed that report in 1456 (Crone 1937: 70). In 1462, Pedro de Sintra reported rice on the Kaloum Peninsula, site of Conakry in modern Guinea (Crone 1937: 80). In 1479, the Fleming Eustache de La Fosse saw "a lot of good rice" that was said to have come from the nearby Iles de Los (La Fosse 1897: 185).

The Portuguese historian Duarte Pacheco Pereira, writing in the first decade of the sixteenth century, confirmed the presence of rice in modern Guinea and was the first to report its cultivation in today's Guinea-Bissau, Sierra Leone, and the San Pedro/Sassandra region of Côte d'Ivoire. He said it grew abundantly all along those coasts (Pacheco Pereira 1937: 91, 92, 94, 97, 99, 106, 107, 115; Fernandes 1951: 47, 49, 55, 57, 59, 73, 81, 95, 99). In 1519, Martín Fernández de Enciso said rice was also eaten on the coast beyond Sierra Leone as far as Cape Three Points (just beyond Axim) in modern Ghana (Enciso 1932: 105). By c. 1544–1545, a French sailor, Jean Alfonce, reported rice as a staple on what would become the Liberian coast (Hair 1976: 30). As early as 1491, Portuguese colonists on São Tiago (Santiago), principal island of the Cape Verdes, were importing rice from the African mainland, presumably to feed the slaves they were bringing from the same region (Lauer 1969: 42; Blake 1942: 1:87). By the next decade, they were growing it (Fernandes 1951: 115; Anonymous Portuguese Pilot 1550: 125–129).

Within a century, and despite the relative paucity of written reports, the presence of rice farming was thus fairly well documented from Senegal almost to the Bandama River, which bisects Côte d'Ivoire and still divides West Africa's rice-eating west from its yam-eating east (Miège 1954: 27–28, 30–32). As we have seen, Enciso reported rice further east, on the Ghanaian coast. This was confirmed at the turn of the seventeenth century by the Dutchman Pieter de Marees (1987: 62, 63, 159) and the turn of the following century by his compatriot Willem Bosman (Bosman 1967: 6–7; Jones 1995: 220). In the gap between the Bandama River and Axim, French visitors to Assinie in the southeastern corner of Côte d'Ivoire reported rice cultivation in 1692 and 1701 (Roussier 1935: 60, 168–169, 184, 191, 213). A modern French scholar, Claude-Hélène Perrot, suggests that the use of rice in basic local rituals, as noted in 1701, points to its longtime presence in the area (Perrot 1990: 13).

The early Europeans identified not only places where rice was grown but coastal peoples who grew it and still do. Between them, Pacheco Pereira and Valentim Fernandes wrote of the Diola (Jola), Fulani (Fula, Peul), Malinke (Mandinka), and Serer of Senegambia; the Balanta (Balante) and Nalu of Guinea-Bissau; the Banyun (Banhun), Biafada (Biafar), Dialonke (Jalonke), Landuma, Susu, and Tiapi of the Republic of Guinea; the Bulom and Temne of Sierra Leone; and the Grebo of Liberia and Côte d'Ivoire (Pacheco Pereira 1937: 92, 94, 97, 99, 107; Fernandes 1951: 47, 49, 55, 57, 59, 81; Rivière 1968: 733; Hair 1967a: 249–257). A French navigational manual of the 1540s contains a vocabulary identified in 1964 as that of the Kra (Kran) of Liberia and including their word for rice (Dalby and Hair 1964: 174, 178, 181–182). Before the end of the sixteenth century, the Kasanga of Senegambia, the Bram (Buramo) and Bijago (Bijogo, Bissago) of Guinea-Bissau, and the Baga of Guinea had also been identified as rice growers by two Cape Verdeans who separately visited the mainland in the 1570s and 1580s (Almada 1984: 72, 82, 92, 99; Donelha 1977: 99). The rice-growing kingdom of Kquoja described by the Dutch geographer Olfert Dapper appears to have spoken Vai (Sierra Leone and Liberia) between the 1630s and 1650s (Dapper 1668: 395–396; Hair 1964: 129–139). The Sape (Sapi) of Sierra Leone were linked to rice by the 1660s and the Papel (Pepel) of Guinea-Bissau by the 1680s (Hair 1982: 59–60; Cultru 1913: 214).

Eltis, Morgan, and Richardson acknowledge the early presence of rice along the West African coast but try to play it down. "Even in…coastal Guinea-Conakry and Guinea-Bissau…," they aver, "which are usually assumed to be predominantly rice-growing areas, rice cultivation was only a part of agricultural activities – and often a small part until well into the eighteenth century." They offer only one source for this, and name only one ethnic group, the Balanta of Guinea-Bissau (Eltis et al. 2007: 1345–1346). Moreover, their evidence is debatable. It is based on the testimony, probably dating to 1612–1613, of Portuguese missionary Manuel Álvares, who was stationed in Sierra Leone and was clearly no expert on the Balanta. In his 90,000-word text, he devoted two paragraphs to the Balanta and 26 chapters to the inhabitants of Sierra Leone. Moreover, while he attributed "little rice" to the Balanta (and the neighboring Biafada, who rated four paragraphs), he credited other peoples from Senegambia to Sierra Leone with abundant rice production (Hawthorne 2003: 35–36, 152–153; Álvares 1990: parts 1 & 2). More than a century earlier, Fernandes included the "Balangas" (whom the late British Africanist P.E.H. Hair equated with the Balanta) among rice-growing peoples of the Casamance (bordering Guinea-Bissau), and, two decades before Álvares, the Cape Verdean merchant André Álvares de Almada reported that the Balanta traded rice to the Biafada (Fernandes 1951: 59, 160 n. 115; Almada 1984: 92; Hawthorne 2003: 51–52 n. 38; Hair 1967a: 251–252, Hair 1967b: 39–40; 63 n. 42–45).

One would think European seamen and traders would have named at least portions of the coast for rice, its primary staple, but they chose other labels instead: Guinea of Cape Verde, Rivers of Guinea, Rivers of the South, Ivory or Tooth Coast, Windward Coast, or Upper Guinea (for almost the whole extent). Liberia was the Malagueta or Pepper or Grain Coast (an appellation that spanned four centuries), the grain not for any cereal but for the spice *Aframomum melegueta* K. Schum. aka grains of paradise. "Rice Coast" does not seem to have been coined until the second half of the eighteenth century, when it appeared in South Carolina publicity for the auction of newly arrived slaves. Eltis, Morgan, and Richardson dismiss it as "mercantile advertising" and "marketing gambits" (EMR 2007: 1348–1349); S. Max Edelson, a historian of Lowcountry colonial plantations, sees it as a belated "indirect recognition of African capabilities" (Edelson 2006: 78).

The rice that Europeans originally found growing from the Saloum Delta to Axim must have been *O. glaberrima*, but its resemblance to *O. sativa*, brought earlier by Arabs to the Mediterranean region (Lauer 1969: 40, 44), kept Europeans from recognizing it as a distinct species. Some did notice differences. An anonymous Dutchman remarked in the mid-seventeenth century that "a large amount of grey rice" was available on the St. Paul River in today's Liberia (Jones 1995: 21). This may have been a way to distinguish it from Asian rice. A few decades later, the Huguenot Jean Barbot bought rice on Liberia's Cess River and reported it was "not as large or as white as that of the Levant," which he later changed to "as that of Milan and Verona," meaning in both instances *O. sativa* (Hair et al. 1992: 1:268, 282 n. 13). A prescient ship captain named Nathaniel Cutting, who, in 1790, supplied Thomas Jefferson with a ten-gallon keg of African upland rice from Sierra Leone, wrote him that the "Red Rice, as this kind is sometimes call'd," was "a distinct species" (Richards 1996: 217). In 1792, Royal Navy Capt. Philip Beaver noticed the difference in the flavor of African and Carolina (Asian) rice in Guinea-Bissau (Beaver 1805: 346). But African rice was not identified botanically until a German specialist, Ernst Gottlieb Steudel, studied samples of *O. glaberrima* collected on the Iles de Los in 1845–1848 by a Frenchman and named the species in 1855. He knew the species grew in West Africa but did not claim it was domesticated there (Steudel 1855: 1:3; Portères 1955a: 536).

Steudel chose the name for the plant's glabrous stem that, unlike *O. sativa*'s, usually lacks bristles or down. But the only obvious difference between the two species is ligule shape and length. Grain shape can also be used to distinguish them, but it takes an expert to recognize the disparity. For the layman, the big difference, as Capt. Cutting indicated, is color. While both African and Asian kernels are white, *O. glaberrima* has bran ranging from red to brown, at least when not fully mature. It has often been called red rice, though some *sativa* varieties also have red bran.

Oryza sativa

The main reason *O. sativa* has replaced *O. glaberrima* in much of West Africa is its higher yield. Moreover, the husk is easier to remove and, in *japonica* varieties at least, the grain is less likely to shatter. *O. glaberrima* tends to have weaker stalks that windstorms can topple, especially when compared to *japonica*. Imperfectly hulled *O. glaberrima* mixed with white *O. sativa* adds red coloring that reduces the cereal's market value (and red *O. sativa* does likewise). *O. sativa* is said to exhibit greater diversity and flexibility than *O. glaberrima*. On the other hand, *O. glaberrima* has significant virtues. It is more tolerant of soil infertility, salinity and excessive iron, fluctuations in water depth, and human neglect. It seems to compete better against weeds. It may be more resistant to some diseases and pests. If it is less genetically diverse than *O. sativa*, it is nevertheless plastic and has adapted to many different soils, terrains, and climates. Some types mature faster than *O. sativa*, providing an emergency food. In the Inland Niger Delta region, farmers developed a floating *O. glaberrima* capable of surviving floods three meters deep. Some Africans favor *O. glaberrima* for its taste, aroma, and even its redness. And to some, traditional rituals calling for rice are meaningless unless the old standby is used.

The best written source I have found for distinguishing African from Asian rice is the National Research Council's *Lost crops of Africa, volume 1, Grains*. The authors, advocates of *O. glaberrima*, even suggest that in practice it is more nutritious because it is harder to polish. *O. sativa* is invariably polished to a greater degree and thus loses more of its nutrients. Another good source is Olga F. Linares, African rice (*Oryza glaberrima*): History and future potential, available online via Wikipedia (NRC 1996: 17, 21–22, 25, 28–29, 33; Linares 2002: 3; Richards, 2009, personal communication; Nuijten, 2009, personal communication).

The Portuguese are thought to have introduced *O. sativa* to western Africa sometime in the sixteenth century, but evidence is slender. The earliest source may be Alfonce who, in a report dating to c. 1536, said people of the kingdom of Benin (southern Nigeria) ate rice. This was well beyond the coastal range of *O. glaberrima*, but, if Alfonce was accurate, it could conceivably have descended from the north since *O. glaberrima* had spread eastward from the Inland Niger Delta as well as to the Atlantic coast (Hair 1976: 35). André Donelha, who visited Sierra Leone c. 1574–1585, said it produced "the best [rice] in all our Guinea,…as attractive as that of Valencia and very white." His modern translator, Hair, suggested it was *O. sativa* (Donelha 1977: 81, 211 n. 48). In 1599, a German soldier, Johann von Lübelfing, reported local rice at Cape Lopez (Gabon), a more likely place than old Benin or Sierra Leone for the Portuguese to have introduced *O. sativa* (Jones 1983a:13). Marees said the rice he saw on the Gold Coast in 1601 had come from India and had "little yellowish skins" and "Grains…whiter than the best Wheat." This sounds like *O. sativa*, but Marees's modern editors think it was too early for Asian rice to have reached the Gold Coast (Marees 1987: 159 & n. 3; Lauer 1969: 43–44). Carney has plausibly identified Marees's illustration of a rice plant as *O. glaberrima* and suggests it had moved eastward from its coastal heartland to the Elmina area (Marees 1987: 158; Carney 2005a: 333–334).

Columbian Exchange

O. sativa was clearly established in western Africa by the mid-seventeenth century, if only on the then Portuguese islands of Fernando Po (Bioko), Príncipe, São Tomé, and Annobón (Pagalu), which would logically have been among the first places to bring seed to from the Indian Ocean (Jones 1995: 58; Dapper 1676: 2:67, 68; Lauer 1969: 42–43). By then, rice was already on the way to becoming one of the major slave-ship provisions, perhaps ranking third after yams and maize, based on records I have kept of provisions mentioned in the slave-trade literature. (Since my records are far from exhaustive, rice may have ranked a bit lower. As the Slave Trade Database has now made clear, the Portuguese achieved an inglorious first place in slaving (Eltis and Richardson 2008: 40–41), and their slave menu from Congo/Angola often featured cassava meal or flour. Another very common provision was broad or horse beans [*Vicia faba* L.] from Europe.)

São Tiago may have been supplying rice to Brazil-bound slave ships from the mid-sixteenth century (Duncan 1972: 167–168). Álvares, who lived in Sierra Leone from 1607 to c. 1616, seems to have been talking of slavers when he described the purchase of rice as ship victuals at the ports of Cacheu and Bissau (Guinea-Bissau) (Álvares 1990: part 1, ch. 5, p. 4; ch. 11, p. 1). Rice bought c. 1616 on the Cess River (Liberia) by a Dutch ship was not meant for slaves – the vessel took malagueta pepper, ivory, and gold back to Holland – but it suggests that the cereal was already provisioning slavers on that part of the coast (Jones 1983a: 78). Cape Verdean Francisco de Lemos Coelho told of rice being sold to European ships, apparently slavers, on the coast from Gambia to Sierra Leone in the 1640s or 1650s (Lemos Coelho 1985: ch. 2, pp. 11, 29; ch. 3, p. 7; ch. 9, pp. 2, 6, 24).

There can be little doubt that these early shipments of rice with slaves from Upper Guinea and the Cape Verdes involved *O. glaberrima*. It of course was never identified as such because the slave trade was just about over when the two rice species were scientifically distinguished. Even as late as 1793, a slave ship loaded one and a half tons of red rice in Guinea-Conakry that may have been unmilled *O. glaberrima* (Mouser 2002a: 86 & n. 282). But the evidence lies in the presence of *O. glaberrima* in the Western Hemisphere.

The earliest transfer may have occurred not on slave ships but on vessels taking Portuguese settlers to Brazil. A 1549 expedition that founded Salvador da Bahia stopped at São Tiago to take on livestock and crops (Duncan 1972: 167). Among the crops transferred from the Cape Verdes, either on that voyage or one not long after, was seed for both wet and dry rice. (Others included sugarcane, coconuts, ginger, and African yams.) The rice was termed by a contemporary landowner "as full-grained and handsome as that of Valencia" (Ribeiro 1962: 153). The comparison with Valencian rice (*O. sativa*) gives pause, but more likely *O. glaberrima* was involved because there is no sign *O. sativa* had reached that far west in Black Africa that early. Modern French plant geneticists have assembled strong evidence for the penetration of African rice into Brazil, including the absorption of *O. glaberrima* characteristics by local wild species and the presence of *Oryza barthii* A. Chev., the wild progenitor of *O. glaberrima*, which must have tagged along when it crossed the ocean (Second and Rouhan 2008: 6, 8, 10–11).

Carney offers evidence from Portuguese-language sources that rice, at least some of it described as red, was long grown in Brazil by slaves as a subsistence crop. The red-coated grain seems also to have been cultivated by others, because in 1772, the Portuguese colonial government decreed a 1-year jail sentence for whites who grew it (and a 2-year term for blacks and Amerindians). This suggests to Carney that *O. glaberrima* was involved because its admixture with *O. sativa* would have lowered the value of rice grown for export (Carney 2001a: 151–152, b: 388–389, 2004: 16–17; 2005a, b: 336–337, 340, 2008: 101). (Against the apparently unanimous view of botanists and agronomists that rice has been domesticated only twice, first in Asia, then Africa, the authors of "Agency and Diaspora" assert [p. 1342] that a red rice grown in Brazil's Maranhão state was "native to the Americas," supporting that claim with a single source.)

Elsewhere in South America, Carney points to reports or hints of slaves-cum-subsistence-rice in Colombia, Ecuador, and Peru (Carney 2001a: 75, 78, 196 n. 51; Crosby 1972: 70, 107), but the case for *O. glaberrima* is much stronger in French Guiana and Suriname. In 1938, an agricultural official named Vaillant came across a sample of *O. glaberrima* in French Guiana that Portères later found to be identical to a variety grown by the Baga of Guinea. The Missouri Botanical Garden Herbarium reportedly contains an *O. glaberrima* specimen described by Portères in 1944 (Portères 1955b: 680; Carney, 2009, personal communication). Vaillant went on to describe a number of apparent *O. glaberrima* varieties grown in the region of the Maroni River that divides French Guiana and Suriname (Vaillant 1948: 520–529). Carney suggests that Dutch planters who left Brazil when the Portuguese reestablished control in 1654 and settled in Suriname brought elements of the Brazilian plantation system with them, notably "the right of a slave to an individual garden plot as well as control over income derived from the sale of produce grown on it," including rice (Carney 2005a, b: 336–337). In 1928–1929, Melville and Frances Herskovits found Saramaka Maroons in Suriname growing what they recognized as an African rice on slopes. They also saw a woman who stamped, pounded, and winnowed rice African style (Herskovits and Herskovits 1934: 100, 214).

There are hints of *O. glaberrima* on Caribbean islands – an early-eighteenth-century report of slaves in Jamaica sowing subsistence rice that was hard to husk (Sloane 1707–1725: 1:103), a specimen collected in Cuba before 1877 and kept in London's Natural History Museum (Hunnex, 2010, personal communication), and a reference to rice growing by slaves in Martinique (Edelson 2006: 67) – but more evidence in Central America. Portères devoted an article to the *O. glaberrima* of El Salvador, which he linked to the Susu of Guinea (Portères 1960: 442–443, 1976: 441). He likewise traced *O. glaberrima* in Panama to the Susu (Portères 1976: 441; Edelson 2006: 67) and is also said to have collected the grain in Costa Rica (NRC 1996: 33). Nicaragua and Honduras may also have been host to African rice (Carney 2001a: 195–196 n. 51).

None of the rice that reached South Carolina after the arrival of English colonists in 1670 was provably *O. glaberrima* – no trace has yet been found in archaeological excavations. But enough circumstantial evidence exists to make its introduction quasi-certain. Before detailing that evidence, I might usefully observe that a slew of other plants either domesticated in Africa or domesticated elsewhere but transmitted from Africa during the slave trade reached the New World largely unnoticed. They include some of the most important additions to the Western Hemisphere's crop inventory: akee, cowpea (black-eyed pea), kola nut, okra, roselle/sorrel, tamarind, and watermelon among African domesticates; cinnamon, ginger, indigo, pigeon pea, and taro of Asian origin; chickpeas and purslane of Mediterranean provenance; and the castor-oil plant and sesame of disputed pedigree (Grimé 1979; Carney 2001b: 377–396, 2003: 167–185).

African crops well documented as slave-ship migrants to the Americas were yams – doubtless preponderantly the African domesticates *Dioscorea minutiflora* Engl. and *D. cayennensis* subsp. *rotundata* (Poir.) J. Miège but including two Asian species – malagueta pepper, millet, and sorghum. Sorghum was never identified as such: it was often confused with millet or called "Guinea corn" or "country corn." In French, it was *gros mil*. Eltis, Morgan, and Richardson cite sorghum (EMR 2007: 1356 n. 58) as an example of an African domesticate that "did not take off as an Atlantic crop" because, though slaves knew all about it, Euro-American entrepreneurs were not interested. This was not a particularly apt choice by the authors, who themselves mistake "Guinea corn" and "country corn" for millet (EMR 2007: 1345), since the United States is now the world's no. 1 producer and exporter of sorghum (mainly as livestock feed). It ranks third among US cereals after corn and wheat. Mexico, Argentina, and Brazil are also major producers. The African oil palm too crossed the Atlantic, but it was the oil itself that was frequently listed as a slave-ship provision – I have found only two references to the nut (Jones 1995: 74, 119). Crops of ultimately Asian origin on slave ships included *Oryza sativa*, plantains and bananas, citrus fruits (oranges, lemons, and limes), coconuts, and sugarcane.

O. glaberrima apparently reached South Carolina in the early years of the colony, brought not by slaves, as the transfer of African crops is sometimes described, but by slavers. The slaves barely had clothes, much less baggage. So whatever the rice grown there, it cannot rightly be put on the slaves' side of the ledger. The only contrary claim is a legend among Maroon descendants in Suriname and French Guiana and slave descendants in Brazil that one or more women on slave ships introduced rice to the Americas by hiding unmilled grains in their hair (Vaillant 1948: 522; Carney 2001a: 153–154, 2005a: 339–340, b: 219; 318 n. 61, 2008: 101). Eltis, Morgan, and Richardson mock the "black rice" school on this point (EMR 2007: 1357): "The possibility of a single enslaved African woman carrying a few grains of rice in her hair can become all that is necessary to sustain this thesis [that] founding migrants have an influence all out of proportion to their numbers." This strikes me as an unwarranted *reductio ad absurdum* of the plausible argument that it did not require great numbers of slaves with rice-growing experience to introduce their knowledge into South Carolina.

Evidence for the presence of African rice there is scattered and ambiguous but on the whole credible. Around 1688, the provincial governor, Nathaniel Johnson, is said to have planted rice, "but it being a small, unprofitable Kind little Progress was made in its Increase." It was contrasted with a variety that arrived from Madagascar (where Asian rice had been introduced many centuries earlier by Indonesian voyagers). "The first Kind is bearded,…and requires to grow wholly in Water. The other is larger, and brighter, of a greater Increase, and will grow both in wet and tolerable dry Land." The author thought one variety turned red because it was sown too often in the same soil (Catesby 1731–1743: 1:xvii).

Letters written in 1690 indicate that various kinds of rice were being tried out, wet and dry, broadcast, or planted in rows (Littlefield 1981: 102). A reference to two types being grown in the 1690s spoke of "one called Red Rice in Contradistinction to the White, from the Redness of the inner Husk or Rind of this Sort, tho' they both clean and become white alike" (Carney 2001a: 144). In the same decade, a Portuguese slave ship reportedly arrived in South Carolina "with a considerable quantity of rice, being the ship's provision: this rice the *Carolinians* gladly took in exchange for a supply of their own produce.—This unexpected cargo was distributed…, but was not sufficient to supply the demand of all those that would have procured it to plant" (Carney 2001a: 145–146, 2005b: 210–211, 2008: 94–95). A visitor in the first decade of the eighteenth century, John Lawson, reported "several sorts of Rice" being cultivated in South Carolina, "some bearded, others not, besides the red and white; But the white Rice is the best." A black sold his party "some small Quantity of Tobacco and Rice," which Edelson suggests came from the slave's "own provisions plot" (Lawson 1967: 81–82; Edelson 2006: 67).

In 1750, a Dutch official in what would become Guyana wrote about a red rice he knew of in Carolina, comparing it unfavorably with rice grown in his colony because "the red husk…gives so much trouble…to get off" (Oka 1961: 20–21). Half a century later, a describer of South Carolina remarked: "Besides the white and gold rice,…there are some others in the State, of little note or consequence; principally cultivated by negros. They are called *Guinea rice, bearded rice, a short grained rice*, somewhat like barley, and a species of *highland rice*" (Drayton 1972: 125). This account, Carney notes, "draws attention to the role of slave provision gardens for growing varieties of rice favored by slaves," subsistence plots where such familiar

African crops as okra, millet, yams, and black-eyed peas were also cultivated (Carney 2001a: 150).

A deliberate French effort in 1718–1719 to bring seed rice and slaves with rice-growing experience from West Africa to the new colony of Louisiana makes one wonder whether the English ever undertook a comparable initiative for South Carolina that did not get recorded in the historical haze of that colony's early years. The captains of the first two French slave ships bound for Louisiana were instructed to buy several blacks familiar with the grain and three or four barrels of seed on the way to pick up their main cargo at Ouidah on the Slave Coast (modern Benin), and we know that one vessel did stop at Cap Lahou (Côte d'Ivoire), where rice was grown. The ships reached Louisiana in June 1719, and the following year, rice was reportedly being cultivated along the Mississippi (Herpin 1718: 201–204; Hall 1992: 59–63, 121–122, 382). This may have been early enough for the rice to be *O. glaberrima*.

In an astonishing generalization coming from recorders of 35,000 slaving voyages, Eltis, Morgan, and Richardson tell us that "women did not mill rice on the Middle Passage." They base this on a sample of 20 Royal African Company vessels that visited Gambia and Sierra Leone between 1679 and 1688 and loaded only husked rice (EMR 2007: 1347). If all rice was already milled when stocked on slave ships, how could it have reproduced itself in the Americas? And why would slavers not have brought plantable African rice to the New World on occasion when they clearly transferred other native African foods? The above evidence suggests otherwise, but there is even more explicit proof that the authors are wrong.

A French official noted in 1685 that on crowded ships, the slaves "are fed only once a day, and *gros mil* [probably sorghum] is cooked for them, hulled by the women then swollen like rice with a little water and salt" (Cultru 1913: 274). In 1694, the English slaver Thomas Phillips got some "cancy-stones" on the Gold Coast "for our slaves to grind their corn upon," a reference to the "canky" or "kenkey" bread made by local women (Donnan 1930–1935: 1:395; Atkins 1970: 76). Many decades later, another English witness to the shameful trade saw "the woemen [sic] slaves in one part [of a ship at the Iles de Los] beating rice in wooden mortars to cleanse it for cooking" (Smeathman 1773).

In 1768, the Danish slave ship *Fredensborg*, before leaving the Gold Coast for St. Croix in the Caribbean, took on board "three 'millet stones'…hollowed out of sandstones, for grinding millet grains" (Svalesen 2000: 112). A 1785 painting of its successor, the *Fredensborg II*, shows two slave women on deck with pestles and a mortar pounding what a modern author believes to be millet (Svalesen 2000: 40–41, 104). A physician who visited a Savannah-bound slave ship when it stopped off in Barbados in 1795 or 1796 said "their food is chiefly rice." He saw several slaves, presumably women, "employed in beating the red husks off the rice, which was done by pounding the grain in wooden mortars, with wooden pestles, sufficiently long to allow them to stand upright while beating in mortars placed at their feet… They beat the pestle in time to…song" (Dow 2002: xxiii–xxiv).

The journal of the slave ship *Mary*, bound for Savannah from the Gold Coast on June 20, 1796, reported "The Women Cleaning Rice and Grinding Corn for corn cakes" (Donnan 1930–1935: 3:376). An early-nineteenth-century lithograph in a museum in Brittany shows women slaves eating out of a big on-deck pot that appears to be set in a wooden mortar (Carney 2005b: 213, Fig. 12.1).

Eltis, Morgan, and Richardson concede that slave ships may have carried mortars and pestles but assert that "whatever pounding…occurred was more likely of millet" than rice. To support this, they cite a single, flawed source, a statement that "the term for a domestic female slave in the Senegal River region was *pileuse* ('pounder of millet')." In fact, *pileuse* means a female pounder of any cereal or root crop that goes into a mortar, and secondly, there was no rice in the lower Senegal River area to pound (EMR 2007: 1347; Searing 1993: 122).

A price differential between milled and unhusked rice may help explain why the above-mentioned 20 English ships bought only the former. African coastal merchants, aware that they could earn much more from cleaned rice, might have withheld unmilled grain from the market in that period and in those regions (Mouser 2002a: 86 n. 282, Mouser 2002b: 356 n. 42). On the other hand, it was said that rice in the husk kept better on board ship than the processed cereal (De Vorsey and Rice 1992: 118).

Lowcountry Slave Role

Though in general a critic of the black-rice school, Edelson acknowledges almost lyrically that slaves may have been the first to grow rice in South Carolina:

> From above the Lowcountry's freshwater swamps, European colonists watched African slaves plant an edible species of grass that distinguished itself from forage in late summer. As its heavy panicle of rice matured, the plants appeared to bow to the ground under the burden of the grain…[C]olonists appropriated the example of African American wasteland farming to focus their plantations on the production of a new market crop…[S]laves may have…[shown first] that it was possible to raise rice in swamps…South Carolina's wasteland agriculture, in which the hoe-wielding slave replaced the horse-drawn plow, set the new-world plantation apart from the old-world farm (Edelson 2006: 53, 54, 73, 76).

How rice reached South Carolina, or whether *O. glaberrima* preceded *O. sativa*, are not at the crux of the matter. The black-rice school agrees that *O. sativa* soon became the rice of choice for Carolina planters because of its greater yield and market appeal. The debate revolves around the third "core element" discerned by Eltis, Morgan, and Richardson in the black-rice argument, the "putative parallels… between rice cultivation in Africa and its counterparts in the Americas." They concede that "[f]rom land preparation through sowing, weeding, irrigating, threshing, milling, winnowing, and cooking, African practices *seemingly* [my emphasis] left a deep imprint on New World ways of growing and processing the crop." A little later they go a bit further, observing that "African slaves were the primary cultivators of rice" and that some of them "introduced Old World customs of sowing, threshing, and winnowing the crop into the New World." Subsequently, Africans get possible credit for "the hollow logs or trunks that regulated the flow of water between rivers and fields" (EMR 2007: 1333, 1335, 1338). Toward the end of their article, the authors add, unreservedly, that "some slaves introduced...a distinctively African sowing style, pressing a hole with the heel and covering the rice seed with the foot; that they hand-processed rice using an African-style mortar and pestle; and that they fanned rice with African-style coiled grass baskets. As slaves planted, hulled, and winnowed – accompanied by their distinctive songs – they incorporated African folkways into their routines. But these 'survivals' do not amount to 'an entire agricultural complex'," as claimed by Carney (EMR 2007: 1356–1357; Carney 2001a: 167).

And then, in a sweeping statement they do not substantiate, and which seems to contradict what they have told us, the authors find "no evidence that the rice culture of South Carolina [and] Georgia...was any more dependent on skills imported from Africa than were its tobacco and sugar counterparts in the Chesapeake, the Caribbean and Brazil" (EMR 2007: 1357). Tobacco was a New World crop domesticated and developed by Amerindians and imported to Europe in the sixteenth century. Sugar was an Asian crop introduced by Arabs to the Mediterranean around the eighth century and grown in Madeira in the fifteenth, whence it was introduced to São Tomé but eclipsed after a few decades in the sixteenth century by Brazilian production. There was nothing to be learned about either crop from Africa.

Before looking further into parallels, we should note that many are attributed by the black-rice school to slaves who came from rice-growing areas of West Africa. Eltis, Morgan, and Richardson maintain that during the early, formative years of South Carolina's rice industry, few slaves met that requirement. "[B]efore 1715 or so," they say, "most slaves arrived in the lowcountry in small shipments from the West Indies...Not only would very few of these slaves have had much prior knowledge of rice production in Africa, but none of them would have grown the crop in Barbados," which they consider the principal source (EMR 2007: 1337). They credit Peter Wood, founder of the black-rice school, for this information, but the pages they cite in his 1975 book do not support them (Wood 1975: 13–34, 143–144). Slaves from Barbados were important in the first 2 or 3 years of the Carolina colony, but soon they were reportedly coming also from the British Isles, Bermuda, the Bahamas, New England, New York, New Jersey, and the Chesapeake region.

In fact, as of 2008, the Slave Trade Database still had no information on slave arrivals in the Carolinas and Georgia before 1701, i.e., 31 years after the founding of South Carolina, and information on only 227 slave arrivals in 1701–1710 (Eltis and Richardson 2008: 48). Yet no one disputes the official figure of 4,100 black slaves reported for South Carolina in 1708, slightly more than the figure for whites (Wood 1975: 144), which suggests that some slave shipments leading up to that year may not have been so small. In their 2005 paper, Eltis, Morgan, and Richardson admitted that "the lowcountry slave trade data for [1700-c. 1720] are too sparse to be meaningfully analyzed" (EMR 2005: 5). This was omitted in the 2007 version but apparently not on the basis of new data. Even by the authors' own statistics, nearly 5,000 slaves from Upper Guinea reached South Carolina and Georgia before 1751 and close to 36,000 more by 1775 (EMR 2007: 1336). Richardson himself, in an earlier article, found South Carolina to be "the largest market for slaves in eighteenth-century North America" and noted that "an unusually large number of slaves entering [the colony] came from regions in Africa such as Gambia and the Windward Coast" (Richardson 1991: 127, 129, 160), i.e., rice-growing areas.

Despite all the historical evidence cited above for longtime and almost uninterrupted rice growing from the Senegal's Saloum Delta to Cape Palmas at the southern tip of Liberia, i.e., practically the entire length of Upper Guinea, Eltis, Morgan, and Richardson conclude, without convincing proof, that "the proportion of Upper Guinea Africans with rice-growing skills was especially minimal during the years when risiculture (a word used 11 times by the authors, presumably a misspelling of the convenient French word for rice growing, *riziculture*, which has begun appearing unitalicized in English-language literature) became established in the lowcountry" (EMR 2007: 1348). This is designed to bolster one of their crucial theses, that there were too few slaves with rice-growing experience to play a decisive role in the genesis and early development of South Carolina's rice industry. Those who think there were, they contend, "must posit a charter-generation argument in which founding migrants have an influence out of all proportion to their numbers" (EMR 2007: 1357). South Carolina planters, they insist, "learned how to use land and water to grow rice…without a critical mass of Africans to teach them" (EMR 2007: 1338).

History, however, is full of examples of relatively small groups who, usually because of technological advantage, had an influence on other societies all out of proportion to their numbers. The British in India, the Spaniards in the New World, and the Portuguese in circumglobal places come quickly to mind. Even if slaves from rice-growing peoples were a tiny minority in South Carolina, they could still have passed along their expertise to their white masters and their fellow chattel. Eltis, Morgan, and Richardson themselves furnish us with a good example: in 1627, about three dozen Arawaks brought from Dutch Guiana taught the English in Barbados how to grow tobacco (EMR 2007: 1335 n. 14). Small groups of anonymous Europeans introduced many new crops to western Africa (Alpern 1992: 13–43, 2008: 63–102); the reverse could well have been true for the Lowcountry and other parts of the Americas (with the logistical assist of slavers), especially since the slaves were (unwillingly) there to stay and not mere transients.

African/American Parallels

And now for the parallels on which the black-rice school builds its case and which its detractors consider overrated. As we have seen, Eltis, Morgan, and Richardson more or less acknowledge a good number of them. Some years earlier, Morgan even wrote: "African slaves undoubtedly contributed strongly to the development of South Carolina's rice economy, perhaps more in the technology of production and processing than in anything else" (Morgan 1998: 183), as if technology were not of the essence.

On the African side, few of the visitors who, in the two and a half centuries before 1700, reported rice in West Africa did more than locate its presence, link it to certain ethnic groups, or identify it as a slave-ship provision. But enough provided details of riziculture to prove that a wide range of techniques had been mastered. Some peoples grew rain-fed upland rice and some irrigated paddy. Some relied for water on river floodplains, some on freshwater swamps, and some on river tides. Some carved fields out of coastal mangrove forests. Trees and bush were cut down everywhere to clear land for planting. Earth was heaped to form dikes, embankments, causeways, and furrows or dug out for canals and ditches. Water was enclosed and drained, and seed was started in beds then transplanted or directly broadcast onto the fields.

As we have seen, in the mid-fifteenth century, Estêvão reported what might be called rice polders in marshland along what was probably the Gambia. In the early sixteenth century, Valentim Fernandes heard that denizens of the Casamance/Gambia border area sowed and harvested two crops a year (Fernandes 1951: 57), and much later, Lemos Coelho reported the same achievement in Sierra Leone (Lemos Coelho 1985: ch. 9, p. 13). An English traveler, Richard Madox, described the staggering of rice plantings and harvests in Sierra Leone in 1582 (Donno 1976: 170). Around the same time, Almada said rice was transplanted from nurseries, and diked fields were flooded for more than three months in Gambia (Almada 1984: 57; Linares 2002: 2; Rodney 1970: 20–21). A contemporary fellow Cape Verdean, Francisco d'Andrade, said Gambian natives "irrigate [rice fields] with fresh water from the river, in the absence of rain, so that they always have [food]" (Boulègue 1967: 83). In 1606, a Jesuit missionary, Baltasar Barreira, described the prolonged annual clearing of vegetation on lands to be sown with rice along the Guinea coast (Hair 1989: section 13, p. 13). Later in that century, a Franciscan missionary, André de Faro, told of forests being cut down in Sierra Leone and the soil being scraped with an "iron rake" to prepare it for rice (Hair 1982: 49, 60). Dapper's informant on Vai-speaking Kquoja also spoke of tree cutting but said field hands *following* the sower used "iron harrows and rakes" to cover the seed with earth. He also reported that swarms of birds had to be chased from fields, which were surrounded with "good, strong hedges" to keep out buffalo and elephants, and that women weeded the crop and beat the harvested rice (Dapper 1668: 395). Englishman Richard Jobson saw lines of short-hoe wielders raising furrows in rice fields along the Gambia in 1620–1621 (Gamble and Hair 1999: 163). In the Casamance in 1685–1687, a French official, Michel Jajolet de La Courbe, saw causeways crisscrossing paddies to keep water in them and described the rice growers' tools as "shovels of wood provided with a flat piece of iron at one end and a long handle" (Cultru 1913: 208–209; Linares 2002: 2). Barbot said women in Sierra Leone "pound the rice in slightly hollowed tree-trunks" and on the Gold Coast beat it "in wooden mortars made from a thick tree-trunk, or in holes hollowed out of a rock…then winnow it" (Hair et al. 1992: 1:186, 2:512). Long after these random observations, when ethnography became a profession and West Africa's rice-growing cultures were studied, their pre-1700 complexity and sophistication could be verified to some degree retroactively, given the conservatism of traditional African societies. See especially studies of the Baga, Diola, Kissi, Kru, Mende, Susu, Temne, and Vai.

On the Lowcountry side, the parallels may be categorized under rice growing, "drivers," rice processing, cooking, other skills, and the Gullah subculture.

Rice Growing

The heel-and-foot planting method described by Eltis, Morgan, and Richardson was still common in the Lowcountry in 1939, when William R. Bascom visited the area. He heard that it was used "especially [for] rice in the old days" and believed it to be "entirely foreign to European tradition" (Bascom 1941: 49; Wood 1975: 61). Melville Herskovits had already filmed the procedure in Dahomey (now Benin) and Haiti (Herskovits 1937: plate opp. 100, 254, 1938: 1:32–33, 1941: 146).

"[W]hen Carolina blacks moved through the rice fields in a row," notes Wood, "hoeing in unison to work songs, the pattern of cultivation was not one imposed by European owners but rather one retained from West African forebears" (Wood 1975: 61; Bascom 1941: 45–46). This communal form of labor, called the "gang system," appears to have coexisted with the individual-oriented "task system" typical of Lowcountry plantations. Morgan has acknowledged that "[m]asters might mix the two [systems]" (Morgan 1998: 179; Joyner 1985: 58–59).

As noted above, the authors concede that Africans may have contributed sluices, "the hollow logs or trunks that regulated the flow of water between rivers and fields," but they say Europeans too employed such logs for irrigation. Earlier, Morgan was less ambivalent. "African slaves," he wrote, "might well have transferred to South Carolina the primary mechanism used to regulate water flow in early rice culture: the hollowed log and plug," and he pointed to "[s]luice valves in West African rice culture" (Morgan 1998: 157). Carney says "plug trunks…placed in an embankment…enabled control over the flow of fresh water for field flooding and the elimination of weeds" in the Lowcountry, and that "[t]his is the exact water management system still used in mangrove rice production in West Africa" (Carney 2001a, b: 95–96). In fact, the authors can find only two aspects of Lowcountry water control, hanging floodgates and irrigation reservoirs, that they consider of definite European origin (EMR 2007: 1338, 1353). But in 1793, slaving captain Samuel Gamble saw what he called reservoirs on Baga rice farms (Mouser 2002a: 75; Paulme 1957: 263–264; Carney 2001a: 87–88; Fields-Black 2008: 50).

Tools employed for rice growing also had African antecedents. West Africans used hoes instead of plows, and so did Lowcountry slaves. "Those slaves who came from intensive wet-rice-producing areas," Morgan tells us, "would also…have known how to handle a shovel," and he cites references to long-handled Senegambian spades (Morgan 1998: 178 n. 49; Murdock 1959: 266; Harlan et al. 1976: 16). He notes a plantation inventory that included "rice hooks" among slave tools. They recall sickles used to harvest rice in West Africa (Morgan 1998: 178 n. 49; Harlan et al. 1976: 16; Phillips 1966: 90).

Parallels even extend to protecting crops from "ricebirds." West African rice was (and still is) targeted by weaverbirds, and Lowcountry rice by migratory songbirds known as bobolinks. African children and women scared away the birds with slingshots, arrows, stones, and noise. Lowcountry slaves did it with guns and also shot bobolinks for food (Phillips 1996: 90; Morgan 1998: 152). Unlike African planters, who also had to cope with monkeys, buffalo, hippopotami, and elephants, Lowcountry planters were plagued just by feathered foes, but the problem was so notorious that South Carolinians were once nicknamed Ricebirds (Webster's Third 1966: 1951).

"Drivers"

Some high achievers among Lowcountry male slaves, counterparts of men who directed rice growing in West Africa, rose to positions of authority and responsibility though fully aware that no personal accomplishment might change their chattel status. These were the "drivers" who effectively managed rice plantations under the general supervision of white "overseers" (who were usually less qualified) or even on their own (Turner 1974: 4). Occasionally, slaves even served as overseers. As managers, Morgan notes, drivers were "valued for their ability to take decisive and independent action" (Morgan 1998: 343). Edelson suggests that "a small contingent of experienced [slave] rice farmers who worked with planters to select land and direct cultivation [may have become] the Lowcountry's first drivers" (Edelson 2006: 84). Ulrich Phillips thought "[t]he chief advantage of the task system lay in the ease with which it permitted a planter or an overseer to delegate much of his routine function to a driver" (Phillips 1966: 247). According to Charles Joyner, "[o]verseers were dependent upon the black drivers to make the major decisions regarding everything from the supervision of the labor force to the regulation of the water flow in the fields" (Joyner 1985: 65).

Since drivers assigned tasks, imposed discipline, and checked whether orders were carried out, they were also responsible for punishing fieldworkers who did not meet performance standards. Inevitably, this power against fellow slaves was abused, but that does not seem to have been common (Joyner 1985: 65–68).

Rice Processing

In the Rio Nunez region of Guinea-Conakry, men thresh rice, i.e., remove grain heads from stalks, "by swinging waist-high batons over their heads and bringing them down onto the piles of harvested rice" (Fields-Black 2008: 41). In South Carolina, "[b]undles of grain were placed on the ground with their heads facing out, while skilled slaves walked down the rows of bundles beating the heads off the grain with flailing sticks" (Joyner 1985: 48). Unquestionably, the latter technique derived from the former (Littlefield 1981: 108–109; Phillips 1966: 90; Rosengarten 1997: 144).

The pestles and mortars seen on slave ships were standard food-processing implements throughout West Africa and would reappear in the Lowcountry. Their use, notes Wood, "became the accepted method for removing rice grains from their husks." (Pestles detached bran as well as husk.) While the technique is worldwide, he says "there was a strikingly close resemblance between the traditional West African means of pounding rice and the process used by slaves in South Carolina," and, as we have seen, the authors would not disagree. "Even the songs sung by the slaves who threshed and pounded the rice," Wood adds, "may have retained African elements" (Wood 1975: 61–62; Joyner 1985: 48; Morgan 1998: 153, 178 & n. 49, 183 n. 58; Edelson 2006: 81–82; Ribeiro 1962: 136).

No Lowcountry artifact dating from eighteenth-century riziculture has been as thoroughly researched as the coiled winnowing or fanner basket, designed to rid grain of chaff. As defined by the researcher, Dale Rosengarten, "[t]he fanner is a wide winnowing tray used to 'fan' rice – that is, to throw the threshed and pounded grain into the air or drop it from a basket held at a height into another basket, allowing the wind to blow away the chaff" (Rosengarten 1997: 63). She calls it "the archetype of all lowcountry baskets" and confirms earlier reports that such coiled basketry "appears most closely related to traditions in Angola, the Congo, and Senegambia" (Rosengarten 1994a: 18–21, 1994b: 146–148, Rosengarten 1997: 4, 14; 2008: 104–110; Morgan 1998: 153; 178 n. 49, 183 n. 58; Edelson 2006: 82; Joyner 1985: 48). As for the rice fanner itself, "a reasonable candidate for a single-source theory…is Senegambia…, where rice and millet still are winnowed in baskets that look just like lowcountry fanners" (Rosengarten 1997: 145).

Cooking

According to culinary historian Karen Hess, "Africans brought the secret of cooking rice with them" to South Carolina. "This is…self-evident…[N]obody else knew how." She explains that they cooked long-grain rice "in such a way as to eliminate any loose starch so that the cooked grains would be perfectly separate and fluffy" (Hess 1992: 27, 31). The method used was (and is) parboiling, whereby steam drives off excess moisture and keeps grains from sticking (Carney 2005b: 218, 2008: 100). Steamed rice is still a "classic staple dish" in West Africa (Wilson 1972: 75).

Hess calls hoppin' John "the signature dish of South Carolina, black and white" (Hess 1992: 104), and another food writer, John Egerton, "the toast of the Carolina coast" (Egerton 1987: 308). It is composed basically of rice and black-eyed peas (cowpeas), which are more closely related to beans than peas. Since both crops have a long history in Africa, and rice-and-bean dishes "exist in all rice lands," Hess assumes they were combined by African cooks before crossing the Atlantic (Hess 1992: 93–96, appendix 60–61; Carney 2005b: 218; Morgan 1998: 141; Wilson 1972: 137; Joyner 1985: 96).

Slaves are also thought to have contributed other important components of Lowcountry cuisine from their subsistence plots and cooking facilities. The African domesticate okra, reported growing in South Carolina in the 1780s (Pace 1983: 171–172) but probably introduced much earlier since it was already common in Jamaica in the early eighteenth century (Sloane 1707–1725: 1:223), has featured in soups, stews, and pilau (Hess 1992: 8, 111–112, appendix 40–41, 46, 56, 57; Egerton 1987:

304; Grimé 1979: 19, 63–64; Carney 2001a: 150, 2003: 179; Pollitzer 1999: 96). Sesame seeds, known by the African-derived name of benne in the Lowcountry and doubtless brought on slave ships, have gone into soup, salads, puddings, bread, cake, and especially wafers (Egerton 1987: 322; Hess 1992: 8, 102; Grimé 1979: 25–26; Morgan 1998: 141; Carney 2001a: 213 n. 48, 2003: 179; GWP 1940: 71, 178; Pollitzer 1999: 96). The African yam *Dioscorea cayenensis*, known as the yellow Guinea yam, reached the New World so early it was mistakenly named for Cayenne. It might lie behind the semantic confusion in the southern states between yams and the Amerindian domesticate sweet potatoes, which are botanically quite different but, according to Egerton, "as far as cooking is concerned…absolutely and completely interchangeable" (Egerton 1987: 48, 306; Grimé 1979: 23; Morgan 1998: 141; Carney 2001a: 150, 2003: 178; Hess 1992: appendix 43). The *white* Guinea yam (*Dioscorea rotundata*) is an alternative candidate for confusion with the sweet potato (Pollitzer 1999: 97). Other food plants that may have gone from Lowcountry slave gardens to the community at large included the African domesticates sorghum and watermelons and the peanut, an Amerindian domesticate that went from Brazil to western Africa, then back to the West Indies and North America. Throughout the eighteenth century, English-language writers thought it was of African origin (Grimé 1979: 19–20, 73; Hess 1992: 95; Chaplin 1993: 156). Sloane, for instance, found peanuts being grown in Jamaica and said they were "brought from Guinea in the Negroes Ships" (Sloane 1707–1725: 1:184). A similar round trip may have brought Amerindian red peppers (*Capsicum*), a seasoning for hoppin' John and rice pilau, via slave ships to Charleston (Hess 1992: 46, 93, 101, 109–110, 112, appendix 55, 61).

Other Skills

Ironically, Philip Morgan provides us with perhaps the most comprehensive survey of slave contributions to Lowcountry life beyond food production in a chapter titled "Skilled Work" of his 1998 book, *Slave Counterpoint* (pp. 204–254). He lists the following crafts and trades and supports his comments with a wealth of sources:

Basketmaking – Lowcountry slaves made a great variety of baskets besides rice fanners. "In Africa," Morgan notes, "both sexes made baskets, with men making the larger work baskets and women the smaller, fancier household baskets. A similar distinction seems to have persisted in early South Carolina…Close stylistic parallels link the coiled rush baskets with parts of Africa, notably Senegambia" (Morgan 1998: 233). Dale Rosengarten, concluding that slaves "were encouraged [by their masters] to practice their ancient art of basket making," lists many basket types that may date to colonial times and are still made in the Lowcountry. They include head-tote baskets, covered work baskets, hats and hat boxes, granaries, bowls, trays, food covers, mats, hampers, sewing baskets, and wastepaper baskets (Rosengarten 1997: 152–155, 192, 203, 212–213, 1987: 7–8, 16–21).

Woodworking – According to Morgan, "woodworking was the trade to which most Lowcountry Africans were put in the eighteenth century…No doubt they were responsible for fashioning even that most utilitarian of objects – the rice mortar – so as to resemble the shape of, and incorporate decorative motifs from, African drums. Blacks in coastal Georgia were famous for a distinctive wood-carving tradition – walking sticks embellished with reptile and human figures" (Morgan 1998: 236). At the other extreme from canes, slaves made dugout canoes out of swamp-cypress logs (Morgan 1998: 236; Wood 1974: 166, Wood 1975: 123–124, 2008: 84–87) and the rice-field sluices mentioned above, plus a gamut of wooden objects in between, including barrels and fences. In fact, as time went on, carpenters apparently became "the elite of the slave craftsmen," building, with new skills, their masters' houses and furniture, rice mills, and barns, not to speak of slave cabins (Joyner 1985: 71–74; Phillips 1966: 248; Pollitzer 1999: 176–178). They also worked in Charleston area shipyards (Morgan 1998: 229).

Pottery Making – Lowcountry slaves made unglazed pottery known as Colono Ware that Morgan believes were "styled on generalized mental images drawn from their [African] homelands." He cites "face vessels, or mask pitchers, that have been linked to sculptural techniques employed in the region of Angola…[B]road artistic principles that lie at the heart of these creations…may well be African in origin" (Morgan 1998: 234–235). Excavations principally by archaeologist Leland Ferguson in recent decades have belatedly brought to light the links between Africa and Colono Ware, once regarded as an Amerindian craft. Marks on pots found in South Carolina, for example, resemble Bakongo cosmograms (Ferguson 1992: 7–32, 82–107, 110–116; Rosengarten 1997: 22–24; Joyner 1985: 75; Pollitzer 1999: 171–173).

Blacksmithing – West Africans were already smelting and shaping iron when the Europeans arrived, and Morgan, for one, thinks "[b]lacksmithing was…one trade into which slaves [in North America] incorporated African influences." He cites a modern-day black ironworker in Charleston whose designs and improvisation reveal "connections to an earlier and broader African American tradition" (Morgan 1998: 232). The "elaborate hand-fashioned iron grill work" of plantation mansions has been attributed to slave smiths (Joyner 1985: 71), and their ordinary output is said to have consisted of "nails, hinges, screws, bolts, rakes, tubs, weights, and all other metal goods" (Pollitzer 1999: 175–176). I imagine the latter included the business ends of hoes, shovels, axes, picks and sickles, as well as horseshoes.

Brickmaking and Bricklaying – Certain West Africans made mud bricks, but in South Carolina, slaves learned how to bake and lay European ones. Morgan tells us that "[e]ven as early as the 1730s, bricklaying slave families had emerged in the Lowcountry," and later in the century, masters boasted of slave brickmakers who could make 5,000–6,000 a day (Morgan 1998: 228–229).

Musical Instruments – African-style drums and African-derived banjos were made by Lowcountry slave craftsmen. Reportedly, such drums were made along the South Carolina-Georgia coast into modern times (Pollitzer 1999: 156; GWP 1940: 62, 100–101, 107, 148–155, 180–194). Banjos became the instrument of choice for slave dances by the mid-eighteenth century (Morgan 1998: 583). Slave musicians soon mastered the Anglo-American fiddle, but I have seen no evidence of local manufacture (Morgan 1998: 593; Pollitzer 1999: 156). Gourd rattles containing seeds recalled African models (GWP 1940: 148, 165, 176, 184).

Other Crafts – There are references in the literature to slave leatherworkers such as tanners and cobblers, tinsmiths, broom, fan and fly-whisk makers, fish casting-net and trap makers, seamstresses and weavers, and to gourds made into receptacles and dippers as in Africa and also used for banjos and rattles (Joyner 1985: 70–71; Pollitzer 1999: 165–167; Rosengarten 1997: 168).

Fishing and Boating – African fishing skills were easily transplanted to the Lowcountry. "By the early eighteenth century," Morgan notes, "an identifiable group of 'fishing Negroes' had emerged in South Carolina, particularly in Charleston…[B]y midcentury, blacks had established a virtual monopoly on town fishing…[It was] observed that slaves dominated the fishing business because they were 'permitted to keep boats, canows, etc.'…Whether in their own or their masters' boats, Lowcountry slave fishermen were successful" (Morgan 1998: 240–241).

Slaves not only built canoes (and rowboats) but manned them to transport people and goods. "Canoe services," says Morgan, "were literally in the hands of slaves, and more than one traveler testified to their proficiency…Schooner services linking inland towns to coastal ports or connecting contiguous coastal towns depended on slaves…[A]ll-slave crews increasingly manned South Carolina's river and coastal boats" (Morgan 1998: 237–239).

Morgan gives examples showing that "[s]ome masters…were aware of their African slaves' familiarity with boats and canoes," that their "boating prowess" was linked to their African background. "The alacrity with which some African slaves took to the water, their familiarity with navigating boats between ocean and inland rivers and swamps, and their penchant for singing…'their plaintive African songs, in cadence with the oars,' are a true measure of the influence of homeland experiences" (Morgan 1998: 243–244; Wood 1975: 122–124, 200–205; Pollitzer 1999: 178–179; Joyner 1985: 15–16, 76–77, 99–100).

Hunting – "[P]robably no group of slaves," Morgan suggests, "could match those of the Lowcountry for the amount of time spent fishing and hunting." Game was caught with traps and hunting dogs but also with firearms. "Lead shot and gunflints have been found at almost all the Lowcountry…slave sites." Morgan quotes an observation from the 1770s that "a dextrous negroe will, with his gun and netts, get as much game and fish as five families can eat" (Morgan 1998: 138–139). According to Wood, slaves hunted for their masters as well as themselves (Wood 1975: 127; Joyner 1985: 100–101).

Closeness to Nature – The French anthropologist Claude Lévi-Strauss observed that "indigenous" peoples in general possess "extreme familiarity with their biological milieu," devote "passionate attention to it" and acquire "precise knowledge of it" (Lévi-Strauss 1962: 10; Wood 1974: 162, 1975: 117, 2008: 79). Earlier, a compatriot, geographer Jean Dresch, remarked that "[t]he adaptation of the [African] peasant to the environment has been underlined many times. One cannot, indeed, fail to be surprised by the extent and precision of his knowledge of vegetation insofar as it is useful to him. In the same way he knows soils, so uniform in appearance, so variable in reality" (Dresch 1949: 295). Slaves apparently brought the African's sensitivity to nature to the Lowcountry. According to Edelson, they "learned to read the landscape with a 'visual acuity' that could well have proved useful [to planters] in identifying viable sites for production" of rice in South Carolina wetlands (Edelson 2006: 78).

In Wood's view, the slaves also inherited the Amerindians' "ability to cope with this particular natural world…[They] assimilated the largest share of their lore and …increasingly took over their responsibilities as 'pathfinders' in the southern wilderness." He cites the slaves' roles as transporters of goods, passengers, and messengers through uncharted forests and across watercourses, and their "general familiarity with the uses of wild plants," such as medicinal species, edible herbs, and even poisons put in streams to drug fish (Wood 1975: 116–122, 2008: 80).

Gullah Subculture

In a feat of omission, the authors completely ignore the Gullah subculture, North America's principal repository of Africanisms, and its intimate links to Lowcountry rice cultivation. Edelson manages the same feat in his *Plantation Enterprise in Colonial South Carolina*. Eltis, in *The Rise of African Slavery in the Americas* (2000), is not only silent on the Gullah (and Georgia) but barely mentions South Carolina. Morgan, on his own in *Slave Counterpoint* (1998), does give due attention to the Gullah language (pp. 465–466, 565–570, 572, 575, 622).

As Joseph A. Opala observes, "the Gullah have been able to preserve more of their African cultural heritage than any other group of Black Americans," and few social scientists, if any, would dispute him. He goes on to explain that the Gullah "speak a creole language similar to Sierra Leone Krio, use African names, tell African folktales, make African-style handicrafts…, and enjoy a rich cuisine based primarily on rice." Siding with the black-rice school, he also maintains that Lowcountry colonists "adopted a system of rice cultivation that drew heavily on the labor patterns and technical knowledge of their African slaves" (Opala 1987: i, 2).

The development and perseverance of the Gullah subculture are generally credited to several factors: demographic, geographic, climatic, and genetic. For most of the eighteenth century, blacks outnumbered whites in South Carolina, a unique situation in the colonies that became the United States. The primary cause was the perceived need for large numbers of slaves to grow rice. This alone would have favored the persistence of African ways, but the demographic advantage was reinforced by the relative isolation of black communities. This was due in large part to the subtropical climate and especially the health risks of the coastal marshlands and islands where the rice was grown. Slave ships apparently brought the mosquito vectors of malaria and yellow fever from the Old World to the New, and the insects settled in the Lowcountry along with the slaves. The inherited sickle-cell hemoglobin trait protected many slaves against malaria, and rarely fatal childhood exposure to yellow fever gave black adults immunity. Whites were more susceptible to the two deadly diseases and avoided that environment if they could (Wood 1975: 64–76, 79–91; Pollitzer 1999: 18–20, 69–74; Cassidy 2008: 19; Edelson 2006: 83, 152, 162–163; Opala 1987: 8–10; Carney 2001a: 147–148; Phillips 1966: 90–91).

The Gullah subculture was not brought widely to the attention of scholars until 1949, when Lorenzo Dow Turner published the first in-depth study of the language in *Africanisms in the Gullah Dialect*. Melville Herskovits had publicized Turner's research earlier, calling Gullah "the most distinctive form of Negro diction in this country" (Herskovits 1941: 276–280). Just the year before Turner's book appeared, H.L. Mencken could describe the language as the "one [American] dialect that stumps a visitor from any other part of the country" (Mencken 1948: 101). As Michael Montgomery has summed it up, Turner found "significant [African] influences in every area of the language" (Montgomery 1994: 170). Specifically, he linked almost all of 3,595 personal names to Africa, plus 251 other words used in conversation and 92 expressions heard only in stories, songs, and prayers (Pollitzer 1999: 109). He related eight words to terms for rice or rice dishes in African languages (Turner 1949: 70, 111, 144, 191, 195, 196, 198, 200).

While Turner's methodology has been questioned – his knowledge of African languages, for instance, was not encyclopedic – even a critical Hair said he had "conclusively proved that Gullah contains more 'Africanisms' than any other Negro dialect of the United States" (Hair 1965: 79), and much more recently, two scholars reported that the African origin of the many personal names Turner collected "is not now substantially in dispute" (Baird and Twining 1994: 27; Wood 1975: 167–191; Cross 2008: 125–147). Beyond the language, it is important to remember that slaves who spoke Gullah were involved in all the cultural parallels discussed above.

***

One of Eltis, Morgan, and Richardson's closing paragraphs (EMR 2007: 1355–1356), mostly an African history lesson, is remarkable in that nearly every sentence invites a rejoinder. "Before the French invasions in the later nineteenth century," the authors assert, "Europeans could never impose their will on the indigenous populations of West Africa." This ignores European forts and trading posts established along the coast beginning in the fifteenth century, the European focus on commerce rather than conquest until the nineteenth century, and the British expansion that competed with the French (plus German, Belgian, and Portuguese outreach if West-Central Africa is included). The proposition that Europeans could never impose their will before the late nineteenth century would, in effect, seem to place primary responsibility for the slave trade on the Africans.

Next, "Africans…already produced most of what they needed and European could offer little that was essential to African lives." If so, why did they sell slaves to Europeans for an increasingly broad range of goods? African needs expanded from the moment the Portuguese first offered them brass basins and glass beads (Alpern 1995: 5–43).

Then we are told that "rice…hinged on an intensely exploitative plantation system that could not exist in Africa." This underestimates the organizational/administrative abilities and hard-nosed entrepreneurial talents of Africans. The rulers of

Asante, Dahomey, and old Benin, and probably of the Oyo Empire too, relied for food supplies on plantations worked laboriously by foreign chattel slaves with no legal rights. T.C. McCaskie tells us such slaves in Asante were subject to "unremitting exploitation that ended in death" (McCaskie 1995: 96, 97–99; Wilks 1975: 52, 70, 93–94, 176–177, 675; Kea 1982: 5, 20, 56, 182–183, 197). Herskovits thought servitude on royal Dahomean plantations was "perhaps more closely akin to the slavery known in America and the West Indies than has been recorded in any other portion of Africa…[S]laves…, under the control of overseers whose duty it was to get the utmost yield from the fields,…worked hard and long" (Herskovits 1938: 1:82, 1: 99–100, 2:97; Law 1991: 66–67, 272–273). R.E. Bradbury says wealth in old Benin "was invested primarily in buying slaves," who were installed in villages and forest camps to farm for their masters, including chiefs as well as the king (Bradbury 1973: 53, 59, 81, 180, 253, 265–266).

Robin Law notes that wealth in Oyo "consisted in the possession of slaves," that the first ingredient of power was "the labour of slaves employed in agricultural production," and that a Hausa slave revolt in Oyo in 1817 pointed to an "exploitative" form of servitude (Law 1977: 205–207, 229, 231, 233, 306).

While these examples all come from highly centralized polities in non-rice-producing parts of Lower Guinea, we have a record of a Sierra Leone rice plantation owned by a Sherbro native called Mr. Tucker by English travelers. Slave-ship surgeon Alexander Falconbridge visited the property near Cape Mount in 1783 and saw male workers clearing the ground for rice (Lambert 1975: vol. 72: 601, 605, 606, 612–613).

The "Agency and Diaspora" paragraph concludes with the observation that unlike Africa, "[t]he American environment – especially, we may hypothesize, the capitalistic mentalité of planters – was essential to a rice-based export sector." That evidently pejorative, bilingual phrase, "capitalistic mentalité," presumably refers to the age-old, ubiquitous, and natural desire of producers and merchants, including West African market women and male traders, to maximize profit.

There is an element of strawmanship in several of the authors' arguments, if I may coin a neologism. They assert early on (EMR 2007: 1334) that the black-rice school's "linkage between the rice-growing regions of Africa and those in the Americas is…tenuous" and rests on three "key claims." "First,…Africans from rice-growing areas are said to have been either a significant minority or even a majority of those slaves arriving in New World regions that specialized in rice." But they support this with only one reference, to two pages in an article by Carney on the introduction of rice into Suriname. What Carney actually says is that a significant percentage of the earliest slaves brought to Brazil came from Guinea-Bissau and neighboring rice-growing areas, that they grew rice in Brazil on subsistence plots, and that Dutch planters expelled from Brazil took subsistence conventions to Suriname (Carney 2005a, b: 336–337; Hawthorne 2003: 67–69). In response, the authors contend that "[t]here are no records of any slaves from African rice-growing regions in Brazil before the Dutch conquest" in 1630, and that between 1574 and 1630, African origins are known for only 16 voyages to Brazil, and they all came from Angola and São Tomé (EMR 2007: 1343). But Portuguese colonists almost certainly brought slaves from São Tiago to grow the crops (including rice) and tend the animals they picked up there when they founded Bahia in 1549, and in the next 25 years, there were doubtless other sailings to Bahia from the Cape Verdes carrying slaves as well as settlers. Most of the slaves on those islands came from rice-growing parts of the African mainland known as the Rivers of Guinea. As Gwendolyn Midlo Hall points out, "just because something is not included in a historical document or a database, that does not mean it did not happen" (Hall 2010: 138–139).

As far as South Carolina is concerned, Peter Wood, pioneer of the black-rice hypothesis, wrote of slaves entering the colony that "many, perhaps even a great majority, had never seen a rice plant" (Wood 1975: 61). Daniel Littlefield noted "the dominance of peoples from Congo-Angola," where rice was not grown, in the early years of the colony, and said "[t]he preponderance of Angolas among their slaves was a fact of which South Carolinians were very much aware" (Littlefield 1981: 124, 126). (In fact, the word "Gullah" probably derives from "Angola.") And Carney remarks that in the decade before the 1739 Stono rebellion, over 20,000 slaves arrived, a more than twofold increase over the slave population of 12,000 in 1720, and that 70% of the newcomers "originated in Angola and its interior" (Carney 2001a: 82; Pollitzer 1999: 43).

The second and third "key claims" have to do with South Carolina's alleged preference for female slaves because of their prominent role in West African riziculture. Again, only one source is cited, Carney's *Black Rice*, p. 107. While the authors make a convincing case, based on slave-trade statistics, against Carney's position, it can hardly be called a key claim of the black-rice school. Nor did it mean that male slaves hailing from rice-growing peoples knew nothing about riziculture. The gender division of labor was not one-sided; the heaviest work in rice cultivation was often done by men. As Africanist Adam Jones cautions, "generalization about the sexual division of labour may in fact conceal a considerable amount of shared involvement" (Jones 1983b: 190). The old stereotype of African males off hunting and warring, or sitting around drinking palm wine or millet beer, while their spouses worked nonstop, was often well off the mark. And even where riziculture was a female monopoly, men still ran the show. The evidence for the male role in West African rice farming is abundant (Almada 1985: 57; Hair 1982: 60; Dapper 1668: 395; Paulme 1954: 23–42, 1957: 266–271; Little 1948: 37–48, 1967: 80–82; Linares 1992: 3–5, 9, 18–23, 58–60; Thomas 1959: 108, 112, 115–117; Pélissier 1966: 726–727, 738–741, 747, 754–757; Massing 1980: 119–144; Langley 1939: 69–70; Johnny et al. 1981: 606–607; Chauveau et al. 1981: 630).

Another black-rice view that the authors seem to regard as a key claim is that the task labor system, ostensibly easier on slaves than the gang system, may have been a *quid pro quo* for the slaves' rice-growing expertise. They raise plausible doubts that it *was* much easier and that slaves were able to use their experience with rice as negotiating leverage (EMR 2007: 1356), but this is not a major plank of their opponents. Neither Wood nor Littlefield single out the task system, and Carney, who does, makes a good but unsubstantiated case for slaves bargaining better terms from their masters (Carney 2001a: 98–101, 105–106).

On the basis of their writings, I doubt very much whether Wood, Littlefield, and Carney, as alleged, see South Carolina and Georgia's major eighteenth-century export crop "as predominantly a creation of Africans" (EMR 2007: 1332). I also doubt they would deny that planters "held the reins of power" and "called the shots" once the coercive plantation regime was established (EMR 2007: 1357). After seeing all the circumstantial and inferential evidence of cultural transfer, I wonder why the authors dismiss Carney's claim that an "entire cultural system" based on rice reached the Lowcountry (Carney 2001a: xii; EMR 2007: 1333). And I find Eltis, Morgan, and Richardson's minimalist concession, that "planters no doubt appropriated some knowledge from their slaves," that "some of [the slaves] contributed ideas about rice cultivation," difficult to accept (EMR 2007: 1356, 1357).

A further example of strawmanship is the authors' contention that "a basic thrust of the 'black rice' hypothesis" is that the slaves were essentially conservative, clinging tenaciously to traditions such as their food preferences, and they ask: "But is it not possible that they were as much improvisers and experimenters as Europeans?" (EMR 2007: 1354). They were indeed, and the black-rice school would surely agree.

Conclusion

From the beginning, the slaves had to adapt to a new environment, both physical and human. The late William Pollitzer noted that "[m]any African ethnic groups made a contribution to the amalgam that arose in the Low Country…The language and culture that developed in the sea islands were more than a retention, more than a mixture; they were a creative synthesis born of memory, necessity and improvisation" (Pollitzer 1999: 198). The slaves had to adjust to whites as well as to each other. Littlefield speaks of "a composite of African and European elements…the meshing of European and African attitudes and attributes…a mutual accomplishment" (Littlefield 1981: 177). Carney speaks of "the hybridized imprimatur of both African and European influences…the technological and agronomic heritage of each knowledge system [that] combined in new ways to shape rice cultivation" (Carney 2001a: 162). Even Eltis, Morgan, and Richardson acknowledge that "[t]he rice regime owed much to improvisation; it was a hybrid, synthetic rather than European or African in character," and at the very end, they talk of "the forging of communities and cultures…the creolization of peoples," a decided shift from the notion that "planters no doubt appropriated some knowledge from their slaves" (EMR 2007: 1354, 1358).

In his groundbreaking study, Peter Wood started from the solid premise that some slaves knew how to grow rice and their early European masters did not (Wood 1975: 35). He sought to correct the impression left by the planter elite and many historians that slaves were generally unskilled, that Africans contributed nothing more than menial labor to South Carolina's early development (Wood 1975: 56, 62). What Eltis, Morgan and Richardson see as a deeply flawed revisionism inspired by "a desire to celebrate an African accomplishment in the New World" – historiography as ego massage – I see as a long overdue and laudable effort to set the record much straighter.

In view of all the evidence presented above of contributions by African slaves to Lowcountry culture, how can one deny that they helped in large measure to give the eighteenth-century rice boom its distinctive character? Pollitzer thought "[t]he material culture of Africa was retained on the coast of Carolina and Georgia because this holdover of memories and talents was useful to blacks and whites alike...[that it] was molded by the customs and economic needs of Europeans into something new and unique" (Pollitzer 1999: 186). I would agree with him.

Acknowledgments For help on this chapter, my thanks go to Frances and Jennifer Alpern, Judith Carney, Philip Cohan, Per Haldbo, Gwendolyn Midlo Hall, Adam Jones, James La Fleur, Joseph Lauer, Philip Lovdal, Dorothy, Hannah and Peter Lubin, Erica Maillart, Susan Keech McIntosh, Edwin Nuijten, Élisabeth Ortunio, Monique Picard, Tania Pos, Anne Pouchoulin, Paul Richards, Dale Rosengarten, Dana Sardet, and Micky Van Tekelenburg.

References

Abbreviations Used:

HA for *History in Africa*
IFAN for *Institut Fondamental d'Afrique Noire* (Fondamental replacing Français in 1966)
JAH for *Journal of African History*
JATBA for *Journal d'Agriculture Tropicale et de Botanique Appliquée*
SA for *Slavery and Abolition*

Almada AA de (1984) Brief treatise on the rivers of Guinea (c. 1594) (trans ed: Hair PEH). Department of History, University of Liverpool, Liverpool

Alpern SB (1992) The European introduction of crops into West Africa in precolonial times. HA 19:13–43

Alpern SB (1995) What Africans got for their slaves: a master list of European trade goods. HA 22:5–43

Alpern SB (2008) Exotic plants of western Africa: where they came from and when. HA 35:63–102

Álvares M (1990) Ethiopia Minor and a geographical account of the province of Sierra Leone (c. 1615) (trans: Hair PEH). Department of History, University of Liverpool, Liverpool

Anonymous Portuguese Pilot (1550) Navigationi da Lisbona all'isola di san Thomè posta sotto la linea dell'equinottiale scritta per un pilotto Portoghese. In: Ramusio GB (ed) Navigationi e viaggi, vol. 1, pp. 125–129. Stamperia de Giunti, Venice

Atkins J (1970) A voyage to Guinea, Brazil, & the West Indies, reprint of 1735 edn. Frank Cass & Co., London

Baird KE, Twining MA (1994) Names and naming in the Sea Islands. In: Montgomery M (ed) The crucible of Carolina. University of Georgia Press, Athens, pp 23–37

Bascom WR (1941) Acculturation among the Gullah Negroes. Am Anthropol 43(1):43–50

Beaver P (1805) African memoranda: relative to an attempt to establish a British settlement on the island of Bulama, on the western coast of in the year 1792. C. & R. Baldwin, London

Blake JW (1942) Europeans in West Africa, 1450–1560, 2 vols. Hakluyt Society, London

Bosman W (1967) A new and accurate description of the coast of Guinea, new edn. of 1705 translation from the 1704 Dutch original. Frank Cass & Co., London

Boulègue J (1967) Relation de Francisco d'Andrade sur les îles du Cap-Vert et la côte occidentale d'Afrique (1582). In: Bulletin de l'IFAN 29, series B, nos. 1–2. Dakar

Bradbury RE (1973) In: Morton-Williams P (ed) Benin studies. Oxford University Press, London

Carney JA (2001a) Black rice: the African origins of rice cultivation in the Americas. Harvard University Press, Cambridge

Carney JA (2001b) African rice in the Columbian exchange. JAH 42(3):377–396

Carney JA (2003) African traditional plant knowledge in the circum-Caribbean region, Journal of Ethnobiology 23, no. 2(fall/winter), 167–185

Carney JA (2004) 'With grains in her hair': rice in colonial Brazil. SA 25(1):1–27

Carney JA (2005a) Rice and memory in the age of enslavement: Atlantic passages to Suriname. SA 26(3):325–347

Carney JA (2005b) Out of Africa: colonial rice history in the black Atlantic. In: Schiebinger L, Swan C (eds) Colonial botany: science, commerce, and polities in the early modern world. University of Pennsylvania Press, Philadelphia, pp 204–220

Carney JA (2008) Rice in the new world. In: Rosengarten D, Rosengarten T, Schildkrout E (eds) Grass roots: African origins of an American art. Museum for African Art, New York, pp 94–101

Cassidy FG (1994) Gullah and the Caribbean connection. In: Montgomery M (ed) The crucible of Carolina. University of Georgia Press, Athens, pp 16–22

Catesby M (1731–1743) The natural history of Carolina, Florida and the Bahama Islands, 2 vols. London

Chaplin JE (1993) An anxious pursuit: agricultural innovation and modernity in the Lower South, 1730–1815. University of North Carolina Press, Chapel Hill

Chauveau JP, Dozon JP, Richard J (1981) Histoires de riz, histoires d'igname: le cas de la moyenne Côte d'Ivoire. Africa 51(2):621–658

Crone GR (ed) (1937) The voyages of Cadamosto and other documents on western Africa in the second half of the fifteenth century. Hakluyt Society, London

Crosby AW Jr (1972) The Columbian exchange: biological and cultural consequences of 1492. Greenwood Publishing Co., Westport

Cross W (2008) Gullah culture in America. Praeger, Westport

Cultru P (1913) Premier voyage du sieur de La Courbe fait à la coste d'Afrique en 1685. Société de l'Histoire des Colonies Françaises, Paris

Dalby D, Hair PEH (1964) 'Le langaige de Guynée': a sixteenth century vocabulary from the Pepper Coast. Afr Lang Stud 5:174–191

Dapper O (1668) Naukeurige beschrijvinge der afrikaensche gewesten. Jacob van Meurs, Amsterdam

Dapper O (1676) Naukeurige beschrijvinge der afrikaensche gewesten van Egypten, Barbaryen, Lybien, Biledulgerid, Negroslant, Guinea, Ethiopiën, Abyssinie…, 2nd edn. Jacob van Meurs, Amsterdam

de La Fosse E (1897) Voyage à la côte occidentale d'Afrique, en Portugal et en Espagne (1479–1480). Revue Hispanique 4:174–201

De Vorsey L Jr, Rice MJ (1992) The plantation South: Atlanta to Savannah and Charleston. Rutgers University Press, New Brunswick

de Zurara GE (1960) In: Bourdon L (ed) Chronique de Guinée. IFAN, Dakar

Donelha A (1977) An account of Sierra Leone and the rivers of Guinea of Cape Verde (1625) (trans ed: Hair PEH). Junta de Investigações Científicas do Ultramar, Lisbon

Donnan E (ed) (1930–1935) Documents illustrative of the history of the slave trade to America, 4 vols. Carnegie Institution of Washington, Washington, DC

Donno ES (1976) An Elizabethan in 1582: the diary of Richard Madox…. Hakluyt Society, London

Dow GF (2002) Slave ships and slaving, reprint of 1927 edn. Dover, Mineola

Drayton J (1972) A view of South Carolina, as respects her natural and civil concerns, reprint of 1802 edn. Reprint Co., Spartanburg

Dresch J (1949) La riziculture en Afrique occidentale. Annales de Géographie 58(4):295–312

Duncan TB (1972) Atlantic islands: Madeira, the Azores and the Cape Verdes in seventeenth-century commerce and navigation. University of Chicago Press, Chicago

Edelson SM (2006) Plantation enterprise in colonial South Carolina. Harvard University Press, Cambridge

Egerton J (1987) Southern food: at home, on the road, in history. Knopf, New York
Eltis D (2000) The rise of African slavery in the Americas. Cambridge University Press, Cambridge
Eltis D, Richardson D (eds) (2008) Extending the frontiers: essays on the new transatlantic slave trade database. Yale University Press, New Haven
Eltis D, Morgan P, Richardson D (EMR) (2005) The African contribution to rice cultivation in the Americas. In: Georgia workshop in early American history and culture, Athens, GA
Eltis D, Morgan P, Richardson D (EMR) (2007) Agency and diaspora in Atlantic history: reassessing the African contribution to rice cultivation in the Americas. Am Hist Rev 112(5):1329–1358
Enciso MF de (1932) A brief summe of geographie (trans: Barlow R, Taylor EGR). Hakluyt Society, London
Fagan BM (1969) Radiocarbon dates for sub-Saharan Africa: VI. JAH 10(1):149 169
Ferguson L (1992) Uncommon ground: archaeology and early African America, 1650–1800. Smithsonian Institution Press, Washington
Fernandes V (1951) Description de la côte occidentale d'Afrique (Sénégal au cap de Monte, archipels) (trans ed: Monod T, Teixeira da Mota A, Mauny R). Centro de Estudos da Guiné Portuguesa, Bissau
Fields-Black EL (2008) Deep roots: rice farmers in West Africa and the African diaspora. Indiana University Press, Bloomington
Gamble DP, Hair PEH (eds) (1999) The discovery of the River Gambra by Richard Jobson (1623). Hakluyt Society, London
Georgia Writers' Project (GWP) (1940) Drums and shadows: survival studies among the Georgia coastal Negroes. University of Georgia Press, Athens
Grimé WE (1979) Ethno-botany of the black Americans. Reference Publications, Algonac
Hair PEH (1964) An early seventeenth-century vocabulary of Vai. Afr Stud 23:129–139
Hair PEH (1965) Sierra Leone items in the Gullah dialect of American English. Sierra Leone Lang Rev 4:79–84
Hair PEH (1967a) Ethnolinguistic continuity on the Guinea coast. JAH 8(2):247–268
Hair PEH (1967b) An ethnolinguistic inventory of the Upper Guinea coast before 1700. Afr Lang Rev 6:32–70
Hair PEH (1976) Some minor sources for Guinea, 1519–1559: Enciso and Alfonce/Fonteneau. HA 3:19–46
Hair PEH (trans ed) (1982) André de Faro's missionary journey to Sierra Leone in 1663–1664. University of Sierra Leone, Freetown
Hair PEH (trans) (1989) Jesuit documents on the Guinea of Cape Verde and the Cape Verde Islands, 1585–1617. Department of History, University of Liverpool, Liverpool
Hair PEH, Jones A, Law R (eds) (1992) Barbot on Guinea; the writings of Jean Barbot on West Africa 1678–1712, 2 vols. Hakluyt Society, London
Hall GM (1992) Africans in colonial Louisiana: the development of Afro-Creole culture in the eighteenth century. Louisiana State University Press, Baton Rouge
Hall GM (2010) Africa and Africans in the African diaspora: the uses of relational databases. Am Hist Rev 90(1):136–150
Harlan JR, de Wet JMJ, Stemler ABL (1976) Origins of African plant domestication. Mouton, The Hague
Hawthorne W (2003) Planting rice and harvesting slaves: transformations along the Guinea-Bissau coast, 1400–1900. Heinemann, Portsmouth
Herpin (1718) Instructions pour le S. Herpin Commandant du Vaisseau L'aurore destiné pour la Traitte des Negres a la Coste de Guynee. Service Archives Nationale, Paris, series COL/B/42 bis, pp 201–204. Instructions to French slave-ship captain, 4 July 1718
Herskovits MJ (1937) Life in a Haitian valley. Knopf, New York
Herskovits MJ (1938) Dahomey: an ancient West African kingdom, 2 vols. J.J. Augustin, New York
Herskovits MJ (1941) The myth of the Negro past. Harper & Brothers, Boston

Herskovits MJ, Herskovits FS (1934) Rebel destiny: among the Bush Negroes of Dutch Guiana. McGraw-Hill, New York

Hess K (1992) The Carolina rice kitchen: the African connection. University of South Carolina Press, Columbia. Appendix includes reprint of a Carolina rice cookbook compiled by Mrs. Samuel G. Stoney at Charleston in 1901

Johnny M, Karimu J, Richardson P (1981) Upland and swamp rice farming systems in Sierra Leone... *Africa* 51(2)

Jones A (1983a) German sources for West African history, 1599–1669. Franz Steiner Verlag, Wiesbaden

Jones A (1983b) From slaves to palm kernels: a history of the Galinhas country (West Africa), 1730–1890. Franz Steiner Verlag, Wiesbaden

Jones A (trans ed) (1995) West Africa in the mid-seventeenth century: an anonymous Dutch manuscript. African Studies Association Press, Atlanta

Joyner C (1985) Down by the riverside: a South Carolina slave community. University of Illinois Press, Urbana

Kea RA (1982) Settlements, trade, and polities in the seventeenth-century Gold Coast. Johns Hopkins University Press, Baltimore

Lambert S (ed) (1975) House of Commons sessional papers of the eighteenth century. 147 vols, vol 72. Scholarly Resources, Wilmington

Langley ER (1939) The Temne; their life, land, and ways. Sierra Leone Stud 22:64–80

Lauer JJ (1969) Rice in the history of the lower Gambia–Geba area. M.A. thesis, University of Wisconsin, Madison

Law R (1977) The Oyo empire, c. 1600-c. 1836: a West African imperialism in the era of the Atlantic slave trade. Clarendon Press, Oxford

Law R (1991) The slave coast of West Africa, 1550–1750: the impact of the Atlantic slave trade on an African society. Clarendon Press, Oxford

Lawson J (1967) In: Lefler HT (ed) A new voyage to Carolina (1709). University of North Carolina Press, Chapel Hill

Lawson A (2001) Recent archaeological research on Gambian Iron Age habitation. Nyame Akuma 55:32–35

Lemos Coelho F de (1985) Description of the coast of Guinea (1684) (trans: Hair PEH). Department of History, University of Liverpool, Liverpool

Lévi-Stauss C (1962) La pensée sauvage. Plon, Paris

Linares OF (1992) Power, prayer and production: the Jola of Casamance. Cambridge University Press, Cambridge

Linares OF (2002) African rice (*Oryza glaberrima*): history and future potential. Online via Wikipedia

Little KL (1948) The Mende farming household. Sociol Rev 40, section 4:37–55

Little KL (1967) The Mende of Sierra Leone, revised edn. Routledge & K. Paul, London

Littlefield DC (1981) Rice and slaves: ethnicity and the slave trade in colonial South Carolina. Louisiana State University Press, Baton Rouge

Marees P de (1987) Description and historical account of the gold kingdom of Guinea (1602) (trans ed: van Dantzig A, Jones A). Oxford University Press, Oxford

Massing A (1980) The economic anthropology of the Kru (West-Africa). Franz Steiner Verlag, Wiesbaden

McCaskie TC (1995) State and society in pre-colonial Asante. Cambridge University Press, Cambridge

McIntosh SK (ed) (1995) Excavations at Jenné-Jeno, Hambarketolo, and Kaniana (Inland Niger Delta, Mali), the 1981 season. University of California Press, Berkeley

McIntosh RJ, McIntosh SK (1981) The Inland Niger delta before the empire of Mali: evidence from Jenne-Jeno. JAH 22(1):1–22

McIntosh SK, McIntosh RJ (1993) Field survey in the tumulus zone of Senegal. Afr Archaeol Rev 11:73–107

Mencken HL (1948) The American language: supplement II. A.A. Knopf, New York

Miège J (1954) Les cultures vivrières en Afrique occidentale. Cahiers d'Outre-Mer 7:25–50
Montgomery M (1994) Lorenzo Dow Turner's early work on the Gullah. In: Montgomery M (ed) The crucible of Carolina: essays in the development of Gullah language and literature. University of Georgia Press, Athens, pp 158–174
Morgan PD (1998) Slave counterpoint: black culture in the eighteenth-century Chesapeake and Lowcountry. University of North Carolina Press, Chapel Hill
Mouser BL (2002a) A slaving voyage to Africa and Jamaica: the log of the Sandown, 1793–1794. Indiana University Press, Bloomington
Mouser BL (2002b) Who and where were the Baga? European perceptions from 1793 to 1821. HA 29:337–364
Murdock GP (1959) Africa: its peoples and their culture history. McGraw-Hill, New York
Murray SS (2004) Searching for the origins of African rice domestication. Antiquity 78(300):1–3 (of online extract)
National Research Council (NRC) (1996) Lost crops of Africa, vol 1, grains. National Academy Press, Washington
Oka H-I (1961) Report of trip for investigation of rice in Latin American countries. National Institute of Genetics, Mishima
Opala JA (1987) The Gullah: rice, slavery and the Sierra Leone-American connection. U.S. Information Service, Freetown
Ozanne P (1966) The Anglo-Gambian stone circles expedition. West Afr Archaeol Newsl 4:8–18
Pace A (trans ed) (1983) Luigi Castiglioni's viaggio: travels in the United States of North America, 1785–87. Syracuse University Press, Syracuse
Pacheco Pereira D (1937) Esmeraldo de situ orbis (trans ed: Kimble GHT). Hakluyt Society, London. Written in early sixteenth century
Paulme D (1954) Les gens du riz: Kissi de Haute-Guinée française. Plon, Paris
Paulme D (1957) Des riziculteurs africains: les Baga (Guinée française). Cahiers d'Outre-Mer 10:257–278
Pélissier P (1966) Les paysans du Sénégal. Fabrègue, Saint-Yrieix
Perrot C-H (1990) Semailles et moissons dans la région d'Assinie vers 1700. J des Africanistes 60(1):9
Phillips UB (1966) American Negro slavery. Louisiana State University Press, Baton Rouge. First published in 1918
Pollitzer WS (1999) The Gullah people and their African heritage. University of Georgia Press, Athens
Portères R (1950) Vieilles agricultures de l'Afrique intertropicale. L'Agronomie Tropicale 5:489–507
Portères R (1955a) Historique sur les premiers échantillons d'*Oryza glaberrima* St. recueillis en Afrique. JATBA 2:535–537
Portères R (1955b) Présence ancienne d'une variété cultivée d'*Oryza glaberrima* St. en Guyane française. JATBA 2:680
Portères R (1960) Riz subspontanés et riz sauvages en El Salvador (Amérique centrale). JATBA 7:441–446
Portères R (1976) African cereals: eleusine, fonio, black fonio, teff, *Brachiaria paspalum*, *Pennisetum*, and African rice. In: Harlan JR, de Wet JMJ, Stemler ABL (eds) Origins of African plant domestication. Mouton, The Hague, pp 409–452
Ribeiro O (1962) Aspectos e problemas da expansão portuguesa. Junta de Investigações Cientificas do Ultramar, Lisbon. Ribeiro quotes from Sousa GS de (1587) Noticia do Brasil
Richards P (1996) Culture and community values in the selection and maintenance of African rice. In: Brush SB, Stabinsky D (eds) Valuing local knowledge: indigenous people and intellectual property rights. Island Press, Washington
Richardson D (1991) The British slave trade to colonial South Carolina. SA 12:125–172
Rivière C (1968) Le long des côtes de Guinée avant la phase coloniale. Bulletin de l'IFAN, series B, Dakar

Rodney W (1970) A history of the upper Guinea coast, 1545 to 1800. Clarendon Press, Oxford
Rosengarten D (1994a) Row upon row: sea grass baskets of the South Carolina Lowcountry, revised edn. University of South Carolina Press, Columbia
Rosengarten D (1994b) Spirits of our ancestors: basket traditions in the Carolinas. In: Montgomery M (ed) The crucible of Carolina. University of Georgia Press, Athens, pp 133–157
Rosengarten D (1997) Social origins of the African-American lowcountry basket. PhD thesis, Harvard University, Cambridge
Rosengarten D (2008) By the rivers of Babylon: the Lowcountry basket in slavery and freedom. In: Rosengarten D, Rosengarten T, Schildkrout E (eds) Grass roots: African origins of an American art. Museum for African Art, New York, pp 104–121
Rosengarten D, Rosengarten T, Schildkrout E (eds) (2008) Grass roots: African origins of an American art. Museum for African Art, New York
Roussier P (1935) L'établissement d'Issiny, 1687–1702. Larose, Paris
Salley AS Jr (1919) The introduction of rice culture into South Carolina. Bull Hist Comm South Carolina 6. Cited by Carney 2001a
Searing JF (1993) West African slavery and Atlantic commerce: the Senegal River Valley, 1700–1860. Cambridge University Press, Cambridge
Second G, Rouhan G (2008) Human-mediated emergence as a weed and invasive radiation in the wild of the CD genome allotetraploid rice species (*Oryza*, Poaceae) in the Neotropics. PLoS ONE
Shaw T, Sinclair P, Andah B, Okpoko A (eds) (1993) The archaeology of Africa: foods, metals, and towns. Routledge, Printed by the author, London
Sloane H (1705–1725) A voyage to the islands Madera, Barbados, Nieves, S. Christophers, and Jamaica, 2 vols. Printed by the author, London
Smeathman H (1773) Letter to Dru Drury, July 10, 1773, MS D.26, Uppsala University Library, Sweden. Smeathman was a naturalist who spent years in Sierra Leone, Drury an entomologist who helped sponsor his travels
Steudel EG (1855) Synopsis plantarum glumacearum, 2 vols. Metzler, Stuttgart
Stewart J (1931) Letters from John Stewart to William Dunlop. South Carolina Historical and Genealogical Magazine 32 (January). Cited by Littlefield 1981
Svalesen L (2000) The slave ship Fredensborg. Indiana University Press, Bloomington
Thilmans G, Descamps C, Khayat B (1980) Protohistoire du Sénégal: recherches archéologiques, tome 1, les sites megalithiques. IFAN, Dakar
Thomas LV (1959) Les Diola. IFAN, Dakar
Turner LD (1974) Africanisms in the Gullah dialect, new edn. University of Michigan Press, Ann Arbor. First published in Chicago in 1949
Vaillant A (1948) Milieu cultural et classification des variétés de riz des Guyanes française et hollandaise. Revue Internationale de Botanique Appliquée et d'Agriculture Tropicale 28:520–529
Webster's Third (1966) Webster's Third Indernational Dictionary of the English Language Unabridged. G & C Merriam Company, Springerfield, MA
Wilks I (1975) Asante in the nineteenth century: the structure and evolution of a political order. Cambridge University Press, London
Wilson EG (1972) A West African cookbook. Flare/Avon, New York
Wood PH (1974) 'It was a Negro taught them': a new look at African labor in early South Carolina. J Asian Afr Stud 9(3–4):160–179
Wood PH (1975) Black majority: Negroes in colonial South Carolina from 1670 through the Stono rebellion. W.W. Norton, New York
Wood PH (2008) They understand their business well: West Africans in early South Carolina. In: Rosengarten D, Rosengarten T, Schildkrout E (eds) Grass roots: African origins of an American art. Museum for African Art, New York, pp 78–93

Chapter 4
African Origins of Sesame Cultivation in the Americas

Dorothea Bedigian

Abstract Contributions from Africa to crop introduction in the Americas have received insufficient recognition. A cultural, ecological, economic, geographical, and historical study of the transfer of sesame from Africa to the New World, this chapter surveys knowledge about sesame in the American colonies concerning medicine, myth, magic, culinary, and industrial use. African and New World usage by subsistence farmers was strikingly similar. Preference for mucilaginous foods, shown here, is widespread throughout Africa. A benefit already known by enslaved Africans, their masters in the American colonies eventually recognized the mucilage properties of the leaves and its value in treating summer dysentery. On both continents, according to folk belief, sesame has attributes as a good luck plant.

Sesame is a valuable crop introduced to Africa from Asia long ago. Language reveals important clues about transmission routes in the absence of written textual evidence. Surviving traces of African languages and customs in modern African-American and Caribbean material culture *vis-à-vis* sesame appear with culinary and healing traditions of Africa and the African diaspora. Eyewitness reports, including correspondence of Thomas Jefferson, provide firsthand data, supplementing herbarium specimens and limited, widely scattered published passages about the dynamic nature of contributions by African-born slaves in the dissemination and use of sesame in the Americas.

Keywords *Benne* • Gullah-Geechee • Kongo • *Vanglo* • *Wangila* • West Africa

---

This is a major revision of the article titled: Introduction and early use of sesame in America: medicine, cookery and folkways (Bedigian 2011b), published in *Sesame: the genus Sesamum*. D. Bedigian, ed. ©2011. Taylor & Francis, CRC Press, Boca Raton, FL.

D. Bedigian, Ph.D. (✉)
Research Associate, Missouri Bot Gard, St. Louis, MO, USA
e-mail: dbedigian@yahoo.com

R. Voeks and J. Rashford (eds.), *African Ethnobotany in the Americas*, DOI 10.1007/978-1-4614-0836-9_4, © Springer Science+Business Media New York 2013

Introduction and Objectives of the Study

The goal of this study is to document the early history of sesame (*Sesamum indicum* L., syn. *S. orientale* L.) in the Western Hemisphere. It links the crop with the transatlantic slave trade and the geography of colonization. Reports about sesame in the Americas appear in diverse sources: agronomic, botanic, culinary and economic histories, letters, linguistic, newspaper clippings, shipping records, and archival manuscripts. Together, these data portray surviving traces pertaining to sesame, from African languages and customs into modern African-American cultures. A culmination of research begun in 1996 with an early draft titled "Precious Cargo," this homage to enslaved Africans connects faraway communities through edible use, medicinal use, and language.

Methodology and Sources

With few written records, researchers must opt for other sources to learn about agricultural history and economic transformation (Bedigian 2004b; Ehret 1977, 2005). Unlike cash crops that are endowed with more detailed records, sesame, which geneticist Jack Harlan dubbed a "poor man's crop" (Bedigian 2011c), lacks such, as it is cultivated on a small scale, often intercropped among other food plants. Sesame's early American history is recorded in disparate fragments of ancillary information, including correspondence, diaries, *materia medica*s and pharmacopoeias, letters in newspapers, ship's logs, and commonplace sources such as recipes, all obscured under various African names, each with several variant spellings. The Americas, as described here, span Brazil and Latin America, Caribbean nations, and the coastal eastern United States. These accounts reveal the history of their ancestral African culinary, healing, and linguistic traditions (Bedigian 2003b, 2004d).

Herbarium voucher specimens occasionally record vernacular names (Bedigian 2004b). Label information was gleaned from 3,600 herbarium specimens of *Sesamum* and 700 of the closely related genus *Ceratotheca* Endl., held at 47 of the world's major herbaria in Africa, Europe, India, and the USA. The Microsoft Access 2003 Relational Database Management System for Windows contains my database of specimens. Searches of standard reference works such as the *Dictionary of American Regional English* (1985); Cassidy (2002), and the definitive *Oxford English Dictionary* (*OED* 2nd ed. 1989; online version June 2011) enabled assembly of unforeseen, diverse sesame names among the earliest sources. A thorough search in every dictionary available in an academic library introduced additional unfamiliar names for sesame. Etymology of each name was traced, and each name was tracked down in the published literature. African historian and linguist Christopher Ehret helped to assess colloquial names.

Distinguishing Crop Domestication from Cultivation

Hundreds of reports from recent centuries have proposed the "origin" of sesame in Africa. These views often draw upon the fact that the Pedaliaceae family, to which sesame belongs, arose in Africa, and many wild species in the genus *Sesamum* occur there. Another argument, proximity of morphological resemblance, fails to fulfill the definition. The fundamental test to demonstrate a crop's origin [domestication] is achievement of fully fertile hybrids from reciprocal crosses between crop and a putative progenitor, as shown for sesame (Annapurna Kishore Kumar 2003; Annapurna Kishore Kumar and Hiremath 2008; Bedigian 1984, 1988, 2000, 2003a, b, 2004a, 2010, 2011a; Bedigian et al. 1985, 1986; Hiremath and Patil 1999). Cultivation of a wild species is a separate matter. That is a practice widespread in Africa, whereby wild species are protected – deliberately and selectively – not weeded out of farmers' fields, unlike undesirable species (Bedigian 2003a, 2004d).

Sesame domestication occurred on the Indian subcontinent. Long ago, it reached Africa along with other traded commodities.

Earliest Record of Sesame in Africa

The earliest unmistakable sesame seeds in Africa come from an archaeological context: a basket of seeds found among the grave goods of King Tutankhamen (Bedigian 1998, 2000, 2004a, 2011c). The seeds are uncharred, slightly smaller than average in size, but within the range of variation for cultivated material. While these seeds do not provide evidence of cultivation in Egypt (they may have reached Egypt by trade), they show that sesame was present in Egypt by 1350 B.C.E. and deemed a fit offering for the royal grave.

Roman Era

Travel and trade on the Indian Ocean were described by an anonymous merchant of the first century C.E. in the *Periplus of the Erythraean Sea* (1912: 27): "Ships customarily fitted out from places across this sea, from Ariaca and Barygaza to Sachalites, bringing to these far-side market towns the products of their own places; wheat, rice, clarified butter, sesame oil, cotton cloth and girdles, and honey from a reed called *saccharum*. Sesame oil was traded along with cloth and wheat, for frankincense." Note that it was the value-added product, oil extracted from the seeds that was exported from modern Broach (Baroda), in Gujarat India, to the southern Arabian Peninsula. According to van Beek (1958: 142), "Claudius Ptolemy explicitly states that Sachalites is located beyond Syagrus… Ras Fartak, Southern Arabia, Bay of Oman, Arabian Peninsula, present day Yemen." Salomon (1991: 732), translating *telasa* as oil, provides a handful of little known fragments of inscriptions in Indian languages and scripts found in Egypt to corroborate the testimony of classical sources about the flourishing trade between India and the Roman Empire.

Cappers (1999) and van der Veen (1999) count sesame among the plant foods recovered in Egypt during the time of the Roman Empire. Their data support the assertion by Kajale (1991) that plants such as cinnamon, black pepper, lemon grass, and sesame were being exported from India into the Roman Empire during the early historical period. Clapham and Rowley-Conwy (2007) found sesame seed at post-Meroitic Qasr Ibrim, Lower Nubia, Sudan (500–550 B.C.E.).

Ancient Egyptian traders also fetched luxury goods from the South, including resins, ebony, ivory, animal hides, gold, and oils (Reisner 1918: 6).

Earliest Historical Report About Sesame Cultivation in Africa

Scholars writing about early sesame in Africa, including Mauny (1953), Busson (1965), Lewicki (1974), Burkill (1997), and Alpern (2008) state that sesame was observed by the Arab traveler al Zouhri (Zuhri) in 1150 C.E. at Kawkao, interpreted as Gao, Mali (***Merriam-*** *Webster's New Geographical Dictionary* 1997). Their source is the Arabic text of al-Zuhrî, Kitâb al-Ja'râfîya (1968: 184), where on numbered paragraph 333, al-Zuhrî wrote about the abundance of *al-simsim*, an Arabic word for sesame, at Kawkaw: "Among the towns of the Habasha in this section is the town of Kawkaw, which is the capital of the Habasha. Caravans from the land of Egypt and from Wāraqlān reach it, and a few from the Maghrib by way of Sijilmāsa. This town is on an island, cut off in the middle of the Nīl. The Nīl surrounds it on all sides and nobody can enter it except by boat. These people live entirely on pulses (*qatānī*) and rice which they sow on the banks of the Nīl. *The rice plant and its ears are exactly like millet (*dukhn*).* They have plenty of sesame and sugar cane and know no other fruit except what is brought to them from Egypt *and the Maghrib* such as dates and raisins." "*Kawkaw*" (i.e., "*Gawgaw*") is a reference to Gao (called "*Gaawo*" in the Songhay language, and "*Gawgaw*" in the Tamasheq language of the Tuareg).

At least as far back as Herodotus, "Ethiopian" referred to people with "burnt faces" and hence was synonymous with Black Africans. Sudan became the preferred term in Arabic sources, but with the same range of meaning. Since many of the Arab geographers were familiar with the Greek sources, it is more than plausible that they might revert to some of the Greek terminology (D. Crummey, personal communication, 2011). The word "Habasha" was used by Arab authors not only for Ethiopia but also for parts of West Africa, and the river Niger (frequently amalgamated by them with the Senegal) was often referred to by those Arab authors as "the Nile," or "the Nile of the Blacks." They thought the Nile of Egypt and the "Nile of the Blacks" were branches of the same river (P. de Moraes Farias, personal communication, 2011).

This detail corrects my earlier accounts (Bedigian 2006a, 2011b; Hopkins and Levtzion 2000) tell us that there are at least nine known al-Zuhri manuscripts and no two are quite identical, which might account for the difference of opinion (Alpern, personal communication, 2011).

Although it is difficult to show where and when the earliest sesame fields grew in Africa, we have genetic indications that its African occurrence has considerable antiquity. The fact that geographically isolated Nuba farmers cultivated a great diversity of sesame landraces (Bedigian 1981, 1988, 1991; Bedigian and Harlan 1983; Nadel 1947) attests to a crop's long evolutionary history in a region (Harlan 1951, 1970, 1976, 1992). Eventually, sesame became widespread across most of Africa (Mkamilo and Bedigian 2007).

Sesame Cultivation in Africa: Earliest Travelers' Accounts

What we know of early sesame cultivation was tracked through travelers' accounts, e.g., Ibn Battuta's (Muhammed ibn abd Allah) *Travels in Asia and Africa*, 1325–1354, and Leo Africanus. Ibn Battuta (1958–2000) witnessed edible sesame oil in Egypt's markets (c1330). These reports do not establish, however, that sesame was cultivated there. Leo's African observations, c1500 C.E. (Leo et al. 1896: viii: 324), state: "This little citie built vpon the riuer of Nilus,.. aboundeth greatly with the graine or seed called *Sesama*."

Significantly, a sixteenth century record indicates recognition of its leaf properties. A Portuguese, Duarte Barbosa, translated an early Spanish manuscript, a description of the coast of East Africa, written ca. 1514. Describing coastal Mozambique, Barbosa's translation indicated (1866: 8): "their rations are meat, rice and oil of sesame." The footnote below designates "*Ajonjo* (*Ajiongoli*) plant with a viscous substance, *Ajonjoli* sesame plant."

Cape Verdean André Álvares de Almada traded along the Upper Guinea coast (modern Guinea, Sierra Leone, and Liberia) in the 1570s and 1580s. Describing the Jalofos of Guinea (i.e., the Wolof of Senegal) (Almada 1594: 18): "The staple foodstuffs are rice, *milho maçaroca*, another grain called 'white *milho*' and sesame – from which oil is made." Regarding the trade and customs of the Kingdom of the Buramos (more commonly known as the Brame or Bram, inhabiting Guinea-Bissau and the Republic of Guinea), Almada (1594: 78) remarked, "This land provides much trade in provisions, in rice, *milho* (maize), and sesame; and in slaves, wax and ivory." Hall (1909: 47), writing of the Portuguese trade of Sabia and Limpopo Districts of Mozambique, 1505–1760, observed: "what trade came from them consisted only of ivory, ambergris, and iron, sesame and other vegetables."

A footnote in the *Papers of Henry Laurens,* August 26, 1730, indicates (1746–1755: 115): "Although the manufacture of oil from the seed was never established in South Carolina, the production of the seeds continued, providing a local delicacy known as *benne* seeds, the *Sesamum* plant being called *benne* by the natives of Sierra Leone in Africa." Thompson's *West Africa* established that sesame grew close to one major Atlantic port city (1858: 198): "*bene*-seed… is produced in large quantities, for its oil. It is bought up in Freetown, and exported to England and France; makes a superior oil… It is very aromatic when parched, and the natives are fond of it, to mix with their sauce." Capus and Bois (1912: 300–301) report: "Among our colonies, we must mention especially Guinea, where the plant was once the subject of quite important culture, and it is found almost everywhere in the loop of the Niger, where the Bambara denote it as *Béné*. Our possessions of the west coast of Africa have exported from 100–200 tonnes annually."

Early African Names of Sesame and Their Sources

While much justifiable attention has focused on the trade of enslaved Africans in the seventeenth and eighteenth centuries, there was also a thriving, complex exchange of foodstuffs that was part of the same regional network. Slavers and traders carried sesame to the Western Hemisphere. Willem Piso (1648: 21) devotes a chapter to sesame, identifying its Congolese name, *Gangila*, and in Portuguese, *Girgilim* [now *gergelim,* derived from Arabic *jiljilan/juljulan* from an Arabic noun *jaljala* "sound, echo," referring to the rattling sound of ripe seeds within the capsule].

Pliny wrote that Egypt obtained a large amount of oil from *gingelly* (*Sesamum indicum*) (Natural History 15.7.31). *Gingelly* is a name for sesame still in use today in India and Europe. It transferred into Spanish as *ajonjoli* and French as *jugleone*, and present-day Arabic medicinal and botanical works employ both *al-juljulan* and *simsim. Juljulan* was used at least as early as the eighth century in a poem by Waddah el Kubani-al Yamani d. 709 (quoted in *Lis n al-'Arab* 1981; Faroukh 1965). *Juljulan* is defined as "sesame before the seeds are removed," i.e., the capsule. *Juljulan* is derived *from juljul,* meaning "small bell" (Zabadi 1970), which gives, as the first definition, "bells worn around the necks of camels," and the second as "the horse *jeljela,*" meaning the "neighing horse rings" (Bedigian and Harlan 1986: 150).

The *Oxford English Dictionary* (1961) gives "jingle" as one definition for the entry "*gingelly*." Sesame flowers are campanulate. The capsule could be described as bell-shaped, and the seeds rattle within the capsule before it opens (Bedigian and Harlan (1986: 150)). The names commonly used in Morocco, *zinz'lan*, *djouldjoulan*, *zelzlane*, and *djyldjylan* (Bedigian 2004a), may well have originated with *gingelly* and *gangila.* A footnote in Flückiger and Hanbury (1879: 474[2]) states: "The word *Gingeli* (or *Gergelim*), which Roxburgh remarks was (as it is now) in common use among Europeans, derives from the Arabic *chulchulân*, denoting sesame seed in its husks before being reaped (Dr. Rice)."

Blacks called it *Chuni* and Portuguese *Pilaon* [this is actually the word for a pestle]. Piso vouched for the Portuguese import of sesame into Brazil from Guinea. A Dutch scientific expedition to Brazil in the 1640s found sesame cultivated in plantations. There, Blacks ate it with millet, replacing butter. A description of agriculture in Brazil in the sixteenth c. mentions sesame as *gergelim* (Hoehne 1937: 331).

A footnote in Flückiger and Hanbury (1879: 474[2]) states: "The word *Benné* is, we believe, of West African origin, and has no connection with *Ben*, the name of *Moringa*." Mathews pointed out (1948: 115–116): "The plant name *benne* is a word of especial interest to philologists. The first time *benne* appeared an English dictionary, so far as I have discovered, was in 1828 when it was included in the first of the long series of Webster's unabridged works. In the 1828 edition, *benne* did not get in as a separate entry, but it occurs under the entry *Sesame*, *Sesamum* where with reference to this plant the statement is made: 'One species of it is cultivated in Carolina, and the blacks use the seeds for food. It is called there, *bene*.' In the next edition of Webster, that of 1841, *bene* was entered in its proper alphabetical place as has been done ever since, but in one of the early editions, perhaps first in that of 1864, the word was provided with an etymology which derived it from Malay. Somewhat incongruously in this same edition of Webster, the word was defined as follows: 'The *sesamus orientale*, or oil plant, of the West Indies, from the seeds of which an oil is expressed called *benne*-oil, used in medicine as a laxative. Who the gentleman was who ventured to derive *benne* from Malay will probably never be known… This error has persisted in all subsequent editions of Webster's dictionary, though in the current edition the plant in question is no longer associated with the West Indies… There is nothing any clearer than that the word *benne* passed into the English language first here in the South, and that it did so as a direct taking over from African slaves of a word they brought with them from Africa where they had made the acquaintance of the plant before they were brought to this country. Bernard Romans, a British traveler writing of Florida in 1775 reported, speaking of this plant, 'The negroes use it as food either raw, toasted, or boiled in their soups and are fond of it; they call it *Benni*.'"

Legacy of the Transatlantic Slave Trade: Sesame in the Americas Retains African Practices and African Names

The earliest mention of *bene* occurs in a letter by Morel of Georgia in presenting sesame seed to the American Philosophical Society, Philadelphia (1769: 239–240): "I send you a small keg of *Bene* or *Bene* Seed, which you will please to present to your Society for their inspection. This seed makes oil equal in quality to Florence [olive], and some say preferable. Some say one hundred-weight of seed will produce ninety pounds of oil, others say less, be that as it will, it certainly makes very fine oil, and produces amazingly. If it is put to the trial, care should be taken to have the press well cleaned, so as leave no tincture from what may have been already pressed; in my opinion, this is an article of consequence, and I believe it will grow in Philadelphia. The way to sow it is in holes about three feet asunder, dropping in each hole about ten grains; when it comes up, thin it to three or four of the most promising, the seeds will appear in pods about September, and should, when full grown, and before dry, be gathered in. The method is as follows: As soon as you perceive about three-fourths or four-fifths of the pods rise on the stalk, and the lower pods begin to lose their seeds, it is then time to take it in; for after that; as much as ripens one day a top, so much falls out of the pod at bottom, you take a sharp hatchet bill or some such weapon, and with it cut off the stock 12–18 inches below any of the seed, holding the stock with the left hand; and when cut, a second person receives it, keeping it upright, till he has his load, for if you turn it downwards the ripe seed will fall out of the pods, you may immediately carry it into a barn, and set it upright on a close floor till you perceive all the pods fully dry and open. (You may, if you chuse, leave it in the field, which must be the case if a large quantity is planted) then thresh it, and run it through a proper sieve, and it is fit for use."

"I am quite unacquainted with the method of expressing the oil, but I believe if it is designed for table use, nothing should be done to the seed, as it might give it an ill taste. The lighter and dryer the soil is in which it is planted, the better."

A detailed early botanical description of *Sesamum indicum* is given by Bryant (1783: 345–346) with added information about its transfer to the Americas: "This plant is not only cultivated in Asia, but in Africa, and from the latter the negroes have carried it to South Carolina, where they raise large quantities of it, being very fond of the seeds, and make soups and puddings with them, as with Rice and Millet. They parch them too over the fire, and with other ingredients, stew them into a hearty food. The seed in Carolina is called *Oily Grain*, it yielding oil very copiously. This when first drawn has a warm pungent taste, and is otherwise not palatable, but after being kept a year or two the disagreeableness goes off, and it becomes mild and pleasant, and is then used in their sallads, and for all the purposes of *Olive Oil*."

Early agricultural reference works described sesame under its African names. Warden indicated (1819: 247): "*Sesamum Orientale*, or *benny-seed*, is now cultivated in Virginia and the Carolinas for domestic purposes. The oil which the seed affords is equal to olive oil of the best quality, and it is difficult to distinguish the one from the other by the taste." An anonymous news report in the *Agricultural Intelligencer, and Mechanic Register* (1820a: 101) with the headline "Cultivation of the grape, olive, &c" stated: "Among the productions raised for the manufacture of oil, is the vine or bush producing beans or pulse, called by the Africans the *beney* or *benny*, and called in Indostan the '*sesamum*'; in botany '*sesamum orientale*'. It has been brought into South-Carolina by the African ships, is raised, there, and its oil has been sent hither and to New-York, as salad oil, in presents, and for sale… The pulse, bean, or pea, when pressed and deprived of much of its oil, is excellent and pleasant food, boiled alone, or with vegetables, which require butter or oil. It would be well for our government confidentially to instruct our consuls and commercial agents to send us this seed from Africa and Indostan." Another anonymous account in the *American Farmer* (1820b: 135) urged: "The *oil* and *the food* for man and beast produced by the vine called the *Sesamum Orientale*, or *Bene* bush, or vine, are well worthy of the attention of the people of the United States. The *bene* vine or bush, has been produced for some time, in small quantities, in the southern states, from seed imported directly or indirectly from Africa, and from Asia. It abounds in the former, and in Bengal. Most of the blacks of the Mississippi, have continued the propagation of the seed of the *Bene*, and make soup of it after parching. The seed may be procured from them and from the blacks in the Carolinas and Georgia."

Bigelow (1822: 338) included *Benne Oil* in the *Materia Medica*: "Its seeds were used by the ancient Egyptians for food, and are still employed by the negroes and Asiatics for this purpose. The plant is now cultivated in the southern parts of the United States." Skinner of Skidaivay [Skidaway] Island GA (1824: 38) presented

"*Bene* Seed": "This plant (pronounced *Binne*) is the *Sesamum* L. and was probably introduced into our southern states, by the negroes from Africa. It abounds in many parts of Africa, and Sonini, and Brown, both late travelers into Egypt, say, it is much cultivated there for the purpose of feeding horses, and for culinary purposes. The negroes in Georgia, boil a handful of the seeds with their allowance of Indian corn. Probably no plant yields a larger proportion of oil." Skinner's note revealed the African origins of the plant.

Quoting from Goodwin's letter titled "*Bene* seed—its culture and use" (1824: 46): "The negroes in this part of the country are well aware of its virtues as a medicine; they likewise cultivate it for food; it is thought by them to be much better in soup than *okra*, and it is used by them in the same manner." Hooper (1829: 127) defined "*Benne* Seed": "Among the negroes, in Georgia, a plant is cultivated which appears to be a species of *sesamum*. They call it *benne*, which is probably its African name… The plant is now cultivated in the southern parts of the United States."

Parrish' essay begins (1863: 245–248): "The *Benne* plant (*Sesamum orientale* Linn.), is believed to be a native of Africa, whence it was probably brought to the United States by the negroes." "The negroes are in the habit of roasting the seeds and infusing them in water to form a drink like coffee… the leaves, which plucked at the time of their maturity, are very mucilaginous, and extensively used in the treatment of some complaints of children." Willis' *Flora* (1894: 196) named *Benne* among its commercial names, indicating pragmatically: "It was taken to the West Indies by Europeans, and *S. Indicum* has found its way to Florida and other Southern States and has crept along as far north as the vicinity of Philadelphia." *Cyclopedia of American Agriculture* (1909–1910: 501) identified *bene*: "said to have been brought to South Carolina by the early slaves. It now runs wild in parts of the extreme South, and is cultivated in small patches, chiefly by the negroes."

An illustrated oral history (Wightman and Cate 1955: 163) records: "Living alone and in a community of Negroes, Charles [Wilson] continued to practice these superstitions which he had learned from his parents long after they had been discarded by other Negroes. He was the last of the old Negroes to plant *bene* (*Sesamum*) at the end of every row of cotton, corn, etc. in the field. *Bene* was called the 'good luck' plant and was said to bring a good crop. When asked where he got his *bene* seed, Charles said he saved it; but when pressed to tell where he got his original start, he said his parents always had it and he was told 'Dey brung it fum Africa.' He said the 'old folks' gathered the seed to use in cooking. They made *bene* cakes and, on special occasions, *bene* candy. Also, it was ground in a mortar to extract the oil, which was used as a shortening in the making of cake, oyster stew, or as a butter substitute." Woods featured the West African method of extracting oil designed for a household level of production and consumption (1995: 38): "It is also evident that *benne* was introduced into the lowcountry by African-born men and women. In the 1930s Rosanna Williams, who lived in Tatemville, southwest of Savannah, recalled that her African-born father used to 'plant mosly *benne* an rice…. He use tuh beat *benne* seed in mawtuh an pestle, sometime wid a lill shugah an sometimes wid a lill salt an make a pase. He eat it on bread aw he eat it jis so.'" Granger (2007: 84) and Morgan (1982: 573) reiterate this account.

During recent decades, progressive contributions have begun to authenticate the introduction of crops from Africa into colonial America (Carney 2001; Chaplin 1993; Cross 2008; Grimé 1979; Harlan 1992; Holloway 2005). These include *benne*, greens, indigo, kola, okra, rice, sorghum, watermelon, and yams. The index entry of "okra" in the *Cambridge World History of Food* (2000: 1824) identifies it as "native of tropical Africa, carried by slaves to the Americas in the late seventeenth c." Carney (1998, 2001, 2003) substantiates this route for African rice, Attfield (1865) for kola nut, and Tobin (1999) for watermelon and yams. Africans came to cultivate rice in the 1526 settlement at the abortive colony at San Miguel de Gualdape that ultimately failed. These Africans were members of the *Mende* [*Minde*] tribe from the region now called Sierra Leone in West Africa. Spanish explorer Lucas Vázquez de Allyón had brought these people specifically because they were experienced in growing rice (Stormer 1993).

There are indications that in their free time, enslaved Africans tended their own vegetable gardens. They were given the use of a plot of land, generally about an acre per family, where they were encouraged to grow rice and vegetables including greens and according to Hess (1992: 8): "presumably African favorites such as okra, sorghum, black-eye peas, eggplant and *benne* seed." These crops were likely imports to the colonial southern states during the slave trade (Carney 1998, 2001; Savannah Unit Georgia Writers' Project Work Projects Administration 1986). Abrahams and Szwed (1975: 28) reckon: "By far the largest number of people entering South Carolina during the colonial period came from West Africa, and, in the course of a century of immigration, items indigenous to parts of that vast region were transported with them." Enslaved Africans in the Carolina Lowcountry grew root crops such as tania, African grains (including millet and sorghum), sesame (making soups and puddings and using its oil for salads), African peppers, and okra (Morgan 1998: 141): "They introduced sesame – what they called '*Benni*' – to the region. A late eighteenth-century history of plants noted that South Carolina slaves made both 'soups and puddings: of sesame and used its oil in salads.' Morgan reported that in 1762, an observer of the 'private fields' of Lowcountry slaves saw their '*beny*-seed' which the bondpeople planted for 'their own use and profit', revealing the enslaved Africans' oversight of seed stock for *benne*. Morgan reported that during travel to the West Indies and North Carolina in the late colonial period, Janet Schaw was so impressed by how Carolina slaves used their 'little piece[s] of land' to grow vegetables, 'rear hogs and poultry, sow calabashes, etc.,' that she thought they cultivated them 'much better than their Master[s].' The harvesting of gourds, picked from creeping vines rather than small trees (which was the common practice in Africa), is a particularly noteworthy activity, indicating that slaves actively produced their own drinking, eating and storage vessels. Schaw also believed that Carolina 'Negroes are the only people that seem to pay any attention to the various uses that the wild vegetables may be put to.'"

Hatley (1984: 19) wrote: "strange sounding names such as *tanniers*, collards and *benne* all grew on small 'provision gardens' which were alternately ignored and encouraged by slave owners." Little (1991: 61) observed: "Thus far, no archaeological evidence of a central plantation kitchen has been reported (citing Singleton

1988: 349). This finding is significant, not because it discredits a written account, but because it suggests that slaves prepared their food to suit their own tastes, perhaps incorporating aspects of their traditional cuisine."

Chaplin proposed that Southern planters would invest commercially in African plants only after crops grew successfully in the West Indies, or after enslaved Africans in America had cultivated them in garden plots. This indicates some autonomy by enslaved Africans. Chaplin observed (1993: 156–157): "Rather than importing African crops, planters more often discovered them in the gardens of their slaves. For these crops, blacks were the true experimenters and relied on a transatlantic network much different from that emanating from the Royal Society. Through the Atlantic slave trade, blacks had gradually transferred African plants (like sesame, guinea corn, okra)." "Southern planters, for instance, found lucrative commercial uses for sesame seed (known locally as *benne* or *binny*), after seeing it grown in slave gardens." "Sesame seed, which blacks called *benne* or *binny*, was another African crop nearly invisible to whites until the market beckoned for it. James Crokatt, a South Carolina merchant who removed to London, sent back in 1747 a model of a mill to press out sesame oil (as well as samples of cochineal and Spanish indigo) for display at the treasurer's office in Charleston. Crokatt intended to lead planters' thoughts toward diversification, and residents did pay some attention to sesame during the depression of the 1740s. Production expanded again during the commercial vicissitudes of the War of 1812; at this time, local planters looked for sales selling three dollars a bushel to cover the cost of growing the seed or (a more interesting possibility) the cost of buying it from their slaves plus a profit."

As they commonly had done in Africa, substantial slave-patch plantings used intercropping that "went unnoted in plantation record books. Judging by surviving documents, planters rarely attempted *benne* production on a scale to create more than a modest, largely local market for oil (the maximum plantings tended to be fifty acres), yet a number of plantations engaged in artisanal production, using sesame in crop rotations with corn, sweet potatoes, and cowpeas, or with rice and sweet potatoes" (Shields 2010: 30). Wilson and Ferris (1989: 21) confirmed: "Techniques of growing vegetables, as well as the plants grown, were another aspect of the African-American tradition. A 19th c. planter on the Sea Islands off the coast of Georgia described… a small patch where arrowroot, long collards, sugarcane, *tanniers* (an edible starchy root found in the tropics, also known as cocoyam), ground nuts, *benne*, gourds, and watermelons grew 'in commingled luxuriance.'"

Registry of Historical Accounts About Sesame in the Americas

United States

P. Miller (1735: pages not given) noted about "*Sesamum*; Oily grain": "promiscuously cultivated in the fields of *Syria*, *Egypt*, *Candy*, &c. where the inhabitants use the seeds for Food; and of late Years these Plants have been introduced in *Carolina*, where they succeed extremely well. The Inhabitants of that Country make an Oil from the Seed, which will keep many Years, and not take any rancid Smell or Taste, but in two Years becomes quite mild."

An account by Rev. William Stukeley, M.D., about Alexander Gordon (d. 1756) of Charles Town reports: "However at one point he took enough interest in the colony to send a description of it to the Royal Society; he wrote of 'its admirable fertility, and wonderful produce of unnumerable curious and useful things – the vine, wine, *sesamum*, oil for soap, cotton, mulberry, silkworms, cochinel, *opuntium* a yellow dye, hemp, flax, potash, &c. &c. But after all this profusion of nature's bounty, the inhabitants, through stupidity or laziness, made no profit or improvement in any one article for commerce, employing themselves wholly in the culture of rice'" (Rutledge 1949: 641).

Mease (1811: 44–45) demonstrated detailed knowledge of *Sesamum* in an essay recommending *bene* cultivation: "It is highly probably that the *Sesamum* plant was introduced into S. Carolina and Georgia, by the African negroes imported at an early period after the settlement of that part of the country; and there can be no doubt of the plant having been continued by them, for the purpose of adding to the various articles of vegetable aliment, as corn, sweet potatoes, and rice, of which their diet chiefly consists. They also parch the seeds, and after bruising them in a mortar, make them into soup, which they season with salt and pepper. The seeds parched and ground with an equal quantity of cocoa, make an excellent chocolate." Toasting intensifies the nutty flavor of sesame seed.

According to Meriam (1846: 970): "The *bene* plant was introduced into the West India islands from Africa, and subsequently in some sections of the United States. South Carolina and Georgia, I believe, are States in which this plant has been cultivated…. The plant is useful as a medicine, and I enclose a small printed sheet which I obtained from Mr Thorburn's seed store, for your information. More than twenty years ago I purchased several casks of the oil of the *bene* seed,… made from seed sent to that house from a correspondent in Georgia, or South Carolina. The seed was sent to a linseed oil mill in this vicinity, where it was crushed and pressed. A portion of the oil was obtained by cold pressing, the residue by hot pressing. the cold-pressed oil was light colored and flavorless; the hot-pressed was a little colored, and had the peculiar smell of nut oil. I purchased this oil at a low price, for the purpose of making soap; but finding it contained too much mucilage, I was induced to make sale of it to the house Clark & Co., druggists, Maiden Lane, New York. The Messrs. Clark paid me two dollars per gallon for the cold-pressed, and 90 cents per gallon for the hot pressed oil. Mr. Wyckoff used the cold-pressed oil for table oil, and found it excellent."

Holmes' compilation of cultivation instructions includes *Bene* (1852: 69), adding that "The seeds of *bene* contain a large quantity of oil, which is said to be of superior quality. *Bene* is also used in cakes and pastry, and the plant for its medicinal virtues." Burr included "*Bene*-Plant. Oily Grain. *Sesamum* sp." among medicinal plants (1874: 548–549): "Said to have been introduced into this country from Africa by the negroes."… "The seeds were at one time used for food; being first parched, then mixed with water, and afterwards stewed with other ingredients. A sort of pudding is made of the seeds, in the same manner as rice, and is by some persons much esteemed." Simmonds recognized (1889: 413): "*Benni* seed, as it is called in parts of Africa, is extensively used in Oriental countries for aromatizing the church bread and for the preparation of the renowned *Chalba*, which is eaten during fasts by all Orientals. It consists of the finely powdered seeds, which are mixed with honey, and oftentimes also with sugar."

Brazil

De Candolle (1884: 422) citing Piso (1648: 211) wrote that the Portuguese transported sesame from the Guinea coast to Brazil, but it is unknown how long it has been cultivated in the rest of Africa.

French Guiana

Aublet (1775: 665–666) observed: "Sesame is cultivated in the gardens of the Negroes. One presses the seeds [to obtain] oil which is good to eat fresh; it is called in the East Indies '*Gingili* oil.' We also roast the seeds with which one makes a kind of nougat, with honey or sugar." The Oxford African American Studies Center database shows a photograph of a woman in French Guiana cleaning sesame stalks.

Jamaica

Sloane (1707: i 161) observed: "It is frequently planted here by the *Negros* in their Gardens, and agrees to the Descriptions of Authors, having a flower like *Digitalis*. The seed is very often beat up in Chocolate." Using its local names, *vanglo*, *zesamum* or *Sesamum Africanum*, Lunan indicated that he found it in Jamaica under the name *Zezegary*. "The first time I saw the plant, it was growing in a negro's plantation, who told me, they ground the seed between two stones, and eat it as they do corn. Their seed-vessels are full of small white seeds, which the negroes call *soonga*, or *wolongo*. The oil that is drawn from it is called *sergilim* oil. The seed is often mixed and ground with coco, to make chocolate" (Lunan 1814: 251–252). Roxburgh and Ford (1898) list *zezegary* or *wanglo* among crops introduced by the British into Jamaica, for cultivation.

Sesame was a prized source of oil, with good reason. Long (1774: 809–810) raved about *vanglo* and *wongala*: "There are few which more deserve to be extensively cultivated into general domestic use, in the room of that abominable rancid butter imported hither from Europe. Nothing but the grossest prejudice, in favor of old habits, can influence the inhabitants to persevere in the importation of that unwholesome, nauseous stuff, and to swallow it every day with their food, when they may supply themselves with so fine, nourishing and wholesome an oil, as the

Sesamum, for an ingredient in their pastry…." It is germane to note that the names *soomgba/somgba* or *soomgwa* are colloquial for sesame in northern Congo's Ubangi region near the Central African Republic, whereas Lingala speakers near Kinshasa say *wangila* (Samson Mutolo, personal communication, 2011).

Trinidad

Warner-Lewis (1991) includes *Bene* in the lexicon of Wolof contributions to the language of Trinidad. Africans in nineteenth century Trinidad had a ceremony of thanksgiving and intercession to ancestors called the *saraka* involving animal sacrifice and offerings of unsalted food. Ritual plant ingredients for the ceremony included kola nut, rum, stout and wine, *bene* (called *ziziwi* there), black-eyed peas and rice, corn cooked with olive oil exclusively, and bread. Correspondingly, a feast that included *bene* made into balls and sugar cake characterized the Yoruba wedding in central Trinidad (Warner-Lewis 1991).

***Sesamum* and *Ceratotheca*: Closely Linked Species with Interchangeable Use of Their Slimy Leaves and Oily Seeds**

Ceratotheca is the genus closest to *Sesamum* (Abels 1975; Bedigian 2004d, 2011a). Several species in *Ceratotheca* and *Sesamum* have edible seeds; their slimy leaves serve as food and medicine in Africa (Bedigian 2003b, 2004d). Comparable use of the leaves of *Sesamum* and *Ceratotheca* suggests that African agency established these species in the colonies; there are clear links between usage in Africa and in colonial America (Manning 1991; Miller 1976). Their presence as plants naturalized in Florida (Manning 1991) offers botanical clues that speak for African peoples who arrived against their will in the Americas. These include medicinal uses of the mucilaginous leaves: for hair, as eyewash, laxative, and treatment of cholera and dysentery (Bedigian 2003b, 2004c, d; Bedigian and Adetula 2004). Knowledge about *Ceratotheca*'s useful mucilage is most likely derived from Africa.

Ceratotheca triloba (Bernh.) Hook.f. is decidedly of African origin. It is reportedly cultivated as a substitute oil source in Africa (Jardin 1967). A voucher specimen of *Ceratotheca triloba*, native to Angola and Namibia (Abels 1975; Mabberley 1997), was collected in the early twentieth c., in Florida. Wunderlin and Hansen (2008) list both species: "*Sesamum indicum* is known in the wild from Washington Co. (west of Tallahassee); *Ceratotheca* from Lake and Highlands Counties, north of Lake Okeechobee." Wunderlin (1982: 336) represented *Ceratotheca triloba*: "Disturbed sites, rare; Polk Co. Native to Africa; *S. indicum. Benne*. Disturbed sites. Rare; Lee Co. Native to East Indies." Batson (1984: 158) depicted *Ceratotheca triloba* as "introduced; borders, S. Florida." It is important to note that the country described as Angola also encompassed the Congo at that date, from which a substantial number of enslaved Africans were shipped (Gomez 1998; Klein 1972, 1978, 1997).

Benne je is the Bambara name of *Ceratotheca sesamoides* Endl. (Bedigian and Adetula 2004; Bedigian 88, MO), a wild relative in the genus closely related to sesame, indisputably from Africa, and cultivated as a substitute oil source (Bedigian 2004d; Jardin 1967). Busson (1965) provides the name *dobobéné* used by the Toma of Guinea. Dalziel (1955: 447–448) found that the designation "black *benniseed*" was used throughout much of West Africa for both *Ceratotheca* and sesame, and he considered "those of pale colour probably commonly an adulterature with seeds of *Ceratotheca*, at least in the Benue region."

Small (1913: 1367) found benne "in cultivated grounds and waste places, Gulf States. Naturalized from the tropics." Subsequently Small (1933: 1243) augmented: "Roadsides, cult. Grounds, and waste-places, Coastal Plain, Fla. to Tex. Nat. of East Indies." Moreover, Small (1933: 1243) included the related African species *Ceratotheca triloba*: "high pinelands and roadsides, peninsular Florida. Nat. of Africa."

Romans (1775 [1962]: 130–131) wrote about sesame in Florida: "the Negroes use it as food either raw, toasted, or boiled in their soups and are very fond of it, they call it *Benni*." Vignoles wrote all too vaguely: "the attention of the farmer will be directed to the *Benne* plant from which … a fine table oil is procured nearly equaling that from the olive; it grows in almost any kind of soil; the peculiar advantages of the *benne* and its history and productions, have been set forth a length in the publications upon Florida, which are already before the community and it is not necessary here to repeat them" (1823 [1977]: 103). Gray and Thompson assessed (1941: 901): "After the cession of Florida to the United States in 1819, a large part of the Spanish population left the State. In 1822, the total population was estimated at 5,000. A large proportion, were runaway slaves. The opportunities of the new territory excited much interest, and there was a considerable immigration of Americans. Very early, the possibility of growing various subtropical products aroused the same interest as at the beginning of British occupation in the colonial period. The imaginations of prospective immigrants were excited by the prospects of growing almonds, oranges, lemons, citrons, olives,… sesame, silk, sea-island cotton, ginger, breadfruit, mangos,… and various drugs and spices."

Irvine (1954) noted the addition of leaves of *Ceratotheca sesamoides* to soups, in the southern states. Manning considered it to be weedy or a naturalized escape from cultivation (1991: 315).

Mucilaginous Leaves Used as Medicine

The eighteenth c. experienced a surge in knowledge, exploration, and rapidly growing technology and expanded record keeping made possible by advances in the printing press. Medical theory and practice of the 1700s developed too, as evidenced by extensive plant collection, along with descriptions of diseases, their conditions, and treatments. Books on science and technology, agriculture, natural philosophy, and even cookbooks all increased at that time. This literature survey demonstrates what Fett pointed out: "This colonial exchange of plants across the Atlantic reorganized the pharmacopoeia that both European and African descendants drew on for medicines in North America" (2002: 63).

There is a preference for mucilaginous foodstuffs, above all, okra, throughout Africa. Today's subsistence practices, still observed in remote villages in Africa, may well reflect remnants of earlier customs that survived intact for hundreds of years. A review of *Sesamum* herbarium labels and published sources reveals widespread medicinal usage of these mucilaginous leaves in Africa and Asia (Bedigian 2003b, 2004c, d). Miller (1976) cites Williamson (1975), who registers the name *benniseed* and documented the use of sesame leaves (*umpeza*), cooked as a side dish in Port Herald, Nyasaland. Bedigian (2004d) described similar examples in Kenya, Sudan, Tanzania, and Uganda.

Among the earliest records of American usage, a number of reports indicate that sesame leaves were recognized officially, as the plant part used for medicine. Sloane (1707 i: 161) referred to use of the mucilaginous plant as well as its oil-rich seeds: "A Decoction of the Plant is used for resolving *Ophtalmie*, of they be applied to the Eyes, for Coughs, Pleuresies, Inflammations of the Lungs, and hard schirrous Tumours. Women use it for hardness of the Uterus, it moving the Menses. They use the same for the Diseases of the Skin and Face from Spots. The Herb and Seed boil'd in Honey, makes a resolving Plaister for hard and sanguineous Tumors, and dried Nerves. The Decoction is good in Clysters. The Seed gives gross Nourishment, and fattens very much. The decorticated Seed fattens; the Oil more; and the Dregs (which are eaten for Food in Ethiopia) more than that. Women ordinarily drink the Oil to be fat, with the Dregs it is given to four ounces in Plurisies and Pains, and in all defedations of the Skin, outwardly, as well as inwardly." Browne (1756: 270) wrote about *Vanglo* or Oil-plant: "A decoction of the leaves, and buds, is looked upon as a good resolutive; and frequently ordered in inflammations of the eyes, where warm fomentations become requisite."

Barham indicated (1794: 121–122): "A decoction of the plant is good for coughs, pleurisies, inflammations of the lungs, hard schirrous tumours, and women use it for hardness of the womb. The herb and seed, boiled in honey, make a good cataplasm or poultice for hard tumours, and dried nerves or shrunk sinews; so doth the oil. A decoction of the whole herb, flowers, and seeds, is good in clysters, soften the belly, and give a stool or two. The juice of the herb or distilled water is good for sore eyes."

Thomas Jefferson (1808) wrote to Dr. Gustavus Horner about the *benne* plant's effectiveness in intestinal disorders: "The leaf is a specific in dysenteries & other visceral complaints. Two or three without being bruised, being put into a pint of cold water, in a few minutes produce a mucilage equal to that of the white of an egg. 5. or 6. pints are taken in the course of the day. The leaves dried under cover retain the property of producing mucilage" (Betts 1944).

Thacher discovered the *bene* plant on a plantation near Savannah, in 1810, where it was cultivated extensively: "It was planted in hills like beans, and the produce of seed was exceedingly abundant." A tribute followed: "Originally an African plant, has become well known by the name of *bene* in South Carolina and Georgia, or the *vangloe* of the West Indies.... Of late years, the seeds have been introduced into the states of Georgia and South Carolina, by the African negroes, where the plant succeeds extremely well, and they boil a handful of the seeds with their allowance of Indian corn, which forms a nourishing food. But the excellency of these seeds consists in their yielding a larger proportion of oil than any other vegetable with which we are acquainted. One hundred weight of seed will produce ninety pounds of oil of an equal and even preferable quality to Florence [olive] oil. It will keep good many years without contracting any rancid smell or taste... It also burns well in lamps. The leaves of this plant, by infusion or decoction are found to afford an excellent mucilage, well adapted to all the intentions of that class of remedies, and in 1809 was used with the most marked good effect, in an epidemic dysentery in South Carolina. Considering therefore, the great utility and importance of the *bene* plant, its cultivation by our patriotic planters cannot be too strongly recommended" (Thacher 1821: 373–374). *Materia Medica* (Ewell 1827: 643) promoted, "*Benne*, *Se Samum Orientale* – Is now cultivated in South Carolina and Georgia. The leaves by infusion afford an excellent mucillaginous [as spelled] drink, which is used with manifest advantage in dysentery, diarrhoea and cholera *infantum*. The seeds yield a pure and pleasant oil, which in doses from one to two wine-glassfuls, acts well on the bowels. It is now generally used at the tables of the wealthy." Thacher extolled the benefits of *bene* mucilage (1835: 40): "*Bene* or *Sesamum* seed... very much used in this country, for all complaints of the bowels; it is particularly efficacious in dysentery... seed is sown in April, and the leaves are fit for use in June. They are steeped in spring water, two or three to a tumbler, and in 10 or 15 minutes the water will be fit to drink."

Mease (1811: 45, 48) demonstrated full knowledge of the utility of *Sesamum* leaves in dysentery: "it is also worthy of attention by reason of the medicinal qualities of its leaves... The dysentery, a disease that frequently ravages our country settlements, yields very readily to an infusion of the leaf in water. In the year 1803, during an epidemic flux, which raged with great violence in the upper country of South Carolina, this remedy was attended with the best effects. Three or four leaves infused in a pint of cold water will in a short time yield a thick mucilage, which may be given to the quantity of five or six pints daily. The infusion of the dried leaves is equally beneficial."

Lunan (1814: 251) wrote: "The leaves are of a very mucilaginous nature, as well as the seeds; and the emulsion of both have been recommended as excellent remedies in dysentery." Hooper (1817: 741) noted that "The seed and leaves are used medicinally in some countries on account of the bland oil the former contains, and for the mucilaginous nature of the latter." Cathcart (1821a, b) reported that "the grain parched makes a pleasant light food, and may be prepared as a substitute for chocolate, and that an infusion of the leaves in water produces a gelatinous drink highly recommended in bowel complaints." Roughley reported: "The *wangola*, a small Indian shrub, whose leaf, when steeped for a short time in water, or simple liquid, makes it of a mucous, glutinous quality, and is deemed an excellent medicine in case of flux or dysentery" (1823: 416–417).

Former US vaccine agent Smith, writing from Baltimore July 14, 1823, pointed out (1823b: 619): "The great value of the *sesamum indicum*, or *bhene* plant, is not yet known as generally as it ought to be… in cases of *dysentery* and cholera *infantum*, I am fully persuaded we have nothing superior to it, that can be used in the treatment of either of these distressing complaints. Young children, who cannot be forced to swallow any other medicine, without great difficulty, are fond of this, which is refreshing and nourishing to them. Those whose stomachs nauseate at the sight of almost every thing that we can present to them, take the clear mucilaginous infusion of the leaves of the *bhene* plant, with the same avidity they would drink cool spring water." Submitted from Georgetown, July 15, 1823, Smith recommended *Bhene* for the "summer complaint" (1823c: 619): "drying of the leaves in the shade, previous to their being touched by frost, and putting them away, in order to be prepared for early attacks of Cholera *Infantum* in the spring… upon administering an infusion of the dried leaves, it was, in the course of the day, entirely relieved… the simple manner in which it is prepared for use - which is, by placing one or more of the green leaves, (according to the size), in a tumbler of cool water, and the mucilage is immediately imparted, and it is fit to drink - Its great advantage is its insipidity. The leaves when dry, require to remain rather a longer time in the water, to deprive them of their mucilage. When green, a full grown leaf is sufficient for three or four tumblers of water." Similar reports appeared in Rhode Island and Richmond (Smith 1823a, d). Smith provided additional details conveying his own case history (1823e: 332): "On the first of September I was attacked myself, in the morning, with an inflammation in my right eye. During the whole day, I was in constant pain, and as the evening approached, the irritation increased to such a degree that I was obliged to retire from the light. I prepared some of the *Bene* mucilage with *warm* water; and applying it directly to my eye, I derived immediately, much greater relief from it than I had anticipated. I next took the leaves out of the warm water, and applied them to my eye. They were exceedingly soft, smooth and comfortable, and procured me a good night's rest; in the morning I found myself entirely free from any appearance of the inflammation that attacked me."

Smith promoted the leaves (1824: 1): "One leaf of the *Bene* plant, immersed in a tumbler of pure water, changes the whole of it immediately, into a perfect mucilage: that is clear, tasteless and inodorous. Sick children take this mucilage with the same avidity they would drink cold water - and as it is perfectly innocent, they may be allowed to take as much of it as they like… The *Bene* mucilage is useful in all cases where other mild and mucilaginous remedies are proper to be recommended… The *sesamum* is one of the most invaluable plants I know… I may safely assert, that if the virtues of the *Bene* leaves were known as generally as they should be, they would save the good people of the United States many thousand dollars annually, that are now spent to procure nauseous drugs of various kinds, that are not only useless, but often unjurious to those who take them."

The *American Farmer* (January 9, 1824: 331–333) cited by Dictionary of Americanisms on Historical Principles (1951). University of Chicago Press, Chicago, IL. described its medicinal use: "The *Bene* mucilage is useful in all cases where other mild and mucilaginous remedies are proper to be recommended." Mills (1826: 93) described the "*Bennè* Oil Nut, *Sesamum indicum*; the seeds of this plant furnish an excellent oil for salads, and every purpose for which olive is used; the grain parched makes a pleasant light food, and may be prepared as a substitute for chocolate; and an infusion of the leaves in water produces a gelatinous drink highly recommended in bowel complaints; one or two wine glasses full of the oil acts well on the bowels." Harrison recommended *benne* leaves above most any other remedy (1828: 1): "The best drink, when the stomach is irritable, is the roasted meal gruel, or a cold infusion of the *benne* leaf (*sesamum orientale*), or slippery elm bark."

The *New Harmony Gazette* (Anonymous 1827: 277) carried a rapturous promotion: "The *Bene* mucilage is useful in all cases where other mild and mucilaginous remedies are proper to be recommended. It is particularly useful in the dysentery and summer complaints of children; and the leaves may be applied in case of inflammation of the eyes, and as a dressing for burns, cuts, bruises, gun-shot wounds, etc." Rafinesque identified *Sesamum* as *benny*, *benne*, *vanglo*, and *zezehan*: "The mucilage of the leaves has long been known as an excellent remedy for diarrhoea, dysentery, cholera *infantum* &c. diseases that will always prevail in the United States in Summer, as long as green fruits are allowed to be sold in the markets, and parents will be so imprudent as to allow their children to eat this rank poison. Dr. Smith of Baltimore was the first to use, recommend and extol this mild and useful property of the *Sesamum* leaves. Therefore, it is nothing new, but not the less useful for all that. It is to be regretted that this plant is not cultivated on a larger scale for the seeds, oil and leaves" (Rafinesque 1831: 3).

Williams (1832: 2) wrote of *Bene* Plant: "The seeds of this plant seem to have been introduced into our southern states by the negroes from Africa, and is cultivated by them in almost every patch, or negro garden, to a limited extent; and is considered by them as a specific in all cases of dysentery, diarrhoea and cholera. For this purpose, about two quarts of cold water are put into a vessel, two green leaves are then taken from the *bene* plant and the water kept stirring with them for about five minutes, by which time the water will have assumed nearly the consistency of starch, perfectly colourless and tasteless. Of this water, the patient is made to drink freely and often, with the most beneficial effects in those complaints.... As it is only cultivated as a medical plant, a few seeds will probably be enough for any one family. Whether it will preserve its medical qualities after it is dried, I am not informed."

Fessenden (1842: 51) praised "*BENE* PLANT – *Sesamum* – This was introduced into the Southern States by the negroes from Africa. It abounds in many parts of Africa. Sonnini and Brown, travelers in Egypt, say it is much cultivated there for the purpose of feeding horses and for culinary purposes. The negroes in Georgia boil a handful of the seeds with their allowance of Indian corn. Probably, no plant yields a larger proportion of oil, which Dr. Cooper of Philadelphia has pronounced equal to the finest oils. But it is worthy of cultivation in the Northern States, principally as a medicinal plant. A gentleman in Virginia has given Messrs. Thorburn & Sons the following account of its virtues: "A few leaves of the plant, when green, plunged a few times in a tumbler of water, make it like a thick jelly, without taste or colour, which children afflicted with the summer complaint will drink freely, and is said to be the best remedy ever discovered. It has been supposed, that (under Providence) the lives of three hundred children were saved by it last summer in Baltimore, and I know the efficacy of it by experience in my own family.""

Directions for using the *bene* plant were published by the Commissioner of Patents (Meriam 1846: 970): "Take eight or ten of the leaves, wash the sand off, and let them lie twenty minutes in a half-pint tumbler of cold water. A think jelly will soon be formed, without taste or color, which children afflicted with the summer complaint will drink freely. A little sirup may be added to tempt the taste. It has been supposed that the lives of a great number of children have been saved by it, wherever it has been used. In some instances it has been administered with perfect safety to infants only a few days old." Miller (1976: 136) documents early use of *benne* leaves at four Ohio locations, ranging from 1847 to 1874.

Porcher (1863: 450–453) showed considerable knowledge about *Sesamum indicum*: "*Bené.* Introduced by the Africans. Fl. July. This is the sesame of the *Anabasis*, mentioned also by Dioscorides, Theophrastus, and others. The seeds contain an abundance of fixed oil as tasteless as olive, and for which it may be substituted; it is said to be used extensively in Egypt and Arabia. Lind. *Nat. Syst.* 280; *U. S. Disp.* 661. Mérat says that in Egypt they drink large quantities of the oil morning and evening, to give them embonpoint. It is also used medicinally as a laxative, and is by some preferred to castor oil; also as an application to furfuraceous eruptions. In India it is regarded as an emmenagogue and as provocative of abortion; employed in cutaneous affections and ophthalmia; a solution is given in colic and dysentery, and used as an application for softening the skin. Mérat and de L. *Dict. de M. Méd.* vi, 332, and the *Supplem.* 1846, 657, according to which it is also becoming an object of considerable commercial importance, being substituted for olive oil in the manufacture of Marseilles soap. See Essay of M. Hardy, *Revue Agricole*, Avril, 1845, 177. In the *Trans. Phil. Soc.*, it is said that one hundred parts of the seed yield ninety of oil. Coxe, *Am. Disp.*, art. *Sesam. orient.*, states it was found beneficial in a dysentery which prevailed in 1803. We have seen it given to some extent, and with great advantage, in New York, in diarrhoea and dysentery, particularly in these affections as they occur in children; two or three of the leaves, thrown in water, are sufficient to render it very mucilaginous. This is taken internally. It also serves as a convenient vehicle for enemata, gargles, collyria, etc. In South Carolina the seeds are largely used by the negroes in making broths. They are also eaten parched, and are made with sugar into a very nice candy. It might be made a source of profit to raise the plant in the Confederate States, as it grows well and the seeds bring a high price. The above was contained in my report on the *Med. Bot. of South Carolina*, published in 1849."

"The oil pressed from the seed will keep many years without acquiring any rancid taste, but in two years becomes quite mild, so that the warm taste of the oil when first drawn is worn off, and it can be used for salads and all the ordinary purposes of sweet oil. In some countries it is used for frying fish, as a varnish, and for some medicinal purposes. Nine pounds of seed are said to yield upward of two pounds of fine oil. The oil may be extracted by bruising the seed and immersing them in hot water, when the oil rises on the surface and may be skimmed off. But the usual mode of extraction is similar to that practised in the expression of linseed oil. The plant is generally sowed in drills about four feet apart, in the month of April. *Am. Farm. Encyc.* I consider, after examination that the sassafras leaf contains more mucilage than the *Bené*, and that both should be gathered and cured for winter use in making mucilaginous teas to be used in dysenteries, pulmonary diseases, etc."

"From a statement of H. M. Bry, of Louisiana, P. O. Rep., 1854, p. 225, sixteen bushels of seed of *Bené* plant (*S. orientale*) was sent to a mill in Cincinnati to be manufactured into oil. It yielded thirty-nine gallons of clear oil and about five quarts of refuse oil, or about two and a half gallons to the bushel. In consequence of the mill imparting the flavor of flaxseed he could not use it as a salad oil, for which purpose he was confident it would be superior, when pure, to the adulterated imported olive oil. It was used however, as a substitute for castor oil. All who used it praised it for its gently purgative effect, and because it was free from the nauseous taste peculiar to castor oil. Twenty bushels is believed to be a moderate estimate of the amount of the seed produced by an acre. It yields a gallon of oil to the bushel more than flaxseed."

"The excellent effect of the leaves steeped in water as a mucilage to be used in diarrhoea and dysentery is testified to by all persons who have used it. For this purpose, two or three leaves are soaked in a tumbler of water and administered repeatedly. This plant will act as a substitute for gum-arabic on account of the mucilage it yields. It should be used in the bowel affections of children and among our soldiers in camp. Planters should collect and cure all the leaves at their disposal. At page 338 of the same volume, another paper on the *Bené* is to be found. It is there stated that the plant will throw out a great profusion of leaves by breaking off the top when it is half grown…[]. The *Bené* probably yields as much oil as any plant we possess, as I am informed by a practical gardener."

The Friend, a religious and literary journal (1855: 387), extracted a letter from a correspondent of the Patent Office dated Monroe, Washita Parish, Louisiana, for its forthcoming agricultural report. It speaks of "The *Bene* Plant, from which oil of a pure quality is produced in great abundance" Washita Union. "In 1813 I sent 16 bushels of seeds of the *bene* plant (*Sesamum orientale*) to a mill in Cincinnati, to be manufactured into oil. It yielded 39 gallons of clear oil, and about 5 quarts of refuse oil, or about 2.5 gallons to the bushel." […] "The leaf of the plant is an excellent remedy for bowel complaints in children, and also in adults. For this purpose, two or three leaves are put in a tumbler of water, which they immediately render mucilaginous, but impart no disagreeable taste. The negroes cultivate it for food, using the parched seeds with their meats. I consider it so useful that a few stalks at least should be raised in every garden. And I believe it will soon be extensively cultivated for manufacturing oil, yielding, as it does, about a gallon to a bushel more than flaxseed oil."

Wood and Bache (1858: 714) add: "*Benne* Leaves.-These abound in a gummy matter, which they readily impart to water, forming a rich, bland mucilage, much used in the Southern States as a drink in various complaints to which demulcents are applicable, as in cholera *infantum*, diarrhea, dysentery, catarrh, acute cystitis, strangury, etc. The remedy has attracted attention also in the North, and has been employed with favorable results in Philadelphia." De Voe (1867: 358) references "The *Bene* plant. This plant is much used in the South for culinary purposes…" Most strikingly, "A few leaves, when green, plunged a few times in a tumbler of water, make it a thick jelly, without taste or color, which children afflicted with the summer-complaint will drink… the best remedy ever discovered." Dunglison (1868: 881) states: "The leaves – *Sesamum* (Ph. U.S.) – afford, by infusion, an excellent mucilage."

Burr included "*Bene*-Plant. Oily Grain. *Sesamum* sp." among medicinal plants (1874: 548–549): "A few of the leaves, immersed a short time in a tumbler of water, give it a jelly-like consistence, without imparting color or flavor; and in this form it is generally used." M'lean (1875: 64) extolled The Dysentery Plant as a "wonder-working plant": "I have grown the plant for the last eight years, and have known many wonderful cures resulting from its use." Maïkere-Faniyo et al. (1789) and Neuwinger (2000) provides contemporary confirmation of its efficacy in dysentery. Codrington (1878: 151) wrote of *Bene* or *Wanglo*: "The leaves are very mucilaginous. One of them placed in a tumbler of water soon turns it quite ropy. It is given to check severe vomiting."

Bentley and Trimen (1880: entry 198) devoted nearly a full page to medicinal use of sesame leaves: "*Sesamum* or *Sesami Folium*. - General Characters. - *Benne* leaves abound in a mucilaginous or gummy principle which they readily impart to cold water and form a thick transparent mucilage. … Medicinal Properties and Uses. - An infusion made by adding one or two of the full-sized leaves to about half a pint of cold water, and occasionally stirred, is much used in the Southern States of North America as a demulcent drink in cholera *infantum*, and other disorders of the bowels in children. It is also employed in catarrh and affections of the urinary organs. The infusion is sometimes made with the dried leaves; tut in this case hot water should be substituted for the col. This remedy has likewise been successfully used in other parts of the United States. The leaves also serve for the preparation of emollient poultices. *Benne* leaves are also much valued in the treatment of dysentery in some parts of Australia; and within the last few years the mucilage obtained from the leaves of the Indian-grown plant has been very successfully employed in the treatment of the milder forms of dysentery in India."

Sweringen (1882: 76) names *bene*, *benne*, and *vangloe*: "The leaves of *Sesamum indicum* and *Sesamum orientale*." Fenner (1884: 473) identifies *Sesamum indicum* as *benne*: "The leaves are demulcent, and are employed for cholera *infantum*, and other intestinal troubles." Bolles (1885: 487/1) indicated: "The *Benné* plant is a native of Africa, and probably also of Asia. The leaves of *Benné* are very mild and mucilaginous."

The *Yearbook of the U.S. Department of Agriculture* (1895: 198) stated: "The leaves of the sesame plant are considered of medicinal value, from the mucilaginous matter they contain…. The negroes near Charleston, S.C, are said to have grown sesame in a small way for two hundred years." Miller's Cyclopedia *Sesamum* entry indicates (1902: 1662): "leaf mucilage, dysentery, *bene*. Runs wild in the extreme South" a reference to sesame or perhaps to *Ceratotheca triloba* (Bernh.) Hook.f. Raubenheimer (1910: 477) recorded: "The leaves of sesame were official in the *U.S.P.* 1830 (1st revis.) to *U.S.P.* 1880 (6th revis.) inclusive." Grieve's *Herbal* (1931 [1971]) included the leaves "which abound in gummy matter when mixed with water form a rich mucilage used in infantile cholera, diarrhoea, dysentery, catarrh and bladder troubles, acute cystitis and strangury. The oil is said to be laxative and to promote menstruation."

Linguistic Residues: Alternative African Loan Words Suggest Independent Origins in Transfer of African Names with Sesame

Dictionary searches and herbarium specimens yielded clusters of alternative colloquial names for sesame from Africa that emerged in the Americas. Examination of each of these appears below.

Benne

Gray's *Manual of Botany* (1848 [1997]: 1302) includes the African name with a botanical description: "*Sésamum* (ancient name) *indicum* L. (of East Indies), Sesame or *Bené* (*S. orientale* L.) of the Pedaliàceae, an erect annual with lanceolate to oblong-ovate leaves, the lower often divided, white to roseate ventricose-campanulate corolla with bilabiate 5-parted limb, linear stigmas and oblong-quadrangular velvety capsule, is sporadic in waste ground but rarely persistent (Adv. from Old World)."

The sesame name spelled variously as *benne/bene/bennie/beny/benney/binny* is of particular importance because it is the preferred idiom used in the Southern states since the Colonial era (Bedigian 2004d: S7). References from archival sources about the southern USA show European colonists' interest in sesame. Thomas Jefferson's letters (Betts 1944) represent sesame by the West African name, *benne* (Bedigian 2000, 2006b; 1958: 278–279; Wilson 1964). Darby (1818: 185) specified its use in Louisiana, "That species of *Sesamum*, called oriental *bhené*." The *Dictionary of American Regional* English (1985: 220) entered *benne* as an official name, as in "Charleston *Benné* Wafers," listed in *American Regional Cookery* (Hibben 1946: 316) as a native regional food.

Turner compiled a comprehensive list of *Africanisms in the Gullah dialect* (1969: 62); there, *bene* is a feminine personal noun, "the sesame" in Bambara (West Africa, particularly Mali) and Wolof (Senegal and Gambia). Gamble (1967: 29) included sesame among Senegambia's "minor crops" indicating: "*Beniseed* (*bene*) is sometimes seen, though this was commoner in former times." Gullah, the Angola-Bantu dialect of South Carolina, may refer either to Angola or to the Gola of modern Liberia.

The Gullah people of the Sea Islands of South Carolina have preserved ways of life and speech from West African slave culture and plantation times. African-American foodways specialist Jessica Harris was amazed to learn about the resemblance (Harris, personal communication, 1995). She was able to recall very few other cases where the original language and loan word are identical. Harris considers this the clearest example she has encountered of the transfer of a crop with its name, from Africa. Harris (1996: 246; in Fra-Molinero and Nero 2007: 295) viewed the word as "a corruption of *bene*, the term for sesame among the Wolof people of the Senegambia region." As many southern cookbooks do, Harris selected "*benne*" for her book index' search term, rather than "sesame."

This remarkable resemblance of the word *benne* in the American south and the countries in Africa that border the "Slave Coast" suggests that the crop arrived with its local name, to the southern colonies of Georgia and South Carolina, from western Africa. Historian Peter Wood shows (1974: 171) that the language families of West Africa reappear in Gullah speech, the roots of Black English. Variations in spelling exist: *benne*/*benni*/*bene* are well-known Wolof, Mandingo, and Bambara words; Dogon use the variant *many-many* (Bedigian, personal obs., 1997). It appears that the crop and its name transferred together, from Africa to the Americas.

Beniseed was designated as a subject heading entry by the Online Computer Library Center (OCLC); it was an (archaic) standardized subject heading for sesame that has now been combined with the latter word. The term *benne* is so well established in Africa that the index of a key reference work (Irvine 1954, 1969) redirects readers searching "sesame" to "see *benniseed*." Busson (1965) reported the name *dobobéné* used by the Toma of Guinea. The name "black *beniseed*" is used throughout much of West Africa for *Sesamum indicum* (Bedigian 2004d, 2006b; Dalziel 1955), and wild *beniseed* is used for *S. radiatum* Thonn. ex Hornem. (Bedigian 2004c, 2006b; Dalziel 1955). Bedigian and Adetula (2004) include *benne je* among names of *Ceratotheca sesamoides*.

Herbarium vouchers provide supporting data. The name *beni* is used in Niger (Barter 266, K), Sierra Leone (Dawe 87, K; Deighton 1972, K), and Nigeria (Johnson 51, K); the variant *benin* among Dioula in Ivory Coast (Ambe 48, BR). That Deighton 1972 specimen designated the names *bene* used by Kono, Koranko, and Mandingo; *beni* among Mande and Mende; and *beno* by Kisi peoples.

Reference to *Benne* in the *Oxford English Dictionary* (*OED*)

The online *OED* (1989) identifies the African source of the American name *benne* thus: "Sesame, the plant *Sesamum indicum*. Chiefly attributed as *benne*-oil, -seed. Also *bene, beni, benni, benny*. [ad. Mende (Sierra Leone) *bene*]." *OED* (1989) cites early examples, e.g., from 1769, in Phillips and Lesley (1884: 44): "On cutting & gathering the *Bene* seed"; Romans (1775: 130): "The negroes use it as food either raw, toasted, or boiled in their soups and are very fond of it; they call it *Benni*"; Mayne (1853): "*Benne* oil;" Flückiger and Hanbury (1874: 425): "*Oleum Sesami*. Sesamé Oil, *Gingeli*,..*Til* or *Teel* Oil, *Benné* Oil. Moloney, Bruce and Shelley (1887: 50) equate "*Beni-seed* (*Sesamum indicum*)."

Vanglo

Sloane (1696) mentioned sesame in Jamaica under the name given to it by enslaved Africans: *vanglo*. Browne (1756: 270) gave further accounts of the Jamaican *vanglo* or oil plant, while Lunan (1814) collected the learning on *vanglo* by Linnaeus, Sloane, Browne, and Barham. Dancer (1801: 362) valued *vangloe* (*Sesamum*) for "emulsion of the seeds." Hart (1957) pointed out that the various importing waves which brought sesame into Jamaica by "Jews, negroes, and from French sources" brought also the varying names of *vanglo*, *wangla*, and *zezegary*, but the African name *bene* or *benne* seed (by which the seed is known in the Turks and Cayman Islands) never stuck in Jamaica, nor the Indian name *simsim*, nor the famous Spanish name (most euphonious of all) *ajonjoli* (pronounced *ahonoli*).

The online *OED* (1989) *vanglo* entry begins with 1756-*. BL 1: "The small climbing *Sesamum* orientals, formerly valued for its oily seeds and other parts." The modifications listed are as follows: "*Wangla/wánggla/* occas./*wánggra/* sb now dial; 1756-to *vanglo*, 1801, 1820 *vangloe*, 1817 *vangla*; [a 1756 (1794) *wolongo*], 1774 *wongala*, 1817-*wangla*, 1823 *wangola*, 1835 *whangra*, 1895 *wanglo*; and Kongo *wangila*. Cf *OED vanglo* 1756--. BL are noted. The small climbing *S. orientale*, formerly valued for its oily seeds and other parts. Also called *hockalenah*, oil plant, *vangloe*."

Long (1774: 809) identified it as *vanglo*, *wongala*, or oil plant. *Curtis's Botanical Magazine* (Sims 1815) listed synonyms *vanglo* and *vangalosa*. S. F. Gray synonymized *gingelly* as "*Vangloe. Sesamum orientale*. Seeds yield an oil which is sufficiently mild to be used for food, and in emulsions as a pectoral" (1821: 63). Loudon (1829: 515) commented: "From the Arabic word *semsem*. Forskahl, p 68. These plants were introduced into Jamaica by the Jews, and are now cultivated in most parts of the island. They are called *vanglo* or oil-plant. The seeds are frequently used in broths by many of the Europeans, but the Jews make them chiefly into cakes… In Japan, China and Cochin-China… they use the oil for frying fish, and in dressing other dishes; as a varnish; and medicinally as a resolvent and emollient. Nine pounds of the seed yield upwards of two pounds of neat oil."

Webster's Dictionary (1844: 83) defined *vangloe* as the popular name of *Sesamum orientale* in the West Indies. A later edition (1913) identified *vanglo* as: "*benne* (*Sesamum orientale*); also, its seeds; so called in the West Indies." Redwood and Gray (1847) described sesame's mucilaginous leaves, correctly identified its East Indies origin, and listed several major vernacular names: *gingelly*, *guiggiohma*, *jugeoline*, and *vangloe*. Dalton (1855: 206) reports *vanglo* or sesame in the British Guiana. The online *OED* (1989) indicates that Simmonds (1858) defines *vanglo*, a West Indian name for the *teel* seeds of the East (*Sesamum orientale*), the name commonly used in India today (*til*). Simmonds (1889: 413) gave the name, "in Demerara as *wanglo*." Demerara (Dutch: Demerary) is Guyana, colonized by the Dutch in 1611. The British invaded and captured the area in 1796. It was located about the lower courses of the Demerara River.

Webster's New International Dictionary of the English Language (1959) elaborates: "*Vanglo*, *vangloe*, n. [also *wangala*, fr. Kongo *wangila*] Sesame. W. Indies." Hirt and Bindanda (1993) indicate the Kikongo vernacular name of sesame is *wangila*. Writing about eighteenth century Jamaica, Morgan (1995: 70) listed "*wangla*, a Kongo word for the small oily seed, *sesamum*, used in cakes and broths." The Dictionary of Caribbean English usage (1996), defines wangla (Belize, Jamaica) as the sesame plant, or sesame seeds, (especially used in flavouring, making sweets, etc.), Kikongo waangila 'sesame'.

Wangila

It seems likely that there were several independent introductions of sesame from Africa to New World colonies, since, in addition to the *benne* cluster of names from West Africa, herbarium specimens from Zaire indicate *wangila* (Pauwels 802, BR; Pauwels 6457, BR; Vanderyst s.n., BR; Verschueren 334, BR), *uwangila* (Vanderyst 77, BR), and *wangkila wu matebo* [Kikongo] (Pauwels 6457, BR) for *S. indicum*, as well as *wangila* for *S. parviflorum* (Pauwels 6458, BR). The variant *wangila wu matebo* [Kikongo] is used for *Ceratotheca integribracteata* Engl. (Delhaye 335, BR; Pauwels 6459, BR); *wangila ya nzanaa* [Kikongo] for *S. radiatum* (Compere 688, BR) and *wangima nseke* for *S. parviflorum* Seidenst (Gillet and Hendrickx 3105, BR). A specimen of sesame from Angola (Gossweiler 8413, K) uses *uanguilla*. A sesame accession in the USDA gene bank named *wanguila* from Angola, PI 278164, strengthens this view.

Adjanohoun et al. (1988) mention the term *bonanguila* for sesame in the Beembé and Doondo languages from Congo-Brazzaville. Lingala speakers near Kinshasa use *wangila* (Samson Mutolo, personal communication, 2011). "This term has a specifically Bantu-looking structure. It has the typical structure of a Bantu noun derived from a verb (–CVC- verb root plus *-il- verb extension) and would best be looked for in northern Angola-Kongo regions. So the Lingala source sounds very much right… But to get to the New World, I would not be surprised to see it in a few other languages of the region" (Ehret, personal communication, 2011).

Herbarium Specimen from Suriname Labeled Bowangala

Lending strong support to these name variants on *wangila* is the long-forgotten Herbarium Paul Hermann (1645–1695) specimen of sesame housed at Leiden, labeled *Bowangala* (Veldman 2012) (Fig. 4.1). Collected by Paul Hermann in 1689, during his tenure as director of the Leiden Botanical Garden, from 1679 to 1695, it is positive proof that people were already growing sesame in Suriname at that time. That local name, now pronounced "*boangila*," is the word for sesame in Aucan (a Surinamese Maroon language). The present day Maroons grow sesame for food and oil production, as well as magic medicines (van Andel, personal communication, 2011). The Latin text below the plant states: "Carries white flowers, looking like *Hyosciamus*, when the seeds are pressed they yield a sweet, soothing oil." This information derives from a review of the entire Surinamese Hermann herbarium, including a translation of the all?? original Latin texts in the collection. The British Museum's Sloane herbarium holds two voucher specimens of sesame (Richardson 191, BM; Sloane 192, BM), the latter named *Sesamum ueterum*, collected at approximately the same time in Jamaica. Pulle (1906: 430) includes sesame in his *Enumeration* of Surinam's plants, citing the collection by Hostmann and Kappler 629, named *Capraria integerrima* Miq., a synonym of sesame, according to Tropicos.org.

Fig. 4.1 *Sesamum indicum* from Suriname in the Hermann Herbarium, ca. (1680). The same local name, *Bowangala* (now pronounced '*boangila*') is used by present-day Suriname Maroons, who grow sesame for food and oil production, as well as magic medicines. Image courtesy Netherlands Centre for Biodiversity Naturalis

Wangla and Related Names, in Medicine and Confectionary

John Quincy assembled an early lexicon of medical terms (Quincy 1802: iv 281) that identified *Sesamum* as *ganglia.* Williamson (1817: 171–172) wrote about sesame: "The *wangla*, or *vangla*, a plant, whose leaf, when agitated between the fingers in fine spring water, or any other, if it cannot be procured, gives a glutinous consistence, without affecting its transparency, and has been much used in the dysentery which prevailed that month. It had been conveyed to the island by its reputed effects at St Vincent and Granada, in some of the black regiments, where it passed under the French name of *zi zigree*."

Roughley (1823: 416–417) also specifies the medicinal benefits of sesame leaf mucilage: "The *wangola*... whose leaf, when steeped for a short time in water,... makes it of a mucous, glutinous quality, and is deemed an excellent medicine in case of flux or dysentery."

Cassidy and Le Page (2002: 462) list numerous variant names: *Wangila*, *wangla*, *wangola*, *wangala*, *wongolo*, *wongala*, *wangra*, *wánggla*, *wánggra*, *whangra*, and the Gullah *waŋga* (Turner 2002 [1949]: 203). Several of those sources use the name *wangla* not only for sesame but for sesame brittle or brickle: "Sullivan (1893:81): The natives make a sugar cake with. A bean called '*wangla*'; *Benne*tt (1942: 10): Candy seller. "Nice young man, come here, what you want? *Pinda*-cake, *Wangla*?" *Pinda* is Dutch for peanut. Gleaner's list (1943) of dialect words, all parishes: Clarendon, Kingston, *Wangla*, a peanut cake containing *wangla* seeds; Parish of St. Thomas, *Wangra*, a kind of candy; Cassidy (1952) Parish of Portland, Parish of St. Mary /*Wànggláɪ*/ cakes of hard sugar with peanuts, formerly with *wangla* seeds."

Sesame brittle, a popular candy in Belize, known as *wangla* seed in Creole or "*jojoli*" in Spanish, is a "thin peanut brittle made with thousands of sesame seeds" (Martinez 2006: 14). Cassidy (2007) elaborates: "The seeds of sesame, most commonly called /*wánggla*/ today – and this seems to have been the early form of the name too, though the *OED* and others list it under *vanglo*e. Native made sugar and sugar concoctions have a number of folk names. New or wet sugar is freshly made, and as the second name suggests, has lots of molasses in it." Cassidy (2007: 346) specified that "The w-- forms are more etymological: Kongo *wangila* means sesame. *Wangla* seeds are burned in some *Obeah* practices." Mufwene and Condon (1993: 145) consider this an African contribution to French-based Creole as *Wāg*, "*Sesamum* sp." from Kikongo, *wangila*. The Bakongo use the name *wangila* for sesame.

It is significant that similar African names and healing properties appear in the Guianas. DeFilipps et al. (n.d.) specify "*Sesamum indicum* L. (*Sesamum orientale* L.) French Guiana: *ouange*, sesame. Guyana: sesame, *wangala*. Uses: Root: emollient. Leaf: mucilaginous. In a decoction for dysentery. Flower: in an emollient tisane. Seed: diuretic; yields an oil for soothing burns."

Soonga

A term used in Jamaica, Cassidy and Le Page (2002: 416) reveal: "*Soonga* sb obs; cf Ngombe *sóngbá*, sesame seeds. = *wangla*. 1726 Barham (1794: 121–122): Oily Pulse, Which is called *sesamum*, or *sesamum Africanum*. The first time I saw this plant, it was growing in a negro's plantation. [They have] seed vessels, full of small white seeds, which the negroes call *soonga* or *wolongo*, which is much like the sago sold in shops, but very oily." Significantly, the names *soomgba/somgba* or *soomgwa* are colloquial for sesame in the northern Congo/Ubangi region near the Central African Republic (Samson Mutolo, personal communication, 2011).

Zezegary

The name *zezegary* is less common for sesame. Dancer (1801: 362) evaluated sesame as follows: "I cannot help thinking that a very undue importance has been given to the boasted one, of the specific virtues of *vangloe* (or the *zizigary* as it has been called), in dysentery." Lunan (1814: 251) distinguished "*Sesamum orientale*... This is a native of East Indies, and lately brought to Jamaica, under the name of *zezegary*, though Browne notices it as common in the island in his time." Williamson (1817: 171) noted its French name, *zi zigree*, and (1817: 178) paid tribute: "In the early stage of the continued epidemic fever... The *wangla*, or *zi zigree*, had been used with singular benefit. Andrew, at Newhall, was cured by it under dysentery."

The index to Grisebach (1864: 789) designates the name *Zezegany* for *Sesamum orientale*. Cundall (1891: 163) states: "The *wanglo* or *zezegany* was sent by Sir Basil Keith in 1802 (Journals Assembly 10: 638)." Cundall continues (1891: 165): "During the years 1791–1807... In order to make the island less dependent on America for supplies every encouragement was given to the cultivation of yams, cocoes, maize, plantain, and such products as the breadfruit, *zezegany* or *wanglo*, nutmeg, clove, cinnamon, *pindars* and coffee, it being believed that the cultivation of these valuable exotics would, without doubt, in the course of a few years lessen the dependence of the sugar islands on North America for food and necessaries; and not only supply subsistence for future generations, but probably, furnish fresh incitements to industry, new improvements in the arts and new subjects of commerce."

These variable names suggest several independent introductions of *benne*, to the Carolinas, from West Africa, and as *wangla* to Jamaica from Angola-Kongo. Cassidy's Jamaica Talk (2007) names *Zezegary* or *zizigary*. Cassidy and Le Page's (2002: 489) indicate: "*Zezegary* sb bot; etym not found. A name under which *wangla* was imported at the end of the... *zizigary* see *Zezegary*." De Ficalho (1947[1884]) indicated that sesame names "*gingeli*, *jinjili*, *gegari* and *gergelim* and also other spellings, are employed by the Portuguese since the discovery of the eastern lands, but do not seem to derive from the languages of these lands, but from a rather obscure source." It seems plausible to this writer that *zezegary* may be a conflation of known prefixes used for sesame, *zizi-*, *sesa-*, or *Zesamum* with *gegari*.

Enslaved Africans Attached Supernatural Subtexts to Sesame

Sesame forms one component of a set of beliefs which today fall under the category of voodoo, including spirits, talismans, lucky and unlucky actions, and omens. African in character, these beliefs survive in small pockets of the USA today (Herskovits 1941 [1958]: 245–251). According to Granger (2007: 202) and Woods (1995: 38), who credited informant Emma Hunter: "a superstition attached to the cultivation of *benne*. Her African-born grandmother Betty used to 'plant *benne* seed. Once you staht plantin *benne*, yuh got tuh plant em ebry yeah aw yuh die.'"

It is worth noting that in accord with prevailing views among Chinese, East Indians (Bedigian 2000), the Nuba of Sudan (Bedigian and Harlan 1983), and populations in Southwest Asia (Bedigian 2004a, 2011c), Lowcountry Carolina residents of African descent believed that sesame seeds are a symbol representing good fortune. A note included with a recipe for *Benné* candy, in a source that used the oral histories of "black domestics" in and around lowcountry Savannah, Georgia (Colquitt 1933: 161), describes *benne* as an East Indian herb which is supposed to be a charm with which to secure entrance and exit through any portal. "They make *benné* candy and *benné* cakes and are said to pound the bean into a kind of paste, which they eat on their hominy. But, most interesting of all, is the fact that although they never heard of Ali Baba and his magic words, they sprinkle it on their doorsteps to bring them luck and ward off *ha'nts*."

Similarly, a recipe book compiled by the Junior League of Charleston (1950) states that "According to legend among descendants of Negro slaves along the coast of Charleston, *benne* is a good luck plant for those who eat thereof or plant in their garden. It was originally brought in by the slaves from West Africa to this Coastal region." The Federal Writers Project, Writers' Program of the Work Projects Administration in the State of South Carolina (1976: 204) says the following in a mention of the slave market: "Typical sweetmeats sold here are made from old recipes, and include peach leather, '*benne* brittle,' made with *benne* (sesame) seed – called 'goodwill seed' by negroes – 'monkey-meat' (coconut) cakes, and 'groundnut' (peanut cakes. The last two are not cakes at all, but round patties of candy." It notes that "*Benne* seed, or sesame, brought from Africa according to tradition, is still grown in South Carolina" (1976: 156).

Genovese (1974: 542–543) wrote: "As for sesame seeds, the effects of which drew compliments from Thomas Jefferson among many others, the slaves regarded them as bearers of good luck and planted them everywhere they could for use in candies, cookies and desserts, - most of which, however, were enjoyed by their masters." That view also appears in *slave recipes* (Anonymous n.d.: pages not numbered): "*BENNE* SEEDS … Planted in slave days as a 'good will crop' *Benne* seed, which is another name for *Sesamum indicum*, suggests an origin in the East Indian archipelago… These rich, spicy seeds have long been a favorite with the Low-Country Negro, who eats them raw with sugar or with milk, and, often, in gravies… In ante-bellum times the seeds were prepared for 'de buckra' in the form of cakes, wafers and brittle." The term "*buckra*" used contemptuously by Black people refers to a White man, probably derived from Efik, *mba-ka-ra*, master (*The Free Dictionary* 2011).

In conjunction with an exhibition titled "Transatlantic Linkage: The Gullah/Geechee-Sierra Leone Connection" at the Museum of Coastal History, St. Simons Island GA, 1995, miniature cloth sacks of sesame seeds were distributed to visitors, bearing the following note: "History of *Benne*. When slaves first came to coastal areas of Georgia and South Carolina, they brought with them, as their most valued possession, a little handful of *benne* seed, which they believed held for them the secret of health and good luck." This matches folk belief in Jamaica (Berlin and Morgan 1993: 38) and Africa (Bedigian unpublished mss., Bedigian and Harlan 1983).

Sesame Superstition: In *Obeah* Practices, Wards Off Evil

During the slave era, *obeah* practice was: "a type of sorcery which largely involved harming others at the request of clients, by the use of charms, poisons, and shadow catching. It was an individual practice, performed by a professional who was paid by his individual clients" (Patterson 1967: 188).

Madden (1835: 62) reported a plant called *whangra*, used by *obeah* men, representing folk magic found among those of African descent in the West Indies derived from central African Kongo or Bantu sources that has a close North American parallel in African-American conjure or v/hoodoo. Beckwith (1929: 129) relates "*wangla*" with *obeah* power: "*Wangla* should be planted in the provision ground and seeds carefully gathered. If a thief robs the field and you can find a fresh footstep, take up earth carefully in a leaf, measure in a spoon, put 4/3 [four thirds] as much '*wangla*' seed with it and put into a pot upon fire. Call the name of the person you think is a thief, and if you are right, as many 'bumps' will appear on his foot as there are seeds that 'pop' or 'his skin will strip off' says Cundall's informant; another says he will die. Falconer says the only thing can save him is to have previously eaten some '*wangla*' seed. Banbury says if one burns '*wangla*' with pepper and salt in a road which thief passes, it will give thief Jamaica leprosy called '*cocobay*' [repulsive sores]. Beating '*wangla*' planted in a field as you call thief's name also effective in bringing swellings on his leg."

Clavel reported a variant from the Bahamas (1904: 37) using *benne* seed to overthrow a spell: "There is another way to catch a hag. If you think you are being hagged, take a pint of *benne* seed (as small as mustard seed) and guinea corn (also a small seed); spill it all in the four corners of your house; that will catch the hag, as she cannot leave the house before she has picked up all the seeds, one by one, during the night. In the morning you will see her raw body, the skin being away; she will be so ashamed that afterwards she will never come near you."

Farnsworth and Wilkie provide some analogies from Africa (2006: 64–65): "*Benne* seed was used in a number of ceremonial contexts as well. Among the Bura peoples [of Nigeria], *benne* seed was thrown over the shoulder to prevent evil or put in a *habtu* (a pot used for ceremonial purposes) to bring good luck (Meek 1931:164). In Sierra Leone, *benne* seed was used as part of a ritual punishment in Poro-Sande cultures. Differing degrees of punishment were inflicted upon those who were found guilty of some wrong by the society. A serious crime perpetuated against the community, such as theft, resulted in the perpetrator being required to pick up a quart of *benne* seed that had been poured on the ground by the elder. The Kilba are known to have incorporated *benne* seed into healing ceremonies, placing the seeds as offerings into pots designated as male and female to bring the two spiritual aspects back into harmony (Meek 1931: 1992)." Creel (1988: 181–182) described how this ritual was carried on within the praise houses of the Sea Island Gullah.

Berlin and Morgan (1993: 37–38), citing historian Barry Higman, assert that gardens and provision grounds permitted the elaboration of African-influenced conceptions of spatial order. Here, boundaries could be fluid and irregular, as against the geometric and rigid European notions of order. Moreover, enslaved Africans infused their gardens and grounds with magical beliefs, many carried from Africa. Saramaka maroons closely examined potential garden sites in order to avoid forest spirits and snake gods and to placate the god who presided over their chosen location. In Lowcountry South Carolina and Georgia, enslaved Africans planted *benne* at the end of rows in their private fields to help ward off intruders. Fett (2002: 81) concurs: "Slave gardens in the Georgia and South Carolina Lowcountry, featured *benne*, or sesame, at the ends of the rows. Sesame had West Indian associations with *obeah*, and its presence in these Lowcountry gardens may have been intended to repel thieves and other intruders." Berlin and Morgan reported (1993: 38) that a visiting absentee owner first saw sesame in Jamaica: "growing in a negro's plantation. Known as *wangla*, it was used in *obeah* practices. Jamaican slaves also planted the cut eye or overlook bean at the borders of their provision grounds as protection from the evil eye. Thus when the Intendant of Saint Domingue described the slaves' provision grounds as '*une petite Guinée*' he was more perceptive than he knew."

Gomez (1998: 50–51), writing about amulet, charm, and talisman, notes that the "*zinzin* charm, so-called in Louisiana Creole used for 'support or power' has the same meaning in Bambara, as *Gris-gris* from Mande, *gerregerys*."… "Another destructive amulet called *wanga* in N. Orleans, is also of Mande derivation."… "Africans arrived in possession of *gris-gris*" (a charm used to harm others and to ward off evil). Read (1931: 125) derived: "*Ouanga* or *wanga*, *m.* A charm; from Southwest African *owanga*, 'witchcraft'; compare Angola *wanga*, 'magic' and western Zambesia *Vuañga*, 'witchcraft.' [Footnote cites Sir H. H. Johnston, Comparative Study I, 343, 358, 369.] Here too, should be noted Congo *mbwanga*, a charm consisting of a bundle of aromatic powder, used as a cure for headache; or a charm made of a bundle of powder, bits of leaves, serpents' heads, birds' beaks, etc., having power to produce or avert disease."

"From *ouanga* has been formed La.-Fr. *ouangateur* (*wangateur*). *m.*, conjurer, 'voodoo-doctor,' with the feminine *ouangateuse* (*wangateuse*). These terms are well known in New Orleans; they are also used in St. Martinville, and perhaps elsewhere in South Louisiana.

"The Congolese use various charms to protect themselves against the witches whom they believe to be responsible for sickness, misfortune, and death. If a man becomes seriously ill, the *nganga a moko*—arms doctor—pretends to find out who is bewitching the patient. Cuban Spanish *uangá*, *m.* 'witchcraft' is likewise of African origin [Footnote cites Ortiz, *Glosario*, p 468]."

The encyclopedia *World of a Slave*'s "*benne*" entry (Twitty 2011: 416) reiterates: "By the height of the trans-Atlantic slave trade, the plant was raised in almost every slave society from the Lower South through the Caribbean to South America. By the late 18th c., enslaved South Carolinians harvested great quantities of sesame with other crops such as gourds, collard greens, and watermelon. This mirrored West African traditions where *benne* seed was companion planted with other crops and used by some groups to give spiritual warning not to steal produce when planted on the borders and edges of the garden." Twitty suggests that to steal from a garden surrounded by this sacred plant was an invitation to bring on wrath, possibly based on Granger's primary source (2007: 179): "*Benne* seed is bad fuh duh witch too an keep um way."

Sesame, Means of Punishment in Gullah Folk Practice, and Belize Belief

Washington (1994: 57–58) reported that enslaved Africans engaged in a peculiar type of penance as punishment for sin: "If the crime was of the first magnitude, the perpetrator was condemned to pick up a quart of *benne* seed (among the smallest of seeds) which had been thrown on the ground by the priest; if of the second, a quart of rice; if of the third, a quart of corn…." If the perpetrator failed to collect every single *benne* seed, the individual was condemned to repeat the process. "It was also a rule among them never to divulge the secret of stealing: and if it should be divulged that one had to go on the low seat or pick up the *benne* seed." Pollitzer showed (1999: 141): "Some practices of the black religious societies of the sea islands met with the disapproval of Christian missionaries. Methodist Thomas Turpin reported an African-derived ceremony in a slave society on Bull Island in 1833 that he considered corrupt. Punishment was inflicted according to the perceived degree of the crime. For a minor infraction, the wrongdoer had to pick up a quart of corn poured on the ground by the priest; for a graver offense, a quart of rice; and for a crime of the first magnitude, a quart of benne seed, a task that could take all night. … Failing to recognize its African origin, Turpin viewed such penitential rites as akin to Catholicism."

A variation of this penalty is found in a folk legend from Belize, the Old Hag Tale (Martınez 2006: 14–15): As in Jamaica (Hart 1957), "*Wangla* is both the name of a popular sesame candy made in Belize, and the name given to *Wangla* Lady, or Old Heg. 'Skinny, skinny, yu no know me?' This is a tortured cry of the vampire-like Ole Heg, popularly depicted as an old woman in stories told throughout rural Belize. Loggers at our old mahogany camps told about her, and she is believed to have African roots but also resembles the witches or *brujas* of Maya lore. Belizean poet James Sullivan Martinez (2006) who spent much of his youthful years with his father at logging camps, was inspired to write about her in a local newspaper: "Ole Heg is said to frequent communities with dilapidated buildings, emerging late at night, in search of victims from whom she can suck blood. She prepares for her attack by uttering an oration or prayer that will allow her to remove her skin, which she hangs on a *Ceiba* tree or places in a mortar (these were kept in the outdoor cooking area in the earlier days). It is said that she will then transform herself into a flying ball of fire and dwindle to a minute ray of light that can enter a home through the keyhole. Some say that she straps two straw mats to her arms in order to fly on to the thatched roof of a house and sits there, peeking in until the moment for her entrance is right. Once inside, she chooses any sleeping person, although children are her favorite victims. She visits on several successive nights, sucking blood from the same person. Her victims become paler and thinner each day and often die as a result."

"Ole Heg can be kept at bay in several ways, one of the easiest being to dress in blue, a color she avoids. At night, parents often put their children to sleep in a blue garment, as protection from her. Another of Ole Heg's enemies is asafoetida, a foul-smelling gum resin that will keep her far away. An up-side down broom will also prevent her from coming into the house. It is common for people to scatter *wangla* (sesame) seeds on the floor in their homes or on their doorsteps, because, on seeing the seeds, Ole Heg becomes obsessed with picking them up. She does this slowly, one by one, and when her hand is full, she pours them on the ground and begins to collect them again, over and over until morning comes. She cannot leave until night falls again, and once Ole Heg has been trapped in this way, she can be beaten to death with a broom. Perhaps because of her association with these seeds she is called *Wangla* Lady."

Durant and Knottnerus (1999: 48) draw attention to Margaret Creels' discussion of the *benne* seed ritual among the Gullah people in the South Carolina low country as "an example of African cultural beliefs that were translated into the religion of enslaved blacks who constructed reality in their own interest. This ritual was derived from the secret men's (Poro) and women's (Sande) societies in Sierra Leone."

"In the *benne* seed ritual, Gullah's dealt with theft in a spiritual context under the guise of the Baptist faith. But the moral judgment was the opposite of Christianity. *Sin* was not upon the slave who stole from the master, but the individual who divulged the act. Gullahs possessed a keen sense of communal morality which, in this regard, was not subverted by dogmatic features of Christianity. While stealing from *Maussa* was a sin, it was wrong - that is against elements of collective interests and values - to steal from a fellow slave and hence betray community confidence" (Creel 1998: 239).

Sesame Featured in Lowcountry Recipes

Characteristic of the South Carolina Lowcountry since colonial times, *benne* stars throughout the South in plantation kitchens where some dishes with African origins persist. Romans (1775 [1962]: 130–131) praised sesame because it "will never go rancid even if left exposed to the air; the Negroes use it as food either raw, toasted, or boiled in their soups and are very fond of it, they call it *Benni*." Other traditional sesame items were *Benne* Bits, biscuits salted with red pepper; *Benne* Wafers, crisp paper-thin cakes; and *Benne* Brittle (Anonymous n.d.).

Originally published in 1847, *The Carolina Housewife* (Rutledge 1979) gave a recipe for *Bennie* soup. According to Paul R. Begley, Reference Services, South Carolina Department of Archives and History, Columbia, SC, (personal communication, June 14, 1995): this was "among the earliest [or the earliest] recipe[s] for *benne* in a Carolina cookbook." The intended audiences of *The Carolina Housewife* were white women of aristocratic southern society. Information collected for the book was rooted in recipes used by friends and acquaintances of Rutledge. It indicates that *benne* soup, a West African dish (Greene 1980), was well established in the homes of plantation owners. Hess (1992: 102) clarifies further: "*Bennie* Soup [sesame, brought from Africa, or rather by way of Africa]."

The enslaved Africans who arrived in the northern segment of the hemisphere, like their "cousins" who arrived in South America and the Caribbean, were initiated into the cultivation end of the food chain and placed in charge of big house kitchens, where they instituted what historian Genovese (1974: 543) has called: "the culinary despotism of the slave cabin over the big house." Genovese (1974: 542–543) wrote: "Sesame seeds and oil, as well as red pepper, came from Africa with the slaves and became central to southern cooking."

Sturtevant (1919 [1972]: 532) described: "use by the negroes of South Carolina, who parch the seeds over the fire, boil them in broths, and use them in puddings." Clinton (1982: 178) described Christmas as most festive: "Women cracked nuts, seeded raisins, cut orange peel, mashed currants, and prepared fruit cakes… they also made a candy called *benne* brittle from sesame seeds (called *benne* seeds by the Africans who first brought them to America)." Cook (1982) describes *benne* seed brittle, a savory that harks back to plantation days: "the sort of sweet that makes dentists blanch. Made with sugar, vanilla and lemon extract mixed with two cups of parched *benne* seed, the brittle is cut into squares while still warm and is frequently devoured on the spot," and delicate *benne* wafers that a friend associates with the Southern funerals of her youth.

Weiss (1948: 44) citing Reed (in fact, Read 1931) notes its use in Louisiana pralines, a sugar or syrup-based candy, "praline de *Bennés*." Read (1931: 63) wrote: "Pralines de *Bennés*, *f.* Candy prepared with syrup or sugar and the seeds of the sesame plant (*Sesamum indicum* L.)… The term *pralines de bennés* is well known at Lafayette, Breaux Bridge, and the vicinity of these towns." "Further information came from Miss Louise V. Olivier, who is conducting an Acadian project, gathering handiwork from the people of the towns of rural Louisiana. She said, 'Years ago, *benne* was planted here in St, Landry Parish, on a small scale. My father's tenants always planted 2 to 4 rows and every season made personal calls to our home to deliver my father's share" (Weiss 1948: 44)

A Guide to South Carolina (Writers' Program of the Work Projects Administration in the State of South Carolina 1941: 155) names: "Peach leather and *benne*-seed [sesame seed] brittle Charleston specialties in the way of confectionery." The *Dictionary of American Regional English*'s firsthand interviews with the native Southern populace found the following, from 1966 to 1968 in South Carolina (1985: 220): "Addit SC, Sesame seeds are *benne* seeds -- (informant has never heard of sesame); ceSC: *Benne* seed – sesame seed." From 1966 to 1970, (Qu. H80, "Kinds of candy often made at home around here), Inf SC62, *Benne* brittle; SC4, *Benne* candy; SC19, *Benne* wafers; (Qu. H82b, cheap candy… sold years ago), Inf SC70, *Benne* [*beni*] candy." The National Council of Negro Women (1993) avoided the question of agency: "*Benne* is a West African word for sesame which is still in common usage in South Carolina." Most early southern recipe books employed "*benne*" over "sesame."

Schoolcraft (1969) recounted: "After packing two trunks full of clothes for each of them, they sewed Up bags of hempen cloth, to be filled with dried chinquapins, and pecans, and *benne* and groundnuts, and walnuts, for them to take with them, as all children delight in nuts; besides, at school, such temporalities, when distributed, make the scholars friends… Every one is a friend to him that giveth gifts." Dalgish (1982) recorded *benne* cake.

The figure legend of a photograph credited to the Carolina Art Association (Whitelaw and Levkoff 1976: 147) features "the groundnut-cake *Mauma* with her tray of sweetmeats and a palmetto fan in one hand to whisk away the visiting flies…. *Benne* cakes and 'monkey-meat,' a coconut and molasses candy." That archival early twentieth century portrait by G. W. Johnson appears on his book cover (Johnson et al. 2005, and p 48) with this explanatory text: "The ground-nut cake huckster plied her trade every day at the corner of St. Philip and George streets, near the College of Charleston. She sold ground-nut cakes, coconut cakes in chocolate, pink or white colors. She also sold *benne* seeds, a Charleston staple, and monkey meat."

Fowler admits (2008: 30): "Introduced into the Lowcountry by the West African slave trade, *benne* (sesame seeds) have become an indelible part of the region's cookery." Dabney raves about *benne* seeds (2010: 181–182): "(Pronounce them *bennie*)… one of the great legacy plants of the Carolina-Georgia coastal plain, having come over with the slaves from Africa during the colonial era. *Bene* means sesame in Gambia and Senegal. They have long been rated as a Lowcountry food icon, nearing the status of the legendary she-crab soup."

The depth to which *benne* permeated South Carolina's African American culture is displayed in a "lyrically written piece of fiction," *Neeny Coming, Neeny Going*, reviewed by Johnson-Feelings (1997: 556) "that is concerned, most of all, with the idea of change." Karen English, its author, "calls attention to the historical moment during the fifties when many sea islanders were displaced because pollution was destroying their oyster beds and thus, part of their way of life."

The tension in *Neeny Coming, Neeny Going* revolves around the strained relationship between two cousins who 'used to live like sisters': Essie, the narrator, and Neeny, who now lives on the mainland but has returned to 'Fuskie [Daufuskie] Island for a visit… Neeny no longer appreciates anything about life on Daufuskie Island… English attaches significance with these archetypes: "Neeny's own 'fingers forget what Dada show them' and so can no longer bundle sweetgrass. She declares that chocolate is superior to the traditional benne candy made by the community's midwife. It is quite significant that she repudiates African American aesthetics…"

Thomas Jefferson's Garden

According to Peter Hatch, director of gardens and grounds at Monticello (2005: 57), "Salad oil was a perennial obsession for Thomas Jefferson. He referred to the olive as 'the richest gift of heaven' and 'the most interesting plant in existence.' When he found domestic olive oil imperfect and imported olive oil too expensive, Jefferson turned to the possibilities of a salad oil extracted from the sesame plant, or *benne*…. Jefferson sowed sesame annually from 1809 until 1824, and purchased or concocted three different sesame oil presses. He was disappointed by the low yield of seed to oil and by the problems of extracting chaff and leaves during the pressing."

Leighton's chapter titled "Every Man His Own Doctor" states that "Jefferson became interested in cultivating *benne* seed, a plant of which much was hoped as a source of oil equal to olive oil. Grown by the Negroes in the south for their own use, it was taken up by their white owners, like other African plants (such as Guinea corn) the slaves had brought with them. Jefferson reports that the seed is eaten parched for dessert and used in soups, but its chief use was as an esculent oil… for dysenteries and other visceral complaints… two or three leaves in a pint of cold water yield a mucilage the consistency of a white of egg" (1976: 181).

Benne in Thomas Jefferson's Garden Book (Betts 1944)

Thomas Jefferson's Garden Book (Betts 1944) is the richest single source on the crop and records plantings of sesame grown at Monticello as long as the book was kept (to 1824). Careful examination of Jefferson's letters reveals that among the full texts that Betts published, from 1807 to 1811, there are 13 references to sesame cultivation, the history of the crop, and praise for its fine qualities.

Betts (1944: 351) includes a letter from William Few to Jefferson dated September 26, 1807, that states: "I take the liberty of sending to you by Mr. Gallatin a bottle of salad oil, the first perhaps that was ever made in the United States. It was pressed from the seed of a plant which grows in the southern States, and is known there by the name of *Bene*, and is cultivated in those States by the Negroes only for their own use, the pod which contains the seed before it is matured, I am told is the part which they use…."

"I have not learned the Botanic term of the plant nor under what class, or order it is arranged. The seed was sent to me from Georgia by Mr. Milledge of the Senate of the United States whose Agricultural and scientific researches have rendered important services to that State."

"Six bushels of the seed produced about six gallons of cold drawn oil, of the quality I send, and about twelve gallons of warm drawn oil that is not quite so pure and well tasted, but it may be used as salad oil, or for painting, or lamps…" (*Jefferson Papers*, l.c.).

Jefferson replied to William Few from Washington on January 3, 1808 (Betts 1944: 361–362):

> "I thank you for the specimen of *Benni* oil which you were so kind as to send me. I did not believe before that there existed so perfect a substitute for olive oil. I tried it at table with many companies and their guesses between two dishes of salad dressing, the one with olive oil, the other with that of *Beni*, shewed the quality of the latter in favor of which the greater number guessed. certainly I would prefer to have it always fresh from my own fields to the other brought across the Atlantic and exposed in hot warehouses. I am therefore determined to go into the culture of it for domestic use, and should be thankful to you for the process of expressing the oil from the seed in which you appear to have succeeded so perfectly. all the minutiae in new processes give aid towards perfecting them. Dr. Mitchell supposes the *Benni* is a *Sesamum*...." (*Jefferson Papers*, l.c.)

Shortly after, Jefferson wrote to John Taylor (Betts 1944: 362) on January 6 (1808):

"… The African negroes brought over to Georgia a seed which they called *Beni*, & the botanists *Sesamum*. I lately received a bottle of the oil, which was eaten with sallad by various companies. all agree it is equal to the olive oil. a bushel of seed yields 3. gallons of oil. I propose to cultivate it for my own use at least…." (*Jefferson Papers*, l.c.)

From Jefferson to Anne Cary Randolph (Betts 1944: 368):

"Washington Mar. 22. 08.

… I inclose you some seed of the *Beny*, or Oriental *Sesamum*. this is among the most valuable acquisitions our country has ever made. it yields an oil equal to the finest olive oil. I received a bottle of it, and tried it with a great deal of company for many days, having a dish of sallad dressed with that & another with olive oil, and nobody could distinguish them. an acre yields 10. bushels of seed, each bushel giving three gallons of oil. an acre therefore, besides our sallad oil, would furnish all kitchen & family uses, most of them better than with lard or butter. you had better direct Wormly to plant these seeds in some open place in the nursery, by dropping two or three seeds every 10. or 12. I. along a row, and his rows 2. feet apart. the plant grows somewhat like hemp. it was brought to S. Carolina from Africa by the negroes, who alone have hitherto cultivated it in the Carolinas and Georgia. they bake it in their bread, boil it with greens, enrich their broth &c. it is not doubted it will grow well as far North as jersey, tho' McMahon places it among green house plants." (*Jefferson Papers*, M. H. S.)

A letter from Jefferson to John Taylor (Betts 1944: 372) repeats his view that slaves introduced sesame:

"Washington, June 23. 08.

… We have lately received from S. Carolina & Georgia the seeds of a plant brought from Africa many years ago by their negroes & by them called *Benney*… you will find a good acount of it in Millar's Gardener's Dictionary under the head of *Sesamum trifolitum*. many persons, from the account of this plant given by the members of S. C. & G. at the last Congress, and the sample of the oil, have sown it this year. it bids fair to supply the place of olive oil, butter, lard & tallow in most cases…." (*Jefferson Papers*, M. H. S.)

Despite their enthusiasm, Jefferson and his correspondents wavered about its spelling; sesame appears variously as *bene*, *beni*, *benne*, *benni*, *beny*, and *benney*. *List of patents for inventions and designs issued by the United States from 1790 to 1847* indicates a patent issued to W. Few, New York, March 1, 1809, for *bene* oil (Burke, comp. 1847: 112, Chemical Processes): "Oil, making *benne* Patent Number X1014" (full text destroyed in 1836 fire).

World Economic Conditions Promote Early Colonists' Interest in Sesame

Enslaved Africans tended sesame as a cash as well as subsistence crop on plantations along the East African coast (Kusimba 2004). Cooper (1977: 80) explained: "Millet was much in demand in Arabia and Zanzibar, while coconuts and sesame - a seed from which oil was extracted - fed an expanding market in Arabia as well as the eagerness of French traders in Zanzibar to obtain oil-producing crops."

Britain presented economic incentives for cultivating edible oils in the American colonies (Wilson 1964). Britain was optimistic that sesame oil might serve as a table oil to replace the olive oil that she was forced to import from her European rivals. In Wilson's view, sesame was one of the products Britain had hoped to obtain by colonizing the New World. Conversely, in nineteenth century France, a government-issued manifesto (Anonymous 1843) urged obstruction of sesame imports, to avoid competing with internal cultivation of the rapeseed crop (Bedigian 2006a). These actions reveal the popularity of sesame in Europe at that time.

Thomas Lowndes of South Carolina sent a sample of sesame oil to the Lords of the Treasury with this letter (August 26, 1730): "My Lords, a planter in Carolina sent me some time ago a parcel of seed, desiring I would try it, and see of what use it would be, for if it turn'd to account South Carolina could with ease produce any quantity of it. By an experiment, I found twenty one pounds weight of seed produced near nine pounds of good oyl, of which more than six pounds were cold drawn and the rest by fire. The name of the seed is *Sesamum* it grows in great abundance in Africa and Asia, and the inhabitants of those parts eat it, as well as use it for several other purposes. Pliny and many other good authors ancient and modern treat of this seed. It rejoyces in the Pine Barren Land (which is generally a light sandy soil).... This seed will make the Pine Barren Land of equal value with the rice land… and is for many purposes even preferable to oyl olive." An early shipping record of sesame aboard the *S.S. Savannah*, registered Samuel Bowen's export of 500 lb *benny* with 20,000 lb sago, 200 gal soy, 200 lb vermicelli, 1,000 lb groundnuts, and a 10 gal keg of sassafras blossom (Lowndes 1730: 261–262; *Records of the States of the U.S.* May 25, 1774, pg 2, column 2).

Henry Laurens lamented in 1747 (1746–1755: 114) that "Our Oyl Manufacture is no more talk'd of, as an Article for Exportation." An accompanying footnote [9] explains: "There had been a hope that oil extracted from the *sesamum* seed would become an important export."

Industrial use also motivated expansion into new crops. Oil for woolen manufacturing was an important item that, according to prevailing view, could be cheaper if produced locally, than imports from foreign countries. A letter signed Thomas Lowndes (1730) reported by Sainsbury (1859: 297) indicates: "The oil may be of great use in woolen manufacture."

Among various vegetable oils tested, Partridge noted: "The *bene* plant is very abundant in Florida, the negroes using the seed in their soup. The seeds of this plant afford about three gallons of oil to the bushel, and the cold expression is fully equal to the best table oil. The raising of this seed was prohibited in France many years since, for fear of its ruining the olive growers. I have no doubt an ample supply of this seed could be obtained from Florida if proper arrangements were made. I obtained my information from Col. John Lee Williams, St. Augustine, Florida" (1834: 2–3). Note that this information connects with herbarium specimens of sesame collected in Florida, reported above.

A Report of Experiments on *Bene* Seed Oil, made by request of Philadelphia's Franklin Institute (1835: 13–14); "In November 1833, several bottles of *bene* seed oil were furnished the Committee on Premiums and Exhibitions, with a view to have determined… the value of this article in the manufacture of wool, and particularly for painting in and out doors work, as a substitute for linseed oil." George Wall tested it on "fine merino wool, which has considerable animal oil in it," with the objective "to make it work more freely in carding and spinning; those who performed that part of the work were of the opinion that it worked as freely as good sperm, or olive oil. In the process of extracting the oil from the cloth, or scouring, as it is usually called, … this oil required more expense of materials and labour than the oils commonly used by manufacturers; this is strong evidence that it contains more glutinous matter, which would be highly objectionable in the manufacture of some kinds of cloth, particularly those of delicate colours." An anonymous note (1844: 13) revealed the broad geographic range of sesame cultivation as a cash crop by mid-nineteenth century: "We find in one of our Louisiana papers, that Mr. McIntyre of Ouschita, is cultivating the *bene* plant for the oil obtained from its seeds."

A summary of early twentieth century sesame culture in the Southern States indicated (Anonymous 1912: 276): "The demand in the United States for *gingelly* oil is constantly increasing, and although the price of the seed is rather low there is undoubtedly a favorable opportunity for developing sesame culture in the Southern States. In case of its production on a scale larger than it is at present it would soon become an economical question, and the price of both seed and oil would increase in value."

Douglas articulates this writer's view (2011: 144): "While the colonial economy's dependence on enslaved labor for agricultural production is recognized, less apparent is the contribution of slaves' technical expertise to making the cultivation of [*benne*] a viable component of the agricultural economy. More obscure are the contributions that enslaved people made to the colony's prosperity by their individual engagement in agricultural endeavors for personal and economic benefit, even if these activities were constrained by the 'peculiar institution.' Many of the African slaves brought to Louisiana in the 1720s for agricultural work also began to grow food as a means of supplementing food allotments and gaining some measure of self-determination, if limited by legal restrictions. Soon they were actively involved in growing familiar things and selling them on the side." The *papers of Henry Laurens* (1778–1782: 268) document this appeal: "that you will procure me as much *Sesamum* or *Bené* Seed as you possibly can, & encourage the planting it by all the Negroes at each plantation."

"Prospecting for Oil" is a review of the "consequential" pursuit of culinary oils and fats by early American farmers, with heavy emphasis on sesame, calling attention to an irony (Shields 2010: 29): "Jefferson had imagined that the olive tree would be first and foremost, a boon for African slaves in the South, providing them a fat that could be produced with less expense and greater volume than lard or bacon grease. […] Yet the slaves had brought with them the source of oil they needed in their diet. *Sesamum indicum* or what the West African Mende called '*Benne*.'"

Participation of Jews in Sesame Introduction

During the 1490s, trade with the New World began to open up. At the same time, the monarchies of Spain and Portugal expelled their Jewish subjects. The Jewish exclusion from Spain coincided with establishment of trading links between Africa, Europe, and the Americas. As a result, many became seafaring merchants, dispersed over critical nodes of the network, transferring assets and information. Jews began participating in all sorts of trade on the Atlantic, including the slave trade.

Early sources indicate that sesame arrived in the Western Hemisphere through the agency of Jews. Browne (1756: 270) was among the first: "These plants were introduced to Jamaica by the Jews, and are now cultivated in most parts of the island: the seeds are frequently used in broths, by many of our Europeans; but the Jews make 'em chiefly into cakes."

Long (1774: 809–810) describes "*Vanglo*, *Wongala*, or Oil-Plant - *Sesamum orientale*" […] "Two species of this plant are cultivated in this island. They are said to have been first introduced by the Jews." and "The Jews use the oil in cakes, instead of butter." Mention of *Vanglo* in Jamaica also comes via Wright (1787: 282): "The *Vanglo* plant (*Sesamum indicum*) was first introduced into Jamaica by the Jews as an article of food. It yields also an expressed oil, which is as clear and sweet as that of almonds… Nine pounds of seed yield two pounds of oil."

Bridges (1828: 587) wrote: "The *Vanglo* (*Sesamum indicum*) was first brought to Jamaica by the Jews, as an article of food. It yields an expressed oil, which is clear and sweet as that of almonds." Loudon (1829: 515) noted: "These plants were introduced into Jamaica by the Jews, and are now cultivated in most parts of the island. They are called *vanglo* or oil-plant. The seeds are frequently used in broths by many of the Europeans, but the Jews make them chiefly into cakes…" Hart (1957) acknowledged the various importing waves which brought sesame into Jamaica by Jews.

Inasmuch as sesame seeds and oil were well known and widely used by Jews in Southwest Asia since the distant past (Bedigian 2004a, 2011c), it seems reasonable that sesame would have been among foodstuffs imported with enslaved Africans brought by Jewish colonists since the sixteenth century, viewed as familiar and favored in the African diet, as well as a productive cash crop. It is likewise congruent that the colonies that have a history with sesame, such as Jamaica, Surinam, (Arbell 2000; Ezratty 1997; Loker 1991; Schorsch 2004), Georgia, and North Carolina (Faber 1998; Friedman 1998; Schorsch 2004) were also those places that had early Jewish populations (Cadoree Bradley, personal communication, July 8, 2011). Diner (2004), Levinger (1949), and Schappes (1950) show that an early population of Spanish and Portuguese Jews arrived in Georgia in 1733 and that Portuguese Jews founded their congregation in Charleston in 1734.

Conclusions and Discussion

The history of sesame in the Americas is a complex question with interlocking underpinnings. The process of sesame domestication materialized on the Indian subcontinent. Sesame reached Africa at some unknown time long ago, where its cultivation became widespread. During the enslavement of Africans, their subjugators looked into crops familiar to their subjects, for provisions. Those captors subsequently adopted the popular oilseed sesame, as a valuable cash crop. Though the Africans arrived under duress and lived, for the most part, segregated from white residents, this record demonstrates the penetration of African expertise about sesame seeds and leaves for medicine and food in the New World, for which they have received insufficient recognition. This report attempts, as Law (1999) wrote, to "bridge the persisting disjunction between the history of Africa and of its trans-Atlantic Diaspora."

There is evidence of sesame derived from Africa assimilated within American cuisine by Africans in the 1600s. Sesame seed grew in South Carolina by 1730 (Gray & Thompson 1941, i: 194) and probably by the 1690s; Sloane saw it before then in Jamaica (1707 and Sloane herbarium specimens, BM). Romans wrote (1775: 130) that sesame was "the best thing yet known for extracting a fine esculent oil… Negroes use it as food either raw, toasted or boiled in their soup and are very fond of it; they call it *Benni*."

A special limitation in the extraction of evidence from the history of enslaved Africans is the paucity of primary sources. Wilson and Ferris reflect (1989: 139): "Linguists have noted that southern speech carries a remarkable load of African vocabulary. This assertion is all the more remarkable when we recall that white southerners have often claimed to have little interaction with blacks. Some regional words have murky origins but there is no controversy for terms such as gumbo, *benne*, *goobers*, okra, jazz."

The term *benne* in the Gullah and Geechee languages persists in the region of the southern United States centered around Charleston South Carolina, and not elsewhere. We know that large numbers came from the rice-growing regions of coastal West Africa to southern coastal states of North America. It appears that enslaved Africans from West Africa – speakers of Bamana, Mande, Mandingo, and Wolof – contributed the name *benne* and brought about a separate introduction of sesame. Cassidy and Le Page (2002) do not include the name *benne* as a name for sesame in Jamaica; instead, they refer to *soonga*, *vanglo*, *wangla*, and *zezegary*.

The data registered here, of edible use and associations between recommended use of sesame leaves in early medical treatises and pharmacopoeias associated with colloquial African names, provide evidence. A herbarium specimen named *Bowangala* at Leiden confirms beyond doubt that sesame grew in Suriname by 1689. That name is one among several clusters of names for sesame found in the Americas, associated with distinct regions of Africa. It is reasonable that since the names *soonga*, *vanglo/wangila*, and *zezegary* are associated primarily with Belize, Jamaica, Suriname, and other parts of the Caribbean, and since speakers of Lingala provided those name variants, and a vast number of enslaved Africans arrived there from the greater Angola-Congo region, that sesame introduction to the Americas occurred from that region also.

Pollitzer (1998: 66) summarizes phenomena that this work documents: "Linguistic, cultural and biological factors tie the Sea Island people of South Carolina and Georgia, to their African ancestors. The early arrival of Bantu peoples from Central Africa, as well as their later infusion, left a permanent mark on vocabulary, religion, magic and many arts and crafts."

Despite several romantic suggestions that portray seeds transferred in the hair of the enslaved (examined by Carney and Rosomoff 2009: 76–77), it would seem more a myth, "given the horrendous conditions under which they were transported to the colonies. The opportunity or possibility of being able to bring anything with them was nonexistent.... *Benne* was a familiar plant to enslaved Africans from West Africa, and upon seeing it here they took it as their own since they were probably more familiar with its uses than were the European colonists" (Begley, personal communication, June 2, 1995).

Yentsch offers a valuable alternative view (1994: 199): "Portuguese ships (which carried free blacks as crew)... carried African foods to the New World - okra, coriander, dasheen, sesame, black-eyed peas, pigeon peas, white yams, kola nuts, varieties of rice and millet, tamarind and watermelon, Guinea fowl, African sheep, calabash gourds and fluted pumpkins, and the West African oil palm. How did the foods arrive? Stories of seeds carried abroad hidden in black ears are legends, part of an African-American mythology; the reality is more complex." "Edible plants were cargo destined to provide the root stock for agricultural experimentation (for example, rice and sesame, sorghum and millet" (Yentsch 1994: 199). Under conditions "created by the slave trade" (Fett 1995), *benne*, yams, and okra were introduced.

The introduction of sesame to the Americas is a solid example of continuity in material culture and medical knowledge carried from the Old World to the New. It is one answer to Herskovits' rhetorical question (1935: 92), "What has Africa given America?" It celebrates the persistence of African heritage in the culture of blacks living on the Carolina-Georgia coast and argues for African contributions to early knowledge about the merits of sesame in the Americas and to America's identity, particularly through food. It also demonstrates the virtues of interdisciplinary research, drawing as it does on anthropology, botany, and history to compile evidence of African expertise.

Acknowledgments Completion of this research has required a decade and a half of persistent searching because the sources assembled here are extremely scarce, and widely scattered. Encouragement from archivists and librarians aided the completion of this manuscript. P. R. Begley, Reference Services, South Carolina Department of Archives and History, Columbia, SC, supplied some local references (June 1995) that helped launch this investigation. M. B. Cadoree Bradley, Reference Specialist, Science, Technology and Business Division, Library of Congress, dipped deeply and unstintingly into the library's rich resources to locate some well-hidden references that answered specific questions about the transfer of African sesame names and practices. C. Ehret, Professor of History, UCLA, provided indispensable aid with linguistic analysis. P. F. de Moraes Farias, Centre of West African Studies, University of Birmingham, settled the location of Gao, in West Africa. A. P. Clark, US National Herbarium Smithsonian Institution; K. D. Perkins, University of Florida Herbarium Gainesville; and T. van Andel, Netherlands Centre for Biodiversity Naturalis, Leiden, provided specimen information. I am grateful to S. Alpern, D. Austin, J. Carney, L. J. G. van der Maesen, Y. Scheven, and R. Voeks for comments on draft versions of this manuscript.

References

Abels J (1975) The genera *Ceratotheca* Endl. and *Dicerocaryum* Boj. (Monographs of the African Pedaliaceae iii–iv). Memórias da Sociedade Broteriana 25:1–358. Coimbra

Abrahams RD, Szwed JF (1975) Discovering Afro-America. E. J. Brill Archive, Leiden

Adjanohoun EI, Ahyi AMR, Ake Assi L, Baniakina J, Chibon P, Cuset G, Doulou V, Enzanza A, Eyme J, Goudote E, Keita A, Mbemba C, Mollet J, Moutsambotè J.-M, Mpati J & Sita P (1988)

Médecine traditionnelle et pharmacopée. Contribution aux etudes ethnobotaniques et floristiques en République populaire du Congo. Agence de coopération culturelle et technique, Paris, p 605
Almada AA (1594) Brief treatise on the rivers of Guinea. Part I. [Liverpool]: Department of History, University of Liverpool, 1984. http://digital.library.wisc.edu/1711.dl/AfricaFocus.Almada01
Alpern SB (2008) Exotic plants of Western Africa: where they came from and when. Hist Afr 35:63–102
Annapurna Kishore Kumar MS (2003) Studies on karyotype, genome size and genome relations in some species of Sesamum L. (Pedaliaceae). Ph.D. thesis, Karnataka University, Dharwar
Annapurna Kishore Kumar MS, Hiremath SC (2008) Cytological analysis of interspecific hybrid between *Sesamum indicum* L. x *S. orientale* L. var. *malabaricum*. Karnataka J Agric Sci 21(4):498–502
Anonymous (1820a, March 31) Cultivation of the grape, olive, &c. Agricultural Intelligencer, and Mechanic Register 1(13):101
Anonymous (1820b, July 21) For the American farmer. *Sesamum* or *Bene*. American Farmer 2(17):135. Republished in *Edwardsville Spectator* Dec. 5, 1820. 2(80):1
Anonymous (1827, May 30) The *bene* plant. New Harmony Gazette 2(35):277
Anonymous (1843) Question du Sésame: Petition adressée aux chambres législatives. La Société Royale et Centrale d'Agriculture, Douai
Anonymous (1844) New product of the south. The Cultivator (N Y State Agric Soc) 1:13
Anonymous (1912, October 5) Sesame culture in the Southern States. Scientific American 107:276
Anonymous (n.d.) Confections delicacies sweets: slave recipes from the Old Slave Mart. 6 Chalmers Street, Charleston
Arbell M (2000) The Portuguese Jews of Jamaica. Canoe Press, Kingston
Attfield J (1865) On the food value of the Kola-nut – a new source of theine. Pharm J 6:457–460
Aublet F (1775). Histoire des plantes de la Guiane Françoise, rangées suivant la méthode sexuelle, avec plusiers mémoires sur différens objets intéressans, relatifs à la culture & au commerce de la Guiane Françoise, & une notice des plantes de l'Isle-de-France, vol 2. P. F. Didot jeune, Londres/Paris
Barbosa D (1866) A description of the coasts of East Africa and Malabar in the beginning of the sixteenth century, by Duarte Barbosa, a Portuguese. Translated from an early Spanish manuscript in the Barcelona library; with notes and a preface, by the Hon. Henry E. J. Stanley
Barham H (1794) Oily pulse. In: Hortus Americanus: containing an account of the trees, shrubs, and other vegetable productions of South-America and the West India Islands, and particularly of the island of Jamaica. Alexander Aikman, Kingston
Batson WT (1984) A guide to the genera of the plants of Eastern North America, 3rd edn revised. University of South Carolina Press, Columbia
Beckwith MW (1929) Black roadways, a study of Jamaican folk life. University of North Carolina Press, Chapel Hill
Bedigian D (1981) Origin, diversity, exploration and collection of sesame. In: Sesame: status and improvement. FAO plant production and protection paper 29. FAO, Rome, pp 164–169
Bedigian D (1984). Sesamum indicum L. Crop origin, diversity, chemistry and ethnobotany. Ph.D. dissertation, University of Illinois, Urbana-Champaign. University Microfilms DA8502071, Dissertation Abstracts International 45, 1985:3410-B
Bedigian D (1988) *Sesamum indicum* L. (Pedaliaceae): ethnobotany in Sudan, crop diversity, lignans, origin, and related taxa. In: Goldblatt P, Lowry PP (eds) Modern systematic studies in African botany, vol 25, AETFAT monographs in systematic botany. Missouri Botanical Garden, St. Louis, pp 315–321
Bedigian D (1991) Genetic diversity of traditional sesame cultivars and cultural diversity in Sudan. In: Oldfield ML, Alcorn JB (eds) Biodiversity: culture, conservation and ecodevelopment. Westview Press, Boulder, pp 25–36
Bedigian D (1998) Early history of sesame cultivation in the Near East and beyond. In: Damania AB, Valkoun J, Willcox G, Qualset CO (eds) The Origins of Agriculture and Crop Domestication. The Harlan Symposium. ICARDA, Aleppo, Syria, pp 93–101. http://www.bioversityinternational.org/publications/Web_version/47/

Bedigian D (2000) Sesame. In: Kiple KF, Ornelas-Kiple CK (eds) The Cambridge world history of food, vol 1. Cambridge University Press, New York, pp 411–421

Bedigian D (2003a) Evolution of sesame revisited: domestication, diversity and prospects. Genet Resour Crop Evol 50:779–787

Bedigian D (2003b) Sesame in Africa: origin and dispersals. In: Neumann K, Butler A, Kahlheber S (eds) Food, fuel and fields – progress in African archaeobotany. Africa Praehistorica. Heinrich-Barth-Institute, Cologne, pp 17–36

Bedigian D (2004a) History and lore of sesame in Southwest Asia. Econ Bot 58(3):329–353

Bedigian D (2004b) Importance of botanical data to historical research on Africa. In: Philips JE (ed) Writing African history: methods and sources. Rochester University Press, New York, pp 163–182

Bedigian D (2004c). *Sesamum radiatum*. In: Grubbin G, Denton OA (eds) Plant resources of tropical Africa [PROTA], vol 2, Vegetables. PROTA programme, Wageningen University, Wageningen, pp 465–467

Bedigian D (2004d) Slimy leaves and oily seeds: distribution and use of wild relatives of sesame in Africa. Econ Bot 58(Suppl):S3–S33

Bedigian D (2006a) Assessment of sesame and its wild relatives in Africa. In: Ghazanfar SA, Beentje HJ (eds) Taxonomy and ecology of African plants, their conservation and sustainable use. Royal Botanic Gardens, Kew, pp 481–491

Bedigian D (2006b) Pedaliaceae. In: Akoègninou A, van der Burg WJ, van der Maesen LJG (eds) Flore analytique du Bénin. Backhuys, Cotonou/Wageningen, pp 846–849

Bedigian D (2010) Characterization of sesame (*Sesamum indicum* L.) germplasm: a critique. Genet Resour Crop Evol 57(5):641–647

Bedigian D (2011a) Cultivated sesame, and wild relatives in the genus *Sesamum* L. In: Bedigian D (ed) Sesame: the genus *Sesamum*. Medicinal and aromatic plants – Industrial profiles series. CRC Press/Taylor & Francis Group, Boca Raton, pp 33–77

Bedigian D (2011b) Introduction and early use of sesame in America: medicine, cookery and folkways. In: Bedigian D (ed) Sesame: the genus *Sesamum*. Medicinal and aromatic plants – Industrial profiles series. CRC Press/Taylor & Francis Group, Boca Raton, pp 389–421

Bedigian D (2011c) Introduction: history of the cultivation and use of sesame. In: Bedigian D (ed) Sesame: the genus *Sesamum*. Medicinal and aromatic plants – Industrial profiles series. CRC Press/Taylor & Francis Group, Boca Raton, pp 1–31

Bedigian D, Adetula OA (2004) *Ceratotheca sesamoides* Endl. In: Grubben G, Denton OA (eds) Plant resources of tropical Africa [PROTA], vol 2, Vegetables. PROTA Programme, Wageningen University, Wageningen, pp 175–177

Bedigian D, Harlan JR (1983) Nuba agriculture and ethnobotany, with particular reference to sesame and sorghum. Econ Bot 37:384–395

Bedigian D, Harlan JR (1986) Evidence for cultivation of sesame in the ancient world. Econ Bot 40:137–154

Bedigian D, Seigler DS, Harlan JR (1985) Sesamin, sesamolin and the origin of sesame. Biochem Syst Ecol 13:133–139

Bedigian D, Smyth CA, Harlan JR (1986) Patterns of morphological variation in sesame. Econ Bot 40:353–365

Bennett L (1942) Jamaica dialect verses. Herald, Kingston

Bentley R, Trimen H (1880 [1981]) Medicinal plants: being descriptions with original figures of the principal plants employed in medicine and an account of the characters, properties, and uses of their parts and products of medicinal value, vol 3, entry 198. International Book Distributors, Dehra Dun/Delhi

Berlin I, Morgan PD (1993) Cultivation and culture: labor and the shaping of slave life in the Americas. University of Virginia Press, Charlottesville

Betts EM (ed) (1944) Thomas Jefferson's garden book, 1766–1824, vol 22. American Philosophical Society, Philadelphia

Bigelow J (1822) A treatise on the materia medica, intended as a sequel to the pharmacopoeia of the United States. Charles Ewer, Boston

Bolles WP (1885) *Benné* oil. In: Buck AH (ed) A reference handbook of the medical sciences embracing the entire range of scientific and practical medicine and allied science, vol 1. W. Wood & Co., New York, p 487
Bridges GW (1828) Annals of Jamaica, vol 1. John Murray, London
Browne P (1756) The civil and natural history of Jamaica. Containing I. An accurate description of that island… with a brief account of its former and present state, government, revenues, produce, and trade. II. An history of the natural productions … native fossils. B. White and Son, London
Bryant C (1783) Flora diaetetica: or history of esculent plants. B. White, London
Burke E (comp) (1847) List of patents for inventions and designs issued by the United States from 1790 to 1847. J. & G. S. Gideon, Washington, DC, pp 112, 432
Burkill HM (1997) Pedaliaceae. In: The useful plants of West Tropical Africa, 2nd edn, vol 4. Families M-R. Royal Botanic Gardens, Kew, pp 414–423
Burr F (1874) Field and garden vegetables of America. William F. Gill & Company, Boston
Busson F (1965) Pedaliaceae. In: Plantes alimentaires de l'Afrique Occidentale. Leconte, Marseille, pp 386–391
Cambridge World History of Food (2000) In: Kiple KF, Ornelas-Kiple CK (eds) vol 2. Cambridge University Press, New York
Cappers R (1999) Trade and subsistence at the Roman port of Berenike, Red Sea coast, Egypt. In: van der Veen M (ed) The exploitation of plant resources in ancient Africa. Kluwer Academic/Plenum Publishers, New York/Boston/Dordrecht/London, pp 185–197
Capus G, Bois D (1912) Le sesame. In: Les produits coloniaux. Libraire Arnaud Colin, Paris, pp 298–301
Carney JA (1998) The role of African rice and slaves in the history of rice cultivation in the Americas. Hum Ecol 26:525–545
Carney JA (2001) Black rice: the African origins of rice cultivation in the Americas. Harvard University Press, Cambridge, MA
Carney JA (2003) African traditional plant knowledge in the circum-Caribbean region. J Ethnobiol 23(2):167–185
Carney JA, Rosomoff RN (2009) In the shadow of slavery: Africa's botanical legacy in the Atlantic world. University of California Press, Berkeley
Cassidy FG (2007) Jamaica Talk: three hundred years of the English language in Jamaica. University of the West Indies Press, Kingston
Cassidy FG, Le Page RB (2002) A dictionary of Jamaican English, 2nd ed. University of the West Indies Press, Barbados/Jamaica
Cathcart JL (1821a, August 4) Miscellany. From the National Intelligencer to the Editor. Columbian Centinel
Cathcart JL (1821b, August 25) Oriental *Bhene*. The Plough Boy, and J Board Agric (1820–1823): 3, 13
Chaplin J (1993) An anxious pursuit: agricultural innovation and modernity in the Lower South, 1730–1815. University of North Carolina Press, Chapel Hill
Clapham AJ, Rowley-Conwy PA (2007) New discoveries at Qasr Ibrim, Lower Nubia. In: Cappers R (ed) Fields of change. Barkhuis & Groningen University Library, Groningen, pp 157–164
Clavel M (1904) Items of folk-lore from Bahama Negroes. J Am Folk-lore 17:36–38
Clinton C (1982) The plantation mistress: woman's world in the South. Pantheon Books, New York
Codrington C (1878) "*Bene*" or "*Wanglo*". South Cultiv 36(4):151
Colquitt HR (1933) Savannah cook book: a collection of colonial receipts. Colonial Publishers, Charleston
Committee on Premiums and Exhibitions (1835) A report of experiments made by request of the Committee on Premiums and Exhibitions, on *Bene* seed oil, furnished at the eighth exhibition of the Franklin Institute, by Wetherill & Brothers. J Franklin Inst 16(1):13–14
Cook J (1982, December 22) Carolina coast's evocative food. The New York Times. http://www.nytimes.com/1982/12/22/garden/carolina-coast-s-evocative-food.html
Cooper F (1977) Plantation slavery on the East Coast of Africa. Yale University Press, New Haven/London

Creel MW (1988) A peculiar people: slave religion and community-culture among the Gullahs. New York University Press, New York

Cross W (2008) Gullah culture in America. Praeger, Westport

Cundall F (1891) The handbook of Jamaica for 1891–92: comprising historical, statistical and general information concerning the island compiled from official and other reliable records. Government Printing Establishment, Kingston

Cyclopedia of American agriculture (1909–1910) Cyclopedia of American agriculture : a popular survey of agricultural conditions, practices and ideals in the United States and Canada. In: Bailey LH (ed) 2nd edn, vol 2, Crops. Macmillan & Company, New York

Dabney JE (2010) The food, folklore, and art of Lowcountry cooking: a celebration of the foods, history, and romance. Cumberland House, Naperville

Dalgish GM (1982) A dictionary of Africanisms: contributions of Sub-Saharan Africa to the English language. Greenwood Press, Westport

Dalton HG (1855) The history of British Guiana: comprising a general description of the colony; a narrative of some of the principal events from the earliest period of its discovery to the present time, together with an account of its climate, geology, staple products, and natural history, vol. 2. Longman, Brown, Green, and Longmans, London

Dalziel JM (1955) Useful plants of West Tropical Africa. Appendix to the flora of west tropical Africa. In: Hutchinson J, Dalziel JM (eds) Flora of west tropical Africa. 2nd reprint of 1937 edn. Crown Agents for Overseas Governments and Administrations, London

Dancer T (1801) The medical assistant; or Jamaica practice of physic: designed chiefly for the use of families and plantations. Alexander Aikman, Kingston

Darby W (1818) The emigrant's guide to the western and southwestern states and territories: comprising a geographical and statistical description of the states. Kirk and Mercein, New York

de Candolle A (1884) Origin of cultivated plants. Kegan Paul, Trench and Co., London

De Voe TF (1867) Market assistant, containing a brief description of every article of human food sold in the public markets of the cities of New York, Boston, Philadelphia and Brooklyn. Hurd and Houghton, New York

DeFilipps RA, Maina SL, Crepin J (n.d.) Medicinal plants of the Guianas (Guyana, Surinam, French Guiana). http://www.mnh.si.edu/biodiversity/bdg/medicinal/medplantsgui3.pdf

Dictionary of American regional English (1985) In: Cassidy FG (ed) vol 1, '*Benne*.' Belknap Press of Harvard University Press, Cambridge, MA/London

Dictionary of Americanisms on historical principles (1951) In: Mathews MM (ed) vol 1, University of Chicago Press, Chicago

Dictionary of Caribbean English usage (1996) In: Allsopp R (ed) University of the West Indies Press, Kingston, Jamaica

Diner HR (2004) The Jews of the United States, 1654 to 2000. University of California Press, Berkeley

Douglas L (2011) Public spaces, private gardens. Louisiana State University Press, Baton Rouge

Dunglison R (1868) Medical lexicon: a dictionary of medical science: containing a concise explanation of the various subjects and terms, with the French and other synonyms, notices of climate, and of celebrated mineral waters, formulae for various officinal and empirical preparations. H. C. Lea, Philadelphia

Durant TJ, Knottnerus JD (1999) Plantation society and race relations: the origins of inequality. Praeger, Westport

Ehret C (1977) East African words and things: agricultural aspects of economic transformation in the 19th century. In: Ogot BA (ed) Kenya in the 17th century. *Hadith* 8. Historical Association of Kenya. Bookwise Ltd./Anyange Press, Nairobi, pp 152–172

Ehret C (2005) Writing African history from linguistic evidence. In: Philips JE (ed) Writing African history. University of Rochester Press, Rochester, pp 86–111

Ewell J (1827) The medical companion, or family physician, 7th edn. Proprietors, Washington, DC

Ezratty HA (1997) 500 years in the Jewish Caribbean: the Spanish and the Portuguese Jews in the West Indies. Omni Arts, Inc., Baltimore

Faber E (1998) Jews, slaves, and the slave trade: setting the record straight. New York University Press, New York

Farnsworth, P, Wilkie LA (2006) Fish and grits: Southern, African and British influences in Bahamian foodways. In: Regis HA (ed) Caribbean and Southern: transnational perspectives on the U.S. South. University of Georgia Press, Athens, pp 34–72

Faroukh O (1965) Tarikh al adab al Arab. Dar ul 'Ilm lil Malayim, Beirut

Federal Writers Project, Writers' Program of the Work Projects Administration in the State of South Carolina (1976) South Carolina: the WPA guide to the Palmetto state. University of South Carolina, Columbia

Fenner B (1884) Fenner's working formulae: a hand-book of the old and the new pharmacopoeias, containing weight and measure standards, working processes, working formulae, the medicinal properties and doses of drugs and their preparations, and being a comparison of the 1870 with the 1880 pharmacopoeia, with remarks, suggestions, original processes and formulae, and other matter of interest and value to druggists and physicians. The Author, Westfield

Fessenden TG (1842) The new American gardener. 15th edn. Otis, Broaders, & Company/Thomas, Cowperthwaite, & Co., Boston/Philadelphia

Fett SM (1995) Body and soul: African-American healing in southern antebellum plantation communities 1800–1860. PhD dissertation, Rutgers University, NJ

Fett SM (2002) Working cures: healing, health, and power on southern slave plantations. University of North Carolina Press, Chapel Hill

Flückiger FA, Hanbury D (1874) *Oleum Sesami*. Pharmacographia; a history of the principal drugs of vegetable origin, met with in Great Britain and British India. Macmillan and Co., London

Flückiger FA, Hanbury D (1879) *Oleum Sesami.* In: Pharmacographia; a history of the principal drugs of vegetable origin, met with in Great Britain and British India. 2nd edn. Macmillan and Co., London, pp 473–476

Fowler D (2008) Savannah cookbook. Gibbs, Smith, Salt Lake City/Charleston/Santa Fe/Santa Barbara

Fra-Molinero B, Nero CF (2007) When food tastes cosmopolitan: the Creole fusion of diaspora cuisine: an interview with Jessica B. Harris. Callaloo 30(1):287–303

Friedman SS (1998) Jews and the American slave trade. Transaction Publishers, New Brunswick

Friend, a Religious and Literary Journal. August 18, 1855. The *Bene* plant 28(49):387

Gamble DP (1967) The Wolof of Senegambia, together with notes on the Lebu and the Serer. International African Institute, London

Genovese ED (1974) Roll, Jordan, roll: the world the slaves made. Pantheon Books, New York

Gomez MA (1998) Exchanging our country marks: the transformation of African identities in the colonial and antebellum South. University of North Carolina Press, Chapel Hill

Goodwin RM (1824) *Bene* seed – its culture and use. Am Farmer 6(1):46

Granger M (2007) Drums and shadows: survival studies among the Georgia coastal Negroes. Forgotten Books, Charleston

Gray SF (1821) A supplement to the pharmacopoeia: being a treatise on pharmacology in general; Including not only the drugs and compounds which are used by practitioners of medicine, but also those which are sold by chemists, druggists, and herbalists… with a collection of the most useful medical formulae. Thomas and George Underwood, London

Gray A (1848 [1997]) Gray's manual of botany: a handbook of the flowering plants and ferns of the central and northeastern United States and adjacent Canada. 8th (centennial) edn. Reprint, 4th printing, 1993. Dioscorides Press, Portland

Greene P (1980) Favourite Sierra Leone recipes. Commercial Printers, Freetown

Grieve M (1931 [1971]) A modern herbal; the medicinal, culinary, cosmetic and economic properties, cultivation and folk-lore of herbs, grasses, fungi, shrubs, & trees with all their modern scientific uses. Dover Publications, New York

Grimé WE (1979) Ethno-botany of the Black Americans. Reference Publications, Algonac

Grisebach AHR (1864) Flora of the West Indian islands. L. Reeve & Co., London

Hall RN (1909) Prehistoric Rhodesia; an examination of the historical, ethnological and archaeological evidences as to the origin and age of the rock mines and stone buildings, with a gazetteer of mediaeval south-east Africa, 915 A.D. to 1760 A.D., and the countries of the Monomotapa, Manica, Sabia, Quiteve, Sofala and Mozambique. T. F. Unwin, London

Harlan JR (1951) Anatomy of gene centers. Am Nat 85:97–103

Harlan JR (1970) Evolution of cultivated plants. In: Frankel OH, Bennett E (eds) Genetic resources in plants – their exploration and conservation. Aldine, Chicago, pp 19–32

Harlan JR (1976) Plant and animal distribution in relation to domestication. Philos Trans R Soc Lond 275:13–25

Harlan JR (1992) Crops and man, 2nd edn. Agronomy Society of America, Madison

Harris JB (1996) Welcome table: African American heritage cooking. Simon & Schuster, New York

Harrison JP (1828) Remarks on cholera *infantum*. Transylvania J Med Assoc Sci (Feb.) 1:1

Hart A (1957) Open sesame. Monthly Comments. Jamaica 2(18): October. http://www.demercado.com/ansellvol2-final.htm#_Toc509546437. Accessed May 2009

Hatch PJ (2005) Thomas Jefferson's favorite vegetables. In: Fowler D (ed) Dining in Monticello: in good taste and abundance. University of North Carolina Press, Chapel Hill, pp 55–64

Hatley T (1984) Tending our gardens. South Change 6(5):18–24

Herskovits MJ (1935, September 4) What has Africa given America? The New Republic 84 92–94

Herskovits MJ (1941 [1958]) The myth of the negro past. Beacon, Boston

Hess K (1992) The Carolina rice kitchen: the African connection. University of South Carolina Press, Columbia

Hibben S (1946) American regional cookery. Little, Brown and Co., Boston

Hiremath SC, Patil CG (1999) Genome homology and the putative progenitor of sesame. J Cytol Genet 34:69–74

Hirt HM, Bindanda M (1993) La Medecine naturelle en Afrique. Editions Centre de Vulgarisation Agricole, Kinshasa

Hoehne FC (1937) Botanica e agricultura no Brasil no seculo XVI. Companhia Editora Nacional, Sao Paulo (in Portuguese)

Holloway JE (2005) "What Africa has given America": African continuities in the North American diaspora. In: Holloway JE (ed) Africanisms in American culture, 2nd edn. Indiana University Press, Bloomington, pp 39–64

Holmes FS (1852) The southern farmer and market gardener. Holmes Book House, Charleston

Hooper R (1817) Quincy's lexicon medicum. A new medical dictionary. M. Carey & Son, Benjamin Warner, and Edward Parker, Philadelphia

Hooper R (1829) Lexicon medicum; or medical dictionary. 4th American, from the last London edn. J. & J. Harper, New York

Hopkins JFP, Levtzion N (2000) Corpus of early Arabic sources for West African history. Marcus Weiner, Princeton

Ibn Battuta (1958–2000) Travels in Africa and Asia, A.D. 1325–1354. Hakluyt Society at the University Press, Cambridge

Irvine FR (1954) West African agriculture, 2nd edn. Oxford University Press, London

Irvine FR (1969) West African agriculture, 3rd edn. Oxford University Press, London

Jardin C (1967) List of foods used in Africa. Food Consumption and Planning Branch, Nutrition Divison, FAO, Rome, and Nutrition Section, Office of International Research, National Institutes of Health, Bethesda

Johnson GW, Bostick DW, Crooks DJ (2005) On the eve of the Charleston renaissance: the George W. Johnson photographs. Joggling Board Press, Charleston

Johnson-Feelings D (1997) Neeny coming, Neeny going. Afr Am Rev 31(3):556–557

Junior League of Charleston (compiler) (1950) Charleston receipts collected by the Junior League of Charleston. Walker, Evans & Cogswell Co., Charleston

Kajale MD (1991) Current status of Indian palaeoethnobotany: introduced and indigenous food plants with a discussion of the historical and evolutionary development of Indian agriculture and agricultural systems in general. In: Renfrew J (ed) New light on early farming: recent developments in palaeoethnobotany. Edinburgh University Press, Edinburgh, pp 155–189

Klein HS (1972) The Portuguese slave trade from Angola in the eighteenth century. J Econ Hist 33:904–918

Klein HS (1978) The middle passage: comparative studies in the Atlantic slave trade. Princeton University Press, Princeton

Klein HS (comp) (1997) Angola slave trade in the eighteenth century, 1723–1771 [machine-readable data file]. DPLS ed. 1970?, Data and Information Services Center [distributor], Madison. "http://www.disc.wisc.edu/slavedata/slaintro6.html" http://www.disc.wisc.edu/slavedata/slaintro6.html

Kusimba CM (2004) Archaeology of slavery in East Africa. Afr Archaeol Rev 21(2):59–88

Laurens H (1746–1755) In: Hamer PM, Rogers GC, Lyles ME (eds) The papers of Henry Laurens, vol 1. University of South Carolina Press, Columbia

Laurens H (1778–1782) In: Hamer PM, Rogers GC, Lyles ME (eds) The papers of Henry Laurens, vol 9. University of South Carolina Press, Columbia

Law R (1999) African cultures in colonial North America. Review: exchanging our country marks: the transformation of African identities in the colonial and antebellum South. J Afr Hist 40(1):132–133

Leighton A (1976) American gardens in the eighteenth century. Houghton Mifflin, Boston

Leo A, Brown R, Pory, J (1896) History and description of Africa, and of the notable things therein contained, written by Al-Hassan Ibn-Mohammed Al-Wezaz al Fasi, a Moor, babtised as Giovanni Leone, but better known as Leo Africanus. Done into English in the year 1600, vol 3. Hakluyt Society. Burt Franklin, New York

Levinger L (1949) A history of Jews in the United States. Union of American Hebrew Congregations, Cincinnati

Lewicki T (1974) West African food in the Middle Ages. Cambridge University Press, London

Lisān al-'Arab (1981) Ibn Mandur. Dar ul Maarif, Cairo

Little BJ (1991) Text-aided archaeology. CRC Press, Boca Raton

Loker Z (comp) (1991) Jews in the Caribbean: evidence on the history of the Jews in the Caribbean zone in colonial times. Misgav Yerushalayim, Institute for Research on the Sephardi and Oriental Jewish Heritage, Jerusalem

Long E (1774) History of Jamaica, vol 3. T. Lowndes, London

Loudon JC (1829) An encyclopaedia of plants: comprising the description, specific character, culture, history, application in the arts, and every other desirable particular respecting all the plants indigenous, cultivated in, or introduced to Britain, combining all the advantages of a Linnean and Jussieuean species plantarum, an historia plantarum, a grammar of botany, and a dictionary of botany and vegetable culture: the whole in English: with the synonyms of the commoner plants in the different European and other languages: the scientific names accentuated, their etymologies explained, the classes, orders, and botanical terms illustrated by engravings, and with figures of nearly ten thousand species, exemplifying several individuals belonging to every genus included in the work/edited by Loudon JC. Longman, Rees, Orme, Brown and Green, London

Lowndes T (1730) Records in the British Public Record Office pertaining to South Carolina 14:261–262

Lunan J (1814) *Vanglo*, or oil plant. *Hortus Jamaicensis,* or a botanical description (according to the Linnean system) and an account of the virtues, &c., of its indigenous plants hitherto known, as also of the most useful exotics. Compiled from the best authorities, and alphabetically arranged, vol 2. St. Jago de la Vega Gazette, Jamaica

Mabberley DJ (1997) The plant book, 2nd edn. Cambridge University Press, Cambridge

Madden RR (1835) A twelvemonth's residence in the West Indies: during the transition from slavery to apprenticeship; with incidental notice of the state of society, prospects, and natural resources of Jamaica and other islands, vol 2. Carey, Lea and Blanchard, London

Maïkere-Faniyo R, Van Puyvelde L, Mutwewingabo A, Habiyaremye FX (1989) Study of Rwandese medicinal plants used in the treatment of diarrhoea I. J Ethnopharmacol 26:101–109

Manning SD (1991) The genera of Pedaliaceae in the Southeastern United States. J Arnold Arbor Suppl 1:313–347

Martinez JS (2006) *Wangla*. San Pedro Sun, pp 14–15. http://www.ambergriscaye.com/sanpedrosun/old/06–06.pdf#search="wangla". Accessed 11 Mar 2009

Mathews MM (1948) Some sources of Southernisms. University of Alabama Press, Alabama

Mathews MM (1958) Of matters lexicographical. Am Speech 33(4):274–279

Mauny R (1953) Notes historiques autour des principales plantes cultivées de l'Afrique occidentale. Bulletin Institut Francais d'Afrique Noire [BIFAN], Dakar 14:684–730
Mayne RG (1853–60 (1860)) An expository lexicon of the terms, ancient and modern, in medical and general science; including a complete medico-legal vocabulary. J. Churchill, London
Mease J (1811) On *Sesamum*, or *Bene* plant. Archives of Useful Knowledge 1:44–48 (July 1810–April 1811). David Hogan, Philadelphia
Meriam E (1846) *Bene* Plant. Doc. No. 140. Appendix No. 19. Jan. 27, 1846. Extracts from the report of the commissioner of patents for the year 1845. A list of patents granted during that year. Ritchie & Heiss, Washington, DC
Merriam- Webster's new geographical dictionary (1997) G. & C. Merriam Co., Springfield
Miller P (1735) The gardeners dictionary: containing the methods of cultivating and improving the kitchen, fruit and flower garden, as also the physick garden, wilderness, conservatory and vineyard, vol 2. Printed for the author, London
Miller W (1902) *Sesamum*. In: Bailey LH, Miller W (eds) Cyclopedia of American horticulture: comprising suggestions for cultivation of horticultural plants, descriptions of the species of fruits, vegetables, flowers, and ornamental plants sold in the United States and Canada, together with geographical and biographical sketches. 2nd edn, vol 4. The Macmillan Company, London, p 1662
Miller AB (1976) Shaker herbs. C. N. Potter, distributed by Crown Publishers, New York
Mills R (1826) Exotic plants. In: Statistics of South Carolina, including a view of its natural, civil, and military history. Hurlbut and Lloyd, Charleston
Mkamilo GS, Bedigian D (2007) *Sesamum indicum* L. In: van der Vossen HAM, Mkamilo GS (eds) Plant resources of tropical Africa [PROTA], vol 14, Vegetable oils. PROTA Programme, Wageningen University, Wageningen, pp 153–158
M'lean P (1875) The "Dysentery Plant.". Madras Revenue Regist 9:64
Moloney CA, Bruce H, Shelley GE (1887) Sketch of the forestry of West Africa with particular reference to its present principal commercial products. S. Low, Marston, Searle, & Rivington, London
Morel J (1769) A letter on the expressing of a fine oil from *bene* seed by Mr John Morel of Georgia. Trans Am Philos Soc 1: 239–240
Morgan PD (1982) The task system and the world of lowcountry Blacks, 1700–1880. William Mary Quart Third Ser 39(4):563–599
Morgan PD (1995) Slaves and livestock in eighteenth-century Jamaica: Vineyard Pen, 1750–1751. William Mary Quart Third Ser 52(1):47–76
Morgan PD (1998) Slave counterpoint: Black culture in the eighteenth-century Chesapeake and Lowcountry. University of North Carolina Press, Chapel Hill
Mufwene SS, Condon N (1993) Africanisms in Afro-American language varieties. University of Georgia Press, Athens/London
Nadel SF (1947) The Nuba. Oxford University Press, London
National Council of Negro Women (1993) Black family reunion cookbook: recipes and food memories, 1st Fireside edn. Simon & Schuster, New York
Neuwinger HD (2000) African traditional medicine. Medpharm GmbH Scientific, Stuttgart
Oxford English Dictionary (1961) In: James AHM, Henry B, Craigie WA, Onions CT, Burchfield RW (eds); being a corrected re-issue with an introduction, supplement, and bibliography of A new English dictionary on historical principles, founded mainly on the materials collected by the Philological society
Oxford English Dictionary (OED). 2nd edn (1989) J A Simpson; E S C Weiner, 1989 2nd ed. / Oxford: Clarendon Press; Oxford; New York
Parrish E (1863) On the culture of *benne* or *Sesamum* in the United States. In: Simmonds PL (ed) The technologist. A monthly record of science applied to art, manufacture and culture, vol 3. Kent & Co., London, pp 245–248
Partridge W (1834) A practical treatise on dying woolen, cotton and silk. Appendix. Published by the author, New York
Patterson O (1967) The sociology of slavery. Associated University Press, London
Periplus of the Erythraean Sea (1912) The Periplus of the Erythraean Sea: travel and trade in the Indian Ocean, by a merchant of the first century

Phillips H, Lesley JP (1884) Early proceedings of the American Philosophical Society for the promotion of useful knowledge. Am Philos Soc. McCalla & Stavely, Philadelphia, p 4
Piso W (1648) Historia naturalis Brasiliae…: in qua non tantum plantae et animalia, sed et indigenarum morbi, ingenia et mores describuntur et iconibus supra quingentas illustrantur. Franciscum Hackium, Lugdun Batavorum, et, Ludgun Elzevirium, Amstelodami
Pollitzer WS (1998) The relationship of the Gullah-speaking people of coastal South Carolina and Georgia to their African ancestors. In: Goodwine ML (ed) Legacy of Ibo landing. Gullah toots of African-American culture. Clarity Press, Atlanta, pp 54–68
Pollitzer WS (1999) The Gullah people and their African heritage. University of Georgia Press, Athens
Porcher FP (1863) Resources of the southern fields and forests, medical, economical, and agricultural. Being also a medical botany of the Confederate States; with practical information on the useful properties of the trees, plants, and shrubs. Steam-Power Press of Evans and Cogswell, Charleston
Pulle AA (1906) An enumeration of the vascular plants known from Surinam: together with their distribution and synonymy. E. J. Brill, Leiden
Quincy J (1802) Quincy's lexicon pysico-medicum improved, or, a dictionary of the terms employed in medicine, and in such departments of chemistry, natural philosophy, literature, and the arts, as are connected therewith containing ample explanations of the etymology, signification, and use of those terms. From the 11th London edn., with many amendments and additions. Printed by T. & J. Swords, and sold at their Book and Stationary Store, New York
Rafinesque CS (1831, July 30) On the *Lobelia* and *Sesamum*. Atkinson's Saturday Evening Post 10:3
Raubenheimer O (1910) Oil of *Sesamum* – its use in pharmacy. Am J Pharm 82:476–481
Read WA (1931) Louisiana-French. Louisiana State University Press, Baton Rouge
Records of the States of the U.S. (1774, May 25) Georgia Gazette. Savannah
Redwood T, Gray SF (1847) Gray's supplement to the pharmacopoeia: being a concise but comprehensive dispensatory and manual of facts and formulae, for the chemist and druggist and medical practitioner. Longman & Co., London
Reisner GA (1918) Outline of the ancient history of the Sudan. Sudan Notes Records 1:3–15
Romans B (1775 [1962]) A concise natural history of East and West Florida. Reprinted by University of Florida Press, Gainesville
Roughley T (1823) The Jamaica planter's guide. Longman, Hurst, Rees, Orme and Brown, London
Roxburgh TL, Ford JC (1898) Handbook of Jamaica for 1898: published by authority, comprising historical, statistical and general information concerning the island. E. Standford, London
Rutledge AW (1949) A cosmopolitan in Carolina. William Mary Quart Third Ser 6(4):637–643
Rutledge S (1979) The Carolina housewife. University of South Carolina Press, Columbia
Sainsbury WN (1859) A list and abstract of documents relating to South-Carolina, now existing in the State Paper Office, London. South Carolina Historical Society. Collections, vol 3. Charleston
Salomon R (1991) Epigraphic remains of Indian traders in Egypt. J Am Orient Soc 111(4): 731–736
Savannah Unit Georgia Writers' Project Work Projects Administration (GWP) Survival Studies among the Georgia Coastal Negroes (1986) [1st ed. 1940]. Brown Thrasher Books. The University of Georgia Press, Athens
Schappes MU (1950) A documentary history of the Jews in the United States, 1654–1875. Citadel, New York
Schoolcraft HR (Mrs.) (1969) Plantation life: the narratives of Mrs. Henry Rowe Schoolcraft. http://aae.greenwood.com/doc.aspx?i=0&fileID=SCP2084&chapterID=SCP2084–290&path=/primarydoc/greenwood/
Schorsch J (2004) Jews and blacks in the early modern world. Cambridge University Press, New York
Shields DS (2010) Prospecting for oil. Gastronomica: J Food Cult 10(4):25–34
Simmonds PL (1858) Dictionary of trade products, manufacturing, and technical terms: with a definition of the moneys, weights, and measures of all countries. G. Routledge, London

Simmonds PL (1889) Tropical agriculture. A treatise on the culture, preparation, commerce, and consumption of the principal products of the vegetable kingdom, New edn. E. & F. N. Spon, Strand, London

Sims J (1815) *Sesamum indicum*. Indian *Sesamum*, or oily- grain. Curtis's Botanical Magazine; or, Flower-Garden Displayed, vol 41: entry 1688. Sherwood, Neely, & Jones, London

Skinner JS (ed) (1824) The American farmer, vol 6. J. Robinson, Baltimore

Sloane H (1696) Catalogus plantarum quae in insula Jamaica: sponte proveniunt, vel vulgò coluntur cum earundem synonymis & locis natalibus, adjectis aliis quibusdam quae in insulis Maderae, Barbados, Nieves, & Sancti Christophori nascuntur, seu prodromi historiae naturalis Jamaicae pars prima. D. Brown, London

Sloane H (1707) A voyage to the islands Madera, Barbados, Nieves, S. Christophers and Jamaica: with the natural history of the herbs and trees, four-footed beasts, fishes, birds, insects, reptiles, &c. of the last of those islands; to which is prefix'd, an introduction, wherein is an account of the inhabitants, air, waters, diseases, trade, &c. of that place, with some relations concerning the neighbouring continent, and islands of America, vols 1 and 2. Printed by B.M. for the author, London

Small JK (1913) Flora of the southeastern United States, 2nd edn. Published by the author, New York

Small JK (1933) Manual of the southeastern flora. Published by the author, New York

Smith J (1823a, July 25) no headline. Rhode Island American, and General Advertiser 15(86):2

Smith J (1823b, c, July 26) *Sesamum Indicum*, or *Bhene* plant. Niles' Weekly Register (1813–1837) 24:619

Smith J (1823d, July 29) *Sesamum Indicum*, or *Bhene* plant. Richmond Enquirer 20(24):4

Smith J (1823e, December 25) *Sesamum Indicum*, or *Bhene* plant. American Farmer 5:331–332

Smith J (1824, May 25) *Sesamum Indicum*, or *Bhene* plant. Trenton Federalist 25:1

Stormer T (1993) St. Helena's Island, South Carolina. Penn Center, South Carolina. www.angelfire.com/sc/jhstevens/penncenter.html

Sturtevant EL (1919 [1972]) Sturtevant's notes on edible plants. Hedrick UP (ed) Dover Publications, New York

Sweringen HV (1882) A dictionary of pharmaceutical science: a guide for the pharmaceutist, druggist, and physician; containing concise explanations of the various subjects and terms of pharmacy, and appropriate selections of the collateral sciences. P. Blakiston, Philadelphia

Thacher J (1821) The American new dispensatory. Containing general principles of pharmaceutick chemistry, chemical analysis of the articles of materia medica, pharmaceutick operations, 4th edn. T. B. Wait, Boston

Thacher J (1835, February 3) *Bene* seed. Farmer & Gardener, and Live-Stock Breeder and Manager 1:40

The Free Dictionary (2011) Collins English dictionary – complete and unabridged. HarperCollins Publishers http://www.thefreedictionary.com/buckra

Thompson G (1858) The palm land; or, West Africa, illustrated. Being a history of missionary labors and travels, with descriptions of men and things in Western Africa; also a synopsis of all the missionary work on that continent. Dawsons, London

Tobin BF (1999) "And there raise yams": slaves' gardens in the writings of West Indian plantocrats. Eighteenth-Century Life 23(2):164–177

Tropicos.org. Missouri Botanical Garden. http://www.tropicos.org/Name/50297261. Accessed 7 Jul 2011

Turner L (1969) Africanisms in the Gullah dialect. Arno Press/New York Times, New York

Twitty MW (2011) Sesame. In: Katz-Hyman MB, Rice KS (eds) World of a slave: encyclopedia of the material life of slaves in the United States. Greenwood, Santa Barbara, pp 416–417

Van Beek GW (1958) Frankincense and myrrh in ancient South Arabia. J Am Orient Soc 78(3):141–152

Van der Veen M (1999) The food and fodder supply to Roman quarry settlements in the eastern desert of Egypt. In: van der Veen M (ed) Exploitation of plant resources in ancient Africa. Kluwer Academic/Plenum Publishers, New York/Boston/Dordrecht/London, pp 171–183

Veldman S, Maas PJM, Thijsse G, van Andel, TR (2012) The forgotten Hermann Herbarium: a historical collection of useful plants from Suriname (ca. 1689). NCB Naturalis http://www.hermann-herbarium.nl/planten/plant26.html

Vignoles C (1823 [1977]) Observations upon the Floridas. A facsimile reproduction of the 1823 edition. Bicentennial Floridiana Facsimile Series. University Presses of Florida, Gainesville

Warden DB (1819) Statistical, political and historical account of the United States of North America from the period of their first colonization to the present day, vol 3. Archibald Constable and Co., Edinburgh

Warner-Lewis M (1991) Guinea's other suns: the African dynamic in Trinidad culture. The Majority Press, Dover

Washington M (1994) Community regulation and cultural specialization in Gullah folk religion. In: Johnson PE (ed) African-American Christianity: essays in history. University of California Press, Berkeley, pp 47–79

Webster N (1844) An American dictionary of the English language: exhibiting the origin, orthography, pronunciation, and definition of words. Harper, New York

Webster N (1913) Webster's revised unabridged dictionary. G. & C. Merriam Co., Springfield

Webster's new international dictionary of the English language (1959) Webster's new international dictionary of the English language. 2d ed., unabridged. Utilizing all the experience and resources of more than one hundred years of Merriam-Webster dictionaries. William Allan Neilson

Weiss CD (1948) Sesame, with emphasis on an unusual use in Louisiana. Herbarist 14:43–45

Whitelaw RNS, Levkoff AF (1976) Charleston come hell or high water. R. L. Bryan Company, Columbia

Wightman OS, Cate MD (1955) Early days of coastal Georgia. Fort Frederica Association, St. Simons Island

Williams RM (1832) From the Genesee farmer. *Bene* plant. Repub Star Gen Advertiser 33(33):2

Williamson J (1817) Medical and miscellaneous observations relative to the West India Islands, 2 vols. A. Smellie, Edinburgh

Williamson J (1975) Useful plants of Malawi. Revised and extended. University of Malawi, Zomba, Malawi

Willis OR (1894) *Pedalineae*. In: A practical flora. American Book Company, New York/Cincinnati/Chicago, pp 195–197

Wilson MT (1964) Peaceful integration: the owner's adoption of his slaves' food. J Negro Hist 49:116–127

Wilson CR, Ferris W (1989) Encyclopedia of Southern culture. University of North Carolina Press, Chapel Hill

Wood PH (1974) Black majority: negroes in colonial South Carolina from 1670 through the Stono Rebellion. Knopf, New York

Wood B (1995) Women's work, men's work: the informal slave economies of lowcountry Georgia. University of Georgia Press, Athens

Wood HC, Bache F (1858) The dispensatory of the United States of America, 11th edn. Lippincott, Philadelphia

Wright W (1787) An account of the medical plants growing in Jamaica. Lond Med J 8(3):282

Writers' Program of the Work Projects Administration in the State of South Carolina (1941) South Carolina; a guide to the Palmetto state. Oxford University Press, New York

Wunderlin RP (1982) Guide to the vascular plants of central Florida. University Presses of Florida, Tampa/St. Petersburg/Ft. Myers/Sarasota

Wunderlin RP, Hansen BF (2008) Atlas of Florida vascular plants. http://florida.plantatlas.usf.edu/

Yearbook of the U.S. Department of Agriculture (1895) Sesame. U.S. Government Printing Office, Washington, DC, pp 197–198

Yentsch AE (1994) West African women, foods, and cultural values. In: A Chesapeake family and their slaves: a study in historical archaeology. Cambridge University Press, Cambridge/New York, pp 196–215

Zabadi A (1970) Qamus al Muhit. Issa al Babi al Halabi, Cairo

Zuhri, Muhammad ibn Abi Bakr, Hajj Sadiq, Muhammad [12th c.] (1968) Kitab al-jarafiyah. al-Mahad al-Faransi bi-Dimashq, Damascus

Part II
Handicrafts and Crafters

Chapter 5
By the Rivers of Babylon: The Lowcountry Basket in Slavery and Freedom

Dale Rosengarten

Abstract This chapter traces the evolution of coiled basketry from its arrival in Carolina, most likely with the first shiploads of African captives, through the twentieth century, when it became the region's emblematic folk art. In the crucible of slavery, basket-making traditions from many places across Africa were forged into a single creole style. Rice winnowing baskets, called "fanners," were commonly made from bulrush sewn with thin splints of white oak or with strips peeled from the stem of the saw palmetto. Essential to the processing of Carolina's staple crop, these hand tools were mass produced by enslaved Africans beginning at least as early as 1690. Once established in South Carolina, the African-American tradition of Lowcountry basketry spread to Georgia, North Carolina, Florida, Alabama, Mississippi, and even Bermuda and the Caicos Islands, as planters moved to new territories with their seasoned rice hands.

What we know about plantation basketry comes mainly from accounts written by masters and their hirelings, such as estate inventories, overseers' records, and runaway slave ads. Basket shapes, however, also tell a tale. Double and triple baskets recall Kongo stacked and stepped-lid forms, and Moses baskets conjure up both the trickster Moses, escaping from Pharaoh in a bulrush basket, and the patriarch, leading his people out of slavery.

Keywords Coiled basketry • Africa • South Carolina • Rice • Plantation slavery • Bulrush • Sweetgrass

---

A version of this essay appeared as a chapter in *Grass Roots: African Origin of an American Art*, edited by Dale Rosengarten, Theodore Rosengarten, and Enid Schildkrout (Museum for African Art 2008).

D. Rosengarten, Ph.D. (✉)
Marlene and Nathan Addlestone Library, Special Collections, College of Charleston,
66 George Street, Charleston, SC 29424, USA
e-mail: rosengartend@cofc.edu

R. Voeks and J. Rashford (eds.), *African Ethnobotany in the Americas*,
DOI 10.1007/978-1-4614-0836-9_5, © Springer Science+Business Media New York 2013

By the rivers of Babylon
There we sat down
And there we wept
When we remembered Zion.

When the wicked carried us away in captivity
Requiring of us a song
Now how shall we sing the Lord's song
In a strange land.

—Adapted from Psalm 137

As they marched from the hinterland to coastal forts and slave ships that would take them across the Atlantic, captive Africans balanced large, coiled baskets on their heads, filled with provisions for the journey. Some number of these baskets likely made the terrible Middle Passage with the people who "toted" them. Arriving on the coast of Carolina, as the British colony was known, many Africans would have noticed similarities with the landscapes they remembered from home. Furthermore, the semitropical vegetation, vast salt marshes, and varieties of palm all offered abundant plant material with which to make baskets.

The Carolina Lowcountry is a country of grasses. From the edge of the high ground, which may be only inches above sea level, salt marsh stretches to the horizon. Along the creeks that drain the marsh grows smooth cordgrass (*Spartina alterniflora* Loisel.), a tough, broad-leafed, green grass equipped with salt glands and specialized cells that allow it to live in soil watered by the tides. Black rush, or bulrush (*Juncus roemerianus* Scheele), occupies slightly higher ground. Bulrush's fiercely pointed tip has earned it the nickname "needlegrass." Its long, round, hollow shafts can tolerate low levels of salinity and withstand intense sun, strong winds, and salt spray. Just out of reach of the tides, in flats between the dunes, or in the transitional zone where marsh meets woods, grow sweetgrass (*Muhlenbergia sericea* (Michx.) P. M. Peterson; syn. *M. filipes*, *M. capillaris* var. *filipes*), salt hay (*Spartina patens* (Aiton.) Muhl.), broomstraw (*Andropogon* spp.), sea oats (*Uniola paniculata* L.), and seaside goldenrod (*Solidago sempervirens* Michx.). Here, southern red cedar (*Juniperus virginiana* var. *silicicola* (Small) A. E. Murray), live oak (*Quercus virginiana* Mill.), cabbage palm (*Sabal palmetto* (Walter) Lodd. ex Schult. & Schult.f.), yaupon (*Ilex vomitoria* Aiton), and wax myrtle (*Myrica cerifera* L.) have pinned down the sand and stopped it from shifting (Porcher and Rayner 2001).

Native Americans found uses for many of these plants. By the end of the seventeenth century, European settlers and their African slaves had adopted Indian practices and had adapted their own Old World customs and tastes to the New World flora. They learned to utilize salt hay to pasture livestock. They made brushes from the fiber of the cabbage palm and ate the heart of the palm as a delicacy. They harvested trees and shrubs for building materials. From longleaf pine (*Pinus palustris* Mill.), they extracted tar to preserve ships' riggings, pitch to caulk wooden boats, and later, spirits of turpentine to make paint and varnish (Porcher 1995, 73–74, 12–13).

In agricultural societies around the world, the baskets people make depend largely on the crops they grow and the basket-making materials that are available.

Fig. 5.1 Rebecca Green winnowing rice, St. Helena Island, SC, ca. 1909 (Photo: Leigh Richmond Miner. Penn School Collection/ Southern Historical Collection, University of North Carolina at Chapel Hill)

On plantations across the American South, Africans and their descendants fabricated a variety of baskets from a diversity of natural fibers. On Lowcountry plantations, two basket types predominated. Using splints of white oak (*Quercus alba* L.), basket makers—usually men—wove large sturdy hampers for harvesting corn and cotton, market baskets with cross-handles, and a range of other forms. Along the rice-growing coast, where Carolina gold was earning its nickname, the escalating demand for winnowing baskets drove up production of coiled "fanners"—wide trays made generally of bulrush sewn with thin oak splints or with strips peeled from the stem of the saw palmetto (*Serenoa repens* (W. Bartram) Small) (Rosengarten 2011, 45–47) (Fig. 5.1).

Lowcountry coiled baskets varied from river to river and between mainland plantations and the Sea Islands, but the range of differences was narrow. Rather than emerge from a long period of experimentation and competition among techniques and forms, African-American coiled basketry developed rapidly and spread up and down the coast as a single creole tradition.

Carolina was a crucible of many cultures. "In no other American colony," wrote geographer D. W. Meinig, "were Europeans, Africans, and Indians so equally and intricately involved during the critical experimental period of working out ways of living in a new plantation region. Nor was this vital process simply an encounter among three peoples," he goes on to say, but a confluence of Europeans from several countries, Indians from various tribes, and Africans from many parts of a vast continent, some of whom were "seasoned" by years in the West Indies (Meinig 1986, 176). Most of these groups converging in the Americas had living traditions of coiled basketry. Europeans from grain-growing regions, for example, made baskets from the straw of their cereal crops, whether rye, wheat, barley, or oats. While this tradition persisted in German settlements of the mid-Atlantic colonies and western North Carolina, white settlers in South Carolina did not make coiled baskets themselves but could readily appreciate the utility of the African-inspired coiled work basket.

As plantation agriculture spread, people kept moving—north, south, west, and even east—and the tradition of coiled basketry that had taken hold in the Lowcountry migrated with them. Rice production expanded into North Carolina and Georgia in the early eighteenth century, as South Carolina planters resettled crews of experienced "hands" to break the new ground and plant the crop (Clifton 1973, 365; House 1954, 20–21). By the mid-1840s, rice operations had extended down the coast of Georgia and into northern Florida and the range of the basket followed suit (Fig. 5.2). Indeed, coiled basketry continues to be made on Sapelo Island, Georgia, where disciples of the island's renowned practitioner, Allen Green (1907–1999), pursue what might be regarded as a hybrid form, using both marsh grass (most likely *Spartina patens*) and sweetgrass, sewn with strips from the stems of saw palmetto (Rosengarten et al. 2008, 256n.24).

Evidence indicates that the Lowcountry tradition moved west into Alabama and Mississippi with convoys of enslaved workers from South Carolina plantations who were relocated to what was then a new frontier (Rosengarten 1997, 282–85, 288–91). Even earlier, refugees from the Lowcountry may have carried coiled basketry to the Bahamas, where some of the planters from South Carolina and Georgia who had sided with England fled after the American Revolution, taking their slaves with them. To this day, on the Caicos Islands southeast of the Bahamas, descendents of this forced migration continue to make coiled grass baskets similar to South Carolina work (Crook 2006).

Rice and Baskets

No Lowcountry baskets from the seventeenth century have survived, nor have scholars found written references to baskets in Carolina before 1700. A historian interested in basketry is handicapped by the perishable nature of the vegetable fibers from which it is made, and by the view of the basket as a humble object, useful but

Fig. 5.2 Map of rivers and Sea Islands of the Lowcountry, with selected rice plantations and towns (Map: Linda Florio. Museum for African Art, New York)

replaceable. For the eighteenth century, documentation of basket production is sparse but informative. Baskets occasionally are mentioned in planters' wills, estate inventories, account books, and overseers' reports where they are described simply as tools and commodities. Fanners, for example, were listed among the personal effects of French Protestant settler Noah Serre on May 18, 1730, and of Carolina-born Joseph Wilkinson on June 10, 1745 (Day 1978, 21).

The earliest surviving artifacts of coiled basketry date to the eighteenth century. A basket fragment from the Revolutionary War era, now on permanent display at

Fig. 5.3 Covered work basket, SC, mid-eighteenth century (Private collection. Photo: Karin Willis. Museum for African Art, New York)

The Charleston Museum, was excavated from the bottom of a privy at the Heyward-Washington House in Charleston (E. Herold, May 10, 1985, personal communication). The new owner of a large, intact work basket that has been in her family for generations found a letter from her grandmother, written in 1955, stating that the artifact was 200 years old. Analysis suggests that a manufacture date of 1755 is a real possibility which, if accurate, would make the straight-sided piece the oldest Lowcountry coiled basket known today (Fig. 5.3).

Baskets were commonly traded between plantations. A Savannah River rice planter recorded in 1774 various amounts "pade for baskets," ranging from 3 shillings to 8 pounds 9 shillings (William Gibbons' Accounts, 1765–1782). Knowing how to make a basket added to the value of a slave. Though not as highly esteemed as competence in carpentry, cooperage, blacksmithing, or dressmaking, the ability to make baskets could be cited as a selling point. "A Negro man" described in a Charleston newspaper in 1791 as "a good jobbing carpenter and an excellent basket maker" was "sold for no fault … but that of having a sore leg" (*Gazette and Advertiser*, February 15, 1791). In 1856, the Reverend C. C. Jones of Savannah described an individual he offered for sale as a "good field hand, basket-maker, and handy at jobs." Jones asked eight hundred dollars for the man, a figure, he told his son, that his wife thought "too low" (Myers 1972, 243–44).[1]

Planters' and overseers' records from the nineteenth century provide details about baskets as tools and their place in the plantation routine. Two or three weeks might be set aside for making baskets before the harvest began. The overseer on Argyle Plantation, in the Savannah River basin, reported two hands "cutting oak for baskets" in 1828 and four hands "sent for rushes" in 1830. The men spent 3 days at the task. The next year the overseer needed more baskets—perhaps the rice crop looked exceptionally good—and for 2 weeks in August, he assigned two workers to making them (James Potter plantation journal, 1828–1831).

[1] For additional references to basket makers on Lowcountry plantations, see Morgan (1998), 232–34, esp. n. 47.

Basket making sometimes stretched into the winter. In February 1836, Charles Drayton II reminded his son and namesake, who managed a plantation on the Satilla River in Georgia, to "have rushes and oak got that you might have some baskets made while you are on the place." Three months later, he instructed him again: "leave strict orders about having baskets made during the summer and fall, and also a quantity of rushes must be laid up for winters work" (Letters dated February 26 and May 12, 1836. Drayton Papers).

After the rice was harvested, it was threshed with a two-handled flail and the stalks were shaken to separate the seed heads from the stems and leaves. Next, the grain was milled or pounded in a mortar with a two-ended wooden pestle to crack the husk, then winnowed by throwing it into the air or dropping it from one basket held high over another, allowing the wind to blow away the chaff. In harvest time, fanner baskets would be issued by the dozens. On Argyle Plantation, beginning in November, more than 50 field hands were engaged in "thrashing [threshing] and winnowing rice." These men and women appear only as numbers on the daily entry but are listed by name under the heading "Disbursement of tools and baskets" at the end of the overseer's report.

By the middle of the eighteenth century, planters had begun moving their operations from the inland swamps to the rivers. The idea was to increase the acreage under cultivation, boost production of the grain, and save labor by having the water do the work. A system of embankments and trunks regulated the flow of tide-driven fresh water onto the impounded fields, transforming each rice plantation into what one planter admiringly called a "huge hydraulic machine" (Quoted in Doar 1936, 8). Not only would water be used to irrigate the crop but also to cultivate, weed, kill pests, hold up the heavy panicles of rice just before the harvest, and drive the threshing machines and mills.

Like the coiled basket, the technology for harnessing the energy of the tides had precedents in Africa and required skill and diligence beyond the brute labor of constructing and maintaining the embankments. Taming the rivers resulted in great profits for plantation owners but no increase in leisure for the workers. Throughout the eighteenth century, laborious hand techniques continued to be used to process ever larger crops of rice. Though the Lowcountry climate and soil were "admirably suited" to the grain, Dr. E. Elliott declared in *DeBow's Review* in 1851, "the planters encountered incredible difficulty in preparing, or dressing, the rice for market." For the first hundred years, Elliott continued, "the grain was milled, or dressed, partly by hand and partly by animal power. But the processes were imperfect—very tedious, very destructive to the laborer, and very exhausting to the animal power" (Quoted in Doar 1936, 18; see Porcher and Judd forthcoming 2013).

As early as 1691, Peter Jacob Guerard patented "a Pendulum Engine," hoping to improve upon the mortar and pestle. Though reputed to perform "much better and in lesse time and labour huske rice" than any other device, Guerard's invention did not catch on (Salley 1967, 69n.2). A "pecker" machine driven by oxen was used until after the Revolution (Doar 1908, 11). Yet on every plantation, workers continued to pound rice by hand—"a very hard and severe operation," wrote botanist Alexander Garden in 1755, "as each Slave is tasked at Seven Mortars for One Day and each Mortar Contains three pecks of Rice" (Quoted in Wood 1974, 79).

Fig. 5.4 The threshing floor with a winnowing-house, Alice Ravenel Huger Smith, ca. 1935. Watercolor on paperboard (Gibbes Museum of Art/Carolina Art Association)

Apart from threshing floors or winnowing houses (Fig. 5.4)—elevated platforms designed to enhance the effect of the wind in blowing away the chaff—no mechanical advances in processing rice were introduced until 1787 when Jonathan Lucas built a successful pounding mill at Peachtree Plantation on the Santee River. Powered by water released from a reservoir, Lucas's mill was the first to employ millstones to remove the hulls of the precious grain. Within 4 years, he "had developed a tide powered mill and, by 1817, one driven by steam." Pounding mills marked a great step forward in rice production. Their impact had "virtually the same significance for the rice kingdom as Eli Whitney's gin for cotton" (Clifton 1978, xxv). Around 1830, a mechanized threshing machine was devised, and an era of prosperity ensued.

Yet the new inventions never replaced the flail, mortar and pestle, and coiled basket. Only the wealthiest planters could manage to build threshing barns and pounding mills, and among those who could afford to mechanize, some chose not to. Nathaniel Heyward, for instance, "the greatest of all rice planters, … long continued to have his crops threshed by hand, saying that if it were done by machines his darkies would have no winter work" (Phillips 1966, 249). Even on rice estates equipped with threshing machines, fanner baskets continued to be used to winnow provision crops and seed grain, and coiled grass baskets of various types

Fig. 5.5 Double basket, Wilcox County, AL, ca. 1850 (McKissick Museum, University of South Carolina. Photo: Karin Willis. Museum for African Art, New York)

remained essential tools in the fields, slave quarters, storerooms, and Big Houses of Lowcountry plantations.

Because large quantities of fanner baskets were produced year after year, many examples have survived in barn lofts and attics and some have made their way into museum collections. Other coiled forms from the plantation era that have been preserved include vegetable baskets, covered work baskets, trays and hot plates, bowl-shaped baskets, as well as a few examples of double or triple sewing baskets, described in detail by Santee River rice planter David Doar (Doar 1936, 33–34). These intricate antebellum forms, reminiscent of traditional Kongo stacked and stepped-lid baskets, are still made by older sewers in Mt. Pleasant, South Carolina, though only on commission (Rosengarten et al. 2008, 59–61, 70–77) (Fig. 5.5).

Coiled basketry thrived along tidal rivers where it was critical to rice processing, yet it also took hold on the Sea Islands where the supply of fresh water was insufficient to support commercial cultivation of rice and long-staple cotton (*Gossypium barbadense* L.) was the chief export crop. Here, rice was grown only for local consumption, for people who came from rice regions of Africa were said to "languish without their favorite food" (House 1954, 53).[2] Also, as in Africa, fanner baskets were used to process other grains and vegetables, such as corn and peas. On the sandy barrier islands off the South Atlantic coast, the dietary preferences of African workers, rather than the profit motives of the masters, drove the production of coiled baskets.

On the mainland, most rush baskets were bound with oak splints. Employing European tools to work a New World material, Lowcountry basket makers split thin, narrow weavers from white oak boles by the same laborious process used in making oak baskets.[3] In place of oak splints, some mainland and all Sea Island basket makers sewed their rows with strips from the stem of the saw palmetto. The way a basket

[2] See also Littlefield (1981), Carney (2001), and Wright (1981), 263.

[3] For a technical description of splitting oak, see Wigginton (1972), 115–18. White oak basket making as a cross-cultural phenomenon is discussed in D. Rosengarten (1997), 296–97, D. Rosengarten (2011), 47, 49.

Fig. 5.6 Santee River knot, bulrush sewn with white oak, from a fanner basket made at the Wedge Plantation, Charleston County, SC, ca. 1890 (Private collection. Photo: Karin Willis. Museum for African Art, New York)

Fig. 5.7 Sea Island knot, bulrush sewn with saw palmetto, from a fanner basket made by Jannie Cohen, Hilton Head Island, SC, ca. 1988 (Private collection. Photo: Karin Willis. Museum for African Art, New York)

is started indicates where it was made. For example, basket makers on mainland plantations usually tied a knot in the middle of the bundle of rush. They would bring together both ends of the bundle, bend it around the knot, and stitch it in place. Sea Island sewers would coil the long length of the grass bundle around a knot tied at one end. After a second or third row was sewn, the free end of the grass that emerged from the center was cut flush with the plane of the basket bottom, facing up as the basket lay in the sewer's lap (Figs. 5.6 and 5.7).

To prepare saw palmetto binders, the outer layer or skin of the stalk was peeled off with a knife and laid across the knees, the smooth, green side resting on a burlap sack or other heavy fabric. Then, the pithy core of the stem was scraped away and the skin split lengthwise into narrow weavers. These were dried briefly in the sun and stored in pails of fresh water to keep them supple. To add a binder, whether of oak or palmetto "butt," the new strip was tied onto the preceding stitch and passed through a stitch in the underlying row. These interlaced stitches curve out from the center of the basket back toward the rim.

Eyewitnesses give us occasional glimpses of the basket makers themselves—gathering rushes, sewing baskets, wearing rush hats. "Jacob and Jim getting stuff for baskets," a Berkeley County, South Carolina, planter wrote—and indexed—in his Journal on August 27, 1836. "Jacob was occupied 3 weeks in making baskets" (Thomas Walter Peyre plantation journal, 1834–1859). Basket making might be assigned to workers no longer fit for field labor. In March 1846, Thomas B. Chaplin of St. Helena Island "put old May to making baskets, 2 a week"—though what kind of baskets Chaplin does not say (Rosengarten 1986, 403). Workers also produced baskets on their own time, either for sale or for personal use. "After one or two o'clock," wrote Daniel Elliott Huger Smith, recalling his youth at Smithfield, a Combahee River rice plantation, "the hands had the rest of the day to themselves and could work their own fields and gardens, or idle at their will. Many of them were expert basket-makers for which on every plantation there was a demand" (Smith 1936, 71).

Though Smith viewed his childhood at Smithfield through a rose-colored lens, his observation nevertheless points to the autonomy that rice hands, working on the task system, were able to turn to their advantage. Under the task system, field workers spent the first part of the day tending the cash crop. When they completed their assigned task—a full task in hoeing, for example, was a quarter of an acre—they could use the remainder of the day for their own pursuits. After slavery, laborers held tenaciously in their contracts with land owners to the privileges they believed were inherent in the antebellum system.

Moses in the Bulrushes

Even in conditions of perpetual servitude, workers were able to use basket-making skills as bargaining chips, to make baskets for their own use, or to sell or carry goods to exchange in the open market or the underground economy.[4] Basketry hats literally protected and financed enslaved people who tried to run away. At the same time, however, the hats and caps runaways wore might be mentioned for purposes of identification in advertisements for fugitive slaves. An ad that appeared in the *South Carolina and American General Gazette* on April 10, 1777, less than a year after the signing of the Declaration of Independence, announced that "a stout black Negro Man, near 6 feet high, named Joe," had absconded. He "had on when he went away, a blue broad cloth coat, a pair of brown German serge breeches, and a rush hat." In 1779, "a Stout young Negro Fellow named Adam," who recently had been brought from Georgia, ran off wearing "a hat made of the same Materials used for Baskets." Would-be slave catchers might also recognize him from the remarkably detailed description of his clothing, down to the color of the buttons on his coat: "a new brown Waistcoat and Trowsers, the Trowsers long and Waistcoat short, double breasted, with yellow Metal Buttons on one Side, except the upper Button which was white: He had with him an old white Coat, and a Pair of Leather Breeches, both much torn" (Windley 1983, 499, 364; for ads referring to straw hats, see 68, 89).

[4] For more on plantation underground markets, see T. Rosengarten (1986), 162.

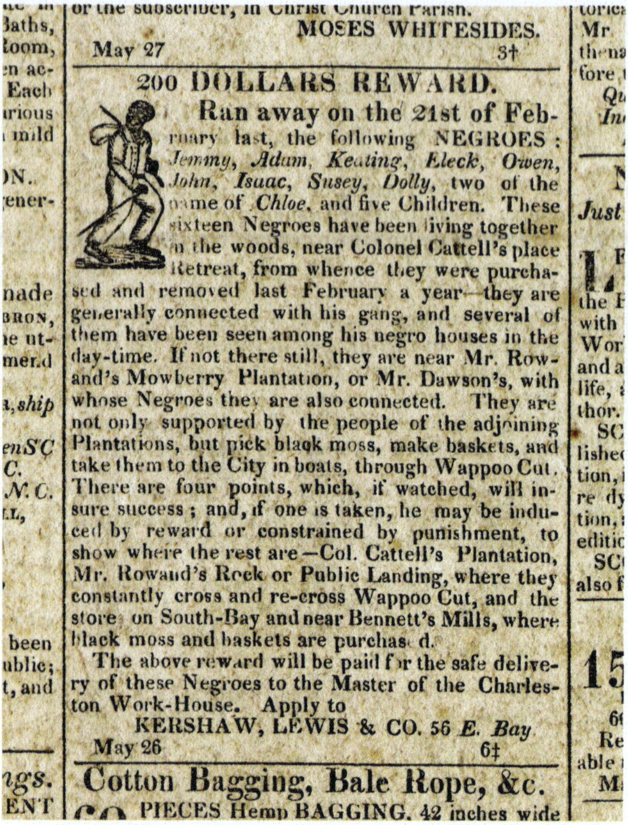

of the subscriber, in Christ Church Parish.
MOSES WHITESIDES.
May 27 3t

200 DOLLARS REWARD.

Ran away on the 21st of February last, the following NEGROES: *Jemmy*, *Adam*, *Keating*, *Eleck*, *Owen*, *John*, *Isaac*, *Susey*, *Dolly*, two of the name of *Chloe*, and five Children. These sixteen Negroes have been living together in the woods, near Colonel Cattell's place Retreat, from whence they were purchased and removed last February a year—they are generally connected with his gang, and several of them have been seen among his negro houses in the day-time. If not there still, they are near Mr. Rowand's Mowberry Plantation, or Mr. Dawson's, with whose Negroes they are also connected. They are not only supported by the people of the adjoining Plantations, but pick black moss, make baskets, and take them to the City in boats, through Wappoo Cut. There are four points, which, if watched, will insure success; and, if one is taken, he may be induced by reward or constrained by punishment, to show where the rest are—Col. Cattell's Plantation, Mr. Rowand's Rock or Public Landing, where they constantly cross and re-cross Wappoo Cut, and the store on South-Bay and near Bennett's Mills, where black moss and baskets are purchased.

The above reward will be paid for the safe delivery of these Negroes to the Master of the Charleston Work-House. Apply to
KERSHAW, LEWIS & CO. 56 *E. Bay*
May 26 6‡

Cotton Bagging, Bale Rope, &c.
60 PIECES Hemp BAGGING. 42 inches wide

Fig. 5.8 Runaway slave advertisement, *Charleston Courier*, May 28, 1825 (The Charleston Library Society)

Baskets themselves could be agents of liberation. Sixteen runaways were able to stay at large by gathering black moss, making baskets, and carrying their handcrafts to town in boats, according to an ad that appeared in the *Charleston Courier* on May 28, 1825 (Johnson 1981, 436) (Fig. 5.8). A basketry boat made headlines in the summer of 1864 when the northern press learned of Jack Frowers's escape to the Union side of Port Royal Sound in a vessel he had made from coarse grass twisted into a rope and "bound round, or, as the sailors would term it, 'served' with other grass." "I got an axe and knife—no matter how—" Frowers told a reporter, "and I cut a lot of rushes, and went to work in the woods and made this boat." He spent two days weaving, then caulked the boat with cotton and waterproofed it inside and out with pitch collected "by cutting into a tree and catching the gum." He fastened three pieces of pine wood to the floor and nailed an old shutter to the bottom, perhaps as a keel. When the boat was ready, Frowers hid out one more day, then paddled to freedom, "too glad to get away" (*Anti-Slavery Reporter* 1864 in Blassingame 1977, 449–54). In the fanfare that followed his escape, the straw boat was sent to

Fig. 5.9 Moses basket, Joseph Foreman, Jr., SC, 2007. (McKissick Museum, University of South Carolina. Photo: Karin Willis. Museum for African Art, New York)

Massachusetts Governor Andrews, who presented it to the Prince Hall Grand Lodge of Free Masons in the South End in Boston, along with Frowers's account of the adventure (Botume 1968, 178–80).

The Civil War marks a decisive shift in written accounts of Lowcountry baskets. While planters had tended to take a strictly pragmatic view, abolitionists and Yankee missionaries were more likely to cite basket making as evidence of the capacity of people struggling to be free.[5] Elizabeth Hyde Botume, appointed in 1864 by the New England Freedman's Aid Society as "a teacher of freed people at Beaufort, South Carolina," mentions baskets three times in her published narrative, first, as a gift from one of the "contrabands," second, as an agricultural tool, and, third, as the boat that carried Jack Frowers to freedom (Botume 1968, 124, 135, 178–80).

Frowers's story is a variant on the biblical account of the Hebrews' exodus from Egypt. The bulrush basket as a vessel of liberation was deeply embedded in the religious imagination of Lowcountry slaves. On the Combahee River in the years before the Civil War, a verse chanted at "shoutings" told the story this way: "Moses in de bull-rushes fas' asleep!/Playing possum in de two bushel basket!" (Quoted in Smith 1936, 76). Indeed, Moses plays a central role in African-American folklore—both as the patriarch who delivers his people from slavery and as the trickster who escapes in a bulrush basket. He is alluded to all the time by contemporary basket sewers because of his association with bulrush. Calling bulrush "the blessed straw … a holy straw," Mary Jane Bennett, who was a midwife and a Pentecostal minister as well as a basket sewer, recalled how Pharaoh's daughter saved the "blessed baby" in a basket made of bulrush (Fig. 5.9). "And that's how come," Mrs. Bennett explained, "the bulrush become to be in the basket" (Bennett 1987). "They made a basket for Moses, when Moses was born," another sewer reported. "Made it of palmetto and 'rushel'—bulrush—just like we use. Some say it started from Africa but I think it started before that, in Moses' time" (Wright 1992).

In November 1861, just 7 months after the Confederate bombardment of Fort Sumter, northern forces occupied the Sea Islands around Port Royal Sound in the

[5] For an excellent guide to published and unpublished materials generated by Yankee missionaries in South Carolina during and after the Civil War, see Rose (1964), 409–33.

Fig. 5.10 Planting sweet potatoes on James Hopkinson's plantation, Edisto Island, SC, 1862 (Photo: Henry P. Moore. The New-York Historical Society)

southeastern corner of South Carolina. Plantation owners fled their estates, leaving behind their personal property and their slaves, whom the Federals designated as "contrabands of war." Almost immediately, an experiment was launched by philanthropic northerners, under the direction of the Federal military, to prepare the newly freed people for participation in post–Civil War society. The goals of the effort, which came to be known as the Port Royal Experiment, were economic independence, literacy, and civil rights.

On April 8, 1862, 5 months after Edisto Island was occupied by Union troops, a number of freed men and women were photographed planting sweet potatoes on James Hopkinson's plantation. Some of the men are wearing Union army caps (Fig. 5.10). Men and women are seated around a pile of potatoes, sorting slips to plant. In the foreground sits a wide vegetable basket and barely visible behind it lays another smaller coiled basket—the first known photograph of Lowcountry forms.

Three months after this historic photograph was taken, Union troops were pulled from Edisto and ordered to Virginia to save General George B. McClellan's army. As a consequence, 1,600 freed people, "with their pigs, chickens, and personal effects," were loaded aboard flatboats and taken to St. Helena Island (Rose 1964, 182–83). Brigadier General Rufus Saxton was charged with organizing production on the liberated plantations. Concerned that the refugees would fall prey to "the evils of idleness" and convinced that there was "a profitable market at the North for the products of their industry," Saxton proposed "to introduce the manufacture of *Rush Baskets* and *Cedar Tubs.*" He ordered his superintendents to "immediately set [people] to work in gathering and curing rushes" and to assign "the more skilful" to make them into baskets (Saxton 1862).

Saxton, a committed abolitionist, thought the baskets and tubs would "find a ready sale" because they were so useful and also because "of the laudable object sought to be attained by their manufacture." Making baskets would keep the freed people busy, demonstrate their willingness to work for themselves, and raise money to support the "experiment" in liberation. Thus, at the moment of transition from slavery to freedom, baskets were still being mass produced under the direction of white overseers—now in the guise of Federal superintendents. In the migrations that accompanied the war, the Edisto basket, along with its makers, had been transplanted 25 miles to St. Helena, where the two island traditions met. Over the next century, under the tutelage of the northern teachers and missionaries who carried the Port Royal Experiment forward, the St. Helena basket, now of mixed lineage, would take on a life of its own as a commodity and an icon of Africa in America.

At the end of the Civil War, rice plantation facilities lay in ruins. Emancipated workers wanted to farm for themselves and demanded land they felt they had earned. Meanwhile, planters lacked the resources to rebuild. Even before 1860, competition from European colonies in southeast Asia and from large tracts in Louisiana and Texas, where rice growers could take advantage of firmer soils and employ animal power and machinery in the fields, had begun to undermine agricultural prices (House 1954, 78; Childs 1999, 41–42; Adas 1974, 31, 58–60). When cultivation in the Lowcountry resumed, it was on a new basis. Postwar plantations frequently were leased or run by managers employed by absentee landlords. The labor was done by free men and women working under contract for wages, for a share of the crop, for land, or some combination of these. Under a common contractual arrangement called the "two-day system," workers exchanged labor for a few acres of land, living quarters, and firewood.

Free and mobile, many black families were able to purchase 10–50-acre tracts. The plantation that had been the milieu of the Lowcountry basket for almost 200 years was now supplanted by the family farm. Half the acreage might be planted in cotton, the other half in foodstuffs, including rice, corn, sweet potatoes, millet, and a host of vegetables (Day 1983, 14–15).[6] No longer under compulsion to mass produce fanners, basket sewers continued to make agricultural work baskets and household forms to meet their needs. In the shift from coerced labor to family farming, growers relied on their old hand tools to process their crops. No longer did schooners dock at plantation levees to load barrels of rice destined for waiting mills, yet rice-growing on a small scale continued. The last rice mill shut down in South Carolina long before the last mortar, pestle, flail, and fanner basket were laid to rest (Fig. 5.11).

[6] For a history of "providence rice" production in the pine belt community of Mars Bluff, see Vernon (1993).

Fig. 5.11 Abraham Herriot pounding rice, Sandy Island, SC, ca. 1930 (Photo: Frank G. Tarbox, Jr. Brookgreen Gardens)

A Lost Civilization and a New Century

Fanners and vegetable baskets continued to be an everyday sight on the streets of Charleston and Savannah. Farmers brought their produce to market in ferries, small boats, or wagons. Seafood was typically transported in sacks or split oak baskets with handles. Vegetables, as a rule, were carried in coiled rush baskets balanced on the head. By the turn of the twentieth century, the figure of a stately woman with a head-tote basket had become a staple in the repertory of a group of artists and writers whose work would collectively come to be known as the Charleston Renaissance. Fueled by nostalgia for a lost civilization, sons and daughters of antebellum rice planters memorialized the world of their parents' childhood in paintings, prints, drawings, prose, poetry, and drama.

Elizabeth W. Allston Pringle, daughter of a South Carolina governor and herself one of the last rice planters on the Waccamaw Neck in Georgetown County, recounted her mother's memories of an upcountry girlhood in *Chronicles of Chicora Wood.* Adele Petigru Allston had admired the beauty of the handwork produced on Badwell Plantation in the South Carolina Piedmont, where African-born "Maum Maria made wonderful baskets and wove beautiful rugs from the rushes that grew along Long Cane Creek." In Pringle's account, Maria had come from royalty in

Fig. 5.12 One or two hands in the barn-yard, Alice Ravenel Huger Smith (From Patience Pennington [Elizabeth W. Alston Pringle], *A Woman Rice Planter*, 1913)

Africa, been taken in battle by an enemy tribe, and sold to a slave ship with her kinsmen, Tom and Prince. These "three quite remarkable, tall, fine-looking, and very intelligent Africans," wrote Pringle, "occupied an important place in my mother's recollections of early childhood" (Pringle 1940, 53–54).

Pringle's first volume, *A Woman Rice Planter*, was a collection of articles that had appeared in the *New York Sun* between 1904 and 1907. Published in 1913 under the pseudonym Patience Pennington, the book was illustrated with line drawings by Alice Ravenel Huger Smith (Fig. 5.12). Smith would go on to paint 30 watercolors that today are the most frequently consulted views of the Carolina rice plantation. Her images faithfully rendered her father's recollections of his boyhood spent at Smithfield Plantation (Severens, 1993, 16). Colorfully dressed, compliant, basket-toting, dark-skinned people evoke a harmonious world of productive laborers working under the benign gaze of genteel masters, their wives, and children (Vlach 2002, 151–78). Suffused with pastel sweetness, her paintings nevertheless provide information about material culture and technology, including basket making and the functions of baskets.

In *The Threshing Floor* (Fig. 5.4), a queue of women and girls with fanners full of rice waits to climb the stairs to the winnowing platform. Fanners also appear in unexpected places. Smith's *Plantation Street or Settlement* (Fig. 5.13) shows Negro babies "sunning in blanket-padded 'fanner baskets,' supervised by a 'mauma,' or nurse." "On fine days," wrote her father, "there might be seen on the open ground" in front of the sickhouse, which doubled as a nursery, "a large number of 'fanner-baskets,' and on each basket a folded blanket, and on each blanket a baby" (Smith 1936, 62).

In *Loading the Rice Schooner* (Fig. 5.14), Smith creates a classical frieze of women with huge baskets on their heads walking along an embankment toward an awaiting ship. "It was always a great pleasure to watch the loading of one of these schooners," wrote D. E. Huger Smith, recalling the line of women "boarding the

Fig. 5.13 The plantation street or settlement, Alice Ravenel Huger Smith, ca. 1935. Watercolor on paperboard (Gibbes Museum of Art/Carolina Art Association)

Fig. 5.14 Loading a rice schooner, Alice Ravenel Huger Smith, ca. 1935. Watercolor on paperboard (Gibbes Museum of Art/Carolina Art Association)

Fig. 5.15 In the shadow of St. Michael's, Charleston, SC, Elizabeth O'Neill Verner, ca. 1928. Etching on paper (Gibbes Museum of Art/Carolina Art Association, on loan from the City of Charleston)

schooner by one plank and going ashore by the other. Close to the open hatchway stood the ten-bushel tub with the captain or mate on one side and the overseer or key-keeper on the other. As each darkey passed the tub," the elder Smith noted, as if describing an automaton pouring gold into his coffers, "she would tilt the basket and pour the golden rice into the tub" (Smith 1936, 63).

Alice Smith may be unique among artists of the Charleston Renaissance in the attention she paid to baskets as tools on Carolina rice plantations, but many other resident and visiting painters and printmakers who wanted to capture the essence of the Lowcountry made use of the coiled grass basket as a prop or symbol. In countless street scenes etched by Elizabeth O'Neill Verner, for example, a faraway figure of a street vendor balancing a head-tote basket provides scale and perspective and signals unmistakably where the picture comes from (Fig. 5.15).

Nothing produced in the Lowcountry has had the power to evoke South Carolina's African inheritance over such a long period of time as has the coiled basket. "One of the most interesting sights of Charleston," touted the caption of a turn-of-the-century postcard, "is the negro vegetable vendor." On the streets of the city, visitors

Fig. 5.16 Vegetable woman, Charleston, SC, ca. 1910 (Photo: George W. Johnson. Gibbes Museum of Art/Carolina Art Association)

could see vendors "of both sexes and of all ages … bearing on their heads enormous round baskets of produce," and hear them "singing their wares in quaint dialect cries that sound to the unfamiliar ear like utterances of a foreign race" (Female Vegetable Vendors, postcard, 1895). In the competition for tourist dollars that were going to watering holes in the Caribbean and pueblos in the Southwest, Charleston promoters were making the point that their city had colorful people, too (Figs. 5.16, 5.17, and 5.18).[7]

The image of the vegetable vendor, however, did not sit well with everyone. When the South Carolina Interstate and West Indian Exposition of 1901 unveiled a sculpture in front of the "Negro Building," educated blacks protested that the work represented "the Negro in a too menial guise." The piece, called the *Negro Group*

[7] On the relation between local crafts, infrastructure, and expanding regional tourism in the Southwest at the turn of the century, see Weigle and Babcock (1996).

Fig. 5.17 Woman with two baskets, Rutledge Avenue, Charleston, SC, 1940s (The Charleston Museum)

(Fig. 5.19), was designed by Mexican-American sculptor Charles Albert Lopez. The original conception by the Exposition's Architect-in-Chief, Bradford Lee Gilbert, called for a cluster of "Negro types" around a "central bale of Cotton or Tobacco," with an "old-time negro," a younger man with a banjo, and "a 'piccaninni'… with a piece of watermelon" on one side, and the figure of Booker T. Washington, "the representative negro of his race today," on the other (Gilbert to McCrady 1901).

Lopez dropped the "piccaninni" but did model a figure after Washington—a muscular man holding a plow and leaning on an anvil. The crowning symbol of the group was a stately black woman, her chin held high, with a basket balanced on top of her head. The basket itself was a mélange of two craft traditions. Shaped like a coiled vegetable basket, its texture appears to be that of splint work.

The sculpture "created quite a commotion among the so-called 'new' negroes of Charleston" (*The Exposition* 1901, 470). The real Booker T. Washington was chief commissioner of the executive committee governing the Exposition's Negro Department. He tried to mollify the dissidents "by pointing out that most members

Fig. 5.18 Female vegetable vendors, postcard, Charleston, SC, 1895 (Collection of Jessica B. Harris)

of the race were tillers of the soil or in menial positions." But the objectors would not relent. Dr. Thomas E. Miller, president of South Carolina State College (the State Colored College at Orangeburg) and an assistant commissioner, urged Washington to look especially at the figure of the boy, whose expression was that of a "blank idiot" (Quoted in Smyth 1987, 218). The protest succeeded in having the sculpture removed from the Negro Building but not from the Exposition. The *Negro Group* was relocated to a more central site in the Court of Palaces, where it was documented by the Exposition's official photographer, G. W. Johnson.[8]

[8] A haberdasher and umbrella manufacturer turned professional photographer, George Washington Johnson (1858–1934) documented the 1886 earthquake and the 1901–1902 South Carolina Interstate and West Indian Exposition. His extensive photo archives of Charleston buildings and street scenes, plantation gardens, and Lowcountry landscapes belong to the Gibbes Museum of Art, Charleston, SC. See also Bostick and Crooks (2005).

Fig. 5.19 The *Negro group*, sculpture by Charles Albert Lopez, at the South Carolina Interstate and West Indian Exposition, with the photographer's daughters Ruth and Lois Johnson, Charleston, SC, ca. 1901–02 (Photo: George W. Johnson. Gibbes Museum of Art/Carolina Art Association)

While Charleston boosters were busy promoting the city as a prime tourist destination, artists Smith and Verner and other luminaries of the Charleston Renaissance, including Alfred Hutty (Fig. 5.20), Anna Heyward Taylor (Fig. 5.21), Prentiss Taylor, and Birge Harrison, established a Lowcountry art scene comparable to such movements as the Harlem Renaissance in New York City and the Southern Literary Renaissance centered in Nashville. Like the "Agrarians," as the Tennessee group was called, the Charleston Renaissance cold-shouldered artists of African descent. In one particularly shameful episode in 1926, The Charleston Museum first offered and then withdrew an invitation to Edwin Augustus Harleston, the city's "sole black academically trained painter," to exhibit his paintings and drawings (Donaldson 2006, 32; McDanniel 2006, 24–25).

Yet for visiting white artists, Charleston's salons and societies, exhibitions and publications made the city a mecca (Severens 1998, 6–7, 11–13). Edward Hopper completed several drawings and 11 watercolors on a visit to Charleston in 1929. Jules Pascin, George Biddle, Rockwell Kent, Charles Smith, Childe Hassam, and Andrew Wyeth all were drawn to the Lowcountry. Photographer Doris Ulmann,

Fig. 5.20 Pounding rice, Alfred Hutty, ca. 1930. Drypoint (South Carolina State Museum Collection)

Fig. 5.21 Cutting rice, Anna Heyward Taylor, 1937. Linoleum block print on rice paper (McKissick Museum, University of South Carolina)

Fig. 5.22 Pounding rice, Sandy Island, SC, ca. 1937 (Photo: Bayard Wootten. Brookgreen Gardens)

working in collaboration with novelist Julia Peterkin, made soft-focus portraits for a volume called *Roll, Jordan, Roll.* Recording what appeared to them as both exotic and distinctly American, Bayard Wootten, Berenice Abbot, Walker Evans, Marion Post Wolcott, and, a decade later, Robert Frank photographed South Carolina street scenes and country roads[9] (Figs. 5.22, 5.23, and 5.24).

Travelers made their way to the Lowcountry by sea, rail, and automobile. In August 1929, a bridge opened over the Cooper River between Charleston and Mt. Pleasant, launching new prospects for basket makers in the vicinity of Boone Hall Plantation. During the plantation era, it was the men who were the chief producers of bulrush "work" baskets. Now, women emerged as makers and sellers of sweetgrass "show" baskets—decorative containers made from sun-dried sweetgrass artfully accented with the rust-colored needles of longleaf pine and sewn together with strips from the unopened fronds of the *Sabal palmetto*, South Carolina's state tree.

[9] For a chronology of events of the Charleston Renaissance and visits of artists from elsewhere, see Severens (1998), 185–92.

Fig. 5.23 Untitled [Baptism in the swamp], 1929–1933 (Photo: Doris Ulmann. Snite Museum of Art, Gift of Milly Kaeser in Honor of Fritz Kaeser)

With the demand for agricultural containers in decline, baskets continued to be made from a different kind of necessity—the need to earn income in a cash economy. The sewers' proximity to Highway 17, then called Route 40, positioned them to take advantage of increased traffic on the major coastal artery. In a stroke of inventiveness, basket makers devised a way to reach this moving market. They began displaying their wares on the edge of the road.

The first basket stands were nothing more than a chair or overturned box, but they quickly evolved into a novel structure consisting of posts or saplings set upright in the ground with thin strips of wood nailed horizontally between them. Nails served as pegs for hanging baskets (Fig. 5.25). On these simple structures extended families invested their labors and their economic hopes. A basket maker could be her own boss, at a time when cleaning house for white people was one of the few alternatives. Behind the scenes, men and women harvested the sweetgrass, pine

Fig. 5.24 A Fourth of July celebration, St. Helena's Island, SC, 1939 (Photo: Marion Post Wolcott. Library of Congress)

Fig. 5.25 Basket stand on Highway 17, Mt. Pleasant, SC, 1938 (Photo: Bluford Muir. Francis Marion National Forest, USDA Forest Service)

Fig. 5.26 Joseph Foreman, Sr., carrying a bundle of bulrush, McClellanville, SC, 1985. (Photo: Theodore Rosengarten)

needles, and palmetto that went into a basket. Children learned to sew at the knee of their mothers and grandmothers and contributed to the family economy by making basket bottoms. The basket stand displayed the end products of their collaboration. Adapting traditional forms and inventing new ones, sewers developed a large repertory of functional pieces—bread trays and table mats, flower and fruit baskets, shopping bags, knitting baskets, thermos bottles or wine coolers, ring trays, and wastepaper baskets. Since the reintroduction of bulrush around 1970 (Fig. 5.26), there is a noticeable trend toward big sculptural forms that feature elaborate surface decoration and meticulous stitching. Now, as then, the stands appear flimsy and makeshift when unattended, but in use, they hold lively exhibitions of original art.

References

Adas M (1974) The Burma delta: economic development and social change on an Asian rice frontier, 1852–1941. University of Wisconsin Press, Madison

Anti-Slavery Reporter (1864) In: Blassingame J (ed) Slave testimony: two centuries of letters, speeches, interviews, and autobiographies, 1977, Series 3, vol 12 (1 September). Louisiana State University Press, Baton Rouge, pp 202–203

Bennett MJ (1987) Videotaped interview in "Row upon row: sea grass baskets of the South Carolina Lowcountry," educators' kit, McKissick Museum, University of South Carolina, Columbia, SC. Additional audio and video interviews with M. J. Bennett are available at: http://www.digitaltraditions.net/html/Artists_Page.cfm. Accessed 25 July 2011

Bostick DW, Crooks DJ Jr (2005) On the eve of the Charleston Renaissance: the George W. Johnson photographs. Joggling Board Press, Charleston

Botume EH (1968) First days amongst the contrabands. Arno Press, New York (Originally published 1893)
Carney JA (2001) Black rice: the African origins of rice cultivation in the Americas. Harvard University Press, Cambridge
Childs AR (ed) (1999) Rice planter and sportsman: the recollections of J. Motte Alston, 1821–1909. University of South Carolina Press, Columbia
Clifton JM (1973) Golden grains of white: rice planting on the lower cape fear. NC Hist Rev 50(4):365–393
Clifton JM (ed) (1978) Life and labor on Argyle Island: letters and documents of a Savannah River rice plantation, 1833–1867. Beehive Press, Savannah
Crook R (2006) "Sweet grass and fanner grass: baskets and connections between the Georgia/Carolina coast and the Bahamas islands," public lecture delivered on 2 February at the College of Charleston, Charleston, SC
Day G (1978) Afro-Carolinian art, towards the history of a southern expressive tradition. Contemporary Art/Southeast 1(5):19–24
Day KY (aka K. P. Young) (1983) My family is me: women's kin networks and social power in a black sea island community. PhD dissertation, Rutgers University
Doar D (1908) A sketch of the agricultural society of St. James, Santee, South Carolina and an address on the traditions and reminiscences of the Parish delivered before the Society on 4th of July 1908. Calder-Fladger Co., Charleston
Doar D (1936) Rice and rice planting in the South Carolina Low Country. The Charleston Museum, Charleston
Donaldson SV (2006) Edwin A. Harleston and Charleston's racial politics of preservation. In: Campbell LP (ed) Edwin Augustus Harleston: artist and activist in a changing era. Avery Research Center for African American History and Culture, Charleston
Drayton Papers (1701–2004) College of Charleston Libraries, Charleston, SC
Female vegetable vendors, an every morning scene on the streets of Charleston, S.C. (postcard), 1895. Collection of Jessica B. Harris
Gazette and Advertiser (1791) Charleston, SC. South Caroliniana Library, University of South Carolina, Columbia, SC, 15 Feb
Gilbert BL to McCrady E (1901) Esq., Charleston SC. Edward McCrady Papers, file #28-307-4, South Carolina Historical Society, Charleston, SC, 18 May
House AV (ed) (1954) Planter management and capitalism in ante-bellum Georgia. Columbia University Press, New York
James Potter plantation journal (1828–1831) Ms 630, Georgia Historical Society, Savannah, GA
Johnson MP (1981) Runaway slaves and the slave communities in South Carolina, 1799 to 1830. William Mary Q, 3rd series 38 July:418–441
Littlefield DC (1981) Rice and slaves: ethnicity and the slave trade. Louisiana State University Press, Baton Rouge
McDanniel MA (2006) Edwin Augustus Harleston: envisioning the talented tenth. In: Campbell LP (ed) Edwin Augustus Harleston: artist and activist in a changing era. Avery Research Center for African American History and Culture, Charleston
Meinig DW (1986) The shaping of America: a geographical perspective on 500 years of history, vol 1: Atlantic America, 1492–1800. Yale University Press, New Haven
Morgan PD (1998) Slave counterpoint: black culture in the eighteenth-century Chesapeake and Lowcountry. University of North Carolina Press, Chapel Hill
Myers RM (ed) (1972) The children of pride: a true story of Georgia and the Civil War. Yale University Press, New Haven
Phillips UB (1966) American Negro slavery: a survey of the supply, employment, and control of Negro labor as determined by the plantation regime. Louisiana State University Press, Baton Rouge (Originally published 1918)
Porcher RD (1995) Wildflowers of the Carolina Lowcountry and Lower Pee Dee. University of South Carolina Press, Columbia

Porcher RD, Judd WR (2013) The market preparation of Carolina rice. University of South Carolina Press, Columbia
Porcher RD, Rayner DA (2001) A guide to the wildflowers of South Carolina. University of South Carolina Press, Columbia
Pringle EWA (1940) Chronicles of Chicora wood. Christopher Publishing House, Boston (Originally published 1922)
Pringle EWA (Patience Pennington) (1992) A woman rice planter. University of South Carolina Press, Columbia (Originally published 1913)
Rose WL (1964) Rehearsal for reconstruction: the Port Royal experiment. Oxford University Press, London
Rosengarten D (1997) Social origins of the African-American Lowcountry basket. PhD dissertation, Harvard University, Cambridge, MA
Rosengarten D (2011) Baskets and basket making. In: Katz-Hyman MB, Rice KS (eds) World of a slave: encyclopedia of the material life of slaves in the United States, vol 1. Greenwood Press, Santa Barbara, pp 45–50
Rosengarten D, Rosengarten T, Schildkrout E (2008) Grass roots: African origins of an American art. Museum for African Art, New York
Rosengarten T (1986) Tombee: portrait of a cotton planter, with the journal of Thomas B. Chaplin, 1822–1890. William Morrow & Co, New York
Salley AS Jr (ed) (1967) Narratives of early Carolina, 1650–1708. Barnes & Noble, New York (Originally published 1911)
Saxton R (1862) Circular by the order of Brig. Gen'l, Commanding, Beaufort, SC. Collection of John Martin Davis, Jr., Beaufort, SC, 2 Aug
Severens MR (1993) Alice Ravenel Huger Smith: an artist, a place, and a time. Carolina Art Association/Gibbes Museum of Art, Charleston
Severens MR (1998) The Charleston Renaissance. Saraland Press, Spartanburg
Smith DEH (1936) A plantation boyhood. In: Smith ARH, Sass HR (eds) A Carolina rice plantation of the fifties. William Morrow & Co, New York
Smyth WD (1987) Blacks and the South Carolina Exposition and West Indian Interstate. South Carolina Historical Magazine 88(4):211–219
The Exposition (1901) Charleston, SC. South Caroliniana Library, Columbia, SC
Thomas Walter Peyre plantation journal (1834–1859) Ms 34/466, South Carolina Historical Society, Charleston, SC
Vernon AW (1993) African Americans at Mars Bluff, South Carolina. Louisiana State University Press, Baton Rouge
Vlach JM (2002) The planter's prospect: privilege and slavery in plantation paintings. University of North Carolina Press, Chapel Hill
Weigle M, Babcock BA (1996) The great Southwest of the Fred Harvey Company and the Santa Fe Railroad. The Heard Museum, Phoenix
Wigginton E (ed) (1972) The foxfire book. Doubleday & Co., Garden City
William Gibbons' Accounts for the estate of Benjamin Farley (1765–1782) item 433, Telfair Family papers, Ms 793, Georgia Historical Society, Savannah, GA
Windley LA, comp. (1983) Runaway slave advertisements: a documentary history from the 1730s to 1790, vol. 3: South Carolina. Greenwood Press, Westport
Wood PH (1974) Black majority: Negroes in colonial South Carolina from 1670 through the Stono Rebellion. Knopf, New York
Wright JL Jr (1981) The only land they knew. The Free Press, New York
Wright L (1992) Interview by Dale Rosengarten, Mt. Pleasant, SC., 11 February. Folklife Resource Center, McKissick Museum, University of South Carolina, Columbia, SC

Chapter 6
Gathering, Buying, and Growing Sweetgrass (*Muhlenbergia sericea*): Urbanization and Social Networking in the Sweetgrass Basket-Making Industry of Lowcountry South Carolina

Patrick T. Hurley, Brian Grabbatin, Cari Goetcheus, and Angela Halfacre

Abstract Despite the visibility of natural resource use and access for indigenous and rural peoples elsewhere, less attention is paid to the ways that development patterns interrupt non-timber forest products (NTFPs) and gathering practices by people living in urbanizing landscapes of the United States. Using a case study from Lowcountry South Carolina, we examine how urbanization has altered the political-ecological relationships that characterize gathering practices in greater Mt. Pleasant, a rapidly urbanizing area within the Charleston-North Charleston Metropolitan area. We draw on grounded visualization—an analytical method that integrates qualitative and geographic information systems (GIS) data—to examine the ways that residential and commercial development has altered collecting sites and practices associated with sweetgrass (*Muhlenbergia sericea* [Michx.] P.M. Peterson) and three other plant materials used in basket-making. Our analysis focuses on the ecological changes and shifts in property regimes that result; we detail the strategies basket-makers have developed to maintain access to sweetgrass and other raw materials. This research highlights how land development patterns have disrupted historic

P.T. Hurley, Ph.D., M.S. (✉)
Environmental Studies Program, Ursinus College, 601 East Main Street, P.O. Box 1000, Collegeville, PA 19426, USA
e-mail: phurley@ursinus.edu

B. Grabbatin, Ph.D.
Department of Geography, University of Kentucky, 1457 Patterson Office Tower, Lexington, KY 40506, USA

C. Goetcheus, M.H.P., B.L.A.
College of Environment and Design, University of Georgia , 108 Denmark Hall, Athens, GA 30602, USA

A. Halfacre, Ph.D., M.A.
Department of Political Science and Sustainability Science, Furman University, 3300 Poinsett Hwy, Greenville, SC 29613, USA

R. Voeks and J. Rashford (eds.), *African Ethnobotany in the Americas*, DOI 10.1007/978-1-4614-0836-9_6, © Springer Science+Business Media New York 2013 gathering practices, namely, by changing the distribution of plants, altering the conditions of access to these species, and reconfiguring the social networking that takes place to ensure the survival of this distinctive art form.

Keywords African-American ethnobotany • Non-timber forest products • Suburban natural resource use • North American political ecology

Introduction

There is growing interest in non-timber forest products (NTFPs) and the gathering practices that characterize diverse NTFPs in the United States (Jones et al. 2002; Emery et al. 2003; Robbins et al. 2008). Much of this research has focused on gathering in the Pacific Northwest and on conflicts between timber management strategies and nontraditional resource users (see Love and Jones 2001; McLain 2002; Robbins et al. 2008). This research also documents harvesting of products, such as fungi (Molina et al. 1993; Liegel et al. 1998), fruits and nuts (Freed 2001), and other plant materials (Butler et al. 2005; Lynch and McLain 2003) in national forests (Emery and Pierce 2005) as well as the collection of plants on private properties (Price and Kindscher 2007; Ginger et al. 2011), and wild foods (Palmer 2000; Cordell et al. 2004) from environments such as city streets (Jahnige 1999; Gabriel 2006). Among these individuals, several minority populations who gather materials for their household economies have come into conflict with other groups over gathering rights at NFTP sites (Richards and Creasy 1996; Emery 1999, 2002; Brown and Marin-Hernandez 2000; Freed 2001; Jarosz and Lawson 2002; Ginger et al. 2011). More recent research suggests that gathering is a practice not necessarily limited to a particular ethnic or racial group (Robbins et al. 2008), and that the same person may collect NTFPs for both economic and noneconomic needs (Emery 1999; Jones et al. 2002; Emery et al. 2003).

Global issues of access for indigenous and rural peoples to culturally important natural resources are well recognized and studied in many parts of the so-called developing world (Zimmerer and Bassett 2003; Robbins 2003) and recently in some developed areas as well (Wehi and Wehi 2010, Ginger et al. 2012). Scholars are also paying greater attention to the ways that changing social *and* biophysical factors influence access to natural resources associated with NTFPs (Ginger et al. 2011). Still, less attention is paid to the ways that development patterns associated with urbanization disrupt NTFP species distributions, uses, and practices. Indeed, increased urbanization in both the United States and around the world is transforming ecosystems (as well as the relationships between people and plants; Theobald 2004; Johnson and Klemens 2005; Head and Muir 2006; Ballard and Jones 2011). The extension of infrastructure and services (e.g., sewer and water) that facilitate urbanization has altered land cover. As denser residential settlements expand, they affect local sites and regional landscapes. Relatively high-density suburban (i.e., ~2–8 dwelling units per acre) and lower-density exurban forms of land transformation (i.e., between

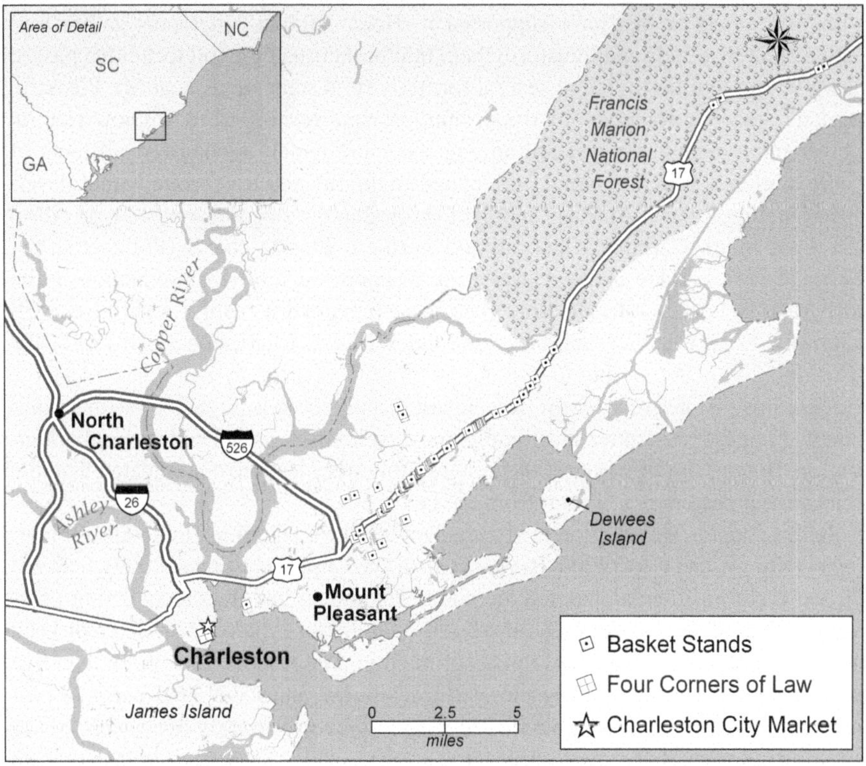

Fig. 6.1 Location of study area within Charleston-North Charleston area, SC, and SE United States. Also pictured are sales locations for basket stands (Figure created by Jeff Levy, University of Kentucky)

1 dwelling unit/acre and 1 dwelling unit/16.2 ha [40 acres]) fragment existing forests, fields, and other nonurban vegetative land covers (Theobald 2004; Johnson and Klemens 2005; Lambin and Geist 2006), dramatically altering ecological systems, including plant and animal species dynamics and ecological functions (see, e.g., Pickett et al. 2001). For example, lawns simplify habitat structure and species composition (Robbins and Birkenholz 2003), and landowners often plant ornamental shrubs, trees, and flowers that differ from the vegetative cover and composition being replaced (Johnson and Klemens 2005; Head and Muir 2007). Similarly, exurbanization (i.e., urban migration to rural areas) alters the plant and animal communities that live there (Theobald 2004; Lambin and Geist 2006). But how does urbanization alter the ecological conditions that support, and social conditions that facilitate, access to plant communities associated with NTFP gathering practices? And how do gatherers and NTFP users negotiate these changes?

Using the case of sweetgrass basket-makers in the greater Mt. Pleasant area, a rapidly urbanizing area within the Charleston-North Charleston Metropolitan area of the South Carolina Lowcountry (Fig. 6.1), we examine the ways that urbanization has altered the political-ecological relationships that characterize this NTFP and the gathering practices that have supported it. Here, African-Americans gather four native plant materials and transform them into baskets using a coil technique passed down within families living in several formerly rural settlements near Mt. Pleasant. Baskets made by forebears of these families were once used to winnow rice on Lowcountry plantations and are among the most widely recognized artifacts of *Gullah* culture, which is based on blended traditions and backgrounds that developed among enslaved Africans and their descendants along the southeastern North Carolina, South Carolina, Georgia, and northern Florida coasts (Pollitzer 2003; National Park Service 2005). Despite the disappearance of rice production in the early 1900s, baskets of different types have remained a regular sight along US Highway 17 in Mt. Pleasant and along sidewalks in Charleston since the 1940s (Coakley 2006). According to popular wisdom, supported by news and magazine articles, this distinctive art form (Rosengarten et al. 2008) is threatened with extinction by rampant residential and commercial development associated with suburbanization. But, as Hurley et al. (2008) suggest, this story is more complicated than the simple disappearance of a plant from the landscape.

In this chapter, we use grounded visualization, the integration of GIS and ethnography (Knigge and Cope 2006; Hurley et al. 2008; Cope and Elwood 2009), to detail the ways that residential and commercial development has altered collecting practices associated with sweetgrass (*Muhlenbergia sericea* [Michx] P.M. Peterson) and other plant materials employed in basket-making. We describe how more than five decades of suburbanization has altered the ecological and social landscapes of basket-making—focusing on ecological changes affecting species presence/absence and shifts in property regimes influencing access—and the strategies basket-makers have devised to maintain access to sweetgrass and other raw materials. While a connection has been drawn between changing gathering practices and the livelihood of basket selling for some time (see, e.g., Hart et al. 2004; Hurley et al. 2008), knowledge about the ways basket-makers accessed specific harvesting sites and analysis of the social networks that connect artists and sellers to gatherers has been limited. This article moves beyond a general consideration of increasing problems associated with access to a direct consideration of patterns of landscape change on historic harvesting sites and adaptive strategies for maintaining supplies from new areas.

Study Context and Methods

Study Background

Enslaved Africans made baskets on South Carolina rice plantations, and their descendents have depended on the basket-making tradition for various cultural uses and economic values. Baskets were one of several technologies adapted to North American environments by enslaved Africans, which helped fuel the early economic growth of Charleston, SC, and its rural surroundings (Carney 2001). The tradition survived as a utilitarian craft during the lean years following emancipation and was transformed into a commodity in the early twentieth century by sewers who promoted the historical, cultural, and aesthetic value to Charleston area tourists and residents (Coakley 2006; Rosengarten 1992, 2008). Despite the disappearance of rice fields and rural landscapes along the South Carolina coast, basket sales remain an important part of the household economy, and basket-making is an essential element of the identities of many Lowcountry African-Americans (National Park Service 2005).

Sweetgrass (*M. sericea*) is the signature resource used in basket-making, but the craft relies on four plant materials that historically occurred in local woodlands and wetlands. Basket-makers bind and sew together rows of sweetgrass (*M. sericea*),[1] longleaf pine needles (*Pinus palustris* P. Mill), and black rush cuttings (*Juncus roemerianus* G. Scheele) with strips made from palmetto fronds (*Sabal palmetto* [Walter] Lodd. ex Schult. & Schult.f) (Twining 1978; Fig. 7a). Sweetgrass typically constitutes the bulk of each basket and currently is the most difficult material to obtain (Hart et al. 2004; Grabbatin 2008). These resources were acquired through personal or familial collecting efforts or purchased from gatherers from one or more of the rural settlements outside the historically small town of Mt. Pleasant (Derby 1980; Hart et al. 2004). Sweetgrass (*M. sericea*) occurs naturally in clumps landward of the second dune line at beaches and in the interface between wetlands and woodlands (Ohlandt 1992; Gustafson and Peterson 2007). Based on herbaria records, published literature, and earlier interviews with basket-makers, it appears that sweetgrass was found in the recent past in a number of places in the greater North Charleston-Charleston metro area (Fig. 6.2; Hurley et al. 2008).

Today, however, the town of Mt. Pleasant is South Carolina's fifth largest city and one of the state's fastest growing metropolitan areas (US Census 2009). Over the past two decades, the greater Mt. Pleasant area has experienced rapid residential growth and geographic expansion through annexation. This growth, however, has been shaped by a geography of historic rural African-American settlements and largely white-owned farmlands (Fig. 6.3). Growth was particularly strong following Hurricane Hugo in 1989. Between 1990 and 2008, the town's population increased from 30,108 to an estimated 65,472 (US Census 2009). Likewise, the town's area has grown to encompass 108.5 km^2 (41.9 square miles) (US Census 2000). Residential subdivisions increasingly encroach on the area's historic rural African-American settlements and, in at least one case, have replaced an entire community (Fig. 6.2 inset). These new subdivisions include numerous upscale planned communities, many with gated entries and extensive walls or fences, and panoramic views of tidal marshlands. Others feature golf courses and/or waterfront parcels with private docks. Thus, development potentially occurs in the places where sweetgrass is most likely to have grown naturally (Hurley et al. 2008). It also has resulted in dramatic changes in patterns of land use and land ownership.

---

[1] Sweetgrass is from the perennial grass *Muhlenbergia sericea* Peterson [synonyms: *Muhlenbergia filipes*, *Muhlenbergia capillaris* var. *filipes*]. It occurs in sandy maritime habitats on barrier islands and coastal woodlands in the southeastern United States (see Gustafson and Peterson 2007 for an overview).

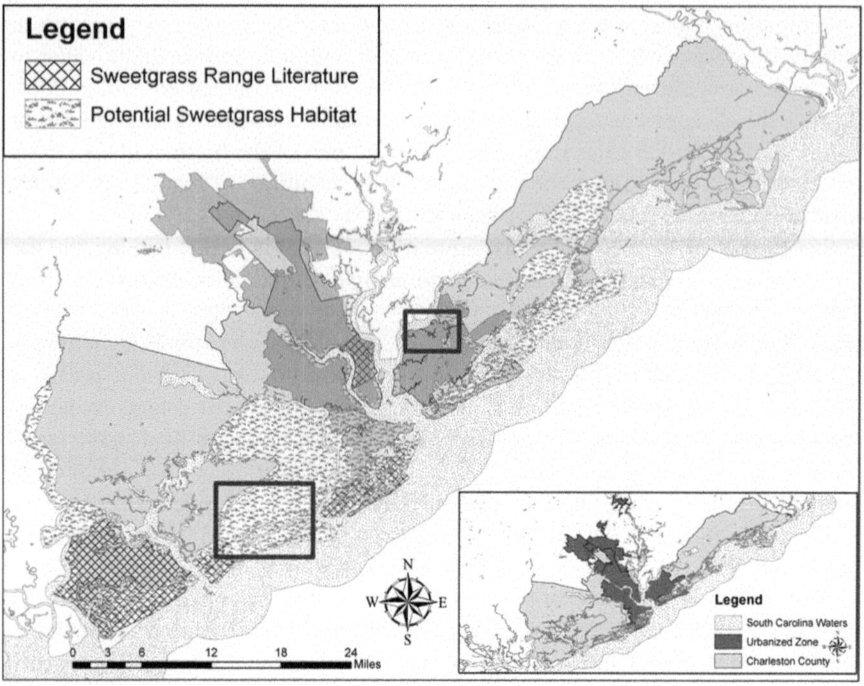

Fig. 6.2 Range of sweetgrass (see Hurley et al. 2008 for methods' discussion; figure created by Norm Levine, College of Charleston)

Growth in Mt. Pleasant also has transformed the area's socioeconomic and racial demographics. Over a century ago, African-Americans comprised roughly 70% and whites just 30% of the population in South Carolina's coastal counties (US Census 2007). By 1990, the percentage of African-Americans in Charleston County and Mt. Pleasant had declined to 34.9 and 15.7%, respectively (US Census 1990). Today, African-Americans comprise 31.9% of Charleston County residents (US Census 2006) and just 7.3% of the population in Mt. Pleasant (US Census 2000). With this increase in development and associated sociodemographic change, Mt. Pleasant land values have increased dramatically, leading to gentrification pressures in historic African-American settlements (Hurley et al. 2008). While gentrification pressures have important implications for sewing and selling baskets (see Hurley and Halfacre 2011; Hurley et al. 2008; Grabbatin 2008), we focus here on the consequences of development for *the gathering* of the key natural resources used in baskets.

Methods

Our project uses grounded visualization, a methodological and analytical approach that integrates ethnographic data with GIS (Knigge and Cope 2006; Hurley et al. 2008; Cope and Elwood 2009), to interrogate the social and ecological processes that

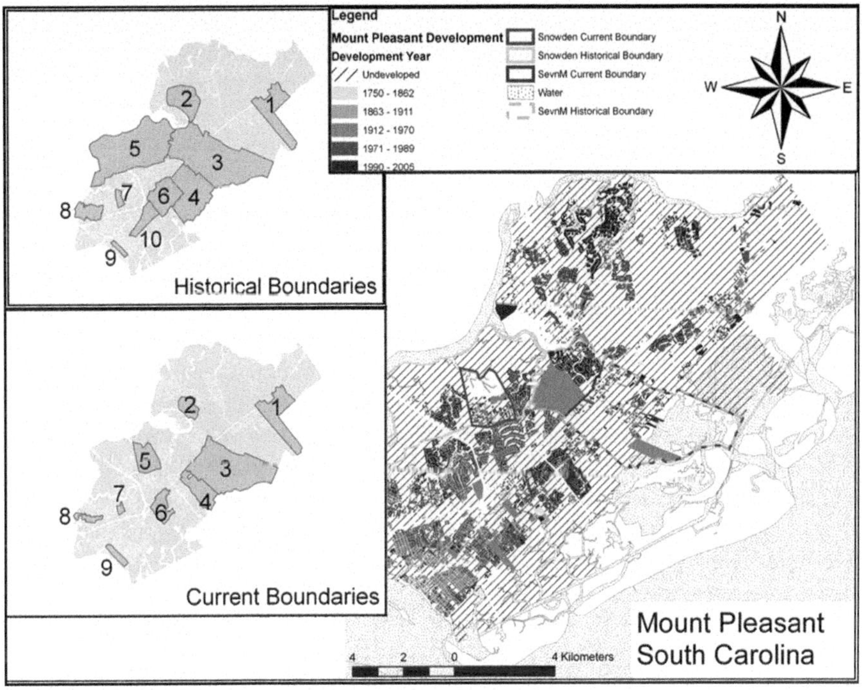

Fig. 6.3 Location of ten African-American settlements within Greater Mt. Pleasant area and residential development trends. *Insets* highlight the comparative cases of Seven Mile/Hamlin and Snowden (Figure created by Norm Levine, College of Charleston)

accompany land-use and land-cover change (Fox et al. 2002; Madsen and Adrianson 2004). First, we interviewed 26 basket-makers (out of 84 contacted) from the four main sales locations in the area (Fig. 6.1). We asked basket-makers how they dealt with the impacts of urbanization on their access to their raw materials, including asking informants to describe where, how, or from whom they had acquired raw materials in the past and how resource acquisition had changed (Grabbatin et al. 2011). These interviews built on data previously gathered between June 2002 and January 2003 for an earlier study of basket-makers' views of and roles in sweetgrass management (Hart 2003; Hart et al. 2004) and 23 interviews (60 invited) in 2003 about collecting strategies. Fifteen respondents from the 2002 study participated in 2003. We used convenience and snowball sampling techniques in each study, with most interviewees identified through visits to basket stand. Given our focus on documenting historic gathering sites, our study does not rely on species vouchers.

Second, between August 2007 and March 2009, we conducted oral history interviews, community tours, community presentations, and field mapping exercises with residents from ten African-American settlements (Fig. 6.3 inset).[2] In our oral history interviews with basket-makers and elderly residents, questions focused on

[2] One of the ten communities in Fig. 6.3 was estimated by the researchers based on historical air photos.

mapping historic resource sites and current settlement boundaries as well as the ways local human-environment interactions have changed over time. We specifically asked residents how basket-making, with particular emphasis on gathering strategies, had changed.

Third, we use historic aerials, ranging from 1946 to 2006, to explore the spatial configuration of historic land uses and land covers associated with sites mentioned during interviews and tours, which identified the location of places where sweetgrass was gathered in the past, to provide further context about the socioeconomic connections among settlements. Using this information, we examine the ways that changes in gathering practices relate to the trajectories of change in land cover, land use, residential and commercial buildings, and property regimes that can be observed in or inferred from air photos overlain with contemporary property boundaries.

Finally, these methods were supplemented with document analysis, participant observation, and attendance at key community events. Over several years, we attended local government meetings, visited basket-makers in their homes, and surveyed numerous newer local subdivisions. We analyzed government land-use documents, reviewed community planning workshops, examined subdivision governance and marketing materials, and followed reporting on basket-making and development in area newspapers and magazines.

Results

Kinship Supply Chains

Generations of basket-makers have relied, in part, on family members and friends to acquire plant material (Twining 1978; Derby 1980; Rosengarten 1986; Hart et al. 2004), but today, more than ever, harvesting has become a business for some collectors (Grabbatin et al. 2011). In our interviews, 18 basket-makers said that family members are instrumental in acquiring materials. They cited husbands, cousins, children, siblings, parents, and in-laws who help supply plant material. Further, 18 of our informants indicated that they currently collect at least one of the four materials. All but two rely on someone else to provide at least one of the materials. For example, basket-maker Vera Manigault collects most of her materials but described how various family members help her out: "my brother, he go out and get it and he give me grass and stuff, you know. My son, all of `em get the material for me, some bring palm." Sweetgrass scarcity in the surrounding area, health problems, and old age prevent many from collecting their own materials. Thus, Emily Johnson described a group of collectors as "mens that are nice enough to try to keep the craft going for different families and the sweetgrass basket weavers." In contrast, Mae Hall described her reliance on "older men who are…probably retired too and they've been doing this all of their lives… It's a way for them to make extra money." Henrietta Snype, a basket-maker who collected materials as a child, said, "[harvesting's] hard work… That's why I don't have any problem paying for it."

Even among basket-makers who say they buy materials, it is clear that networks of kinship and friendship create loyalties that determine which basket-makers are able to get grass and how much they are charged for it. For example, Robena Blake occasionally has to pay cash for her material but described the advantages of having relatives help out: "if it's relatives I don't have to pay, if it's somebody I don't know then, you know, I try to buy some sweetgrass from them." These examples indicate that family members can cut down on costs of obtaining sweetgrass and provide a safety net for basket-makers who still collect on their own. It is important to note, however, that while family and community loyalties can connect basket-makers to resources, these networks can also exclude some basket-makers. One basket-maker, who is not from Mount Pleasant and has no family ties to the area, described how she is shut out of sweetgrass distribution networks, making it difficult for her even to find collectors who will sell to her.

Changing Ecologies, Changing Property Regimes

During earlier decades, mostly prior to the onset of widespread urbanization, collection took place within several of the surveyed settlements' historic community boundaries, including Snowden, Seven Mile/Hamlin, Remley's Point, Six Mile, and Four Mile (Fig. 6.3). Importantly, our informants indicated during community tours and in their oral histories that they specifically found sweetgrass much nearer to their communities than is apparent from Fig. 6.2.

> We used to go to Boone Hall and collect sweetgrass, Brickyard—we call that place The Point. I don't know where the point is between Brickyard and Boone Hall in the back, but that's where we used to go and collect sweetgrass. *Interview with basket-maker, August 4, 2007*

Comments such as these identify places in and around African-American settlements where sweetgrass and the other materials were found and gathered (Fig. 6.4). When viewed within the historical context of land-cover and land-use change, they point out the importance of once extensive woodlands and forest habitats in areas adjacent to or surrounded black settlements characterized by cleared smallholder agricultural fields (Fig. 6.5). Indeed, this spatial relationship became a recurrent theme among informants, highlighting the extent to which current communities represent an ecological shadow of their former selves.

To better understand the historic spatial relationship between community woodlands and land uses, let us briefly consider historic land-use/cover patterns relative to the historic boundaries identified by our informants (Figs. 6.4 and 6.5). These patterns highlight the relationship of cleared land to forests and woodlands. First, nine of our ten study communities were characterized by a historic core dominated by agricultural fields, widely dispersed trees, and varying degrees of small forest clusters (Seven Mile/Hamlin in Fig. 6.4). In comparison to large fields used by white "truck" farmers (i.e., trucked to more distant urban markets), fields in rural African-American settlements are characterized by much smaller fields, landholdings, and

Fig. 6.4 Transformation of areas where sweetgrass was historically collected. Snowden community in the *upper panel*, Seven Mile/Hamlin, is shown in the *lower panel*. *Dark green* and *red boundaries* (Snowden and Seven Mile/Hamlin, respectively) indicate the historic extent of communities, while the *brighter red* and *green boundaries* highlight their current extent. Orange ovals highlight areas where sweetgrass was historically collected

potentially more diverse crops (Fig. 6.5). Second, where these communities did not directly border waterways or white-owned truck farms, they were surrounded by marsh, forested woodlands, and what appear to be areas of scrubby "unmanaged" vegetation. In some cases, these woodlands represent buffers between our study communities and nearby truck farms. Likewise, buffers sometimes separate one African-American settlement from another, while in other cases one buffer type separates the settlement from a nearby marshland or watercourse (Fig. 6.4, Seven Mile/Hamlin, and Fig. 6.5, Remley's point, respectively).

Many community woodlands were critical resource sites during earlier periods. They included diverse plant communities, such as those associated with live oak (*Quercus virginiana*) and sabal palmetto, which are characteristic of Lowcountry maritime forest. There were also wetter areas and patches of longleaf pine forests that could be found in close proximity to rural settlements where basket-makers lived. In the former, sweetgrass could be found and in the latter pine needles easily gathered. Areas of maritime forest provided quick access to palmetto. Likewise, bulrush was frequently found and easily accessed in tidal marshes either directly adjacent to communities, such as the creek in the eastern side of the Snowden

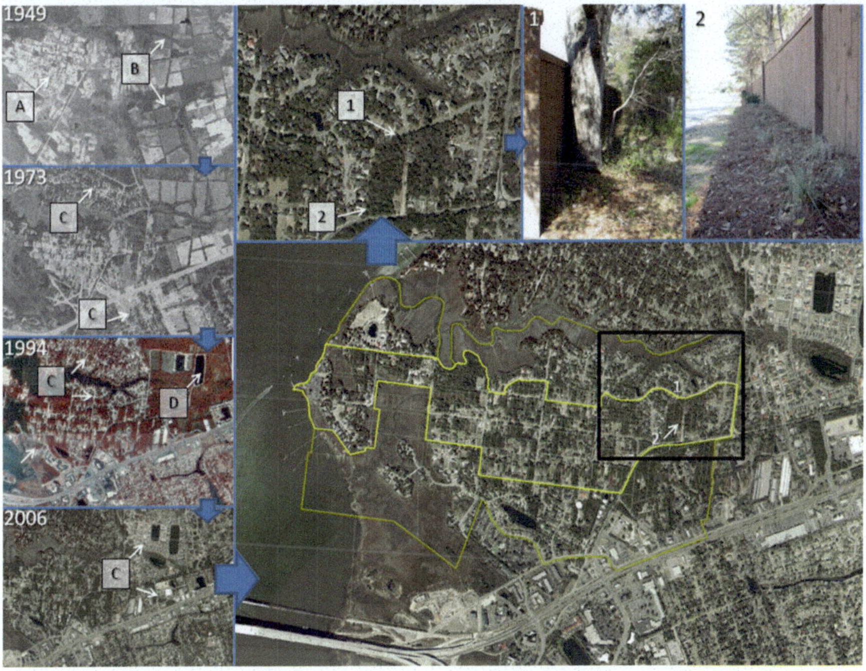

Fig. 6.5 Changing development trends in Remley's Point, including community woodlands. (*A*) Small community farm fields. (*B*) Nearby white-owned, truck farms with larger fields. (*C*) Ongoing residential development around the community. (*D*) Changes to agricultural lands include borrow-pits. (*) Community areas now cut off from nearby waterways, by development and associated fencing (*1*), while sweetgrass appears in subdivision landscaping (*2*)

community (Fig. 6.4), or easily reached through woodlands that separated the settlement from the water. Informants also reported finding sweetgrass growing in close proximity to many of the areas on truck farms where they may have worked picking crops.

While it is unclear how widespread gathering once was, our informants made it clear that basket-making families and others in these settlements also took part in collecting a number of wild medicinal plants, wildflowers for sale in local markets, as well as nuts from trees and shrubs in neighborhood yards, dispersed throughout their communities, and in the surrounding community woodlands. For example, family compounds and individual yards historically included apple, pear, and pecan trees, while muscadine grapes (*Vitis rotundifolia* Michx), wild blackberries (*Rubus cuneifolius* Pursh), huckleberries (*Gaylussacia frondosa* [L.] Torr. & A. Gray), and the seeds of the chinquapin tree (*Quercus muehlenbergii* Engelm) could be found in the nearby woods. Others talked about the collection of "life everlasting" (*species unknown*), "snake root" (*Ageratina aromatic* [L.] Spach), and other plants for use as home remedies against common colds and other ailments, while several described the collection of "magnolia blooms" (*Magnolia grandiflora* L.) and other wildflowers by "flower ladies" who would carry them to Charleston to sell them alongside baskets (see Coakley 2006).

Often community woodlands and forests may not specifically have been owned by African-Americans. Instead, these lands were owned by whites in the area—including some absentee landowners—and access to NTFP resources appears to have been an accepted local practice, including through access arrangements that were approved by the owners of particular properties or because access was not monitored. In other cases, communities believed they owned these lands. Regardless of their specific ownership, community woodlands and forests largely functioned as de facto resource commons.

Today, however, urbanization is transforming these resource commons and thus sweetgrass collection. By using contemporary and past community boundaries to examine air photos (Figs. 6.4 and 6.5), we are able to make better sense of how residential growth has altered the extent and spatial configuration of community woodlands and forested areas, as well as access to nearby marshlands. Let us briefly consider the specific cases of Snowden and Seven Mile/Hamlin, which represent divergent development patterns. A key aspect is the transition from farmland to suburban housing. For example, though their development histories are different, Snowden and Seven Mile/Hamlin share similar histories of NTFP use and a loss of access to community woodlands (Fig. 6.3). By 2006, Snowden was surrounded by predominantly upscale subdivisions (Figs. 6.3 and 6.4), while in the Seven Mile/Hamlin area expansive residential subdivision is relatively new. Here, residential developments are being built both within the core and on the margins of the community. In both Snowden and Seven Mile/Hamlin, there is a recurring theme: new housing developments are built on properties identified as woodlands that were *within* historic community boundaries and from which any number of NTFPs—not just sweetgrass and basket-making materials—were previously collected. While some harvesting still occurs in Mt. Pleasant, most harvesters have branched out into other counties, such as sites in Beaufort and Colleton, and other states, such as Georgia and Florida, to find this locally disappearing resource.

At the same time, these developments do not entirely eliminate woodlands and forests within their boundaries (Fig. 6.4). Still, the proliferation of new subdivisions often includes fences and gates that physically—or socially—separate African-Americans from places once associated with community woodlands, their resource commons, and harvesting (Hurley et al. 2008). This is the case in Remley's Point (Fig. 6.5), where one new subdivision—complete with a street named "Overseer's Retreat"—built a fence along the entire length of its boundary with the historic community (Fig. 6.5 insets). Thus, access to sweetgrass governed by new property regimes may be a more important issue than the plant's disappearance. We now turn to a discussion of current gathering practices and supply chains that illustrate how much access to NTFPs, particularly to sweetgrass, has changed in response to this emerging property regime.

Changing Plant(ing) Regimes and the Importance of Social Networking

New Niches for Sweetgrass

As Mt. Pleasant has transitioned from a largely rural to suburban area, another transformation has occurred, which complicates the NTFP story. Sweetgrass is reappearing throughout the South Carolina Lowcountry as a popular landscaping cultivar, promoted for its showy purple flower (Fig. 6.6d, f) (Hurley et al. 2008; Grabbatin et al. 2011). The proliferation of cultivated sweetgrass would seem to be a solution to the "problem" of reduced abundance of this plant, but these plantings are not necessarily accessible or useful as raw material for basket-making. The cultivars and management techniques used by landscapers often produce plants that are of marginal quality—difficult to harvest and less than ideal to work with. In one instance, 2,000 sweetgrass seedlings were planted on James Island's McLeod Plantation in an effort to "farm" sweetgrass for basket making (Nixon 1993). According to basket-makers and the Historic Charleston Foundation, the sweetgrass grew for the first few years and then died off. Harvesters raised doubts about the cultivation methods and appropriateness of the site, saying the use of fertilizer and poor soils negatively affected the quality of the grass and their ability to harvest effectively.

> The weeds started to set in [at McLeod]… [In its natural habitat] the pine needles regulate the weed overgrowth, the sweetgrass can get through 'cause it's so skinny…most other vegetation can't. So it sort of regulates itself, it don't get overgrown. *Interview with basket-maker, August 25, 2006*

More recently, basket-makers were invited to harvest sweetgrass planted along the roadways of Kiawah Island and along the connector between Spring and Callawassie Islands. At Kiawah Island, landscape plants were good enough that harvests have taken place a second time (Fig. 6.7c). However, Nakia Wigfall referred to the plants along the connector as "sweetgrass on steroids," indicating that they were generally very tough, difficult to harvest because fertilizers were applied, and thus of poor quality for use in baskets.

While grass that is planted as landscaping is typically heavily fertilized, causing the stems to become dry and brittle, often making it of little use for basket making, landscape ecologist Karl Ohlandt has restored grass to maritime wet grasslands on Dewees Island and has plans to do the same on Spring Island (Hunt 2006).

Basket-makers also are growing sweetgrass on their own properties. Ten respondents have tried to grow their own sweetgrass, but for those who have been successful, the few plants they have in their yards provide an inadequate, and often laughable, amount of materials to meet their year-round needs. For example, Elijah Ford said, "Some people got a few hills, but it takes more than a few hills if you want to stay in business… [Laughs] You know what I'm saying." Henrietta Snype referred to the "one little clump" she has growing in her yard as, "nothing to hoop and hooray about. [Laughs] Probably can't make even a pair of earrings with that, but it's

Fig. 6.6 Ecological conditions associated with sweetgrass distribution in the study area: (**a**) in the wild at the interface of a wetland and forest, (**b**) seed collection from plants within a woodland area, (**c**) as a landscaping plant at a local nursery, (**d**) planted along walkways on the Charleston peninsula, (**e**) cultivated garden in basket-maker backyard, and (**f**) ornamental in a commercial parking lot (Photos by authors, inset *A* by Julia Carter)

growing, so far." This discrepancy between the amount of sweetgrass cultivated in yards and the amount needed to make baskets reflects the disappointment of most basket-makers in their efforts to grow a meaningful supply of material for their craft.

A small number of basket-makers have successfully grown sufficient amounts of sweetgrass on their own property. For example, Barbara McCormick lives in a longleaf pine savanna on the edge of Francis Marion National Forest and has been able to grow sweetgrass in relatively large quantities. She describes her cultivation methods:

> Basket-maker: See some people they plant it and then they keep it clean.
> Interviewer: Keep it clean?
> Basket-maker: I just let mine go wild like in the woods, it's one area I don't cut it, I don't go in there and clean it out, nothing… I just let nature take its course, and mine is doing pretty good. I get me a nice size bundle last year…So my husband said he's gonna get some more plot and stick it…I said, I wish I could get the whole area planted, then I wouldn't have to travel [*Laughs*]

This basket-maker attributes her success to the fact that she is growing sweetgrass in its natural habitat (Fig. 6.6a). Other basket-makers living in communities where development patterns have severed their historical relationship with nearby

Fig. 6.7 Links between harvesting and basket-making: (**a**) bulk materials and the creation of a basket, (**b**) finished baskets on a roadside basket stand along Highway 17, (**c**) harvesting landscaped grass at Kiawah Island, (**d**) delivering pine needles to a basket-maker at her stand (Photos by authors)

woodlands maintain small cultivated gardens of sweetgrass within their yards (Fig. 6.6e). Based on community tours and interviews, this trend, however, is not widespread.

These concerns about cultivation and the usefulness of grass are compounded by concerns among scientists over the potential long-term impacts of introduced cultivars on biological diversity. Both Karl Ohlandt and Danny Gustafson have concerns about landscaped sweetgrass that is non-native, given "it can break down genetic combinations that make native plants adaptable to local environments" (Hunt 2006). In response to this concern and the wider need for resource supplies for basket-makers, there now is an effort underway with the Dewees Island Property Owners Association to collect seeds from native plants in order to propagate a supply for basket-makers and landscapers (Fig. 6.6b).

Connecting Peoples and Plants: (Re)Negotiating Access to Plants

The development boom since the 1970s, particularly during the 1990s, has transformed former plantations and farms into residential neighborhoods and resort

Table 6.1 Responses from 26 basket-maker interviews in 2006 and 2007 concerning the people and organizations who were involved in providing access to the *four* basket materials

Land manager	Number[a]
City of Charleston	7
Hunting clubs	5
Federal land managers (USFS and USFWs)	10
Private neighborhood associations	2
Private businesses	2
Friends/family	9
Unsanctioned access	2

Responses for federal land managers speak to pine needle collection in Francis Marion National Forest, where no sweetgrass is obtained. Some basket-makers decided not to share this information, and others gave more than one answer
[a]$N=26$; however, some respondents did not answer and others gave more than one answer

communities (National Parks Service 2005) reconfiguring gathering practices and strategies. Harvesters have had to renegotiate access to historic sites, negotiate access to new sites, or branch out to new locations, with varying degrees of success (Table 6.1). For example, the harvests at Kiawah Island in 2008 and 2009 discussed above (Fig. 6.7c) have reopened access to plants within a private, gated island community. In contrast, one respondent explained how access to historic harvesting areas on Seabrook Island was renegotiated but then discontinued when gathering (or gatherers) was perceived as a disturbance to life on the island. Basket-maker Elizabeth Mazyck describes her experience:

> We used to go to [Seabrook] Island and the people live in great big houses. I guess one day we started laughing and having fun—and one morning this man come out the house… I say, "Oh he must've had a bad night." [*Laughs*] So the next time we couldn't go back out there cause they don't want you in the front of their house…they stop us to go in there, and there was some nice grass, there was beautiful grass, when it dries it's golden beautiful grass. We miss that.

Other communities where grass was not harvested historically have opened their "gates" to collecting. On Dewees Island, basket-makers have gained access to sweetgrass restoration and preservation sites through the efforts of the community association and the help of the island's landscape ecologist Karl Ohlandt. Harvesting is permitted in these areas because plant health is bolstered by this practice. Basket-makers suggest that their harvesting methods are good for sweetgrass plants, helping them to grow back fuller and healthier. Ohlandt agrees, "as the plant grows, it dies and the accumulation of dead growth can choke out the plant. Cutting it gets rid of some of the material, but pulling it out is the best way to manage the growth" (Ohlandt in Johnson 2007). Through Ohlandt's efforts, this ecological understanding of harvesting and plant health was translated into a management strategy on this private barrier island. Similar efforts were undertaken at Spring Island, a gated community on a sea island near Beaufort, when Ohlandt joined the staff there. Basket-makers also harvest in nonresidential areas, such as a hunting club in Hardeeville, where the longleaf pine forest understory is managed with prescribed burning and sweetgrass harvesting. Here, harvesting is open to the entire basket-making community during the summer months. While permits are required, they are easy to obtain and basket-makers can collect large quantities of grass in a single trip at this large acreage site.

More broadly, just as family and community ties play a role in connecting harvesters and basket-makers, these bonds also play an important role in gaining access to new harvesting sites. Sometimes this includes family involvement in the development process itself, albeit if only temporarily. For example, Vera Manigault explains how family members locate and arrange access to new harvesting areas: "Like for instance my brother and them they work, you know, clearing off land and stuff and they find it." Other times, community members pick up resource supplies on their travels through areas with easy roadside access. For example, during field work for the project, a community member walked up to a basket stand and handed a bag of pine needles to the sewers he had gathered on a drive home through the nearby national forest (Fig. 6.7, inset d). Likewise, new friendships are providing links to new harvesting sites outside the region. One basket-maker described how a friendship has opened up access to a steady supply of sweetgrass and pine needles; "I do have a friend who lives on the outskirts here in Savannah [who] normally allows our family to go and retrieve those materials… On his plantation, he has a lot of pine trees, and also the sweetgrass grows on the forested land. So I retrieve that myself." While this basket-maker acknowledged the wider problem with accessing sweetgrass, this arrangement provided a sufficient supply.

Area business and customers are contributing to a new network of personal agreements that supply grass. Elizabeth Mazyck described one business owner who allowed her to harvest and a customer who heard how difficult sweetgrass is to find and decided to plant some to supply her favorite local artist:

> …there's an insurance company up there, [the owner] had a lot [of grass] around their office building, and they was gonna move from there and she gave me all the big hills and I bring it and plant it in my yard, but it doesn't grow as nice as where she had it at… And one of my other customers she lives up in Buck Hall, she's got some hills growing there. She just moved here from some other state and she found out the shortage we have and so she's going to try to grow some on her property.

These arrangements may not be unique; at least one other basket-maker indicated that a customer has planted sweetgrass and plans to allow her to harvest it in the near future. Regardless of how widespread they are, for some basket-makers, these types of relationships are an important link in the supply chains that support their craft and sales.

Discussion and Conclusions

Our analysis reveals the relationship of basket-making communities to a particular set of economic conditions that regulate land use, the distribution of associated land covers, and a wider set of *rural* gathering conditions prevalent in the past.

Our community tours revealed a portrait of historic gathering practices that relied on sweetgrass, longleaf pine, and palmetto fronds found in community woodland commons. In fact, gathering these NTFPs was just one of several practices commonly associated with woodlands and one component of household production. In these rural African-American settlements, economies were also characterized by subsistence gardens, small-scale farming, seasonal labor on larger white farms, and low-wage jobs in nearby cities, such as Charleston.

Today, while sweetgrass appears to be available on larger parcels in more distant parts of the county (see, e.g., Hurley et al. 2008), it still exists within Mt. Pleasant and specifically in areas identified as historic collection sites. However, these lands now are controlled and managed by a greater number and diversity of landowners (Hurley et al. 2008). These landowners may or may not maintain and/or plant sweetgrass within forested areas, in their lawns, or as part of their landscaping. Indeed, land development patterns and the associated property boundaries that result in former collection sites, particularly where sweetgrass may still be present, underscore the extent to which an entirely new property regime—one characterized by both greater numbers of landowners, often from outside the area, *and* collectively owned subdivision commons—is emerging across the study area (Fig. 6.5).

Sweetgrass harvesting persists both in the study area and beyond, despite the fact that rapid land development has displaced its habitats, enclosed sweetgrass within new property regimes, and largely resulted in an absence of "publicly owned or protected" harvesting areas. On the one hand, collecting has expanded from a primarily local activity to a regional one. Today, sweetgrass is gathered from a number of other southeastern states. On the other hand, basket-makers have (re) negotiated access to both historical and new collection locations, where they can gather wild sweetgrass and other key materials. These areas are found within Mount Pleasant and on nearby barrier islands with upscale gated communities, such as Kiawah and Dewees islands, in Charleston County.

Despite the popular wisdom that sweetgrass is disappearing, basket-makers are still finding the plants they need to continue this tradition. Networks of exchange based on friendship, kinship, and entrepreneurship continue to connect basket-makers with gatherers and resource supplies. Since there are difficulties finding and accessing sweetgrass in the Mt. Pleasant area, gathering has expanded beyond the community woodlands and waterways near the neighborhoods where harvesters and basket-makers live. Traditional relationships between people who specialize in collecting sweetgrass and people who only make baskets remain important because harvesters bring sweetgrass *home* to Mt. Pleasant.

Land development patterns in the greater Mt. Pleasant area have disrupted historic gathering practices associated with sweetgrass, by changing the distribution of plants and by altering the conditions of access associated with the species' distribution in the area. These political and ecological changes associated with urbanization mean that those gathering NTFP resources must navigate a complex set of changing land uses and property regimes, characterized by diverse landowners, development patterns, and also new plant types. Only in doing so can users and this distinctive art form—and the special people-plant relationships this represents—persist within this rapidly urbanizing area.

The case of sweetgrass basket-making in greater Mt. Pleasant, SC holds key lessons for studies that focus on NTFP practices in urbanizing areas elsewhere. First, the traditional and/or cultural connections of groups long associated with an area may be actively recognized and carried on, in spite of urbanization and associated forms of economic and cultural change associated with this process. Greater attention by planners and policy-makers to the presence of these links and the continuation of land-based activities by long-time residents in rapidly urbanizing areas is needed. Second, the NTFP and rural livelihoods perspective might facilitate more complex understandings of political-ecological changes brought on by suburban and exurban development. Indeed, this case study highlights the uneven and complex changes brought on by these types of development, particularly in areas where there is a strong "sense of place" among locals and developers seek to market aspects of this sense of place.

References

Ballard R, Jones G (2011) Natural neighbors: indigenous landscapes and eco-estates in Durban, South Africa. Ann Assoc Am Geogr 101:131–148

Brown BA, Marin-Hernandez A (2000) Voices from the forest: the Lives and experiences of non-timber forest workers (Voces de bosque adentro: Vidas y experiencias de trabajadores forestales no-maderables). The Jefferson Center for Education and Research, Wolf Creek

Butler B, Leatherberry E, Williams M (2005) Design, implementation, and analysis methods for the national woodland owner survey. USDA Forest Service Northeastern Research Station, Newtown Square

Carney J (2001) Black rice: The African origins of rice cultivation in the Americas. Harvard University Press, Cambridge, Mass

Coakley JV (2006, July–August) The flower ladies: mothers of invention. Charleston Magazine, pp 158–167

Cope M, Elwood S (2009) Qualitative GIS: a mixed methods approach. Sage Publications, Boston

Cordell K, Betz C, Green G, Mou S, Leeworthy V, Wiley P, Barry J, Hellerstein D (2004) Outdoor recreation for 21st century America. Venture Publishing, State College

Derby D (1980) Black women basket-makers: a study of domestic economy in Charleston County, South Carolina. Dissertation, University of Illinois at Urbana-Champaign, Urbana

Emery MR (1999) Social values of specialty forest products to rural communities. In: Josiah SJ (ed) Proceedings of the North American conference on enterprise development through agroforestry: farming the forest for specialty products. Center for Integrated Natural Resources and Agricultural Management (CINRAM). University of Minnesota, Minneapolis, pp 25–32

Emery MR (2002) Space outside the market: implications of NTFP certification for subsistence use. In: Shanley P, Pierce A, Laird S, Guillen A (eds) Tapping the green market: management and certification of NTFPs. Earthscan, London, pp 302–312

Emery MR, Pierce A (2005) Interrupting the telos: locating subsistence in contemporary U.S. forests. Environ Plan A 37:981–993

Emery MR, Ginger C, Doble S, Giammusso MRB (2003) Special forest products in context: gatherers and gathering in the Eastern United States. NE-GTR-306. USDA Forest Service, Northeastern Research Station, Newtown Square

Fox J, Rindfuss R, Walsh S, Mishra V (2002) People and the environment: approaches to linking household and community surveys to remote sensing and GIS. Kluwer Academic Press, New York

Freed J (2001) Non-timber forest products in local economies: the case of Mason County, Washington. In: Emery MR, McLain RJ (eds) Non-timber forest products: medicinal herbs, fungi, edible fruits and nuts, and other natural products from the forest. Food Products Press, Binghamton, pp 67–69

Gabriel N (2006) Urban non-timber forest products in Philadelphia. Master's thesis, Temple University, Philadelphia

Ginger C, Emery M, Baumfleck M, Putnam D (2011) Access to natural resources on private property: factors beyond right of entry. Soc Nat Resour. doi:0.1080/08941920.2011.633596

Ginger C, Emery M, Baumfleck MJ, Putnam D (2012) Access to natural resources on private property: factors beyond right of entry. Soc Nat Resour. 25:700–715

Grabbatin B (2008) Sweetgrass basketry: the political ecology of an African American art in the South Carolina Lowcountry. Master's thesis, the College of Charleston, Charleston, SC

Grabbatin B, Hurley P, Halfacre A (2011) "I still have the old tradition": The co-production of sweetgrass basketry and coastal development. Geoforum 42:638–649

Gustafson DJ, Peterson PM (2007) Re-examination of *Muhlenbergia capillaris*, *M. expansa*, and *M. sericea* (Poaceae: *Muhlenbergiinea*). J Bot Res Inst Tex 1:85–89

Hart Z (2003) Stakeholder participation in management of Lowcountry South Carolina sweetgrass (*Muhlenbergia filipes*) supplies. Master's thesis, College of Charleston. Charleston

Hart Z, Halfacre A, Burke M (2004) Community participation in preservation of lowcountry South Carolina sweetgrass (*Muhlenbergia filipes* [m. A. Curtis] J. Pinson and W. Batson) basketry. Econ Bot 58:161–171

Head L, Muir P (2006) Suburban life and the boundaries of nature: resilience and rupture in Australian backyard gardens. Trans Inst Br Geogr 31:505–524

Head L, Muir P (2007) Backyard: nature and culture in suburban Australia. University of Wollongong Press, Wollongong

Hunt S (2006, October) Save our sweetgrass. Charleston Magazine, pp 138–145

Hurley PT, Halfacre A (2011) Dodging alligators, rattlesnakes, and backyard docks: a political ecology of sweetgrass basket-making and conservation in the South Carolina Lowcountry. GeoJournal. doi:10.1007/s10708-009-9276-7

Hurley PT, Halfacre AC, Levine NS, Burke MK (2008) Finding a disappearing nontimber forest resource: using grounded visualization to explore urbanization impacts on sweetgrass basketmaking in greater Mt. Pleasant, South Carolina. Prof Geogr 60:556–578

Jahnige P (1999) Non-timber forest product uses and values in Baltimore, Maryland. Community Resources, Baltimore

Jarosz L, Lawson V (2002) "Sophisticated people versus rednecks": economic restructuring and class difference in America's West. Antipode 34:8–27

Johnson EA, Klemens MW (2005) Nature in fragments: the legacy of sprawl. Island Press, Washington, DC

Jones E, Mclain R, Weigand J (2002) Nontimber forest products in the United States. University of Kansas Press, Lawrence

Knigge L, Cope M (2006) Grounded visualization: integrating analysis of qualitative and quantitative data through grounded theory and visualization. Environ Plan A 38:2021–2037

Lambin EF, Geist H (eds) (2006) Land-use and land-cover change: local processes and global impacts. Springer, Berlin

Liegel L, Pilz D, Love T (1998, September) The biological, socioeconomic, and managerial methods and results in the MAB mushroom study. Ambio special report number 9, pp 26–33

Love T, Jones E (2001) Why is non-timber forest product harvesting an "issue"? Excluding local knowledge and the paradigm crisis of temperate forestry. J Sustain For 13:105–122

Lynch K, Mclain R (2003) Access, labor, and wild floral greens management in Western Washington's forests. Pacific Northwest Research Station, Portland

Madsen LM, Adriansen HK (2004) Understanding the use of rural space: the need for multi-methods. J Rural Stud 20:485–497

McLain RJ (2002) Business as usual: the exclusion of mushroom pickers in wild mushroom management in Oregon's national forests. In: Jones ET, McLain RJ, Weigand J (eds) Nontimber forest products in the United States. University Press of Kansas, Lawrence, pp 375–384

Molina R, O'Dell T, Luoma D, Amaranthu M, Castellano M, Russell K (1993) Biology, ecology, and social aspects of wild edible mushrooms in the forests of the Pacific Northwest: a preface to managing commercial harvest. Pacific Northwest Research Station, Portland

National Park Service (2005) Lowcountry Gullah culture special resource study and environmental impact statement. National Park Service Southeast Regional Office, Atlanta

Nixon R (1993) Cultures in conflict: sea island communities are fighting for their survival: stirring new hopes along the coast of South Carolina. South Expo 21:53–56

Ohlandt KF (1992) Where the sweetgrass grows: the restoration of a maritime wet grassland, incorporating harvesting of *Muhlenbergia filipes*. Master's thesis, University of Georgia, Athens

Palmer J (2000) Recreational use of wetlands in Juneau, Alaska. USDA Forest Service, Northeastern Research Station, Newtown Square

Pickett S, Burch W, Dalton S, Foresman T, Grove J, Rowntree R (2001) A conceptual framework for the study of human ecosystems in urban areas. Urban Ecosyst 1:185–199

Pollitzer W (2003) The Gullah people and their African heritage. The University of Georgia Press, Athens

Price DM, Kindscher K (2007) One hundred years of *Echinacea angustifolia* harvest in the smoky hills of Kansas, USA. Econ Bot 61:86–95

Richards RT, Creasy M (1996) Ethnic diversity, resource values, and ecosystem management: Matsutake mushroom harvesting in the Klamath bioregion. Soc Nat Resour 9:359–374

Robbins P (2003) Political ecology: a critical introduction. Blackwell Publishing, Malden

Robbins P, Birkenholz T (2003) Turfgrass revolution: measuring the expansion of the American lawn. Land Use Policy 20:181–194

Robbins P, Emery MR, Rice JL (2008) Gathering in Thoreau's backyard: nontimber forest product harvesting as practice. Area 40:265–277

Rosengarten D (1986) Row upon row, sea grass baskets of the South Carolina Lowcountry. McKissick Museum, University of South Carolina, Columbia

Rosengarten D (1992) Lowcountry basketry: folk arts in the marketplace. South Folk 49:240–255

Rosengarten D (2008) Missions and markets sea island basketry and the sweetgrass revolution. In: Rosengarten T, Schildkrout E (eds) Grass roots: African origins of an American art. Museum for African Art, New York, pp 128–171

Rosengarten D, Rosengarten T, Schildkrout E (eds) (2008) Grass roots: African origins of an American art. Museum for African Art, New York

Theobald D (2004) Placing exurban land-use change in a human modification framework. Front Ecol Environ 2:139–144

Twining M (1978) Harvesting and heritage: a comparison of Afro-American and African Basketry. South Folk Q 42:159–174

U.S. Census Bureau (1990) Census fact finder. www.census.gov. Last accessed 16 Dec 2007

U.S. Census Bureau (2000) Census fact finder. www.census.gov. Last accessed 16 Dec 2007

U.S. Census Bureau (2006) 2006 American community survey: Charleston county, SC. Last accessed 10 Dec 2007

U.S. Census Bureau (2007) Historic census of housing and population: decennial censuses. www.census.gov/prod/www/abs/decennial/index.htm. Last accessed 16 Dec 2007

U.S. Census Bureau (2009) Census fact finder. www.census.gov. Last accessed 7 Oct 2007

Wehi PM, Wehi W (2010) Traditional plant harvesting in contemporary fragmented and urban landscapes. Conserv Biol 24:594–604

Zimmerer KS, Bassett TJ (2003) Political ecology: an integrative approach to geography and environment-development studies. Guilford Press, New York

Chapter 7
Marketing, Culture, and Conservation Value of NTFPs: Case Study of Afro-Ecuadorian Use of *Piquigua, Heteropsis ecuadorensis* (Araceae)

Maria Fadiman

Abstract This chapter explores the pros and cons of non-timber forest products (NTFPs) in relation to three major aspects: (1) economics, (2) culture, and (3) conservation, through a case study of the Afro-Ecuadorian use of a fiber plant, *piquigua, Heteropsis ecuadorensis* (Araceae) Kunth, in the Mache-Chindul Ecological Reserve. The reserve is located in northwestern Ecuador, in the province of Esmeraldas. This chapter addresses the collection, management, use, and marketing of the plant, with particular emphasis on basketry. For these particular forest residents, the plant has minimum economic value and high cultural value, both of which lend themselves to the importance of this plant in terms of conservation. Afro-Ecuadorians, although a minority in the reserve, represent a significant group actively using and managing the forest. The role that *piquigua* plays in their lives and for the forest can help inform the NTFP discussion involving rural areas in the developing world.

Keywords Afro-Ecuadorian • Basketry • Ecuador • Fiber • Hemiepiphyte • *Heteropsis* • Non-timber forest product (NTFP) • *Piquigua*

Introduction

Understanding how people live on the land is crucial for any kind of conservation success, and numerous researchers support the inclusion of local knowledge for effective conservation practices (e.g., Redford et al. 2003; Young and Zimmerer

---

M. Fadiman, Ph.D. (✉)
Department of Geosciences, Florida Atlantic University,
777 Glades Road, Boca Raton, FL 33431, USA
e-mail: mfadiman@fau.edu

R. Voeks and J. Rashford (eds.), *African Ethnobotany in the Americas*,
DOI 10.1007/978-1-4614-0836-9_7,

1998). One way to look at the larger issue of land use and ecosystem health is through non-timber forest products (NTFPs), which are biological materials other than timber which humans extract from forests (Arnold 2010; Balick and Cox 1997). Considering both the benefits and drawbacks of NTFPs provides a framework in which to look at people's resource use in northwestern Ecuador.

There is much debate about NTFPs in relation to ecology, conservation, culture, and economics. While many advocate these organisms' use as a way to promote positive forest use and conservation (e.g., Nepstad and Schwartzman 1992; Peters et al. 1989; Schmidt et al. 2007; Shanley et al. 2002), people also recognize weaknesses of NTFPs (e.g., Arnold 2010; Belcher et al. 2005; Bennett 2002). In order to recommend proper governance, case studies are necessary to better understand numerous situations (Laird et al. 2010). The goals of this chapter are to (1) explore the use of a fiber plant as an NTFP in economic development, (2) address resource use in terms of culture, and (3) understand how the use and collection of this plant relates to conservation. This study addresses the larger NTFP discussion through a case study of Afro-Ecuadorian collection, preparation, weaving, use, and marketing of the fiber plant *Heteropsis ecuadorensis* (Araceae) Kunth, commonly called *piquigua*. This plant was chosen for the study primarily because the people themselves refer to *piquigua* as one of their most important forest plants.

This work is significant in part because it provides previously unrecorded information from a particular region of Ecuador, the northwest. Fewer studies have been conducted in this region compared to the Amazon and the Andes. Furthermore, it addresses a less well-studied group of people, Afro-Ecuadorians. The emphasis of ethnobotanical studies has been focused on indigenous groups in the Amazon (e.g., Bennett and Husby 2008; Byg et al. 2006). However, recognition of Afro-Ecuadorian plant use is growing. Studies such as Afro-Ecuadorian forest use in reserves (Rudel 2000) and sustainability and management practices (Rival 2007) demonstrate the growth in the literature. However, although Afro-Ecuadorians make up a large percentage of the population in Esmeraldas (Johnson 2009), many consider them a group still excluded from national identity (Dixon 2006; Foote 2006). Further exacerbating this dilemma is that Afro-Ecuadorians are a distinct minority in the forested area of the Mache-Chindul Ecological Reserve. Here, the *mestizo* residents have a strong voice as they dominate most communities in the reserve, where there are 25 *mestizo* communities. The indigenous people, the Chachi, also have an influential presence, due to their ethnic status. The government recognizes indigenous groups' rights because they are considered first nation people (INEFAN 1999). The Afro-Ecuadorians however, with only three communities, are often overlooked as a distinct group (INEFAN 2006). Although other groups in the area also do artisan work with forest materials, this chapter gives attention to the Afro-Ecuadorian utilization and conservation practices surrounding *piquigua*.

Piquigua is significant not only for the Afro-Ecuadorian use but also because it is in the widely used genus, *Heteropsis*. Because this genus has species in other parts of Latin America that are also widely used, scholars are interested in how to best manage this genus (Salick 1992). Furthermore, an important and understudied aspect of plant use, according to Toledo (1995), is craft making. This can be particularly illuminating with basketry (Joyal 2003).

This study is significant environmentally because of its geographical location. The reserve includes land considered part of one of the global hot spots of biodiversity (Dodson and Gentry 1991). These forests contain high degrees of endemism, and only about 5% of this kind of ecosystem remains in Ecuador (Neill 2006). The area has been under heavy logging pressure, leading to the creation of the preserve in an effort to protect this ecosystem. Additionally, since numerous people live in and around this region, protection including sustainable resource use by the local residents is the aim of both conservation and development organizations (Aguirre et al. 2000).

In the 1970s, timber became an important commodity in northwest Ecuador (Carrasco 1983), and logging has since been steadily increasing (Sierra 1999, 1999; INEFAN 1998). Local households do much of the tree cutting in the Esmeraldas region. This activity provides one of the few opportunities to earn money. Interested parties are exploring ways to preserve landscapes through a balance of conservation and utilization (Pérez 1998). Many conservation and development organizations are now searching for NTFPs that can provide economic incentive for long-term preservation of ecosystems (Arnold 2010; Nepstad and Schwartzman 1992).

The general trend in the 1970s and 1980s of linking development and conservation added impetus to NTFPs (Schmink et al. 1992). The economic value of NTFPs then began to be studied seriously, with one of the initial important works being conducted by Peters et al. (1989), looking at the economic value of NTFPs from a 1-ha plot in lowland Peruvian rain forest. Today, the importance of NTFPs has gone beyond economics and ecology. Plants provide the basis for forest dwellers' material culture and are intrinsically linked to these people's culture. This is crucial when looking at conservation. With any effective kind of preservation, conservationists need to understand the people living on the land with which they are working (Bennett 2002; Garibaldi and Turner 2004).

The concept of NTFPs has both advantages and disadvantages. In one way, the concept has appeal because interested parties can potentially gain what they need from the forest. Typically, marginalized groups in rural communities can collect, use, and sell their own goods, thus playing an active role in the forest and their own conservation (Hall and Bawa 1993; Prance 1997). Because many of these people already rely on forest goods, the concept of working with NTFPs is familiar to local residents (Balée 1994; Duke 1992; Godoy and Bawa 1993).

As well as having conceptual and cultural value, NTFPs can offer economic incentive, which is a powerful agent. The maintenance of forest functions such as carbon storage, maintenance of water and nutrient cycles, diversity, and conservation of cultural heritage is somewhat abstract, but market value is a concept more readily understood by most people (Belcher et al. 2005; Ruíz-Pérez et al. 2004; Nepstad and Schwartzman 1992). NTFPs show economic potential, with numerous studies proving that it can be more economically viable to use a forest for NTFP collection instead of cutting it down for timber (e.g., Balick and Mendelsohn 1991; Peters et al. 1989; Shanley et al. 2002).

Although NTFPs show promise, there are drawbacks to this concept as well. Long-term commercial harvest has ecological, social, and economic limitations (Pendelton 1992; Tickton 2005). One of the main concerns is the danger of exploitation by overharvesting target species (e.g., Coomes 2004; Godoy and Bawa 1993). Furthermore, the impact of overcollection can affect not only the resource but the surrounding forest as well (Tickton 2005). Although economic benefits increase with high demand, so does the potential for overharvesting (Belcher et al. 2005; Ruíz-Pérez et al. 2004). Thus, NTFP exploitation may only be sustainable in areas where there are low human population densities. Clearly, this limits the strength of NTFP use as a widespread solution to deforestation (Homma 1992).

As well as affecting plant communities, NTFP extraction can have dramatic results on human populations. A product can become so popular that the market encourages its cultivation in plantations. Crop plantations can work well. However, they can also lead to pest infestation, soil depletion, and the clearing of forested land in order to plant the crop (Berry 1978; Southgate and Whitaker 1994). If a group becomes dependent on plantation products that then fail, they not only lose income, but they lost forest from which they could have gathered other products.

Furthermore, success of an NTFP can also lead to a bust through synthetic substitution. The product may become so popular that companies learn to produce the material themselves. In this situation, they then would not have to purchase the materials from forest collectors. The people, who had been gathering, would then no longer be able to rely on that NTFP as a source of income. When people have developed an economy surrounding the collection of an item, there is economic disruption when that avenue is no longer available (Belcher et al. 2005).

Similarly, this livelihood can attract so many people to collect and process the product that it encourages colonization into the very areas that conservation groups are trying to protect (Belcher et al. 2005; Laird et al. 2010). Even if the area colonized is not one of environmental interest, it can still cause complete community dislocation. This kind of social disruption occurred with the rubber trade in the Amazon (Padoch and de Jong 1989).

To increase NTFP success, scholars in the biological and social sciences discuss a need to pursue case studies that focus on extractive economics and the existing and potential uses of tropical forest and fauna (e.g., Plowden et al. 2003; Schmink et al. 1992). Studies should determine if a material is useful, marketable, and can be collected sustainably (e.g., Godoy and Bawa 1993; Shanley et al. 2002).

Relatedly, scientists also need to carefully assess the reality of the economic potential of each NTFP (Bennett 1991; Plowden et al. 2003). Looking at the market includes analyzing transportation, availability, seasonality, and the role of middlemen (Padoch 1987; Ruíz-Pérez 2004). Of these, location and transportation are crucial to the success of an NTFP. If a location is close enough to a market, or there is efficient transportation, then there is much more potential that item will do well. However, if the distance is too far, and transportation is not adequate, then there is little chance of success (Belcher et al. 2005; Lincoln and Orr 2011; Ruíz-Pérez 2004; Runk 1998). Furthermore, it is equally important to discover if a plant is not a viable NTFP in terms of marketing, social importance, or conservation, so that energy and time are not wasted on poor candidates (Sheil and Wunder 2002).

Study Area and Ethnohistory

The Mache-Chindul Ecological Reserve is located in northwestern Ecuador principally in the province of Esmeraldas with its southernmost part extending into the province of Manabí (Fig. 7.1a). Esmeraldas is humid, with 2,000–3,000 mm of precipitation falling in the region each year (Aguirre et al. 2000; Gavilanes et al. 2000). Topographical

Fig. 7.1 (**a**) Map of Ecuador with the Mache-Chindul Ecological Reserve *highlighted*, (**b**) map of study communities and market towns, (**c**) *piquigua* leaves growing out from the stem, (**d**) Afro-Ecuadorian collecting *piquigua*, (**e**) material preparation

relief ranges from 0 to 800 m, with undulating topography. The forest type for most of this area is lowland humid tropical rainforest, with lush vegetation (Gavilanes et al. 2000; Neill 2006).

Along the coastal region west of the Andes, from southern Panama through northern Ecuador, Afro-Ecuadorians comprise a large segment of the population (West 1952; Whitten 1974). The ancestry of this group is linked with the Africans whom the Spaniards brought to Latin America through the slave trade. Beginning in the 1500s, traders brought people of African heritage to Latin America both directly from Africa and from Europe where the enslaved Africans had already worked (Novoa 2001; Whitten 1974).

In the Esmeraldas province, the written history of the local Afro-Ecuadorians begins with an account by the traveler Miguel Cabello Balboa in 1583 (e.g., Novoa 2001; West 1957). Cabello Balboa wrote that the first Africans to arrive in the region came on a slave ship that wrecked on the Esmeraldas coast in 1553. There were 20–23 enslaved Africans, 17 men and 3–6 women, who escaped into the forest. They mixed with local indigenous people, believed to be the Campaces or the Tiguas. Alonso de Illescas, a powerful individual who had already worked as a slave in Europe, led these individuals to develop their own independent group. This free mixed community began to lose power in the 1800s, and the Afro-Ecuadorian population in the area gradually became comprised of descendents of Illescas group, enslaved Africans, and ex-enslaved Africans (Speiser 1991; Villaquirán 1990).

The migration of escaped enslaved Africans from the local and Colombian mines, *haciendas* in highland Ecuador, and those freed by their masters increased the population of Afro-Ecuadorians in Esmeraldas (Novoa 2001; Whitten 1974). The growth of the Afro-Ecuadorian population accelerated beginning in1821 when Spain passed a law gradually emancipating enslaved Africans. In 1851, slavery was officially abolished and larger migrations of freed enslaved Africans coming from the mining areas in Colombia entered Ecuador. Some ex-enslaved Africans continued to work in the local mines in order to have jobs, while others were still virtually enslaved, working off their own debts or those of their parents (West 1952, 1957).

Joining the Chachi who had migrated earlier to the lowlands and have been living in the area since the 1930s, the Afro-Ecuadorians were the second group to come to what eventually became the Mache-Chindul Ecological Reserve. They arrived in the 1940s, from the Canton of Esmeraldas (Alarcón 2000; INEFAN 1996, 1998). Afro-Ecuadorians, as well as the other proximal groups, practice subsistence farming, fishing, hunting, and gathering from the surrounding forest.

Methodology

This study took place from December 2001 to December 2002 and in August 2006, 2008, and 2009. For this study, interviews were conducted with 22 Afro-Ecuadorian weavers. Utilizing snowball sampling, subjects with weaving knowledge were identified and then asked to refer others with similar knowledge and skills (Berg 2001). While many people can do simple weaving, a subset is particularly skilled. The Afro-Ecuadorian informants were from the community of Chiva. Using convenience sampling (Berg 2001), the community was chosen because there was an expression of interest in this project, and it was located within 2 days' walking distance from the base location, Bilsa biological station. Additionally, to evaluate the current and possible market situation, sellers and buyers were also contacted. Three raw material buyers and sellers were interviewed in Quinendé. Also in Quinendé, two finished product vendors were interviewed as well as three more in Esmeraldas, four more in Monte Cristi, and one more in Borbón (Fig. 7.1b).

Most interviews were semiformal or informal, working principally with the weavers within the families. However, as many people do some artisan work, or at least are usually present when the activity takes place, whoever was present in the home or on the walk frequently joined in the discussion. Weaving details and processing were discussed in the home, and collection methods in the forest. Interviews with sellers and merchants took place in their business locations. Although the main informant in these situations was the person who owned the business, usually there were spouses, workers, and friends who also joined in the discussion. This aided in giving consensus about the information received (Reyes-García et al. 2003; Heinrich et al. 1998). Also, the researcher engaged in participant observation by weaving and processing material and accompanying local citizens who collected material (Bernard 2006).

Interviews were carried out in Spanish, all with informed consent. Throughout the study period, the researcher lived with various families. This added casual discussion and observation about the entire process, enhancing the information gathered from the more structured interviews. The names of the communities and informants have been changed to ensure the privacy of the villages and individuals who participated in this study and to protect against potential negative repercussions. Collection methods can become controversial, especially since the Mache-Chindul Ecological Reserve is still struggling with establishing resource use protocol. This technique of protecting privacy is discussed by Bernard (2006) and Martin (2004). A voucher specimen was collected (MF #0046) and is located in the Herbario Nacional (QCNE), Quito, Ecuador.

Study Plant: *Piquigua, Heteropsis ecuadorensis*

Heteropsis (Kunth) is a monophyletic genus of evergreen hemiepiphytes with woody fibrous roots. Although there have been shifts in classification since Engler's (1878) description, Mayo et al. (1997) and Tam et al. (2004) place it in Araceae, subfamily Monsteroideae, and it is sister to *Spathiphyllum*. The neotropical genus contains 17 species, all of which occur in Latin America (Croat 1988; Dodson and Gentry 1978). People use *Heteropsis* throughout its range (e.g., Bennett et al. 2002; Goncalves 2005; Plowden et al. 2003).

Heteropsis ecuadorensis, commonly known as *piquigua*, has subcoriaceous oblong-lanceolate, alternate simple leaves arranged in flat rows along both sides of the stem (Ray 1987) (Fig. 7.1c). It grows in the Pacific Coastal and Amazonian rain forests (Valencia et al. 2000) and has recently been recorded as also growing in Colombia and Costa Rica (TROPICOS 2006). *Piquigua* grows in elevations from 1 to 2,000 m (Jørgensen and León-Yánez 1999; Valencia et al. 2000). Requiring low light and high moisture, it grows in mature forest.

Piquigua is a secondary hemiepiphyte. It germinates on the forest floor with initially skototropic (darkness seeking) stems. When reaching a vertical surface (a tree), the stems, sometimes referred to as root climbers (Knab-Vispo et al. 2003), produce leaves and begin to grow up trees toward the light (Ray 1992; Strong and Ray 1975). When juvenile stems and roots deteriorate, hemiepiphytes produce a few short anchor roots, attaching the plant to the tree. They then send down long slender aerial feeder drop roots, which grow down to the forest floor and reestablish contact with the soil (Croat 1988; Knab-Vispo et al. 2003; Madison 1977; Putz and Holbrook 1986).

Piquigua Collection, Preparation, and Management

Piquigua roots grow in patches throughout the forest, and the Afro-Ecuadorians know specific collection areas. Most families identified a minimum of four collecting locations, sometimes even searching out a particular plant. They locate areas where *piquigua* grows while pursuing other activities in the forest, returning to those places to collect at a later date (Fadiman 2007). Most Afro-Ecuadorians have a few *piquigua* plants on their own land, at the minimum a half-hour walk away. It is common to walk between 2 and 4 h to a collection site. Since they collect only when they need a root, not all collecting trips are the same. People complain that they have to walk farther than in the past to find *piquigua*. A farmer mentioned that he was waiting for a certain root to mature on his property, only to discover that someone had taken it. Another person wrote his name on a root in an effort to claim the material, even though the plant was not growing on his property (Fadiman 2008a).

To collect *piquigua* roots, the harvester cuts the aerial root at ground level with a machete and then with both hands, grabs and pulls until the root separates from the plant and falls to the ground. Collectors yank the root at an angle so that the plant itself remains in the tree (Fig. 7.1d). The harvesters exhibit notable conscientiousness about conserving this resource, carefully ensuring that the growing plant does not become dislodged and fall out of the tree. Collectors also leave a sufficient number of roots growing from each plant, so that the *piquigua* sustains connection with the soil and can continue to produce new roots. On one occasion, while collecting, a mother reprimanded her inexperienced children for pulling the roots too hard. She wanted to prevent them from pulling the entire plant out of the tree. On a different collecting trip, a man made a ladder out of surrounding trees so that he could reach up higher on the root and better control the connection to the stem (Fadiman 2008a).

Plant management has become an increasingly important aspect of ethnobotanical studies (Turner 2001). In this case, villagers responded that they did not manage *piquigua*. Respondents would typically answer questions regarding *piquigua* management by saying: "It is born in the forest, it grows in the forest, and then we collect it from the forest." Generally, this phrase reflects their actions. However, disproving the consistency of these statements, a few specific instances occurred in which people actively managed the plant. They promoted the continued growth of certain roots. For example, one man, seeing a *piquigua* root growing straight down toward a creek, pulled the growing tip to an angle and buried the root tip where it could grow in the soil away from the water. In another instance, while walking through a *piquigua* patch, a woman carefully removed a stick that was weighing down a climbing stem as it grew between two trees. Later, this same woman noted a stem dangling from a tree and secured the wayward stem to the trunk of a tree with a vine (Fadiman 2004). A different kind of management involves the protection of trees when clearing fields that have *piquigua* growing on them. Farmers allow these trees to remain standing, so that the *piquigua* will continue to grow, albeit only for a few years without the surrounding trees (Fadiman 2007).

Most of the families with whom I spoke mentioned the importance of collecting during certain times each month, in reference to the lunar cycle. All three groups emphasized that they must collect during the young moon, and possibly at the full moon, but not when the moon is waning. They say if collected when the moon is receding, the root will be difficult to pull for collection, break easily when weaving, and will quickly dry out. If collected during the correct phase of the cycle, the epidermis is softer and easier to remove. To collect weaving materials in relation to the lunar cycle appears with various ethnic groups. Examples include palm collection in Costa Rica (Joyal 1994) and Mexico (Joyal 1996). Despite the importance placed on the timing, most people admitted that they collect whenever they need the material.

Piquigua preparation begins with the removal of the epidermis in lengthwise strips. The weavers flex the root and peel the epidermis away by pinching it with their fingers and pulling strips off. Occasionally, people will leave the epidermis on the root for weaving, but many people say that this is rough and ugly.

The second step in the preparation of the material is to split the root longitudinally into at least two lengthwise strips by pulling it apart from one end toward the other. With a knife, the weaver makes a small cut at one end of the root. He then takes hold of the end of each cut half and pulls them apart so that the root is torn from end to end. If it is a thick root, he can divide it into more than two pieces (Fig. 7.1e). He will cut these in slices along the diameter and then across subsequent parallel lines.

The information from this study, which is similar to Plowden et al.'s (2003) findings that *Heteropsis* spp. roots had a growth rate of 76.6 + 5.3 cm per year and Hoffman's (1997) data that *Heteropsis flexuosa* can grow at about 3.1 cm per month, provides an example of the growth rate of *Heteropsis ecuadorensis*. Thus, in terms of collection, growth rates need to be kept in mind to ensure that the roots have a chance to establish themselves before more are harvested.

Basketry

The most important items made in Mache-Chindul from *piquigua* aerial roots are baskets, which are found in every home. Afro-Ecuadorians weave baskets in two styles. One type of basket (and weave) in Spanish is called "*regular*" (regular) (Fig. 7.2a). The weaver creates a simple plaited weave by passing an active fiber alternately over one passive fiber and under the next. The second type of basket, a "*chalo*" or "*canasta de ojo*" (eye basket), involves a more open weave, with large hexagonal spaces (Adovasio 1977; Barrett 1994) (Fig. 7.2b).

Almost all weavers in the region are accomplished at making *regular* baskets, which are small and circular, measuring less than 15 cm in diameter. The villagers use these containers primarily to store foods such as eggs, potatoes, beans, and onions. In an effort to protect their food from rodents, the locals hang the baskets from the rafters (Fadiman 2007).

The baskets most utilized are the *piquigua chalos*, used for carrying heavier items for longer distances. These containers serve in harvesting crops, for carrying items between homes and between villages. Furthermore, locals sometimes place a plastic container inside the basket to facilitate water portage from the river. Because

Fig. 7.2 (**a**) Regular basket hanging in a resident's kitchen, (**b**) *chalo*-style basket woven with the hexagonal spaces in the weaving design, (**c**) hats made from *piquigua*, (**d**) initial stage of weaving a *chalo*

the *chalo* weave is strong, stiff, and more durable than the simple weave, the basket maintains its shape, and porters can strap them onto their horses without bruising what they carry. However, the most common way to carry the baskets is on people's backs. The Afro-Ecuadorians tie plastic rice sacks onto the basket for shoulder straps, carrying them like backpacks.

When asked, the villagers consider women to be the weavers. However, despite this image of themselves, in the Afro-Ecuadorian communities, more men than women actually do the weaving with *piquigua*. Although women are those who most often weave with softer fibers, people mentioned that because the material is hard, rough on the fingers, and the initial weaving position is uncomfortable, it is an easier task for men than for women.

Starting with 23 pieces, each about 1 m long, weavers lay the prepared roots in a series of crisscrossing Xs, holding the pieces in place with their toes. They refer to each extending piece as a "foot." Each foot consists of two, three, or four roots laid together side by side, each group acting as a single passive element (Adovasio 1977; Barrett 1994). The more numerous the ply, the stronger the basket. The flat inside part of the split root faces the ground. The weaver will eventually turn these over as he weaves so that the curved parts make up the outside of the finished basket. Weavers then take the active piece, called the "mother," weaving under and over the passive pieces two times around the Xs (Fig. 7.2d). At the end of the second time around, they flip the woven part over and pull up so that the basket starts to have an upward curve. Weavers then take the basket off the floor and sit up to weave. Eventually, he ties off the top in the same way as with the regular weave.

Piquigua brooms, although less prized than the baskets, sit against the wall of all Afro-Ecuadorian homes in the reserve. Each broom lasts 2–3 years, even when used with water. To make brooms, weavers measure a root four hands length long, split it down the middle, and string it between two walls. They then fold smaller pieces, about a foot long each, over the long section, tying each one on with the piece itself. After they have filled the long piece, they then spiral the fringe around a stick, which serves as the broom handle, and tie it at the bottom. Hat making is a craft to which only certain skilled Afro-Ecuadorian *piquigua* artisans devote themselves (Fig. 7.2d). Finally, another aspect of *piquigua* invaluable to the residents of the Mache-Chindul Ecological Reserve is its use as lashing material. The villagers tie together balsa log rafts, fences, sugarcane presses, house structures, livestock, and fish traps (Fig. 7.3a).

Piquigua Market: Current and Future

The residents of the Mache-Chindul Ecological Reserve generally collect and weave the root for their own personal use. However, there are those who demonstrate more skill and interest in weaving than others. Thus, between the Afro-Ecuadorians, there exists a limited amount of buying and selling of *piquigua* articles between members of the same community and sometimes with neighboring villages as well. Various

Fig. 7.3 (**a**) Sugarcane press tied with *piquigua*, (**b**) *piquigua* brooms for sale, (**c**) *piquigua* furniture made for sale in the town of Monte Cristi, (**d**) raw *piquigua* ready for sale in town of Quiñende

weavers will even buy baskets from someone else to save themselves the trouble of collecting and weaving. Still, they sell few articles. One of the main weavers sold five *chalos* in 1 year. When weavers do sell to each other, most often someone has commissioned the article. Bartering is common within the communities, and people will usually trade a chicken for a basket. If there is a money exchange, a large *chalo* costs from USD $3.00 to $4.00, and small simple plaited baskets can be sold for USD $1.00–2.00. However, weavers are more likely to give the latter as gifts or informal repayment for favors. They also sell brooms from USD $1.00 to $2.00 (Fig. 7.3b). However, these are more likely to be given as gifts as well. A small market exists for the large *chalo*s in a few stores in the cities of Quiñende and Esmeraldas, with each basket costing USD $5.00. In areas such as Borbón, a predominantly Afro-Ecuadorian city, one store sells articles almost exclusively made from *piquigua*, mostly bags, small baskets, and hats, costing from USD $1.00 to $5.00.

A successful market for the raw material of *piquigua* exists in various cities. In Quiñende, merchants sell the raw material for commercial furniture and broom making. Furniture makers construct pieces by wrapping the root around metal or wooden frames in decorative patterns (Fig. 7.3c). *Piquigua* furniture is durable and popular throughout Ecuador, and a set of high-quality furniture can cost USD $600, with some large chairs costing $150 each. The high-quality furniture root sells for USD $17.00 to $24.00 per 100 lb. The root sellers buy it from the farmer for about USD $15.40 to $16.00. Roots with knots sell for less, about USD $14.00, with the buyer purchasing these from farmers for USD $12.00 to $13.00 (Fig. 7.3d).

Piquigua sold in western Ecuador comes from various areas, depending on the town in which it is sold. In Quiñende and Monte Cristi, buyers purchase from the collectors north of Esmeraldas: San Mateo, Borbón, Montalvo, Muisne, Atacames, Chontadura, and San Francisco. *Piquigua* also grows in eastern Ecuador, in the Amazon region. However, the distance to that source is too great for people to travel from Esmeraldas and Manabí. Collectors tend to bring in the raw material on the weekends, and seasonally when they are not harvesting other crops. These farmers bring more of the material during July and January than they do in other months. From February through June, they bring less because *piquigua* will keep in the forest and not rot, and they have to tend to other crops as soon as these more perishable items are ready. Partially because of this seasonal nature, the raw material vendors do not deal exclusively with *piquigua. Piquigua* sellers also vend other products such as corn, rice, and coffee in order to earn their livelihood. In terms of conservation, *Piquigua* buyers are also conscious that the collectors need to leave the stem intact and preserve growing roots. Buyers say that when collectors sell the roots, they usually collect little by little until they have enough to bring to the buyer, preventing denuding an area all at one time.

Although the market for *piquigua* is currently limited, opportunities could increase due to changes in and around the reserve. Important factors in marketing are the spatial relations, particularly distance to market (Bennett 2002). In this case, since the roads become almost rivers of mud for half of the year, deterioration of woven articles is a concern. In the past, Peace Corps Volunteers tried to work with selling local artesania, but the items became damaged before they arrived at market. However, looking at markets in closer proximity could address some of these spatial restrictions. El Lago, a small village located where three important roads converge connecting most colonists of the region, is starting to attract tourists. Two lodges have been built, and tourists and student groups from Ecuador and abroad come to spend from one night to a month by the lagoon. Furthermore, volunteers and ecotourists that trek into the Bilsa Reserve pass the first night at the lagoon cabins, before entering the road to the reserve. These travelers can contribute to the region by being potential purchasers of *piquigua* items (Fadiman 2008b). Location is not the only factor that needs to be addressed when looking at market possibilities. Clearly, the amount of resources available is crucial for success. Although there is probably not enough *piquigua* growing in this region to become part of the raw material market, weavers could make small tourist items using little material. All groups already weave small baskets. Some people in the area are already experimenting with miniatures. One person wove a gift of a miniature *chalo*, complete with tiny shoulder straps. Although this would be altering the emphasis that the weavers currently have on weaving functional size containers, it could provide some income without having to cut many more roots.

Discussion and Conclusions

This chapter looks at the case study of the Afro-Ecuadorian use of *piquigua, Heteropsis ecuadorensis*, as a way to add to the NTFP discussion in terms of three main aspects: (1) economics, (2) culture, and (3) conservation. This plant and its uses in part promote the discussion of how useful the species can be as an NTFP. However, in other aspects, it falls short, thus exemplifying some of the weaknesses of the NTFP concept.

In looking at the economic benefits of NTFP marketing, some proponents envision moderate amounts of goods being sold, generating relatively small economic returns (Gentry 1992: Shanley et al. 2002). Others, however, see larger-scale profits at the country level (Padoch 1987), and some even look internationally (Bennett 2002). Marketing and balancing income distribution for NTFPs can be complicated. There are many people involved, and the power structure is not equal. The person earning the least is generally the collector and, in many cases, even the middlemen earn little (Hall and Bawa 1993; Padoch 1992). Furthermore, assessments by Godoy and Bawa (1993) about how much money can be earned raise doubts that participants are gaining as much as various studies have estimated.

There are numerous reasons for these economic weaknesses: the basic inadequacies of the market (Bennett 2002; Southgate 1998), the trends of booms and busts (Homma 1992), and the difficulties in transporting materials and goods (e.g., Lincoln and Orr 2011; Padoch and de Jong 1989). Also, because tropical forests are so heterogeneous, many plants of the same species grow at a considerable distance from each other. Because of the time and energy required to travel between plants, collectors can find it difficult to obtain enough material to make it worth their effort. Collectors are usually so economically destitute that they will choose the activity that generates the most income for their effort, often regardless of destructive results. Thus, in certain circumstances, local people may not choose to collect NTFPs (Browder 1995).

In principle, *piquigua* can be used sustainably as a commercial NTFP. However, although there is a market for the baskets and raw material, given the distance to market, a concept discussed by Belcher et al. (2005) and Ruíz-Pérez et al. (2004), the business possibilities are less viable. This same spatial barrier was also found with two groups selling woven items made from forest fiber in Panama: the Ngöbe women (Lincoln and Orr 2011), and the Wounaan and Embará (Runk 2001).

However, this spatial obstacle can be lessened, as there is increased market potential with an increase in local tourism. On the other hand, most of these tourists are students and volunteers who will be interested in purchasing small portable items for little investment. Thus, in order to properly tap this market, the weavers would need to modify their main form of basketry. At this point, *piquigua* as an NTFP does not generate a substantial amount of income for the Afro-Ecuadorians in the region. They use the material mostly for their own consumption, weaving baskets for themselves, sometimes for barter and rarely for selling. However, they do benefit economically in terms of the fact that they do not need to purchase baskets or twine. Although domestic use does not produce income, it can lessen the need for disposable cash; locals can make certain articles for themselves.

In addition to economics, the nonmarket value of NTFPs is an integral aspect of their overall worth (Garibaldi and Turner 2004; Nepstad and Schwartzman 1992). The importance of learning about the ethnobotany of plant material should not be limited to its economic contributions. The importance of a plant or plant-derived product to a people's own culture can be far more significant than what could be expressed in monetary terms. As Bennett (2002: 294) writes, "While the natural biota yields material provisions for market economies, it has even greater significance for traditional cultures, who collect and directly use biological resources."

The Afro-Ecuadorians self-identify *Piquigua* as one of their most important fiber plants. Their own recognition of this plants' role in their culture constitutes a central aspect of being considered a cultural keystone species, and thus a notable NTFP (Garibaldi and Turner 2004). *Piquigua* in many ways is a central species to the local culture. The fiber plays a role in everyday life through weaving, using baskets, and tying items. Although some individuals are more adept at weaving, all people utilize the finished product.

In terms of weaving, the data show that men are more likely to collect and weave the material. This information aids in creating an understanding of the role men play in the household and with NTFPs. When conservationists, governments, and/or development organizations approach the locals about resource use, they can work with knowledge of the importance of *piquigua*. In terms of gender, these same groups can approach the men about this NTFP. They might have initially began with the women, as weaving is often considered a female sphere of influence. And, for other fibers, such as *Astrocaryum standleyanum* (Fadiman 2008b), women indeed are the major artisans.

Understanding these facets of a community proves crucial when promoting conservation (Bennett 2002). Thus, in looking at the case study of the Afro-Ecuadorians in the Mache-Chindul Ecological Reserve, and their use of *piquigua*, a better understanding is gained about how residents in a protected rain forest area utilize plants in terms of material culture. Understanding the plant can help to contextualize why many villagers protect the resource, and thus value the forest in which the plant grows.

The plant can be collected sustainably when done with precision. However, there is overharvest potential. In order to overcome the barriers that NTFPs face and to take advantage of the benefits, collection and processing need to occur within a system that integrates local plants and products with good ecological management (Bennett 1992). It is important to govern collection (Arnold 2010), as not only individuals but whole plant populations can be affected (Tickton 2005). In many cases, the potential for NTFP commercialization resulting in conservation is limited. Successful NTFP marketing can be contradictory as it often hurts biodiversity (Belcher et al. 2005). NTFPs can become unsustainable if people switch to a cash economy (Ezebilo and Mattsson 2010). Thus, the flip side of the fact that *piquigua* has minimal marketing possibilities for the residents of Chiva is that overcollection or intensive management is not probable (Belcher et al. 2005; Ruíz-Pérez et al. 2004).

Furthermore, daily use of a resource can create an incentive to maintain the health of an ecosystem in which it lives. According to Garibaldi and Turner (2004), as people connect with the landscape through certain species that are important to them, they will be more active in protecting the ecosystem. The hemiepiphyte grows in intact forest, and these people make a conscious effort to collect the material sustainably. In order to overcome the barriers of using NTFPs, the collection and processing need to occur within a system that integrates utilization with good ecological management (Bennett 2002). In this case, it appears that the Afro-Ecuadorians living in the Mache-Chindul are harvesting and using the plant in such a way. Afro-Ecuadorians, although a minority in the reserve, represent a significant group actively using and managing the forest. Bringing attention to these people and their forest activities continues to be an important endeavor (Sanchez 2008). Attention to the community practices in Chiva can promote this group's inclusion into conservation plans made for Mache-Chindul. In looking at the case study of the Afro-Ecuadorian use of *piquigua* in terms of the NTFP discussion, because the plant has little economic potential and high cultural importance, this combination can lead to an increased incentive for conservation of the plant and the ecosystem in which it lives.

References

Adovasio JM (1977) Basketry technology: a guide to identification and analysis. Aldine Publishing Company, Chicago

Aguirre NM, Gavilanes MA, Hofstede R, Sevink J (2000) Sistemas forestales en la costa del Ecuador: Una propuesta para la zona de amortiguamiento de la Reserva Mache-Chindul. Ecopar, Quito

Alarcón HP (2000) Diseño Final del subproyecto de capacitación y comercialización de artesanía Chachi de la zona norte de la provincia de Esmeraldas. FECCHE, Esmeraldas

Arnold JE (2010) Foreword. In: Wild product governance: finding policies that work for non-timber forest products. Earthscan, Washington, DC, pp xxiii–xxiv

Balée W (1994) Footprints of the forest. Columbia University Press, New York City

Balick MJ, Cox P (1997) Plants, people, and culture: the science of ethnobotany. Scientific American Library, New York City

Balick MJ, Mendelsohn R (1991) Assessing the economic value of traditional medicines from tropical rain forests. Conserv Biol 6:128–130

Barrett SA (1994) Los indios Cayapas del Ecuador. Abya-Yala, Quito

Belcher B, Ruíz-Pérez M, Achdiawan R (2005) Global patterns and trends in the use and management of commercial NTFPs: implications for livelihoods and conservation. World Dev 33(9):1435–1452

Bennett BC (1991) Aspectos económicos y sociológicos de la etnobotánica. In: Ríos M, Pedersen HB (eds) Las plantas y el hombre. Abya Yala, Quito, pp 361–365

Bennett BC (1992) Plants and people of the Amazonian rainforests: the role of ethnobotany in sustainable development. Bioscience 42:599–607

Bennett BC (2002) Forest products and traditional peoples: economic, biological and cultural considerations. Nat Res Forum 26:293–301

Bennett BC, Husby CE (2008) Patterns of medicinal plant use: an examination of the Ecuadorian Shuar medicinal flora using contingency table and binomial analyses. J Ethnopharmacol 116:422–430

Bennett BC, Baker M, Andrade PG (2002) Ethnobotany of the Shuar of eastern Ecuador, vol 14. The New York Botanical Garden, Bronx

Berg BB (2001) Qualitative research methods for the social sciences. Allyn and Bacon, Needham Heights

Bernard HR (2006) Research methods in anthropology: qualitative and quantitative approaches, 4th edn. Altamira Press, Walnut Creek

Berry W (1978) The unsettling of America: culture & agriculture. Sierra Club Books, Avon

Browder JO (1995) Redemptive communities: indigenous knowledge, colonist farming systems, and conservation of tropical forests. Agric Human Values 12:17–30

Byg A, Vormisto J, Balslev H (2006) Using the useful: characteristics of used palms in southeastern Ecuador. Environ Dev Sustain 8(4):495–506

Carrasco E (1983) El pueblo Chachi. Abya-Yala, Quito

Coomes OT (2004) Rain forest 'conservation-through-use'? Chambira palm fibre extraction and handicraft production in a land-constrained community, Peruvian Amazon. Biodiversity and Conservation. 13(2):351–360

Croat TB (1988) Ecology and life forms of Araceae. Aroideana 11:4–55

Dixon K (2006) Beyond race and gender: recent works on Afro-Latin America. Lat Am Res Rev 41(3):247–257

Dodson CH, Gentry AH (1978) Flora of the Rio Palenque science center, Los Ríos, Ecuador. Selbyana 4(1–6):1–628

Dodson CH, Gentry AH (1991) Biological extinction in eastern Ecuador. Ann Mo Bot Gard 78:273–295

Duke JA (1992) Tropical botanical extractives. In: Plotkin M, Famolare L (eds) Sustainable harvest and marketing of rain forest products. Island Press, Washington, DC, pp 53–62

Engler A (1878) *Heteropsis* (Kunth). In: De Candolle AC (ed) Monographie Phanerogamarum, vol 2., pp 98–100

Ezebilo EE, Mattsson L (2010) Contribution of non-timber forest products to livelihoods of communities in southeast Nigeria. Int J Sustain Dev World Ecol 17(3):231–235

Fadiman M (2004) Management, cultivation and domestication of weaving plants: *Heteropsis* and *Astrocaryum* in the Ecuadorian rain forest. Calif Geograph 44:1–19

Fadiman M (2007) Exploring conservation: *Piguigua, Heteropsis ecuadorensis,* in Ecuador. Pap Appl Geograph Conf 30:427–436

Fadiman M (2008a) Resource stewardship: rain forest use among three ethnic groups of Ecuador. Pap Appl Geograph Conf 31:310–318

Fadiman M (2008b) Use of *Mocora, Astrocaryum standleyanum* (Arecaceae), by three ethnic groups in Ecuador: differences, similarities and market potential. J Ethnobiol 28(1):92–109

Foote N (2006) Race, state and nation in early twentieth century Ecuador. Nation Nationalism 12(2):261–278

Garibaldi T, Turner N (2004) Cultural keystone species: implications for ecological conservation and restoration. Ecol Soc 9(3):1. Open acess http://www.ecologyandsociety.org/vol9/iss3/art1. 2 June 2011

Gavilanes MR, Delgado SM, Hofstede R, Ronquillo JC (2000) Caracterizaci6n de los bosques y zonas de influencia. Ecopar, Quito

Gentry AH (1992) New non-timber forest products from western South America. In: Plotkin M, Famolare L (eds) Sustainable harvest and marketing of rain forest products. Island Press, Washington, DC, pp 125–136

Godoy R, Bawa KS (1993) The economic value and sustainable harvest of plants and animals from the tropical forest: assumptions, hypotheses, and methods. Econ Bot 47:215–219

Goncalves E (2005) Araceae from central Brazil: comments on their diversity and biogeography. Ann Mo Bot Gard 91(3):457–463

Hall P, Bawa K (1993) Methods to assess the impact of extraction of non-timber tropical forest products on plant populations. Econ Bot 47:234–247

Heinrich M, Ankli A, Frei B, Weimann C, Stitcher O (1998) Medicinal plants in Mexico: healers consensus and cultural importance. Soc Sci Med 47:1859–1871

Hoffman B (1997) The biology and harvest of Nibbi, Heteropsis flexuosa (Araceae): an aerial root fiber product in Guyana. Thesis, Florida International University, Miami

Homma AKO (1992) The dynamics of extraction in Amazonia: a historical perspective. In: Nepstad DC, Schwartzman S (eds) Non-timber products from tropical forests-evaluation of a

conservation and development strategy, vol 9, Advances in Economic Botany. The New York Botanical Garden, Bronx, pp 23–31
INEFAN (1996) Estudio de alternativas de manejo para las montañas de Mache, provincia de Esmeraldas. INEFAN, Quito
INEFAN (1998) Reserva Ecológica Mache-Chindul Remach: Estudio de tenencia de la tierra. Resumen ejecutivo. INEFAN, Quito
INEFAN (1999) Reserva Ecológica Mache-Chindul (Remach): Estudio complementario de tenencia de la tierra sector Y de la Laguna. INEFAN, Quito
INEFAN (2006) Reserva Ecológica Mache-Chindul Remach: Estudio de tenencia de la tierra. Resumen recente. INEFAN, Quito
Johnson E (2009) Schooling as a regime of equality and reproducing difference in an Afro-Ecuadorian region. Ethnograph Educ 4(2):147–164
Jørgensen PM, León-Yánez S (eds) (1999) Catalogue of the vascular plants of Ecuador. Monographs in Systematic Botany from the Missouri Botanical Garden, 75:1181–1182
Joyal E (1994) Palm ethnoecology in the Saripiqui region of Costa Rica. J Ethnobiol 14:161–172
Joyal E (1996) The Palm has its time: an ethnoecology of *Sabal uresana* in Sonora Mexico. Econ Bot 50(4):446–462
Joyal E (2003) University of Hawaii, building bridges conference abstracts. http://www.botany.hawaii.edu/traditionalknowledge/events/sessions/presenta/abstract/Joyal. Accessed 22 Mar 2003
Knab-Vispo C, Hoffman B, Moermond T, Vispo C (2003) Ecological observation on *Heteropsis* spp. (Araceae) in Southern Venezuela. Econ Bot 57(3):345–353
Laird SA, Wynberg RP, McLain J (2010) Recommendations. In: Wild product governance: finding policies that work for non-timber forest products. Earthscan, Washington, DC, pp 367–384
Lincoln K, Orr B (2011) The use and cultural significance of the *Pita* plant (*Aechmea magdalenae*) among Ngöbe women of Chalite, Panama. Econ Bot 65(1):13–26
Madison M (1977) Vascular epiphytes: their systematic occurrence and salient features. Selbyana 2:1–13
Martin G (2004) Ethnobotany: a methods manual. Earthscan, Sterling
Mayo SJ, Bogner J, Boyce PC (1997) The genera of Araceae. Royal Botanic Gardens, Kew
Neill D (2006) Missouri Botanical Garden Ecuador vegetation page. www.mobot.org/MOBOT/research/ecuador/vegetation.shtml. Accessed 13 Mar 2006
Nepstad DC, Schwartzman S (1992) Conclusions and recommendations: the challenge of non-timber forest product extraction. In: Nepstad DC, Schwartzman S (eds) Non-timber products from tropical forests-evaluation of a conservation and development strategy, vol 9, Advances in economic botany. The New York Botanical Garden, Bronx, pp 143–146
Novoa RR (2001) Zambaje y autonomía: Historia de la gente negra de la provincia de Esmeraldas siglos XVI-XVIII. Abya-Yala, Quito
Padoch C (1987) The economic importance and marketing of forest and fallow products in the Iquitos region of Brazil. In: Denevan WM, Padoch C (eds) Swidden-fallow agroforestry in the Peruvian Amazon. New York Botanical Garden, Bronx, pp 74–89
Padoch C (1992) Marketing of non-timber forest products in western Amazonia: general observations and research priorities. In: Nepstad DC, Schwartzman S (eds) Non-timber products from tropical forests-evaluation of a conservation and development strategy, vol 9, Advances in economic botany. The New York Botanical Garden, Bronx, pp 43–50
Padoch C, de Jong W (1989) Production and profit in agroforestry: an example from the Peruvian Amazon. In: Browder JO (ed) Fragile lands of Latin America: strategies for sustainable development. Westview Press, Boulder, pp 102–113
Pendelton LH (1992) Trouble in paradise: practical obstacles to non-timber forestry in Latin America. In: Plotkin M, Famolare L (eds) Sustainable harvest and marketing of rain forest products. Island Press, Washington, DC, pp 252–262
Pérez FL (1998) Human impact on the high páramo landscape of the Venezuelan Andes. In: Young KR, Zimmerer KS (eds) Nature's geography: new lessons for conservation in developing countries. The University of Wisconsin Press, Madison, pp 147–183

Peters CM, Gentry AH, Mendelsohn RO (1989) Valuation of an Amazonian rainforest. Nature 399:655–666
Plowden C, Uhl C, Oliveira F (2003) The ecology and harvest potential of titica vine roots (*Heteropsis flexuosa:* Araceae) in the eastern Brazilian Amazon. For Ecol Manage 182(1–3): 59–73
Prance G (1997) Valuation and sustainable use of tropical rainforest. J Biol Educ 31(4):258–262
Putz FE, Holbrook NM (1986) Notes on the natural history of Hemiepiphytes. Selbyana 9:61–68
Ray TS (1987) Leaf types in the Araceae. Am J Bot 74:1359–1872
Ray TS (1992) Foraging behavior in tropical herbaceous climbers (Araceae). J Ecol 80: 189–203
Redford KH, Coppolillo P, Sanderson EW, Da Fonseca G, Dinerstein E, Groves C, Mace G, Maginnis S, Mittermeier RA, Noss R, Olson D, Robinson JG, Vedder A, Wright M (2003) Mapping the conservation landscape. Conserv Biol 17(1):116–131
Reyes-García V, Goday R, Vadez V, Apaza L, Byron E, Huanca T, Leonard WR, Pérez E, Wilkie D (2003) Ethnobotanical knowledge shared widely among Tsimane' Amerindians, Bolivia. Science 299(5613):1707
Rival L (2007) Alliances for sustainable forest management: lessons from the Ecuadorian Chocó rain forest. In: Ros-Tonen MAF (ed) In collaboration with H. Hombergh and A. Zoomers. Partnerships in sustainable forest resource management: learning from Latin America. Koninklijke Brill NV, Leiden, pp 37–62
Rudel TK (2000) Organizing for sustainable development: conservation organizations and the struggle to protect tropical rain forests in Esmeraldas, Ecuador. Ambio 29(2):78–82
Ruíz-Pérez M, Belcher B, Achdiawan R, Alexiades M, Aubertin M, Caballero J, Campbell B, Clement C, Cunningham T, Fantini A, de Foresta H, Fernández CG, Gautam KH, Martínez PH, de Jong W, Kusters K, Govindan Kutty M, López C, Fu M, Alfaro MAM, Raghavan Nair TK, Ndoye O, Ocampo R, Rai N, Ricker M, Schreckenberg K, Shackleton S, Shanley P, Sunderland T, Youn YC (2004) Markets drive the specialization strategies of forest peoples. Ecol Soc 9(2): 4. Open access: http//www.ecologyandsociety.org/vol9/iss2/. /4 June 2011
Runk J (1998) Productivity and sustainability of a vegetable ivory palm (*Phytelephas aequatorialis*, Arecaceae) under three management regimes in northwest Ecuador. Econ Bot 52:168–182
Runk J (2001) Wounaan and Emberá use and management of the fiber palm *Astrocaryum standleyanum* (Arecaceae) for basketry in eastern Panama. Econ Bot 55(1):72–82
Salick J (1992) Amuesha forest use and management: an integration of indigenous use and natural forest management. In: Redford KH, Padoch C (eds) Conservation of neotropical forests: working from traditional resource use. Colombia University Press, New York City, pp 305–319
Sanchez AJ (2008) Multiethnic nations and cultural citizenship: proposals from the Afro-descendent movement in Ecuador. Souls 10(3):215–226
Schmidt IB, Figueiredo IB, Scariot A (2007) Ethnobotany and effects of harvesting on the population ecology of *Syngonanthus nitens* (Bong.) Ruhland (eriocaulaceae), A NTFP from Jalapão region, Central Brazil. Econ Bot 6(1):73–85
Schmink M, Redford KH, Padoch C (1992) Traditional peoples and the biosphere: framing the issues and defining the terms. In: Redford KH, Padoch C (eds) Conservation of neotropical forests: working from traditional resource use. Colombia University Press, New York City, pp 3–13
Shanley P, Luz L, Swingland IR (2002) The faint promise of a distant market: a survey of Belém's, trade in non-timber forest products. Biodivers Conserv 11:615–636
Sheil D, Wunder S (2002) The value of tropical forests to local communities: complications, caveats and cautions. Conserv Ecol 6(2):9. http//www.ecologyandsociety.org/vol6/iss2/art9/. 3 June 2011
Sierra R (1999) Traditional resource-use systems and tropical deforestation in a multi-ethnic region in north-west Ecuador. Environ Conserv 26:136–145
Southgate D (1998) Tropical forest conservation: an economic assessment of the alternatives in Latin America. Oxford University Press, New York City
Southgate D, Whitaker M (1994) Economic progress and the environment: one developing country's policy crisis. Oxford University Press, New York City

Speiser S (1991) Tenencia de la tierra en los cantones Eloy Alfaro y San Lorenzo provincia de Esmeraldas. FEPP, Esmeraldas

Strong DR, Ray TS (1975) Host tree location behavior of a tropical vine (*Monstera gigantea*) by skototropism. Science 190:804–806

Tam SH, Boyce PC, Upson TM, Darabé D, Bruneau A, Forest F, Parker JS (2004) Intergeneric and infrafamilial phylogeny of subfamily Monsteroideae (Araceae) revealed by chloroplast <011>trnL-F sequences1. Am J Bot 91:490–498

Tickton T (2005) Applying a metapopulation framework to the management and conservation of a non-timber forest species. For Ecol Manage 206:249–261

Toledo V (1995) New paradigms for a new ethnobotany: reflections on the case of Mexico. In: Schultes R, von Reis S (eds) Ethnobotany: evolution of a discipline. Dioscorides, Portland, pp 75–92

TROPICOS (2006) Specimen list for *Heteropsis ecuadorensis*. http://mobot.mobot.org/cgi-bin/search_vast. Accessed 7 July 2006

Turner IM (2001) The ecology of trees in the tropical rain forest. Cambridge University Press, Cambridge

Valencia R, Pitman N, León-Yánez S, Jørgensen P (eds) (2000) Libro rojo de las plantas endémicas del Ecuador. Hojas y Signos, Quito

Villaquirán ML (1990) Alonso de Illescas, estratega, militar y político. In: Savoia P (ed) El negro en la historia: Aportes para el conocimiento de las raices en América Latina. Centro Cultural Afroecuatoriano, Cayambe, pp 193–198

West RC (1952) Colonial placer mining in Colombia. Louisiana State University Press, Baton Rouge

West RC (1957) The pacific lowlands of Colombia: a negroid area of the tropics. Louisiana State University Press, Baton Rouge

Whitten NE Jr (1974) Black frontiersmen: a South American case. Schenkman Publishing Company, Cambridge

Young KR, Zimmerer KS (1998) Introduction: the geographical nature of landscape change. In: Young KR, Zimmerer KS (eds) Nature's geography: new lessons for conservation in developing countries. The University of Wisconsin Press, Madison, pp 3–34

Chapter 8
Berimbau de barriga: Musical Ethnobotany of the Afro-Brazilian Diaspora

James Sera and Robert Voeks

Abstract Enslaved Africans and their descendants were active agents of biocultural innovation and change in the Americas. In this chapter, we examine the cultural and ethnobotanical significance of Brazil's *berimbau de barriga*, a one-stringed musical bow of West African origin. Methods include archival research and semistructured interviews with *berimbau* master craftsmen. Appearing first in early nineteenth-century illustrations by European travelers, the contemporary *berimbau* was originally employed by African-descended street touts to attract customers. Over time, it gained popularity as the iconic musical instrument of the Afro-Brazilian martial art-dance of capoeira. Today, as capoeira increasingly symbolizes Brazil's African cultural heritage, artisans are producing greater numbers of *berimbaus* to meet the demands of tourists and distant capoeira practitioners. Craftsmen in the past took great care in fashioning *berimbaus*, but they now increasingly mass produce the instrument to satisfy the burgeoning commercial market.

The *berimbau* is constructed almost entirely of plant products. The fundamental components include the *verga* (wooden staff), the *cabaça* (bottle gourd), the *arame* (wire), and is often accompanied with the *caxixi* (small fiber rattle). There is no direct evidence of what species were used to build antecedent musical bows in Africa prior to arrival in Brazil. The *verga* is currently fashioned most often from biriba (*Eschweilera ovata*), a native tree of the highly endangered Atlantic coastal rainforests. Extraction of biriba to meet the increasing demand for commercial *berimbau* is leading to conflict between those concerned with the region's cultural patrimony and those who seek to protect the region's unique biological heritage.

J. Sera, Ph.D. (✉)
Department of Anthropology, California State University-Los Angeles,
5151 State University Drive, Los Angeles, CA 90032, USA
e-mail: jsera@hotmail.com

R. Voeks, Ph.D.
Department of Geography, California State University–Fullerton,
800 N. State College Blvd, Fullerton, CA 92834, USA
e-mail: rvoeks@fullerton.edu

R. Voeks and J. Rashford (eds.), *African Ethnobotany in the Americas*,
DOI 10.1007/978-1-4614-0836-9_8, © Springer Science+Business Media New York 2013

Keywords African Diaspora • *Berimbau* • Capoeira • Brazil • Ethnobotany

Introduction

Over the course of 350 years, roughly 10.7 million forced African immigrants reached the shores of the Americas. Immigration figures varied dramatically from region to region; some received several thousands of slaves, others several millions. With 4.8 million enslaved people arriving between 1538 and 1851, Brazil witnessed the arrival of vastly more Africans than any other colony or country in the Americas (Eltis and Richardson 2008). As a result of this tremendous human influx, Brazil's coastal zone was dominated numerically throughout most of its colonial and national history by people of African descent. Yet, given the horrors of the Middle Passage and the often miserable living conditions of chattel slaves, there was surely no diaspora in Earth's history more hamstrung in their ability to introduce their cultures and lifeways to their newfound lands. Nevertheless, from language to foodways, music to dance, and healing traditions to belief systems, Africans contributed mightily to what would emerge over time as Brazil's unique cultural identity.

The myriad contributions of Africans and their descendants to the cultural landscape only began to be acknowledged in a positive sense in the late twentieth century (Carney and Voeks 2003). Indeed, some of the more visible African-derived elements, such as participation in the Candomblé religion, or in the martial art-dance form capoeira, were harassed and apprehended by the police until the 1930s. Long maligned, many of these African cultural contributions today constitute formidable symbols of Brazilian national patrimony and have expanded geographically throughout the Americas. But the recreation and reformulation of these and other African-derived traditions and beliefs in the bosom of a repressive Portuguese civilization was no serendipitous process. For reasons demographic and geographic, enslaved and freed blacks in the Portuguese Atlantic World, particularly in the latter eighteenth through mid-nineteenth centuries, maintained a tighter connection with their homeland than any of the other African diaspora. Inhumane working conditions and consequent lack of replacement level fertility among Brazilian plantation slaves guaranteed the continual importation of new waves of replacement laborers. In the colonial cities of Rio de Janeiro, Salvador, Recife and elsewhere, newly-arrived Bantu from Angola, Yoruba, Hausa, and Igbo from Nigeria, Fon from Benin, and many others, carried desperately sought after news of family and friends from a distant and physically unreachable homeland. "The slave ship," as noted by Costa e Silva (2004: 22), "functioned both as a newspaper and as a courier." This communication corridor between the enslaved and their homeland was facilitated as well by the ubiquity of blackjacks, slave, and free Africans serving as mariners and deckhands on the slave ships, who were ideally placed to pass news and gossip to those at each end of the triangular trade route—Africa, Portugal, and Brazil (Bolster 1997; Silva 2005).

Consequently, to the continuing frustration of colonial and imperial authorities, there existed throughout this period a vast informational and material network of communication between colonial Brazil, West Africa, and Portugal. African dance forms, foodways, spiritual beliefs, liturgical plants, and crafts flowed back and forth across the Atlantic, borrowing and blending and innovating as they adapted to and molded local cultures and conditions. Examples include the common dance forms that were noted among slave and free blacks in the seventeenth- and eighteenth-century port areas of Iberia, Angola, and Latin America. Hybrid music and dance developed from the mixing of Africans of different cultural backgrounds and was facilitated by the large numbers of enslaved sailors and dockworkers (Fryer 2000). Silva (2005) notes, in this respect, that of the 609 sailors working at the port of Recife (Brazil), 403 were slaves. Moreover, for enslaved and freed blacks, material products imported from the "*costa*" (coast of Africa) represented much more than poignant reminders of a distant life; they constituted irresistible touchstones of resistance and topophilia, in many cases attaining sacred status as cultural markers of their distant African provenance. The African *bolsa de mandinga*, for example, an herb-filled leather talisman, was used and traded by Africans in Brazil, Madeira, Portugal, Angola, and even Goa, India. Freed and enslaved paid dearly for the special powers of those *bolsas* hailing from the spiritually-charged shores of Brazil and West Africa (Sweet 2009). In Brazil, liberated and repatriated Africans, recognizing the commercial possibilities of these and other items to the black diaspora, established import-export businesses between the west coast of Africa and Brazil. José Francisco dos Santos, for example, returned to Whydah, Dahomey (Republic of Benin), in the 1840s and began shipping *obí* (divination seeds from *Cola acuminate* (P. Beauv.) Schott & Endl.), atarê (magical pepper from *Aframomum melegueta* K. Schum.), *pano-da-costa* (West African cloth), *palha-da-costa* (the fiber of an African raffia palm), and other materials in demand by northeastern Brazil's black population (Verger 1952; Voeks 2009). Part and parcel to the everyday world of West Africans, these valuable items "from the coast" were incorporated into evolving Afro-Brazilian traditions and came to represent over time powerful cultural symbols of pride and resistance to Portuguese hegemony (Voeks 1997). These and many other cases of biocultural ebb and flow in the Atlantic World remind us that for enslaved and free blacks, Africa was "no vague mythical land… [but rather] a living reality, whence many of the objects they [Afro-Brazilians] use in their rituals are imported, where people they know have visited and… where their fathers or grandfathers came from" (Herskovits and Herskovits 1943: 266).

Among the most successful of all African-derived crafts, the *berimbau de barriga*, a one-stringed percussion instrument, not only survived and morphed over the centuries but became iconic in many ways for the Afro-Brazilian experience. Its evolution over time into a living patrimony followed the trajectory of many other New World African traditions. Tracing its origin(s) to one or several equatorial African musical instruments (Graham 1991; Lewis 1992), the *berimbau* became creolized in colonial Brazil's tropical coastal zone, assuming new meaning and expanded cultural identity as the heart and soul of Brazil's national martial art—capoeira (Anunciação 1971). Indeed, the *berimbau* for many is inseparable from the "game" of capoeira. The nineteenth- and early twentieth-century predecessor of capoeira—*capoeiragem*—was distinctly a fighting tradition, but over time it evolved into a dance-martial art performed to an orchestra of African-inspired percussion instruments, including the *berimbau*, *pandeiro* (tambourine with goat skin as a head), *reco-reco* (scraper), *agogô* (double bell), and *atabaque* (barrel style drum). Musicians bring energy and excitement into the game by singing songs filled with proverbs, life riddles, historical anecdotes, and humorous observations. The *berimbau* is personified by capoeira players as the "commander," who calls the participants to "come and play" (Lewis 1992: 3). In many respects, capoeira does not exist without the *berimbau*, and vice versa.

The florescence of appreciation for Brazil's African cultural heritage over the past few decades has come at a price, as tourism authorities and private entrepreneurs have sought to convert culture to cash. Several million international tourists pour into Salvador, Recife, and Rio de Janeiro each year, many seeking what is widely viewed as the most authentic expression of African culture in the Americas. That these traditions—samba, capoeira, Candomblé, *berimbau*, West African cuisine, and others—persevered the brutality of enslaved existence only adds to their cultural tourism appeal. Capoeira demonstrations appear daily on the streets as well as in the upscale hotels and resorts. And as the material symbol of capoeira, the *berimbau* is now widely sought after as a tourist handicraft, with many thousands per year returning to distant lands with their new owners. In addition, beginning in the 1960s, Brazilian musicians began to incorporate the sound of the *berimbau* into jazz and samba. Brazilian expatriates such as Pernambucan percussionist Naná Vasconcelos introduced this new sound to Europe and the United States, so that by the 1970s this humble African-inspired instrument was being played on six of the seven continents (Fig. 8.1). Not surprisingly, in order to meet the demands of tourists and international musicians, the local production of *berimbau*s has grown dramatically, creating as a result a new set of social and environmental challenges.

The objective of this chapter is to explore the historical development and ethnobotanical significance of the Afro-Brazilian *berimbau*. We begin by describing the various types of *berimbaus* reported by observers in Brazil beginning in the early 1800s. This includes the geographic locations where the musical bow was first reported, the context in which it was being played, and the materials utilized in its construction. We then explore the cultural and ethnobotanical knowledge of *berimbau* master craftsmen, including the provenance and ecological status of the plant materials utilized in its construction. Finally, we consider the growing influence of tourism on the quality of *berimbau* craftsmanship, as well as the question of sustainable harvest of *biriba* (*Eschweilera ovata* Mart. ex Miers), the principal botanical component used in the fashioning of the *berimbau*.

Methods

Archival research was carried out between 2008 and 2010 in Salvador at the Universidade Federal da Bahia and the Instituto do Patrimônio Histórico e Artístico Nacional, in Rio de Janeiro at the Biblioteca Nacional and Instituto

Fig. 8.1 Percussionist Naná Vasconcelos with his *berimbau* (Courtesy of Naná Vasconcelos)

Moreira Salles, in Petrópolis at the Arquivo Histórico do Museu Imperial, and in Recife at the Arquivo Público Estadual Jordão Emerenciano (APEJE), with the assistance of Hildo Leal da Rosa, Fundação Joaquim Nabuco, Comissão Pernambucana de Folclore, and Instituto Arqueológico, Histórico e Geográfico Pernambucano. Additional primary material was provided from the personal libraries of Maurício Lemos de Carvalho (Mestre Vermelho), Frederico Abreu, and Roberto Benjamin.

Interviews were carried out with five *berimbau* master craftsmen in August 2008. Beginning with the well-known craftsman (Valmir) in Salvador, we identified four additional participants via snowballing. We used face-to-face questionnaires consisting of 86 open-ended questions. Inquiries were divided into nine general categories: personal information, the art of *berimbau* making, types of wood used, harvest locations and methods, types and sources of other plant materials used, financial aspects, musical aspects, the role of *berimbau* in capoeira, and general knowledge of the history of *berimbau*. Interviews were carried out in Portuguese, lasted 1–3 h, and were conducted in the *berimbau*-making shop or the home of the craftsperson. We spent time on Itaparica Island observing *biriba* (the primary wood species) being harvested and prepared by Valmir, as well as at various shops observing *berimbaus* being made and painted, at open markets where they are sold to tourists, and at capoeira *rodas* (performances) where *berimbaus* are played. One of us (JS) has been making and playing *berimbaus* in his capoeira academy for over 10 years. We are especially grateful for an interview in 2010 with Pernambucan world renowned percussionist Naná Vasconcelos.

History of the Berimbau

The Afro-Brazilian *berimbau* is clearly of West African origin. There is nothing similar described either for early Portuguese musical instruments or those reported for the indigenous populations of Brazil. The term *berimbau* comes from the Bantu (Kimbundu) speakers of Angola (Schneider 1991: 40). In early Portuguese vernacular, the term referred to an instrument known as a mouth harp, Jew's harp, or jaw harp (Oliveira 1958). The mouth harp was brought to Brazil in the sixteenth century from the Iberian Peninsula and is classified as a lamellophone, or plucked idiophone, which has a metal tongue attached to a metal frame. This mouth harp *berimbau* disappeared over time in Brazil, but the lexeme became attached to various musical bows including the *oricongo* (Debret 1989), *rucumbo* (Ramos 1935), *urucungo* (Araújo 1964), *gunga*, *gobo*, and *bucumbumba* (Alvarenga 1950; Oliveira 1956).

The African-derived instrument was likely developed from a number of Congo-Angolan musical bow prototypes (Graham 1991). Gerhard Kubik, who carried out research in Angola and in Brazil, believes that the *berimbau de barriga* represents a synthesis of several Angolan and Congo gourd bows. One that he recorded in southwest Angola, the *mbulumbuma*, is "virtually identical with the Brazilian *berimbau*" and is "identical in construction and the playing technique" (Kubik 1979: 34). He did not, however, note any musicians in West Africa holding a *caxixi* (small fiber rattle), although he likewise suggests that the origin of this word appears Angolan. Fryer (2000) also cites a number of nineteenth-century observers in Angola and Congo who describe musical bows that are very similar to the *berimbau*. According to Henrique Augusto Dias de Carvalho, the *rucumbo* he saw among the Congo-speaking people near the border of Zaire and Angola is fashioned by taking "a stick of a particular pliable wood and bend it to a bow, tightening the ends with a thick string which is made beforehand from cotton fibre which is kept taut" (1890: 370).

The first accounts of instruments similar to the contemporary *berimbau* being played in Brazil by enslaved and freed blacks are by early nineteenth-century foreign travelers and artists. At that time, there were four types of *berimbaus* reported: the *birimbao* (a mouth harp), the *berimbau de boca* (musicians play the musical bow with their mouth on the string), the *berimbau de bacia* (musical bow using a hole in the ground as a resonator), and the *berimbau de barriga* (the contemporary *berimbau*) (Shaffer 1977). The latter instrument, according to Graham (1991), is quite similar to the *rucumbo* described in 1890 by Henrique Augusto Dias de Carvalho. Whether it was the *rucumbo* that inspired the development of the *berimbau* in Brazil, or one of the many other stringed African instruments that must have arrived over the course of the slave trade, is unknown.

Fig. 8.2 African street vendor in Rio de Janeiro using an *urucungo* to sell his merchandise. The first image of the *berimbau de barriga* in Brazil. Painting entitled *Vendedores Ambulantes* by Joaquim Cândido Guillobel (By permission of the Museu Imperial, Rio de Janeiro/IBRAM/ Ministério da Cultura)

Perhaps the first depiction of a one-stringed musical bow in Brazil is the 1814 painting *Vendedores Ambulantes* (Walking Salesmen) by the Portuguese artist Joaquim Guillobel (Fig. 8.2). He depicted an African street vendor in Rio de Janeiro with his merchandise in a basket on his head, while attempting to attract customers with the sound of his "*urucungo*" (Dos Santos 1941). The player is shown using his left index finger to hold the string against his thumbnail, exactly as capoeira players do today when they play the *berimbau* without a stone or coin. Although the type of *verga* (wooden staff) and resonator are not clear, the latter is held against the player's belly just as it is today. Similarly, in 1815, the Englishman Henry Koster (1816: 241) recounted from Recife having seen "a large bow with one string, having half of a coco-nut shell or of a small gourd strung upon it. This is placed against the abdomen, and the string is struck with the finger, or with a small bit of wood". Unlike Guillobel's depiction of urban blacks, Koster's account appears to be based on rural Africans attached to the Jaguaribe (Pernambuco) sugar mill. This description resembles the contemporary *berimbau* in form but not in the fashion in which it is presently played. The "small bit of wood" in his description is likely the *vaqueta* (thin percussion stick), used to strike the wire of the contemporary *berimbau* as well. A year later, French traveler Louis-François Tollenare (1905: 137) arrived in Recife and described a musical instrument used by "*negros*" having a "cord of tripe extended over a bow and put over a cavity formed by a gourd; they make a sound from a bow and produce tuned and harmonious tones; I did not observe if its music could be dance to, and I say the same about the berimbau". The string in this musical bow is made of animal gut, making this a prototype of the contemporary *berimbau*.

Fig. 8.3 An African or African-descendant male vendor holding a *berimbau* with his *left* hand while supporting a stick with dangling chickens with the *right* hand. This photograph entitled "Negro com berimbau" by Marc Ferrez is to date the earliest photograph of the *berimbau de barriga* in Brazil (By permission of the Acervo Instituto Moreira Salles, Rio de Janeiro)

However, Tollenare does not provide the name of this bowed instrument, and indeed the description suggests that the bowed instrument and the *berimbau* are two different instruments. In 1823, British traveler Maria Graham (1824: 199) witnessed a man in Rio de Janeiro playing an instrument "composed of a crooked stick, a small hollow gourd, and a single string of brass wire". Graham's is the first historical account of a metal wire being used on the musical bow. Regarding how the instrument is played, she noted that "the mouth of the gourd must be placed on the naked skin of the side; so that the ribs of the player form the sound-board, and the string is struck with a short stick" (1824: 199). Again, the short stick is most likely the *vaqueta* used to play the contemporary *berimbau*. The brass wire, according to Richard Graham (1991: 8) would have allowed the musician to produce a "louder, more sustaining tone than a natural string".

Africans or their descendants playing a musical bow are also depicted in a nineteenth-century sketch by Henry Chamberlain, displayed in the Museu Nacional in Petrópolis. The musician is using the musical bow to hawk merchandise on the street (Chamberlain 1822). During the same general period, Marc Ferrez may have taken the first photograph of an African or African descendant holding a *berimbau* while selling chickens in the countryside of Brazil (Fig. 8.3). Based on what we have learned thus far concerning the use of the *berimbau* in the nineteenth century in Brazil, it would not be unreasonable to assume that the vendor used the instrument to attract customers as others did in Rio de Janeiro. Although no date or location is associated with the image, we guess that it was taken in the late 1800s, since that is when the majority of Ferrez's other photographs were taken.

The first suggestion of the wood used in the construction for the *verga* comes from the British Reverend Robert Walsh, who traveled throughout the states of Rio de Janeiro and Minas Gerais from 1828 to 1829. During a trip into the hills near the city of Ouro Preto (Minas Gerais), he encountered a "poor black minstrel boy" playing what he called a "simple" instrument. "It consisted of a single string stretched on a bamboo, bent into an arc, or bow. Half a cocoa nut, with a loop at its apex, was laid on his breast on the concave side; the bow was thrust into this loop, while the minstrel struck it with a switch, moving his fingers up and down the wire at the same time. This produced three or four sweet notes, and was an accompaniment either to dancing or singing" (Walsh 1830: 176). Walsh believed the *verga* to be made of bamboo, and like Koster, he described the resonator as comprised of half of a coconut. Though Walsh comments that the musical bow accompanied dancing and singing, he does not offer any details.

A particularly detailed description of the sound of the *berimbau* is provided by French artist Jean-Baptiste Debret, who lived in Brazil from 1816 to 1831. He made a sketch entitled *O Negro Trovador* (The Black Musician) showing a blind man holding a musical bow, which he refers to as an *urucungo*. The following is Debret's description of the musician and his instrument in Rio de Janeiro (1989: 164):

> That instrument consists of a half gourd attached to a bow formed by a small bent stick with a brass wire of which is struck lightly. One can at the same time study the musical instinct of the player who with the hand supports the front of the gourd with the hope of obtaining a deeper and more harmonious sound by the vibration. This effect, when lively, can only be compared to the sound of the hammered dulcimer string, since it is produced by striking the cord string lightly with a small stick that one holds between the index and middle fingers of the right hand.

A British traveler James Wetherell described in 1860 what appears to be a contemporary *berimbau de barriga* as a "long stick… made into a bow… a thin wire… [and]… half a gourd to serve as a sounding board is attached to this bow by a loop. The bow is held in the left hand, the open part of the gourd pressed upon the body" (1860: 106). The instrument is held "between the fingers and thumb of the right hand is held a small stick with which the wire is struck, producing a tinkling sound; on the other fingers is hung a kind of rattle as the hand moves to strike the string" (Wetherell 1860: 107).

When and by what means the *berimbau* came to be associated with capoeira is unknown. Other than Walsh's observation of "dancing and singing," there are no clear descriptions of the instrument associated with the dance-martial art until the early twentieth century. Robert Farris Thompson claimed that a *berimbau* appears in Johann Moritz Rugendas' 1835 sketch "Jogar Capoëra ou danse de la guerre," which depicts two men involved in capoeira play to the accompaniment of a log drum (Fig. 8.4) (Thompson 1989). Closer examination of the image shows this claim of early *berimbau* use to be unfounded (Price 2006: 98–99). It may well

Fig. 8.4 Africans or African descendants playing capoeira. This is the earliest image of capoeira in Brazil. The original design entitled "Jogar Capoëra ou danse de la guerre" was designed by Johann Moritz Rugendas in 1835. The lithographer Isidore Laurente Deroy engraved the lithography in stone so that it could be published in Rugendas's book (By permission of the Museu Castro Maya—IBRAM/MinC, Rio de Janeiro)

have been introduced into capoeira by happenstance as a *berimbau* player accompanied a drummer during a capoeira match. By whatever means, the *berimbau* along with the *caxixi* was clearly introduced into capoeira sometime between 1860 and 1916, and we believe that Kay Shaffer is correct in that this synthesis occurred in Bahia (Shaffer 1977).

Brazilian scholar Manuel Querino provided a clear description of the fusion of these two elements in Salvador (Fig. 8.5). Querino reported that practitioners of the art "danced to the rhythm of the *berimbau*" and described the *berimbau* as an instrument "composed of a bow of flexible wood, set at its ends by a cord of fine wire, a small gourd connected to the cord, [and] copper coin" (1955: 75). Querino noted that the musician held the *berimbau* in his left hand, while in his right hand he held a "small basket containing pebbles, called *gongo*, along with a thin vine with which [he] would strike the cord producing the sound" (Querino 1955: 75). Later, in the 1950s–1980s, Argentine painter Carybé (Hector Barnabó), who spent most of his professional life in the city of Salvador, made numerous sketches of capoeira and

Fig. 8.5 Three *berimbaus* (*gunga*, *medio*, and *viola*) are played in the capoeira Angola *roda*. The late Mestre Diogo and Mestre Adó play capoeira at an event promoted by the Associação de Capoeira Angola Corpo e Movimento in Salvador under the direction of Mestre Angola (Photo taken by James Sera)

the *berimbau*. He wrote about two types of *berimbau—berimbau de bôca* and *berimbau de barriga* (Carybé 1955). The *berimbau de bôca* he described as an "arc with a vine cord 'timbó', a resonance box is the mouth and the percussion aspect is done with a knife on the cord" (Carybé 1955: 5). The *berimbau de barriga*, on the other hand, was described as a "stick of *pombo* wood in which a wire of steel is attached with tension" (Carybé 1955: 5). Carybé states that "the resonator is a small gourd tied to the wire with a cord. A *vareta* produces the sound and the modulations are produced with a *moeda de vintém* (old copper coin) and with the greater and lesser proximity of the gourd's mouth to the stomach" (1955: 5). The "*pombo*" noted by Carybé is a local vernacular, usually referring to the widespread tree *Tapirira guianensis* Aubl. (Voeks 1995).

Berimbau Craft and Ethnobotany

The *berimbau* is a one-stringed musical bow fashioned largely from local plant products, some cultivated, others growing wild (Fig. 8.6). It is built around a flexible and durable *verga* approximately 1.1–1.2 m in length and 2.5 cm in diameter. The *verga* is set by bending it while a piece of steel wire with a cotton string tied to its end is tautly extended over a piece of leather on top of the staff. The wire is wrapped

Fig. 8.6 The ethnobotanical components of the *berimbau* are displayed for the *verga*, *cabaça*, and *caxixi* (Photos by Robert Voeks and James Sera. Courtesy of Luiz Badaró)

around the staff in a downward direction and then secured by slipping the string on its end string that is already wound. The wire comes from used car tires and is the only significant part of the *berimbau* that is not botanical. When this tradition began is unknown. All the informants indicate that they acquire these from tire stores themselves and extract the *aro* (metal ring). A cut *cabaça* (bottle gourd) is attached with a string approximately 20 cm from the bottom of the staff. The "*boca*," or "mouth" (opening) of the gourd, faces toward the musician's belly and acts as a resonator when struck with the *vaqueta*. Musicians balance the *berimbau* on the pinky of one hand under the string that connects the gourd to the *verga*. They grasp the stick with the middle and ring fingers so that it rests in a vertical position. A smooth rock or a *dobrão* (an old Brazilian copper coin) is pressed against the wire to play higher notes. Many capoeira practitioners also use a *caxixi* by placing their middle and ring finger through the handle with palm of the other hand touching the body of the small instrument.

The *verga*, the most important component of the *berimbau*, is fashioned from locally cut and finished hardwood trees. It is prepared by scraping off the bark, thinning it with a knife, smoothing it with sandpaper, giving it a clear protective coat of polyurethane (or burning or painting, in the case of instruments destined for tourists) (Fig. 8.7), and nailing a piece of raw leather to the top. The master *berimbau* craftsmen were asked to free list the species of trees that they used to fashion

Fig. 8.7 A variety of *vergas* are available to purchase. From *left* to *right*: burned, painted, and recently sanded or natural *vergas* (Photo taken by Robert Voeks)

the *verga*. The results (Table 8.1) show that *biriba* (*Eschweilera ovata)* is the preferred ethnospecies, being named first by all five informants. Several noted that if given the choice, they would always use *biriba*. This preference was followed by *taipoca*, which was named by all five informants, *canduru* (4 of 5), and *camaçari* (3 of 5). They did not include *pau pombo* (*Tapirira guianensis*), the species noted by Carybé, because being brittle, it is employed only to make tourist instruments. Though none of the artisans in our project had ever heard of using bamboo to fashion the *verga*, one of the authors (J.S.) has witnessed *berimbaus* constructed from a species of bamboo known locally in Belém as *cana da índia*. This suggests that the bamboo berimbau tradition cited by Robert Walsh (1830: 176) may be related to the lack of appropriate hardwoods for this bowed instrument. Most of the other ethnospecies were used by only one craftsman and then only rarely. Valmir, who collects *vergas* for himself and to sell wholesale to other craftsmen, could identify three types or "qualities" of *biriba* in the field. Although he believed they were all the same species, he noted nevertheless that the varieties exhibited different morphological properties depending on their specific habitat, that is, sandy or clayey soils, or open or closed canopies.

In spite of the importance of the *verga* to the quality of the instrument, *berimbau* artisans provided surprisingly limited narrative in regards to why the other tree species are less sought after. It became clear in fact that most (4 of 5) informants knew little about the wood properties of the local flora and even less about the identities of trees in the field. Only one actually enters the forest to harvest trees on a regular

Table 8.1 Free-listed ethnospecies used to fashion the berimbau verga

Chupa Molho	Bôbô	Zé do Lenço	Ze Carlos	Valmir das Biribas
Biriba	Biriba	Biriba	Biriba	Biriba roxa
Taipoca	Taipoca	Taipoca	Taipoca	Biriba branca
Canduru	Canduru	Selva de leite	Candeias	Biriba cabeludo
Camaçari	Camaçari	Canduru		Massaranduba
Araça		Vassourinha		Massaranduba-branca
				Taipoca
				Selva de leite
				Canduru
				Camaçari
				Inaíba
				Cabelo de cotia
				Piqia amarelo
				Cafezeiro
				Pindaíba araça
				Pau d'arco

basis; the other 4 generally purchase them from middlemen at markets and from landowners on Itaparica Island. All the informants provided suggestions for the best habitat in which to find *biriba*—"virgin forest on the island of Itaparica", "caatinga and cerrado", "on the coast", and "in the sertão". Most (4 of 5) of the *berimbau* artisans noted that they were told in their youth by their capoeira masters that *biriba* is the best type of wood for making a *berimbau*, and so they continue to use it. Valmir argues, however, that there is a wealth of species that work perfectly well for *vergas*, and that it is simply ignorance that keeps most *berimbau* makers and players demanding *biriba*. According to Valmir, the obsession with *biriba* comes "from the tradition of the great masters like Bimba and Waldemar…and my father. They passed this on to the students—*biriba*, *biriba*, *biriba*". Another suggests that it was the influence of Valmir's father, who is "responsible for the fame of *biriba*".

All of the informants use bottle gourds (*Lagenaria siceraria* (Molina) Standl. *cabaça*) for the gourd resonator (Fig. 8.8). The early observations of a coconut resonator by Robert Walsh (1830) and Henry Koster (1816) were either incorrect identifications or the tradition has been abandoned. We have never seen these used in present day *berimbaus*, and none of our informants had seen this or could imagine it. They note that there is nothing about the coconut's composition that would lend it to giving off the proper sound. None of the informants cultivate *cabaças*, due to space requirements in the city, but rather travel to arid regions in the interior of the state to purchase them in bulk. They also buy smaller numbers at local markets. Although craftsmen agreed (5 of 5) that the most important quality was that the gourd's exocarp be thin, their agreement stopped there. One said that the appropriate properties depended on the needs of the customer, whereas another noted that it should be round but with a straight bottom. Another noted that it depended upon whether the *berimbau* would be used as an accompaniment for regional or Angola capoeira. One of our informants (Bobó) has seen the American calabash fruit (*Crescentia cujete* L., known

Fig. 8.8 Bottle gourds being prepared in Valmir Coutinho Lima's shop, Salvador (Photo by Robert Voeks)

locally as *coite*) used on several occasions as a resonator by contemporary capoeira practitioners. The calabash produces a sound quite similar to a bottle gourd and is readily accessible along the coast of Brazil. According to Budasz (2007), local gourds, whether calabash or bottle gourds, were also used as resonators in seventeenth-century Bahia on four-stringed instruments, or "gourd-guitars".

The last two botanical components are the *caxixi* and the *vaqueta*. The *caxixi* is a small fiber rattle that is usually, but not always, sold and played with the *berimbau*. Three of the five craftsmen always purchase them; two always or sometimes make their own. In either case, they are fashioned from an imported forest vine, woven into a small basket, with seeds in the interior to provide the rattle. According to a *caxixi* maker, the unidentified vines are imported from the northern state of Pará and northeastern state of Maranhão. The sound of the rattle is produced by small dried seeds, especially the seeds from the bananeira do mato (*Heliconia* sp.) or lagrimas de nossa Senhora (Job's tears, *Coix lacryma-jobi* L.), or sometimes just by small pebbles. The *vaqueta* is a small stick, about 0.5 m, used to strike the wire. Three of the five *berimbau* makers use a piece of *biriba* for the *vaqueta*, another uses bamboo, and another uses *ticum*, a local palm. None of the informants felt there was any difficulty procuring material for the *vaqueta*, and most said that they collected it at the same time they were harvesting *biriba* for the *verga*.

Although the specific botanical components that went into fashioning the West African prototype(s) of the Brazilian *berimbau* are unknown, it is evident that the fundamental elements—flexible wooden staff, gourd, stick, vine, and seeds—were available in abundance in Brazil. As the master craftsmen indicated, in spite of the popularity of *biriba* wood, there is a wealth of species that could be substituted for the *verga*. The bottle gourd (*L. siceraria*), although native to the Old World, had arrived in the Americas well before the arrival of African forced immigrants, being the only known early Holocene-cultivated species grown in Asia, Africa, and the Americas (Fuller et al. 2010). The optional *cabaça*, the American calabash (*C. cujete)*, was widespread as well. Columbus noted the exocarp being used to bale water from Indian canoes the day he "discovered" America, making calabash the first New World ethnobotanical observation (Sauer 1993). The species is common in West Africa today and may well have been used in prototypes there, although the earliest positive notation was not until 1841 on the Gold Coast (Alpern 2008). There are myriad lianas and vines that could be used to fashion the *caxixi* shaker, and the seeds were available from native *Heliconias* or from the ubiquitous cultivar *C. lacryma-jobi,* which surely has a long history of cultivation in Brazil. Overall, the native and exotic species composition of Brazil's Atlantic forest would have in no way limited Africans from recreating the one-stringed musical bow. Handcrafted from species or analogues inhabiting both shores of the Atlantic, the presence of the *berimbau* suggests the degree to which trans-Atlantic floristic homogenization facilitated the ebb and flow of botanically based cultural traditions between the Old and the New World.

Berimbau Commercialization

The incorporation of the *berimbau* into capoeira changed the cultural significance of the instrument in many ways. Over time, this humble slave instrument emerged both as the material symbol of the national martial art and as the icon for the Afro-Brazilian experience. Used by nineteenth-century merchants to tout their products, the *berimbau* today is played by vendors in order to sell *berimbaus.* Tourist items that display the image of the *berimbau* are in considerable demand, adorning T-shirts, earrings, necklaces, key chains, pens, and bottle openers. It has become so popular that today telephone booths throughout Salvador are made to imitate a painted *berimbau.*

As demand for *berimbaus* grew, diversification of the instrument followed. One of the pioneers of these changes, and also a great early craftsman of the *berimbau*, was Mestre Waldemar Rodrigues da Paixão of Salvador. According to Shaffer (1977), Waldemar began the commercialization of the *berimbau* in Bahia, and he taught the art of crafting *berimbaus* to one of our informants. Waldemar's works were in considerable demand; to purchase one of his *berimbau*, capoeira players needed to go directly to his neighborhood, or to the old Mercado Modelo in the lower city, or the historic Pelourinho district. One informant (Chupa Molho) recalls that many early capoeira masters purchased *berimbaus* from Waldemar, even the late Mestre Bimba. According to Zé do Lenço, another capoeira master and *berimbau* craftsman, Waldemar baptized and even named his instruments—*Ace de Ouro*, *Ouro Fino*, and *Campeste*. After reviewing a number of photographs of Mestre

Waldemar taken by Brazilian photographer Marcel Gautherot, we noted additional names that Waldemar applied to his berimbaus, such as *Rei do Campeste*, *Campo*, *Flor do Campo*, and *Cacique*. Zé do Lenço suggested that Waldemar used the nicknames as propaganda to sell his *berimbaus*. He noted further that Mestre Waldemar's personal *berimbaus* were made of *taipoca* wood, which is light, flexible, and resistant. Waldemar stated that he was the first to develop the painted *berimbau*, which is now far and away the most popular with tourists (Abreu 2003).

Waldemar eventually had a shop of 15 local boys, including Chupa Molho, assisting to make *berimbaus*. He recalls watching Waldemar make *berimbaus* and play capoeira at the *barracão*, the same open-air room used for his wife's *Candomblé* celebrations. Each child worked on one part of the *berimbau* construction process, for example, washing the bottle gourds, scraping the rubber off the wires cut from car tires, or painting the gourds or *vergas*. At 53 years of age, artisan Chupa Molho continues to make *berimbaus* near Mercado Modelo.

Until recently, capoeira and *berimbau* playing was largely a male domain. Accordingly, most of the *berimbau* makers presently are men, although most (4 of 5) informants were familiar with notable women artisans. Most women are involved in preparing tourist rather than professional *berimbau*. Chupa Molho spoke of two women who make tourist *berimbaus* from *pombo* wood, which is softer and easier to prepare than *biriba*. Bobó, another informant, mentioned a woman who also makes tourist *berimbaus* in the Mercado Modelo. Valmir cited a woman who was particularly good with the finishing stage of preparing the *berimbau*.

As the *berimbau* became increasingly associated with capoeira, especially in response to increasing tourist demand, artisans began to construct a range of quality of *berimbaus*, from children's toys to tourist models to professional quality. Toy *berimbaus* are made with thin, short *vergas* so that young people can easily hold and play them. Tourist *berimbaus* are built with *vergas* that are easy to ascertain and can be mass produced. They are weaker than professional models and offer inferior sound quality. These include the burned *verga* model, the "Olodum" model, and the Bahia model, which is painted with bright colors. Prices vary depending on the size, quality, and paint job. A small tourist *berimbau* sells for about US$3, a medium size costs US$6.50–9.50, and a large tourist *berimbau* costs from US$6.50 to US$12.50.

The highest quality professional *berimbaus* are purchased and played almost exclusively by *capoeiristas* (capoeira practitioners) and musicians. Artisans construct them with the best materials to ensure they produce the finest sound. All of the informants agree that the essence of a quality *berimbau* is to "*casar*," or "marry" the correct *verga* with the correct *cabaça*. This includes choosing the right species of tree, the thinness of the gourd, bending the stick correctly, the correct cut for the *boca* of the gourd, the proper position of the gourd on the stick, and of course a good player. Professional *berimbaus* cost from US$15.00 to US$25.00. Some artisans include a *caxixi*; others do not. Most artisans buy *caxixis* that are already made, so they charge a separate fee for them.

What once consisted of a few *berimbau* musicians constructing their own individual instruments has now become a profession and profitable business. Artisans produce large numbers of *berimbaus*, many destined for the tourist market, others for professional musicians. The livelihoods of the collaborators in this project (4 of 5) depend completely on the manufacture and sales of *berimbau*. One has hired two assistants; another now employs five workers. Most (4 of 5) sell them in open markets, and all sell them in the downtown tourist district, especially the historic Mercado Modelo. Some (2 of 5) send most of their production to the city of São Paulo for sale by others.

The increasing demand for *vergas* fashioned from *biriba* (*E. ovata)* is creating a conservation conflict between those concerned with the region's cultural patrimony and those who seek to protect the region's unique biological heritage. Although other habitats for the collection of *biriba* were mentioned by informants, most trees are harvested from Brazil's Atlantic coastal forest, especially nearby Itaparica Island. This is one of the most biologically rich and acutely threatened biomes in the world (Ribeiro et al. 2011). With over 90% of the original Atlantic coastal forest lost, the Brazilian agency charged with forest use and protection (IBAMA—The Brazilian Institute of Environment and Renewable Natural Resources) has placed a total moratorium on tree harvest on public and private lands. According to Resolution 240/1998 of CONAMA (National Council of the Environment), the extraction, transport, and use of Atlantic forest trees is completely banned. Because of the mounting harvest pressure by *berimbau* craftsmen to meet the burgeoning tourist trade, there have been calls for stepped up enforcement of the harvest ban. At the same time, on June 15, 2008, the Instituto do Patrimônio Histórico e Artístico Nacional (IPHAN) voted to recognize capoeira as a cultural patrimony and to protect the tropical woods used in the construction of the *berimbau*. How the demands of these seemingly competing interests will be met in the future remains to be seen. The artisans censused in the course of this research are adamant that they personally only use sustainable methods to extract *biriba* from the local forests. However, they cannot account for how others collect the trees, and it is clear from walks in the forest that many are being harvested unsustainably. Clearly, a forest management plan that recognizes the biological and cultural values of these forests is needed.

Conclusions

The Brazilian *berimbau* is a one-stringed musical bow of West African origin. Early utilized by street vendors to sell their goods, the *berimbau* became the principal instrument of the martial art-dance capoeira sometime in the late nineteenth century. The instrument is fashioned by master craftsmen using mostly botanical components. Diffusion of this cultural tradition was facilitated in part by the pantropical distribution of cultivated species. Capoeira was officially banned in 1890, and there have even been accounts of police confiscating *berimbaus* during the early twentieth century in Salvador. Remarkably, the once outlawed activity of capoeira became a national Brazilian patrimony in 2008, and the *berimbau* became the only musical instrument to receive national protection as a result. Today, the humble *berimbau* is not only an icon of Afro-Brazilian culture but a symbol of Brazilian nationality.

Acknowledgments We wish to thank Hildo Leal da Rosa, Lindinalva Costa dos Santos, Telma Vasconcelos, Frederico Carvalho, and staff at the Arquivo Público Estadual Jordão Emerenciano (APEJE) in Recife; Maria Paula Fernandes Adinolfe and staff at the Instituto do Patrimônio Histórico e Artístico Nacional (IPHAN) in Salvador; Maria Inez Turazzi, Rosana Carvalho de Oliveira Miranda, and Neibe Machado da Costa at the Museu Imperial in Petrópolis; Cristina Zappa and Vera Lúcia Ferreira da Silva at the Instituto Moreira Salles in Rio de Janeiro; Glaucia Abreu at the Museu Castro Maya in Rio de Janeiro; Rosane Nunes at the Biblioteca Nacional in Rio de Janeiro; and the staff at the Instituto Arqueológico, Histórico e Geográfico Pernambucano (IAHGP). We also thank Frederico Abreu and Roberto Benjamin for opening up their libraries to us, and Maurício Lemos de Carvalho (Mestre Vermelho de Pastinha) and Naná Vasconcelos for their valuable insights. And a special thanks to Antônio Albino Soares (Mestre Angola) for his assistance in this project.

The help of our friends Steve Antonson, Michael Goode, Priscila Cardoso Ferreira, Luiz Badaró, Lenilda Maria de Abreu, Alípio Durans, Eduardo Pereira, and Cristina de Fatima Mendonça during the last 4 years is greatly appreciated, as is the illustration drawn by Kelly Donovan.

Most importantly, we thank the berimbau artisans José Carlos Melo (Zé Carlos), Jaime Sobral dos Santos (Bobó), José Alves (Zé do Lenço), Valmir Coutinho Lima, and Egberto Emilio da Silva (Chupa Molho). We also thank Mestre Marinheiro, Mestre Liberino, Mestre Adó, Mestre Lua Rasta, Contra-Mestre Boi Manso, and the late Valdovino Lima.

References

Abreu F (2003) O barracão do Mestre Waldemar. Organização Zarabatana, Salvador

Alpern SB (2008) Exotic plants of western Africa: where they came from, and when. Hist Afr 35:63–102

Alvarenga O (1950) Música popular Brasileira. Editôra Globo, Rio de Janeiro

Anunciação L (1971) O berimbau da Bahia. Revista brasileira de folclore 11(29):24–33

Araújo AM (1964) Folclore nacional, vol II. Companhia Editoria Nacional, São Paulo, pp 432–433

Bolster WJ (1997) Black jacks: African American seaman in the age of sail. Harvard University Press, Cambridge

Budasz R (2007) Black guitar players and early African-Iberian music in Portugal and Brazil. Early Music XXXV:3–21

Carney J, Voeks R (2003) Landscape legacies of the African diaspora in Brazil. Prog Hum Geogr 27:139–152

Carvalho H (1890) Etnografia e historia tradicional dos povos do Lunda. Imprensa Nacional, Lisboa

Carybé HB (1955) O jogo da Capoeira: 24 Desenhos de Carybé. Coleção Recôncavo n.° 3. Salvador, Bahia: Livraria Progresso Editora

Costa e Silva A (2004) Africa-Brazil-Africa during the era of the slave trade. In: Curto J, Lovejoy P (eds) Enslaving connections: changing cultures of Africa and Brazil during the era of slavery. Humanity Books, New York, pp 21–27

de Tollenare LF (1905) Notas Dominicaes. Empreza do Jornal do Recife, Recife

Debret JB (1989 [1834–39]) Viagem pitoresca e histórica ao Brasil. Editoria Itatiaia, Belo Horizonte

Dos Santos FM (1941) O ambiente artístico Fluminense a chegada da missão Francesa em 1816. Revista do Serviço do Patrimônio Histórico e Artistico Nacional 5:213–238

Eltis D, Richardson D (2008) A new assessment of the transatlantic trade. In: Eltis D, Richardson D (eds) Extending the frontiers: essays on the new transatlantic slave trade data base. Yale University Press, New Haven, pp 1–60

Fryer P (2000) Rhythms of resistance: African musical heritage in Brazil. Wesleyan University Press, Hanover

Fuller D, Hosoya L, Zheng Y, Qin L (2010) A contribution to the prehistory of domesticated bottle gourds in Asia: rind measurements from Jomon Japan and neolithic Zhejiang, China. Econ Bot 65:260–265

Graham M (1824) Journal of a voyage to Brazil and residence there, during part of the years 1821, 1822, 1823. Longman, London
Graham R (1991) Technology and culture change: the development of the Berimbau in colonial Brazil. Lat Am Music Rev 12(1):1–20
Herskovits M, Herskovits F (1943) The negroes of Brazil. The Yale Review 32:263–279
Koster H (1816) Travels in Brazil. In: Harvey GC (ed) Southern Illinois University Press, Carbondale
Kubik G (1979) Angolan traits in black music, games and dances of Brazil. A study of African cultural extensions overseas. Centro de Estudos de Antropologia Cultural, Lisboa
Lewis JL (1992) Ring of liberation: deceptive discourse in Brazilian capoeira. University of Chicago Press, Chicago
Oliveira AM (1956) Berimbau o arco musical da capoeira na Bahia. Revista do Instituto Geographico e Historico da Bahia 80:224–251
Oliveira AM (1958) *Berimbau*. Comissão Bahiana de Folclore, Salvador. (Coleção Antônio Viana:1)
Price S (2006) Seaming connections: artworlds of the African diaspora. In: Yelvington K (ed) Afro-Atlantic dialogues: anthropology in the diaspora. School of American Research Press, Santa Fe, pp 81–112
Querino M (1955) A Bahia de outrora. Editora Progresso, Salvador
Ramos A (1935) O folk-lore negro do Brasil. Civilização Brasileira, Rio de Janeiro
Ribeiro M, Martensen A, Metzger J, Tabarelli M, Scarano F, Fortin M (2011) The Brazilian Atlantic forest: a shrinking biodiversity hotspot. In: Zachos F, Habel J (eds) Biodiversity hotspots: distribution and protection of conservation priority areas. Springer, Heidelberg, pp 405–434
Sauer J (1993) Historical geography of crop plants. CRC Press, Boca Raton
Schneider J (1991) Dictionary of African borrowings in Brazilian Portuguese. Buske, Hamburg
Shaffer K (1977) The berimbau-de-barriga e seus toques. Monografias Folcóricas. Fundação Nacional de Arte, Rio de Janeiro
Silva LG (2005) Escravos das aguas. Nossa Historia 15:66–71
Sweet J (2009) Slaves, convicts and exiles: African travelers in the Portuguese Atlantic World, 1720–1750. In: Williams C (ed) Bridging the early modern Atlantic World. Ashgate, London, pp 193–202
Thompson RF (1989) Capoeira: the Afro-Brazilian martial art. (A on the back cover of LP by Mestre Jelon Vieira)
Verger P (1952) Cartas de um Brasileiro estabelicido no século XIX na Costa dos Escravos. Anhembi 6:212–253
Voeks RA (1995) Candomblé ethnobotany: African medicinal plant classification in Brazil. J Ethnobiol 15:257–280
Voeks RA (1997) Sacred leaves of Candomblé: African magic, medicine, and religion in Brazil. University of Texas Press, Austin
Voeks RA (2009) Traditions in transition: African diaspora ethnobotany in lowland South America. In: Alexiades M (ed) Mobility and migration in indigenous Amazonia: contemporary ethnoecological perspectives. Berghahn, London, pp 275–294
Walsh R (1830) Notices of Brazil in 1828 and 1829, vol II. Frederick Westley and A. H. Davis, London
Wetherell J (1860) Brazil: Stray notes from Bahia. In: William H (ed) Webb and Hunt, Liverpool

Part III
Medicinal and Spiritual Ethnofloras

Chapter 9
Trans-Atlantic Diaspora Ethnobotany: Legacies of West African and Iberian Mediterranean Migration in Central Cuba

Erica S. Moret

Abstract This chapter explores variance in ethnobotanical knowledge based on parallel documented uses between Cuba and its two main centers of mass migration: West Africa and the Iberian Mediterranean. Based on 12 months' fieldwork, it explores the link between medicinal plant-use knowledge, land use, and migrant history at selected sites in central Cuba. Employing natural and social science methods, it seeks to offer a geographical perspective to the study of trans-Atlantic diaspora ethnobotany and New World religious syncretism. Results demonstrate that second-generation residents of selected sites with extensive African immigration (sugar cultivation zones) demonstrate a more detailed knowledge of West African-derived medicinal pharmacopoeia than those in areas with stronger Mediterranean immigration (tobacco cultivation zones). Knowledge derived from the Iberian Mediterranean is more widespread between both agricultural zones and marginally stronger in tobacco cultivation zones of smaller settlement types. This chapter concludes that various strands of diaspora plant-use knowledge thrive in distinct ways in different parts of Cuba, which runs counter to official discourse that portrays traditional medicine as largely derived from Spanish communities. In reexamining religious syncretism in the context of ethnobotany, it advocates a stronger recognition of the important role that African-derived ethnobotanical knowledge has played in Cuba's colonial and postcolonial history.

Keywords Afro-Cuban religion • Syncretism • Mediterranean • West Africa • Medicinal plants

E.S. Moret, Ph.D., M.Sc., B.Sc. (✉)
Jesus College, Oxford University Centre for the Environment, University of Oxford, Oxford, UK
e-mail: erica.moret@sant.ox.ac.uk

R. Voeks and J. Rashford (eds.), *African Ethnobotany in the Americas*,
DOI 10.1007/978-1-4614-0836-9_9,

Introduction

Political, economic, and demographic change has played a crucial role in the Caribbean in shaping the natural environment and the ways in which people use natural resources (Klak and Jackiewicz 2002; Bryant and Bailey 1997). Cuba is a prime example of a Caribbean island where nature and society alike have been influenced by the country's rich migrant history and various waves of geopolitical change over the centuries (Moret 2008a). Ethnobotanical practices in Cuba serve as a metaphor for the island's rich migrant history, incorporating a combination of medicinal and ritual practices from around Africa, Europe, Asia, and the New World, together with those retained from Amerindian groups. Throughout its colonial and postcolonial history, the Caribbean island's rich plant-use traditions have played an important, though as yet poorly recognized, role in helping Cubans adapt and survive in the face of physical hardship, disease, and material scarcity.

Although an extensive body of literature exists on the Spanish influences on plant-use practices in the Americas, scholarly attention has remained more limited with regard to African contributions, and even more so concerning those from Asia (Hammer and Esquivel 1992). Further, while many studies have successfully documented the ways in which plants are used for medicinal and ritual purposes in the New World, less have paid attention to the wider political, economic, or historical context (exceptions include Carney 2001, 2003a, b; Carney and Voeks 2003; Voeks 1997 and Rashford 1993). This is particularly marked in Cuba, where researchers face significant localized constraints. These include a severe deficiency in material resources, a lack of tradition in conducting transdisciplinary research, and difficulties encountered by social and natural scientists – Cuban and foreign alike – in conducting politically focused studies that may be perceived as critical of the government and its policies.

With the aim of contributing to the existing literature by incorporating a geographical and historical focus to the study of Cuban diaspora ethnobotany, this chapter explores variance in medicinal-plant knowledge according to the agricultural and migrant history of selected sites in central Cuba. It draws on works couched in poststructural political ecology and environmental history frameworks and examines the notion of religious and ethnobotanical syncretism to highlight the (as yet) poorly recognized contribution of African-derived systems to modern-day Cuban ethnomedical systems.

Diaspora Ethnobotanical Studies in the New World and Cuba

European contributions to ethnobotanical knowledge and practices in the New World have been well documented since colonial times (Dunmire 2004; Barrera 2002; Guerra 1966). Mass Spanish migration to Cuba occurred between 1861 and 1953 (Girardi 1994), when around one million migrants arrived to the island (Thomas 1971). These migrant workers came largely from Andalusia, Castile, Galicia, and the Canary Islands (Hammer et al. 1992), often bringing with them knowledge of medicinal plant use. Studies on medicinal ethnobotanical practices in Cuba in a non-ritual context (commonly known as *medicina verde* or green medicine) include those by Vasquez Davila (2001), Martínez Callis et al. (1997), Fuentes and Granda (1997), Roig y Mesa (1959, 1988), Cabrera (1984), and Seoane Gallo (1988).

Medicina verde is typically (but somewhat erroneously) connected in the Cuban psyche to the island's Spanish immigrant communities (Moret in press) and thought to be more widely practiced in rural areas. The field has resurged in recent decades, whereby traditional medicine has transcended from the prohibited and discouraged sphere to being actively embraced into the country's national healthcare system (Moret 2008b). This has largely been a consequence of the dire medicinal shortages associated with Cuba's post-Soviet era (since 1989). The US trade embargo, tightened in the 1990s, is also to blame, however. In contravention of a range of international conventions and agreements, it represents the most stringent and longest-standing set of economic sanctions in the world, extending to food and medicines (Garfield and Santana 1997).

Susan McClure (1982) was one of the first scholars to illustrate the importance of African contributions to New World ethnobotanical systems, in seeking to highlight how species such as *Citrus aurantifolia* (Christm.) Swingle, *Abrus precatorius* L., and *Ricinus communis* L. have similar uses in West Africa and the Caribbean (all non-English words, botanical scientific names, and Cuban common plant names are italicized throughout this chapter). More recent studies have emphasized the importance of African knowledge to traditional medicine, agricultural practices, and ritual plant use in the New World (Carney 2001, 2003a, b; Carney and Voeks 2003; Rashford 1987, 1993; Esquivel et al. 1992; Fuentes 1992; Brandon 1991). In Cuba, a range of authors have documented the ways in which practitioners of the island's various African-influenced religions, secret societies, and cults use plants for a combination of medicinal, ritual, ceremonial, culinary, poisonous, and other uses (Volpato et al. 2009; Bolívar Arostegui 2001; Cabrera 1984, 2000; Pérez Medina 1998; Betancourt Omolófaoró Estrada 1997; Moreno Rodríguez et al. 1995; Fuentes 1992; Brandon 1991; Bascom 1950; Volpato & Godinez 1994).

African-influenced religions are widespread throughout Cuba, brought to the island by approximately one million enslaved Africans who arrived in Cuba from 1510 until 1886 (Mayor 2003), and most have rich ethnobotanical traditions. Followers of one faith will often belong to a number of other faiths simultaneously, signifying that they may draw their ethnobotanical knowledge from a range of sources. Most prominent in Cuba are the Yoruba and Spanish Catholic influenced *santería* (or *regla ocha*) and cult of *ifá*. Also prevalent are the Bantu-derived *reglas de congo* (including *palo monte*), alongside *abakuá* (or *ñáñiguismo*), based on similarly structured organizations in the Cross River region of modern-day Nigeria and Cameroon. *Espiritismo*, practiced largely in urban areas, was introduced to Cuba in the early twentieth century from Europe via the United States and employs African-derived techniques borrowed from other syncretic Cuban cults. Also popular in localized parts of the island are *regla arará*, influenced by the Ewe/Fon cultures of what is now Benin, and Haitian *vodún* (voodoo), brought to Cuba through various waves of Haitian immigration.

Studies on Amerindian contributions to modern-day Cuban medicinal plant-use practices are less widespread and merit further study. On Columbus' arrival in Cuba in 1492, Amerindian groups inhabiting the island totaled an estimated 100,000 and included Ciboneys, Guanajatabeyes, and Tainos (Mayor 2003; Barnet 1986). By 1524, less than a 1,000 survived. In today's Cuba, many argue that no true indigenous populations exist, though Amerindian influences persist in tobacco cultivation methods, alongside some medicinal plant-use practices, language, rural building techniques, cookery, and religious rituals (Cabrera 1984).

The waning of African slavery in Cuba marked the start of the immigration of indentured laborers to Cuba, largely originating from China. Hammer and Esquivel (1992) provide a useful account of botanical introductions by Chinese immigrants to Cuba, though wider studies on Asian contributions to Cuban ethnobotanical systems remain limited and represent a worthwhile area for further study.

This chapter examines the ways in which plant-use knowledge derived from Cuba's two principal areas of mass migration (Spain and West Africa) continue to persist in modern-day Cuba. Accordingly, the author selected six sites in central Cuba split between two agricultural zones and differing conurbation sizes. Primary data collection was obtained in sugar cultivation areas (with their long-standing links to African forced labor) and tobacco regions (historically based on Spanish migrant work).

Political Economy of Sugar and Tobacco Cultivation

A central thesis of Cuban scholar Fernando Ortiz's seminal 1940 text, *Contrapunteo cubano del tobaco y el azúcar*, was that all aspects of Cuban life – from politics and economics to nature and society – must be understood in the context of the all-pervasive influence of tobacco and sugar cultivation. Although no longer the most financially significant industries on the island (superseded by tourism, remittances, and nickel mining as of 2006) (Ernst and Young 2006), sugar and tobacco cultivation continue to dominate industrial life and land use in much of the island.

Tobacco cultivation is carried out in small-sized plantations in Cuba, known as *vegas*, which, historically, have been run without the use of extensive or complex machinery or infrastructure. Based on Amerindian practices, tobacco farming and its commercialization have been largely the industry of Spanish-descended peasants, known as *monteros*, *sitieros*, or *guajiros*. Tobacco farming typically provided a continuous harvest throughout the year and regular, though modest, incomes for those involved. This signified that a relatively comfortable lifestyle and stable health conditions were afforded to residents in tobacco cultivation zones (Ortiz 1940).

In contrast to tobacco, the highly capitalistic and territorially expansive venture of sugar production, carried out at sites known as *centrales*, has required masses of workers, largely derived from labor forces made up of enslaved African or African-descended laborers (Ortiz 1940). By 1810, some 20,000 enslaved Africans were being transported to Cuba each year, driven by the labor demands of the sugar revolution (Mayor 2003). A complex infrastructure was required, run by city-dwelling *hacendados* and generating enormous wealth to a powerful minority (Ortiz 1940). By the late nineteenth century, although all provinces of the island were engaged in sugar production, significant urban areas had developed around principal sugar plantation areas in the cities of Havana, Matanzas, and Santa Clara (Boyer 1939). Native to tropical Southeast Asia, sugar is harvested seasonally (Schwartz 2004). This formerly resulted in mass unemployment for parts of the year. As a consequence, this factor, alongside poor living conditions in *centrales*, meant that sugar cultivation areas have been long-standing sites of intense poverty, disease, and malnourishment, in contrast to those of tobacco production (Ortiz 1940).

These factors are taken as a basis for this study, which assumes that residents of areas of long-term sugar and tobacco cultivation production (with their differing migrant and socioeconomic pasts) would have sought solutions to their health problems in historically distinct ways and drawing on different, but often overlapping, ethnomedical knowledge systems. Before proceeding to describe methods and results employed, this chapter first provides a background to the ways in which African and European diaspora ethnobotanical knowledge and botanical taxa were first introduced into Cuba (bearing in mind that they did not necessarily occur simultaneously).

African and European Diaspora Ethnobotany in Cuba

Similarities at the species level between Cuba and either Africa or Europe are rare, and any parity that does exist is thought to be due largely to dispersal by wind and water, alongside introductions by humans or animals. Many similarities that exist at the plant family level between Cuba and West Africa are attributable largely to the fact that parts of the independent land blocks that now make up Cuba were, like Africa, formerly part of the pre-Cretaceous landmass of West Gondwana (Santiago-Valentin and Olmstead 2004).

A range of actors are thought to have contributed to the introduction of new plant species to Cuba over the centuries, combined with knowledge on how they could be used for medicinal purposes. This is likely to have included colonial authorities, spice traders, missionaries, and agricultural migrant workers (from Europe); captive slaves, deckhands, and medicinal plant traders (from Africa); and indentured laborers (from China).

Prehuman Introductions

A number of medicinally useful nonnative plants are thought to have arrived in Cuba through wind and water dispersal, long before the colonial era. The majority of these taxa originate from tropical South and Central America, although African and North American contributions have also been significant (Siegel 1991) (see the methods section, below, on difficulties in categorizing a plant's place of origin).

The *güiro amargo* or *güiro cimarrón*/bottle gourd (*Lagenaria siceraria* (Molina) Standl.) is an example of a precolonial ruderal (or weedy) species, which is found on both sides of the Atlantic. Thought to be native to Africa, the bottle gourd is well adapted to long-distance sea transport. Taking advantage of oceanic surface currents and trade winds in the tropical Atlantic region flowing from east to west, it is believed to have floated across the Atlantic Ocean to the Americas (Voeks 1997). With many thousands of years of cultural history both in Africa and the New World (*ibid.* 1997), *L. siceraria* is used to treat parasites, headaches, and toothache in Cuba (Roig y Mesa 1959) and is also used as a sacred vessel to hold ritual herb concoctions (the *osaín*).

Colonial Agency in Plant Introductions to Cuba

Some species that were commonly used in Africa and Europe are believed to have been introduced accidentally to Cuba and elsewhere in the Americas. Seeds mixed in with cereals for cultivation and found in the straw bedding on slave ships, for example, were inadvertently transported across the Atlantic, where they rapidly spread throughout the Americas (Voeks 1997). The majority of botanical introductions of nonnative species to Cuba are thought to have occurred through deliberate human agency, however. As imported medicines were prohibitively expensive in the colonial Americas (*ibid.* 1997), the economic incentive was strong to investigate the efficacy of locally used medicinal species and to import those already used in Europe and from elsewhere for cultivation.

Species native to the Mediterranean, which were introduced to the region from as early as the sixteenth century, include *hinojo*/fennel (*Foeniculum vulgare* Mill.), *mejorana*/marjoram (*Origanum majorana* L.), *mostaza negra*/black mustard (*Brassica nigra* (L.) K. Koch), *salvia de castilla*/sage (*Salvia officinalis* L.), and *ruta*/rue (*Ruta graveolens* L.) (Dunmire 2004). Medicinal and nutritional species that had been cultivated in Europe for a prolonged period of time before the advent of colonialism, but had originated elsewhere, were also cultivated to the wider from an early stage. From Asia, these included *canela de China*/cinnamon (*Cinnamomum* spp.) and *caña fistula*/cannafistula (*Cassia fistula* L.) and from Africa include *albahaca cimarrona*/African basil (*Ocimum gratissimum* L.) and *albahaca blanca*/sweet basil (*O. basilicum* L.) (Dunmire 2004; Watlington 2003; Voeks 1997).

Spanish settlers also introduced numerous species of global distribution to Cuba. *Llantén mayor*/common plantain (*Plantago major* L.), for example, is a perennial ruderal herb, native to temperate Western Europe but widespread in many parts of the world with similar climates. *P. major* is used in both Cuba and Spain for the treatment of diarrhea and to reduce swellings on the skin (Torres Jiménez and Quintana Cárdenes 2004). *Higuereta*/castor bean (*R. communis*) is thought to have originated from Africa, but diffused to Asia around 3,000 years ago (Carney 2003b), and from there entered Europe. It is used for similar medicinal applications in Africa, Asia, the Caribbean, North America, and Europe (Carney 2003b; McClure 1982) including as a laxative and to treat skin complaints. It has been used as a key medicinal and ritual species in the Caribbean since at least the mid-nineteenth century by a range of actors (Dunmire 2004), and its uses in Cuba extend to the treatment of skin complaints and swellings, respiratory problems, and the warding off of malevolent witchcraft.

African crops and other culinary species were also imported by colonial authorities to provide cultivars for slave communities. Enslaved Africans were generally expected to tend to their own health problems under Spanish rule in Cuba (Ortiz 1975) and were permitted to cultivate land, known by the Amerindian-influenced word, *conucos* (Barnet 1986), from which surplus products could be sold in the marketplace. These included those that continue to be important to modern-day Cuban medicinal and culinary practices, such as *gandul*/pigeon pea (*Cajanus cajan* (L.) Millsp.), *quimbombó*/okra (Abelmoschus esculentus (L.) Moench), *seso vegetal* or *akee de Africa*/akee (*Blighia sapida* K.D. Koenig), and *ataré*/melegueta pepper (*Aframomum melegueta* K. Schum.).

African Agency in the Development of Cuban Ethnobotany

The ethnic groups most significantly represented as slaves in Cuba, particularly those of Yoruba and Dahomean decent, were also those that had among the world's most developed ethnomedical systems at the time of colonial rule. The savanna-rainforest zone spanning Cameroon and Nigeria, from where many of these groups originated, was one of three African centers, or "cradles," of agricultural domestication, where over 2,000 native plant species were being cultivated for food and magico-medicinal use at the time when the slave trade began (NRC 1996 in Carney and Voeks 2003). Yoruba communities, in particular, were widely credited for their healing skills by the various colonial authorities in the Caribbean (Carney 2003b).

Scholarly debate remains divided over whether enslaved Africans would have been capable of importing useful botanical species to the Americas given the brutal conditions of imprisonment and transportation (Miller 1988 in Voeks 1997). Emerick (in Williams 1932: 202), for example, describes the common practice of the shaving of enslaved Africans' heads before the trans-Atlantic journey for the fear that they would hide plants under their hair that may be used for witchcraft or poison. Voeks (1997: 198) also notes that "[a]lthough unnatural concentrations of alien plants and animals may occur along religious migration routes the advertent transplantation of exotics for religious motives is rare."

While deliberate introductions of useful species by enslaved Africans are unlikely to have occurred on a frequent basis, the fact that useful ethnobotanical knowledge was imported is somewhat less contentious. On one side of the debate, Morgan (1997) argues that enslaved Africans in their teenage years and twenties, when sold as captives to European traders, would have been too young to have acquired substantial ritual knowledge. This would have rendered them unable to transmit religious and plant-use knowledge on arrival in the Americas. Nevertheless, Reyes-García et al. (2006b) show that most acquisition of ethnobotanical knowledge takes place before the age of 15. This supports the idea that enslaved Africans, even from a young age, would have been capable of carrying and transmitting plant-use knowledge to places like Cuba.

The sale of ritually and medicinally useful African species to the New World began at an early stage of the slave trade, which would have facilitated the presence of species that would be familiar and useful to enslaved Africans. African healers and deckhands who were employed to tend to the health needs of captive Africans in slave ships are also likely to have transported, and subsequently introduced, seeds of important medicinal value to places like Cuba. Encounters with floristically similar habitats[1] and the presence of pantropical ruderal species on both sides of the Atlantic would also have contributed to the opportunity for experimentation and adoption of naturalized species in Cuba by enslaved Africans and their descendants.

On the West African coast, European traders and freed slaves alike were running lucrative plant-export businesses to the Americas by the nineteenth century (Voeks 1997). This trade included *orogbo*/bitter kola (*Garcinia kola* Heckel), the aforementioned *ataré*/melegueta pepper (*A. melegueta*), *kola*/kola nut (*Cola acuminate* (P. Beauv.) Schott and Endl.), and *obí*/kola nut (*C. nitida* (Vent.) Schott and Endl.).[2] Kola nuts, for example, have been prized for many centuries on both sides of the Atlantic as stimulants and for their use as protection against witchcraft (Voeks 1997).

In sum, the trading and translocation of plants throughout Europe and its colonies resulted in a massive multidirectional spread and mixing of plant species – like its peoples and diseases – between the Americas, Africa, Europe, and Asia.[3] As Enlightenment ideas began to spread throughout the world, diffused by the rapidly expanding European colonial powers, the era simultaneously also saw "a global circulation of the hidden knowledge" that enslaved Africans carried with them on their forced journeys (Geschiere 2006: 120), which extended to ethnomedical techniques.

---

[1] Yorubaland was a principal source of slaves brought to Cuba. The area, which is now part of Nigeria and Benin, spans a range of similar ecosystems to those that make up the Cuban landscape, which include rainforest, savanna, marine, coastal, lagoon, mangrove swamp, and estuarine habitats. Encounters with floristically similar habitats on arrival to Cuba are likely to have provided enslaved Africans with the opportunity for experimentation and adoption of new plants to be used for medicinal purposes.

[2] Many species used in Cuba which have strong cultural ties with West Africa continue to be known in everyday parlance by their African lexical names or some form of references to West Africa (indicated by the postfixes *de Guinea* and *de Africa*). These include *eru* or *pimienta de Guinea*/African Guinea pepper (*Xylopia aethiopica* (Dunal) A. Rich.), *quimbombó*/okra (*Abelmoschus esculentus* (L.) Moench), *corojo de Guinea*/oil palm (*Elaeis guineensis* Jacq.) *aleluya roja de Guinea*/roselle (*Hibiscus sabdariffa* L.), *yerba de Guinea*/ Guinea grass (*Panicum maximum* Jacq.), *akee de Africa*/ackee (*Blighia sapida* K.D. Koenig), *ñame*/yam (*Dioscorea alata* L.), and *fufú*/plantain (*Musa* × *paradisiaca* L.) (Roig y Mesa 1988). Further, plants used in a religious context in Cuba are known largely by lexicons employed in ritual languages including the Yoruba-derived *lucumí* (in *santería*) and Bantu-derived terminologies (in *palo monte*) (Cabrera 2000). See Palmié (2005) for an interesting example of how the tracing of plant name linguistic origins can be a complex task.

[3] Multidirectional flows of ethnobotanical knowledge and practices followed not only the path of West African captive migrants in their establishment in the Caribbean but also a range of other complex networks of human interactions produced through the slave trade between Europe, Asia, and the New World (Argenti and Rochenthaler 2006). A botanical side effect of colonial trade, for example, was that New World species like pineapples, cashews, capsicum peppers, and papaya were rapidly being cultivated in parts of Asia and Africa (Voeks 2004). Many were also incorporated into local healing, hunting, and ritual practices in these regions. Poisons extracted from *Capsicum* and *Nicotiana*, which were introduced to Africa in the seventeenth century from the Americas, have been recorded as fish poisons in parts of West Africa and also serve important ritual functions (Neuwinger 2000).

The following section proceeds with a description of the epistemological approach guiding research for this chapter, followed by details on methods employed. Survey results are then presented followed by a discussion of the implications of West African and Mediterranean contributions to current-day traditional medicine and religious practices in Cuba.

Epistemology and Methodology

Theory

By emphasizing the diversity of ways in which different actors interact with, and make use of, the biophysical world, recent studies focusing on environmental history have successfully highlighted the relationship between memory, natural resource use, and peopled landscapes, often drawing from a political ecology framework (Carney 2003a; Fairhead and Leach 2003; Voeks 1997). Research for this chapter draws on poststructural political ecology, environmental history, and social anthropology studies to explore how distinct forms of knowledge linked to the natural world (in this case West African- and Mediterranean-derived ethnobotanical systems) may be linked to questions of power, identity, and resource access.

Political ecology has been developing since the early 1990s and is suited to trans-disciplinary studies that aim to incorporate social and natural sciences techniques (Peet and Watts 1996). Carney and Voeks (2003), for example, employ such a focus in their study focusing on the ethnobotanical legacies of the trans-Atlantic slave trade and associated acts of agency, resilience, and creativity in the Americas. The burgeoning field, which merits further scholarly attention, is described by Zimmerer and Bassett (2003: 15–16) below:

> [a pioneering] geographic and historical framework for diasporic, North–south political ecology…[creating] a historical vision that encompasses the integration of political-social relations of power *and* the dynamics of ecological-biogeophysical processes…[which can] reveal the prominence of hybrid cultural ecologies of indigenous, mestizo, Creole, and African American peoples.

This chapter also seeks to offer a geographical perspective to the study of trans-Atlantic diaspora ethnobotany through exploring notions of "syncretism" in connection with medicinal-plant knowledge acquisition. In order to do so, it draws on a range of social anthropology works, including Palmié (2005), Apter (2004), and Pels (1997).

Ethnobotanical Methods and Survey Design

The author employed a combination of botanical and ethnographic techniques during 12 months of fieldwork in Cuba carried out between June 2005 and September 2006. Research was conducted as part of a wider social and natural sciences study on political, economic, and environmental change in post-Soviet Cuba (Moret 2007). The survey forming the basis of this chapter was carried out in the central Cuban provinces of Cienfuegos, Villa Clara, and Sancti Spíritus.

During the first stage of research, the author carried out semi-structured interviews with 21 medicinal plant experts in the areas under study to identify their knowledge of West African and Spanish diaspora plant use. This enabled the author to compile an inventory of medicinal plant species used in the region and narrow the list down to those with clear parallels in plant-use traditions with either West African or Iberian Mediterranean centers of early plant cultivation (including native or non-native species of the two regions).[4] Plants that are believed native to Cuba were not included in the inventory as the main question being asked by the survey specifically related to knowledge of nonnative plants. A combined focus on native and nonnative plants could represent a more comprehensive approach to the study of diaspora plant-use knowledge in future studies, however.

The author proceeded to select from the inventory those species derived from either Iberian Mediterranean or West African centers of early medicinal plant use (i.e., those thought to be native to Africa or southern Europe, alongside those that have strong histories of cultural or medicinal use in either of the two regions). The resulting list included a combination of species of pantropical, Old World, and Asian distributions (loosely following Carney's classificatory system 2003b: 169).[5] Sixty-four species in total were selected for inclusion in the survey based on these criteria (serving as the principal tool for data collection for this chapter). The survey list included 46 species originating from West African centers of early plant cultivation and 27 species originating from Mediterranean centers of early plant cultivation. This included nine species shared between the two regions.[6] The author then constructed a Microsoft Access database,

[4] The categorization of a plant as native or nonnative is a problematic one and depends on when, and the way in which, it was introduced to a given region. Information on the ecological, geographical, and genetic history of a given plant is typically used to define its region of origin, though this information is not always readily available. Furthermore, early plant introductions that occurred during the colonial period were not always recorded, and when cited, it was frequently without clear illustrations or reference to voucher specimens (Voeks 1997).

[5] Many plant species from Asia, sub-Saharan Africa, and the Middle East came to be cultivated in the Mediterranean before the advent of colonialism, particularly during the era when the Iberian Peninsula was under Moorish rule (Dunmire 2004). A similar situation is also true in West Africa, which counted on Arab and European trade for centuries before the advent of the trans-Atlantic slave trade (Voeks 1997).

[6] The nine species that have strong histories of medico-cultural use in both Africa and Europe (according to cited literature in the main text) happen also to be those that are listed in Canary Island ethnobotanical texts. These species were included in order to take account of the more "Africanized" ethnobotanical practices originating from these Spanish islands from where many tobacco workers in Cuba originated. These islands, lying just 108 km off the coast of northwest Africa and 1,129 km from mainland Spain, are far closer to the African continent in geographical terms and therefore share some floristic similarities with mainland Africa together with mainland Europe. Furthermore, an African presence in the Canary Islands was strong from an early stage of the colonial era (Schwartz 2004). Sugar cultivation on the island also depended upon African slave labor, and it also served as a stopover for enslaved Africans en route to the New World.

which was used as a basis for entry of the inventory and subsequent results (Appendix 1). The two main groups that make up the 64 survey species will be referred to, hereafter, as "West African" and "Mediterranean" species, for the sake of clarity, though this is not to suggest that they are necessarily considered native to these historically and geographically linked regions.

The second step in survey preparation involved literature research, designed to gather information on as many as possible of the listed uses of each of the 64 species, in their respective areas of early plant use. The author cross-referenced a range of ethnobotanical sources for West Africa and the Western Mediterranean to build a comprehensive and detailed list of medicinal uses for each species (Aiyeloja and Bello 2006; Neuwinger 2000; Verger 1995; Gill and Akinwumi 1986; Buckley 1985; Ademuwagun 1979; Ayensu 1978; Bascom 1972; Ainslie 1937 and Dalziel 1937 [for West Africa] and Dunmire 2004; Torres Jiménez and Quintana Cárdenes 2004; Vázquez et al. 1997; González-Tejero et al. 1995; Alcazar et al. 1990 and Darias et al. 1986, 1989 [for the Iberian Mediterranean]) (Appendix 1). Listed plant use in the available source literature (i.e., focusing on African and southern European practices) does not necessarily represent the full range of possible uses of each plant in the two world regions. Further, not all species with African origins or connotations were included in the survey detailed in this chapter as they were not all commonly used by surveyed participants and not typically available on a commercial basis (as based on central Cuban market studies in Moret 2007).

Medicinal use categories were kept deliberately broad, in line with the more general terminology used in the ethnobotanical literature (such as "headache," "skin complaint," "sore throat," and "aphrodisiac"), without describing medicinal applications or ailments in more detail. Although plants used in a medicinal context typically also serve ritual functions in Cuba, the scope of the study was narrowed to focus only on the medicinal applications of diaspora plant-use practices. Further study to investigate diaspora ethnobotanical knowledge in the ritual sphere in connection with land use practices would be a worthwhile addition to the literature as would a study that incorporated a wider selection of study sites and participants throughout Cuba.

The third step of survey design involved the preparation of visual cues to enable a more thorough opportunity for plant identification by survey participants. Specimens of each of the 64 listed plants were purchased from herb vendors in the town of Cienfuegos and identified to the species level by local botanists. The author further corroborated these identifications through consultations with existing herbaria collections at the National Herbarium of Havana (where specimens from this study were later deposited). The author pressed the specimens in accordance with standard ethnobotanical practices (Alexiades 1996) and subsequently laminated and mounted them in a large scrapbook. Where possible and relevant to the given plant's use, a range of plant parts were included, including leaves, branches, seeds, fruit, flowers, bark, bulbs, and roots. When the author was unable to find the species for sale in herb markets, photographs were used as an alternative. Of the 64 species in the survey, the author recorded 27 (42%) on sale as fresh herbs, branches (*palos*), seeds, roots, or fruit in urban markets in the cities of Cienfuegos and Trinidad during the same time period.

Site and Participant Selection

The fourth stage of the survey design involved site selection. An initial problem encountered was that optimal geological, topographic, and climatic conditions for tobacco and sugar cultivation differ, which results in distinct types of floristic variation in each zone. The Cuban landscape is made up of grasslands, savannas, subtropical rain forests, subtropical thorn forests, Atlantic pine barrens, and mangrove swamps. All but the latter two categories have been successful sites of sugar cultivation, though former subtropical forest is considered the most productive and suitable type of land. The most favorable tobacco cultivation sites are situated on sandy loam soils, in contrast to the clay soils preferred for sugar cultivation (Boyer 1939). Despite this variation, central Cuba counts on a more mixed composition of agricultural zones and floristic diversity, making it a more suitable region in which to carry out the survey in providing a better control for extraneous variables.

Other extraneous variables, such as knowledge acquisition through the media, formal education, and literature, were harder to control for in this study. For this reason, information on the main way in which participants had gained their ethnobotanical knowledge was also gathered during survey implementation. Results were as follows: "from a grandparent" in 30% of cases, "from a parent" in 23%, "from another family member" in 23%, "through religious training" in 17%, and "through state-run education courses" in 7%.

The author selected areas with a long-standing history of sugar and tobacco cultivation according to advice from representatives of Cuba's National Center for Protected Areas (*Centro Nacional de Areas Protegidas*) and the Ecology and Systematics Institute (*Instituto de Ecología y Sistematica*), both in Havana. This information was further corroborated through consultation of historical economic maps and atlases (Instituto Geográfico Nacional 1989; CIA 1977a, b; Automobile Blue Book 1919; Canet Álvarez 1911; Carey and Lea 1826) and other secondary sources (Ayala 1995; Thomas 1971; Ortiz 1940; Boyer 1939). The author selected six sites in central Cuba based on this background research – half in sites of long-term sugar cultivation and half in sites of long-term tobacco cultivation. Each agricultural zone was divided among city, town, and village categories, in order to test not only variation in knowledge according to each area's migrant history but also in relation to settlement size. For the purpose of this study, cities are defined as urban areas with populations of 70,000–170,000, towns of 15,000–70,000, and villages up to 15,000.

The selected sites were Cienfuegos (sugar city, Cienfuegos province), Sancti Spíritus (tobacco city, Sancti Spíritus province), Palmira (sugar town, Cienfuegos province), La Sierrita (tobacco town, Cienfuegos province), Refugio (sugar village, Villa Clara province), and Guaracabulla (tobacco village, Villa Clara province). Resource and bureaucratic restraints prevented repetition of the survey in more areas of each agricultural zone. This signifies that results described in this chapter cannot be used to generalize about the wider context at the national level and exclusively relate to these six study sites alone. Further study is recommended at the national level and with larger sample sizes to build on these initial results.

Forming the fifth stage of research, participants were contacted and invited to take part in the social survey based on plant use inventories, described above. As earlier attempts to test ethnobotanical knowledge among both urban and rural Cuban populations suggest that few nonspecialists had more than rudimentary knowledge of medicinal plant use (Wezel and Bender 2003; corroborated by the author's own experience), survey respondents were limited to ethnobotanical specialists. These included herb sellers and plant cultivators (*yerberos*), plant harvesters (*osainistas* in a ritual context and *monteros* in a commercial context), ritual experts (including adherents to *santería*, *palo monte,* and *espiritísmo*), and those who practice *medicina verde* in a nonreligious context. Surveyed participants were contacted through snowball sampling.

Interviews for the survey were carried out in adherence to ethical guidelines developed by the UK's Economic and Social Research Council (ESRC). Participants were only included if they, and their parents, had been born and brought up in the location where the survey was conducted. This was due to findings of earlier studies that individuals who have lived in a region for extensive periods of time are more likely than those who recently migrated to an area to have a more in-depth knowledge of useful plants and are more likely to use them more often (Nesheim et al. 2006; Byg and Balslev 2006; Voeks 2004). It is worth noting, however, that revolutionary (post-1959) agricultural, educational and healthcare programs and internal migration have meant that an individual's knowledge of ethnobotanical practices is not necessarily defined by their place of residence.

The author surveyed five individuals in each of the six sites (30 individuals, divided between 16 women and 14 men), with a median age of 64 (ranging from 42 to 89). Ethnicity was not included as a factor in the selection of survey participants. This was largely due to the very mixed ancestry of the majority of Cubans, together with the focus of this study being put on ethnobotanical knowledge in relation to place, rather than knowledge in relation to ethnic ancestry. A future study that analyzes ethnobotanical knowledge in relation to self-proclaimed ethnic and religious identity would represent a worthwhile addition to the literature.

Data Collection and Analysis

The author engaged selected survey participants in structured interviews, during which they were shown the plant specimens/photos and given each plant's locally used common name(s). For each of the 64 species, the author asked the respondents if they knew the plant and, if so, to name as many medicinal uses as they could recall for that species. Following completion of the survey, each response was given a score of 1 or 0: the former representing a plant use corresponding to any of the uses listed in the author's diaspora plant-use database and the latter equating to an answer that differed from the literature (Appendix 1). For the six species that are derived from both West African and Iberian Mediterranean centers of early cultivation, the score of 1 represents a response in agreement with both the "African" and "Mediterranean" literature.

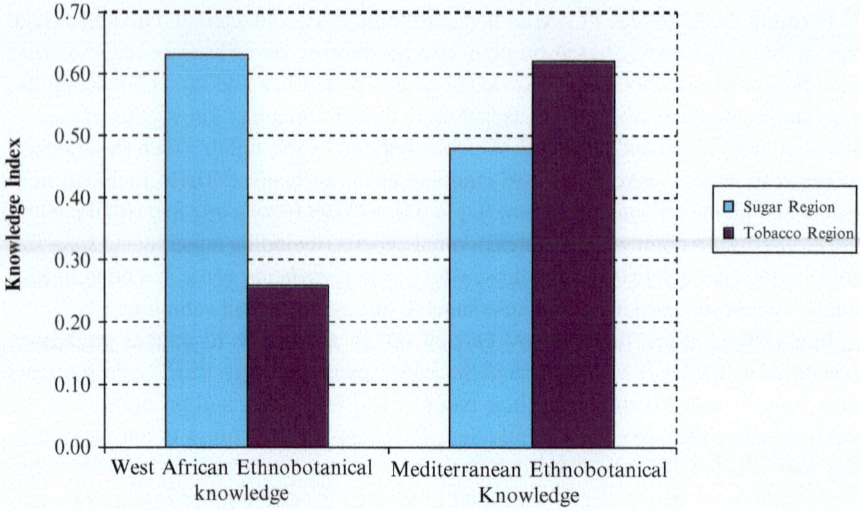

Fig. 9.1 West African diaspora plant knowledge is stronger in sugar regions as compared to tobacco regions. The reverse applies for Mediterranean ethnobotanical knowledge

From these survey results, the author calculated a knowledge index for each of the six regions, representing the percentage of respondents who gave answers in agreement with the relevant literature (or, in other words, receiving a score of "1") for the full inventory of plants in each region. In order to confirm these results, a general linear model (GLM) was employed to analyze the data in terms of the importance of agricultural zones in relation to diaspora ethnobotanical knowledge.

Research Findings

Differences between West African and Mediterranean ethnobotanical knowledge in sugar and tobacco zones are shown in Fig. 9.1.

These results suggest that ethnobotanical knowledge of West African plant use is stronger in the selected sugar cultivation sites and that of Iberian Mediterranean plant use is more widespread between both agricultural zones and marginally stronger in tobacco cultivation zones. A greater disparity is evident between indices of West African ethnobotanical knowledge in sugar and tobacco regions in comparison to those of Iberian Mediterranean ethnobotanical knowledge. Figure 9.2 shows differences in diaspora ethnobotanical knowledge according to settlement type.

These figures suggest that knowledge of West African plant use is stronger than that of Mediterranean plant use in sugar zones, with the reverse being true in tobacco areas. Knowledge of West African species appears to be more marked in larger urban areas over smaller rural settlements in the sugar cultivating zones. Conversely,

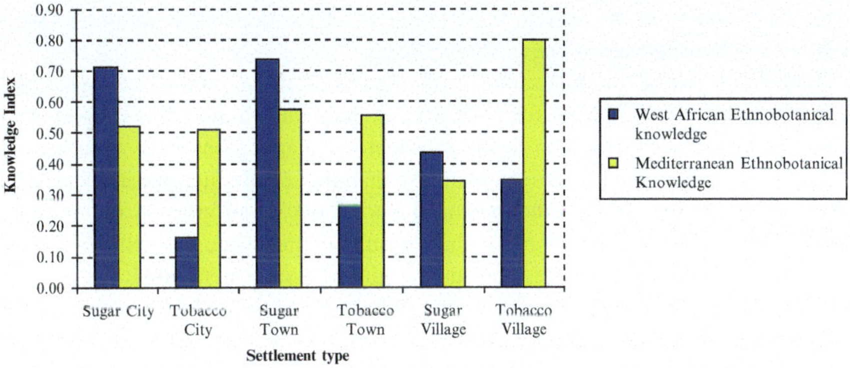

Fig. 9.2 West African diaspora plant knowledge is stronger in the sugar city and town than in the sugar village. Conversely, Mediterranean ethnobotanical knowledge is most marked in the tobacco village category

Table 9.1 GLM results showing statistical significance of sugar and the city/town categories for West African ethnobotanical knowledge

Dependent variable	Source	*df*	*MS*	*F*	*P*
Mediterranean plant knowledge (arcsine transformed)	Sugar	1	125.23	2.16	0.155
	City	1	40.23	0.69	0.413
	Town	1	3.19	0.06	0.816
	Sugar × city	1	1069.04**	18.43	<0.001
	Sugar × town	1	1019.78**	17.58	<0.001
	Error	24	58.02		
West African plant knowledge (arcsine transformed)	Sugar	1	4109.10**	95.29	<0.001
	City	1	11.64	0.27	0.608
	Town	1	197.30*	4.58	0.043
	Sugar × city	1	1083.07**	25.12	<0.001
	Sugar × town	1	711.25**	16.49	<0.001
	Error	24	43.12		

Note: *denotes statistical significance at 95% confidence level
**denotes statistical significance at 99% confidence level

knowledge of Mediterranean plant use is stronger in the tobacco village in comparison to larger settlement types.

In order to confirm these results, general linear model analysis of the data was carried out (Table 9.1). This approach consists of a regression using the knowledge indices of West African and Mediterranean diaspora plant-use knowledge as the dependent variables and the following regressors: sugar dummy (=1 if respondent lives in sugar region and 0 otherwise), city dummy (=1 if respondent lives in a city and 0 otherwise), sugar and city dummy (=1 if respondent lives in a city located in a sugar region and 0 otherwise), and sugar and town dummy (=1 if respondent lives in a town located in a sugar region and 0 otherwise). To overcome the problem of nonhomogenous variance (heteroskedasticity) in the model, the arcsine transformation of the dependant variables was used. The significance tests reported in Table 9.1

were generated from the arcsine transformation of the dependant variables (West African and Mediterranean knowledge indices).

Table 9.1 confirms the first set of results described above, in that the sugar category is statistically significant for West African plant knowledge. This suggests that ethnobotanical specialists who reside in selected sugar regions have significantly greater knowledge of West African plant use than that of Mediterranean species. In other words, West African ethnobotanical knowledge is stronger in the selected sugar cultivation sites. Additionally, these results demonstrate that ethnobotanical specialists who reside in the selected sugar regions have greater knowledge of West African plant use than those inhabiting tobacco regions. The conclusion that Mediterranean medicinal plant knowledge is equivalent between sites cannot be rejected because sugar alone is not statistically significant. Finally, knowledge of West African species is significant in the city, while not in the village, in the sugar-cultivation zone.

Discussion

The geographical role played by the global expansion of what were once highly localized and place-bound religious, ethnobotanical, and healing systems is yet to be well understood (Carney and Voeks 2003). With the aim of contributing to this gap in the literature, this chapter seeks to highlight the importance of migration history and place to Cuban ethnobotanical knowledge.

This study suggests that knowledge of West African-derived plant-use practices is more strongly associated with sugar cultivation zones, where African immigration has been stronger. In contrast, Iberian Mediterranean ethnobotanical knowledge appears more widespread between agricultural zones, and a markedly lower knowledge of African-derived plants is evident in tobacco cultivation zones, which are sites of stronger Spanish immigration. This suggests, at least in the case of the sites under study, that the mixing of ethnobotanical knowledge and techniques has been more profound in areas where African migration has been stronger.

Although further study is required to corroborate these findings and permit broader generalization, it seems logical that those communities historically most in need of a broad range of plant-use practices (i.e., enslaved Africans and their descendants) would possess a richer knowledge of ethnobotanical applications. In contrast to those residing in tobacco cultivation zones, workers in Cuban sugar plantation economies have suffered from greater levels of poverty, malnutrition, disease, inhumane conditions, and poor access to healthcare provisions (Ortiz 1940). As a consequence, the need to find alternative nutritional and medicinal sources in these areas is likely to have been elevated in sugar regions over tobacco production areas. A fluid process of ethnobotanical knowledge transmission is also likely to have taken place, as indicated by the more mixed composition of diaspora techniques cited in areas with stronger African immigration.

Despite its remaining a highly contested term (Argenti and Rochenthaler 2006; Palmié 1995), the notion of "syncretism" is one that can serve as a useful lens through which to explore the ways in which culture, religion, and associated ethnobotanical practices have transformed under diverse pressures of change. The continuous arrival of new slave groups into the Americas, often of priests and traditional healers, meant that religious practices were in a constant process of renewal and erosion throughout the colonial era. As has been demonstrated elsewhere (Nesheim et al. 2006), "modern" forms of ethnobotanical knowledge are likely to have arisen through a combination of ancestral practices, experimentation, and acquisition from other indigenous or imported cultural practices and also from official or scientific sources.

Anthropologist Melville Herskovits is widely credited for his contributions toward the recognition of the importance of African heritage in the New World (Apter 2004). His attempts to describe the African origins of Afro-Caribbean "cultural traits" were criticized widely in the social sciences, however, for their perceived influence on "anthropologically pre-constructed versions of an African past" (Palmié 2005: 94). In 1930s Brazil, Nina Rodrigues was among the first to demonstrate that syncretism between African and European belief systems in the New World was an ongoing, rather than static, process (Carney and Voeks 2003). Cuban scholar Fernando Ortiz (1940) went on to emphasize that a process of mutual influence and multidirectional transformation occurs in the transmission of religious knowledge and practices, something which can also be seen in the learning and sharing of ethnobotanical techniques.

Theoretical stances have now moved away from the idea that religious syncretism arose mainly as a means to avoid punishment from dominating colonial forces. Since the 1970s, Afro-American religions have increasingly been seen as complex symbols of resistance, planning, and proactivity by Africans in the New World (Pels 1997). This has also been shown to be the case with regard to ethnobotanical knowledge linked to syncretic religions (Carney 2003a; Carney and Voeks 2003). Instead of being viewed as cultural systems that have been preserved intact over the centuries, Afro-American religions are also now understood to be diverse melting pots of ritual practices and beliefs that were both reinforced and altered by each new wave of African migration and subsequent political and economic change (Bastide 2004; Cros Sandoval 1975).

The adaptability of ethnomedical systems in Cuba is particularly relevant to medicinal plant-use practices that are employed in modern-day Cuba. The economic crisis that began in 1990s Cuba has seen a drastic restructuring of the economy, accompanied by a notable rise in the commercialization of Afro-Cuban religions (Holbraad 2004) and rise in the use of plant-based remedies to treat ailments in both a secular and ritual context (Moret 2008b). During this time, Afro-Cuban ritual practices and medicinal plant knowledge are two areas that have proved to be highly profitable, albeit contested, fields (Moret in press). A time of intense change has ensued, not only for ritual practitioners in general but also for the plants that are employed in new ways, as novel forms of health, economic, and ideological pressures present themselves under Cuba's recent era of intense material and medicinal shortages.

Although Cuban culture is celebrated in official discourse as a fusion of many traditions, official discourse, including the media and education, presents the island's history as that of a past dominated by Europeans – undervaluing the protagonist role of African migrants to the island (Girardi 1994). Similarly, state actors overwhelmingly portray the island's *medicina verde* as a knowledge system derived largely from Cuba's rural-dwelling Spanish communities (Moret in press). Overt manifestations of African-based culture are more visible in today's Cuba than in Soviet times but remain restricted to income-generating "folkloric" activities and merchandise (focusing on music, dance, and song) rather than as a system of values, philosophies, or healthcare contributions (Moret 2008b; Palmié 2002; Hagedorn 2001; Matibag 1996).

African-influenced traditional medicine – formerly preserved and transmitted through highly secretive and marginalized knowledge systems – has now begun to undergo a shift to more hegemonic, "scientific," and public spheres (Moret in press). In recognition of studies that have highlighted conflicts that might arise in response to the "repackaging" of subaltern practices and knowledge (Fairhead and Leach 2003), this research aims to reveal the ways in which African-influenced ethnomedical systems continue to be relevant in Cuba. In doing so, it aims to encourage a stronger recognition of African contributions to Cuban culture, both in the modern day and throughout history.

Conclusion

This study suggests that West African- and Iberian Mediterranean-derived strands of diaspora plant-use knowledge thrive in distinct ways in different parts of central Cuba, according to a given area's land use and migrant history. Ethnobotanical specialists residing in selected areas associated with high levels of African immigration (sugar cultivation zones) appear to have retained a more detailed knowledge of West African-derived ethnobotanical practices than in areas characterized by stronger Spanish immigration (tobacco cultivation zones). In contrast, plant-use knowledge for species originating from the Iberian Mediterranean is more widespread in both types of agricultural areas. Poor healthcare provision and adverse economic conditions that have characterized sugar cultivation zones in Cuba are proposed reasons for the higher level of both types of diaspora plant-use knowledge in areas of strong African immigration. The author provides a number of suggestions for further study to build on these initial findings.

By exposing the importance of place to Cuban ethnobotanical knowledge, this chapter aims to contribute a geographical perspective to ongoing debates on New World syncretic religions and trans-Atlantic diaspora ethnobotany. Highlighting the important role played by African-derived medicinal plant-use systems, in particular, can represent a valuable tool with which to introduce a stronger recognition of what have been vital, but often under-recognized, intellectual contributions to Cuban culture and healthcare.

Acknowledgments The author would like to thank all those in Cuba who generously gave their time to contribute to this study and to the Cuban Ministry of Science, Technology and Environment (CITMA) for fieldwork permits. Professor Colin Clarke and Dr. Nick Brown at Oxford University provided expert joint doctoral supervision and Yuri Sánchez gave invaluable statistical assistance. Dr. Jan Salick and members of the Oxford Ethnobiology Research Group provided helpful comments on earlier drafts. Generous funding was provided by grants from the UK's Economic and Social Research Council (ESRC) and the Natural Environment Research Council (NERC); Jesus College, Oxford; and London University's Institute of Latin American Studies (ILAS). All views presented in this chapter are those of the author.

Appendix 1: West African/Iberian Mediterranean Diaspora Plant-Use Inventory

Latin name	Common name	West African use	Mediterranean use	Listed West African uses	Listed Mediterranean uses
Abelmoschus esculentus (L.) Moench	*Quimbombó*	1	0	Syphilis, antispasmodic, sudorific	x
Abrus precatorius L.	*Peonia*	1	0	Laxative, urinary complaints, paralysis, rheumatism, sore throat, nervous system, blood, purgative, tuberculosis, syphilis, cough	x
Acacia nilotica (L.) Delile	*Aroma amarilla*	1	0	Astringent, dysentery, tonic, ulcer treatment	x
Aframomum melegueta K. Schum.	*Atare/pimienta de la costa*	1	0	Carminative, fever, stimulant, dysentery, inflammation, tonsils, toothache, abortifacient, headache, sores, constipation	x
Aloysia citriodora Palau.	*Yerba Luisa*	0	1	x	Anti-spasmodic, stomach complaints, sedative, tonic, aphrodisiac
Amaranthus spinosus L.	*Bledo*	1	0	Diarrhea, mouth ulcers, wounds, menstruation	x
Artemisia abrotanum L.	*Incienso/Artemisa/Ajenjo*	0	1	Rheumatism, parasites, stomach problems	x
Arundo donax L.	*Caña de castilla*	0	1	x	Bronchitis, diarrhea, diuretic, antiinflammatorial
Blighia sapida K.D. Koenig	*Seso vegetal*	1	0	Stimulant, piles, angina, ulcers, conjunctivitis	x
Brassica nigra (L.) K. Koch	*Mostaza negra*	0	1	x	Stimulant

(continued)

Appendix 1 (continued)

Latin name	Common name	West African use	Mediterranean use	Listed West African uses	Listed Mediterranean uses
Cajanus cajan (L.) Millsp.	*Gandul*	1	1	Mouthwash, sore throat, abortifacient, burns, dizziness, stomach pain, epilepsy, diarrhea, earache, gonorrhea, measles, syphilis, toothache, wounds	Inflammation, fever
Cardiospermum halicacabum L.	*Farolito*	1	0	Stomach, cathartic, rheumatism, lumbago, swellings, pulmonary complaints, debility, eye infections, laxative	x
Cassia fistula L.	*Caña fistola*	0	1	x	Laxative, vermifuge, skin complaints
Cassia occidentalis L.	*Yerba hedionda*	0	1	x	Respiratory complaints, laxative, skin complaints, antiinflammatory, liver
Ceiba pentandra (L.) Gaertn.	*Ceiba*	1	0	Febrifuge, laxative, rheumatism, colic, stimulant	x
Chenopodium ambrosioides L.	*Apasote*	0	1	x	Stomach pain, vermifuge, pain reduction, asthma
Chrysopogon zizanioides (L.) Roberty	*Vetiver*	1	0	Deodorant, skin complaints, tension	x
Cinnamomum spp.	*Canela de China*	0	1	x	Astringent, stimulant, carminative, antiseptic, antifungal, antiviral, blood purifier, digestive aid
Coffea arabica L.	*Café*	1	0	Diuretic, cardial and renal stimulant, headache, asthma	x

Cola acuminata (P. Beauv.) Schott & Endl. & *Cola nitida* (Vent.) Schott & Endl.	*Nuez de cola/obi*	1	0	Tonic, astringent, aphrodisiac, venereal disease, wounds, colic, diarrhea, ulcers, abscesses, rashes, heart problems, cough, teeth cleaning, childbirth pain	x
Conocarpus erectus L.	*Yana*	1	0	Fever, gonorrhoea, catarrh	x
Cordia spp.	*Ateje* (various)	1	0	Backache, purgative, fever, cough, weakness	x
*Croton lobatus*L.	*Frailecillo cimaron*	1	0	Diuretic, stomach pain	x
Cupressus sempervirens L.	*Cipres*	0	1	x	Slow bleeding, healing of wounds, hemorrhoids, rheumatism, vasoconstrictor
Cymbopogon citratus (DC.) Stapf	*Caña santa/caña de limon*	0	1	x	Fever, digestion, cough, tranquillizer
Desmodium triflorum (L.) DC.	*Amor seco*	1	0	Dysentery, laxative	x
Drymaria cordata (L.) Willd. ex Schult.	*Matemaco*	1	0	Laxative, febrigue, skin complaints, stimulant	x
Eclipta alba (L.) Hassk.	*Eclipta blanca*	1	0	Diarrhea, liver, spleen, purgative, vomitive	x
Elaeis guineensis Jacq.	*Corojo de guinea/ palma africana*	1	0	Excessive menstrual bleeding, headache, gonorrhea, boils, wounds	x
Eleusine indica (L.) Gaertn.	*Pata de gallina*	1	0	Diuretic, laxative, liver, sudorific	x
Evolvulus spp.	*Aguinaldito*	1	0	Antimicrobial, depression	x
Foeniculum vulgare Mill.	*Hinojo*	0	1	x	Diuretica, gas, antiinflammatory, antiseptic

(continued)

Appendix 1 (continued)

Latin name	Common name	West African use	Mediterranean use	Listed West African uses	Listed Mediterranean uses
Garcinia kola Heckel	*Orogbo*	1	0	Tumors, cancer, gonorrhea, closing of wounds, purgative, aphrodisiac, cough, dysentery, aches, vermifuge	x
Heliotropium indicum L.	*Alacrancillo*	1	0	Gonorrhoea, sores, stings, acne, inflammation, ulcers	x
Indigofera spp.	*Añil/Añil cimarron*	1	0	Lice, kidney and spleen complaints, cough, epilepsy, convulsions, ulcers, inflammation	x
Lacmellea utilis (Arn.) Markgr.	*Lechero/pegojo*	1	0	Skin complaints, wound healing, erectile dysfunction	x
Lagenaria siceraria (Molina) Standl.	*Guiro amargo*	1	0	Headache, pectoral, purgative, anthelmintic	Diuretica, diabetes
Leonotis nepetifolia (L.) R.Br.	*Baston de San Francisco*	1	0	Ringworm, scabies, rheuma-tism, sciatica	x
Mirabilis jalapa L.	*Maravilla*	1	1	Purgative	Purgative
Momordica charantia L.	*Cundeamor*	1	1	Bleeding, piles, dysentery, febrifuge, breast cancer, anthelmintic, stomach complaints, diarrhea, burns, ulcers, wounds, aphrodisiac, gonorrhoea, abortifacient	Diabetes, stomach upsets, skin complaints
Moringa oleifera Lam.	*Paraiso frances*	1	0	Heart, circulation, anti-inflammatory, ulcers, antibacterial	x

Mucuna pruriens (L.) DC.	*Ojo de buey/pica pica*	1	0	Cholera, diuretic, piles, anthelmintic, syphilitic ulcers, sores, prevent miscarriage, headache, purgative	x
Nerium oleander L.	*Adelfa*	1	1	Swellings, ulcers, skin disease, diuretic	Heart, sedative, skin complaints
Ocimum basilicum L.	*Albahaca blanca*	1	1	Febrifuge, ringworm, vermifuge, gonorrhoea, urinary infections, diarrhea, dysentery, post-birth pain	Digestive, sedative, tonic, antispasmodic
Ocimum gratissimum L.	*Albahaca cimarrona/albaca de clavo*	1	0	Antibacterial, antifungal, stomach upsets	x
Origanum majorana L.	*Mejorana*	0	1	x	Digestive complaints, respiratory complaints, antiseptic, headache
Phyllanthus spp.	*Yerba de la niña*	1	0	Kidneys, liver complaints, inflammation, digestion	x
Plantago major L.	*Llanten*	0	1	x	Tumors, diarrhea, slows bleeding, antiinflammatorial, repair of scar tissue
Portulaca oleracea L.	*Verdolaga*	1	1	Diuretic, slow blood flow, vomitive, burns	Anti-inflammatory, diuretic, migraine
Punica granatum L.	*Granada*	0	1	x	Astringent, diuretic, digestive, intestinal worms, dysentery, vermifuge

(continued)

Appendix 1 (continued)

Latin name	Common name	West African use	Mediterranean use	Listed West African uses	Listed Mediterranean uses
*Raphanus sativus*L.	*Rabano*	0	1	x	Rheumatism, stimulant
Rauwolfia spp.	*Amorcillo*	1	0	Reduce blood pressure, hypnotic, sedative, indigestion, scabies, tonic, purgative, venereal disease, sexual stimulant, blood pressure, sores, measles, herpes, mental illness	x
Rhizophora mangle L.	*Mangle rojo*	1	0	Slows bleeding	x
Ricinus communis L.	*Higuereta*	1	1	Increase milk flow in breast-feeding, parasites, purgative, sores, burns, ulcers, abortive	Laxative, hair growth, antiheminitic, eyes, anti-inflammatory, allergic reactions
Ruta chalepensis L.	*Ruda*	0	1	x	Cough, stomach ache, abortifacient, slows bleeding, emmenagogue
Salvia officinalis L.	*Salvia de Castilla*	0	1	x	Stimulant, antiseptic, antibacterial, rheumatism
Smilax aspera L.	*Sarsaparilla*	0	1	x	Urinary antiseptic, sudorific, diuretic, purative, laxative, high blood sugar level
Spondias mombin L.	*Jobo*	1	0	Purgative, gonorrhea, infertility, aphrodisiac, tumors, febrifuge	x
Tamarindus indica L.	*Tamarindo*	1	1	Tonic, laxative, diarrhea, jaundice, burns, febrifuge, conjunctivitis	Laxative, purgative

Triumfetta spp.	*Guizazo de caballo/ guizazo de cochino*	1	0	Diarrhea, stops bleeding, gonorrhea, dysentery, facilitate childbirth	x
Verbena officinalis L.	*Verbena*	0	1	x	Sedative, emetic, digestive, disinfectant
Vernonia spp.	*Rompezaraguey*	1	0	Stomach complaints, antispasmodic, parasites	x
Vetiveria zizanoides	*Vetiver*	1	0	Deodorant, skin complaints, tension	x
Xylopia aethiopica (Dunal) A. Rich.	*Eru/Pimienta de guinea*	1	0	Constipation, tonic, fever, mucus, bronchitis, gonorrhea, dysentery, backache, abortifacient, coughs, stimulant, to purify water, fertility, lumbago, sores, headache, vermifuge, vomitive	x
Zanthoxylum spp.	*Ayua*	1	0	Toothache, digestion, rheumatism	x

References

Ademuwagun ZA (ed) (1979) African therapeutic systems. Crossroads Press, Los Angeles
Ainslie JR (1937) A list of plants used in native medicine in Nigeria. Imperial Forestry Institute paper 7. University of Oxford, Oxford
Aiyeloja AA, Bello OA (2006) Ethnobotanical potentials of common herbs in Nigeria: a case study of Enugu State. Educ Res Rev 1:16–22
Alcazar MD, Garcia C, Rivera D, Obon C (1990) Lesser-known herbal remedies as sold in the market at Murcia and Cartagena (Spain). J Ethnopharmacol 28:243–247
Alexiades MN (ed) (1996) Selected guidelines for ethnobotanical research: a field manual. New York Botanical Garden, New York
Apter A (2004) Herskovits's heritage: rethinking syncretism in the African Diaspora. In: Leopold AM, Jensen JS (eds) Syncretism in religion: a reader. Routledge, London, pp 160–184
Argenti N, Rochenthaler U (2006) Introduction: between Cameroon and Cuba: youth, slave trades and translocal memoryscapes. Soc Anthropol 14(1):33–47
Automobile Blue Book (1919) Cuba – Central Cuba. [online map]. Perry Castañeda Library Map Collection. Accessed 12 July 2006. Available from http://www.lib.utexas.edu/maps/cuba.html. Internet
Ayala CJ (1995) Social and economic aspects of sugar production in Cuba 1880–1930. Lat Am Res Rev 30(1):95–124
Ayensu ES (1978) Medicinal plants of West Africa. Reference Publications, Algonac
Barnet M (1986) Biografia de un cimarrón. Ed. Ciencias Sociales, Havana
Barrera A (2002) Local herbs, global medicine: commerce, knowledge, and commodities in Spanish America. In: Smith PH, Findlen P (eds) Merchants and marvels: commerce, science and art in early modern Europe. Routledge, New York, pp 163–181
Bascom W (1950) The focus of Cuban santería. Southwest J Anthropol 6(1):64–68
Bascom W (1972) The Yoruba of Southwestern Nigeria. Holt, Rinehart and Winston, New York
Bastide R (2004) Problems of religious syncretism. In: Leopold AM, Sinding Jensen J (eds) Syncretism in religion: a reader. Routledge, London, pp 113–141
Betancourt m lòfaorò Estrada V (1997) El babalawo: Médico tradicional (Yorubás y santería afro-cubana). Editorial 21, Havana
Bolívar Arostegui N (2001) Plantas medicinales y aromáticas. Revista Bimestre Cubana de la Sociedad Económica de Amigos del País 15:47–63
Boyer HM (1939) Distribution of sugar cane production in Cuba. Econ Geogr 15(3):311–325
Brandon G (1991) Light from the forest: how Santería heals through plants. Blue Unity Press, Newark
Bryant RL, Bailey S (1997) Third World political ecology. Routledge, London
Buckley AD (1985) Yoruba medicine. Clarendon, Oxford
Byg A, Balslev H (2006) Palms in indigenous and settler communities in Southeastern Ecuador: farmers' perceptions and cultivation practices. Agrofor Syst 67:147–158
Cabrera L (1984) La medicina popular de Cuba: Médicos de anta o, curanderos, santeros y paleros de hogaño. Coleccion del Chicherekú, Miami
Cabrera L (2000) El monte: igbo, finda, ewe orisha, vititi nfinda. Notas sobre las religiones, la magia, las supersticiones y el folklore de los negros criollos y el pueblo de Cuba. Ediciones Universal, Miami
Canet Alvarez GA (1911) Atlas de Cuba. Harvard University Press, Cambridge, MA
Carey HC, Lea I (1826) Geographical, statistical and historical map of Cuba and the Bahamas Islands. In: H.C. Carey and I. Lea Publishers (ed) A complete historical, chronological and geographical American atlas. H.C. Carey and I. Lea Publishers, Philadelphia, pp 31–32
Carney JA (2001) Black rice. The African origins of rice cultivation in the Americas. Harvard University Press, Cambridge, MA
Carney JA (2003a) African traditional plant knowledge in the circum-Caribbean region. J Ethnobiol 23(2):167–185

Carney JA (2003b) Agroenvironments and slave strategies in the diffusion of rice culture to the Americas. In: Zimmerer KS, Bassett TJ (eds) Political ecology: an integrative approach to geography and environment-development studies. Guilford Publications, New York, pp 256–273
Carney JA, Voeks RA (2003) Landscape legacies of the African diaspora in Brazil. Prog Hum Geogr 27(2):139–152
CIA (1977a) Cuba – sugar. [online map] Perry Castañeda Library Map Collection. Map 502988. Accessed 12 July 2006. Available from http://www.lib.utexas.edu/maps/cuba.html. Internet
CIA (1977b) Cuba – economic activity. [online map] Perry Castañeda Library Map Collection. Map 502988. Accessed 12 July 2006. Available from http://www.lib.utexas.edu/maps/cuba.html. Internet
Cros Sandoval M (1975) La religión afrocubana. Playor, Madrid
Dalziel JM (1937) The useful plants of West Africa. The Crown Agents for the Colonies, London
Darias V, Bravo L, Barquin E, Martin Herrera D, Fraile C (1986) Contribution to the ethnopharmacological study of the Canary Islands. J Ethnopharmacol 15:169–193
Darias V, Bravo L, Rabanal R, Sánchez Mateo C, González Luis RM, Hernández Pérez AM (1989) New contributions to the ethnopharmacological study of the Canary Islands. J Ethnopharmacol 25:77–92
Dunmire WW (2004) Gardens of New Spain: how Mediterranean plants and foods changed America. University of Texas Press, Austin
Ernst and Young (2006) A business guide to Cuba. [document online]. Accessed 15 Mar 2006. Available from http://www.walterlippmann.com/cuba-business-guide.pdf. Internet
Esquivel M, Fuentes V, Martínez C, Martínez J, Hammer K (1992) The African influence from an ethnobotanical point of view. In: Hammer K, Esquivel M, Knüpffer H (eds) Origin, evolution and diversity of Cuban plant genetic resources. Institut für Pflanzengenetik und Kulturpflanzenforschung, Gatersleben, pp 83–109
Fairhead J, Leach M (2003) Science, society and power: environmental knowledge and policy in West Africa and the Caribbean. Cambridge University Press, Cambridge
Fuentes V (1992) Plants in Afro-Cuban religions. In: Hammer K, Esquivel M, Knüpffer H (eds) Origin, evolution and diversity of Cuban plant genetic resources. Institut für Pflanzengenetik und Kulturpflanzenforschung, Gatersleben, pp 110–137
Fuentes V, Granda M (1997) Conozca las plantas medicinales. Editorial Cientifico-Técnica, Havana
Garfield R, Santana S (1997) The impact of the economic crisis and the US embargo on health in Cuba. Am J Public Health 87(1):15–20
Geschiere P (2006) Afterword. Soc Anthropol 14(1):119–121
Gill LS, Akinwumi C (1986) Nigerian folk medicine: practices and beliefs of the Ondo people. J Ethnopharmacol 18(3):257–266
Girardi G (1994) El ahora de Cuba: Tras el derrumbe del comunismo y tras el viaje de Juan Pablo II. Nueva Utopia, Madrid
González-Tejero MR, Molero-Mesa J, Casare-Porcel M, Martínez Lirola MJ (1995) New contributions to the ethnopharmacology of Spain. J Ethnopharmacol 45:157–165
Guerra F (1966) Drugs from the Indies and the political economy of the sixteenth-century. Analecta Medico-Historica 1:29–54
Hagedorn KJ (2001) Divine utterances: the performance of Afro-Cuban santería. Smithsonian Institution Press, Washington, DC
Hammer K, Esquivel M (1992) The role of ethnic minorities – the East Asiatic case. In: Hammer K, Esquivel M, Knüpffer H (eds) Origin, evolution and diversity of Cuban plant genetic resources. Institut für Pflanzengenetik und Kulturpflanzenforschung, Gatersleben, pp 138–146
Hammer K, Esquivel M, Knüpffer H (1992) A note on European plants in Cuban agriculture. In: Hammer K, Esquivel M, Knüpffer H (eds) Origin, evolution and diversity of Cuban plant genetic resources. Institut für Pflanzengenetik und Kulturpflanzenforschung, Gatersleben, pp 75–79
Holbraad M (2004) Religious "speculation": the rise of ifá cults and consumption in post-Soviet Cuba. J Lat Am Stud 36:643–663
Instituto Geográfico Nacional (1989) Nuevo atlas nacional de Cuba. Academia de Ciéncias de Cuba, Havana

Klak T, Jackiewicz E (2002) What has happened to social development policy in the era of neoliberal globalization and emerging markets?: a look at Latin America in global context. Revista de Ciencia Politica 22(1):112–129

Martínez Callis CR, Herrera Oliver P, Olvieto Prieto R, Moreno Rodríguez E (1997) Importancia de la etnobotánica en Cuba. In: Reportaje del Instituto de Ecología y Sistemática. Academia de Ciencias de Cuba, Havana

Matibag E (1996) Afro-Cuban religious experience: cultural reflections in narrative. University of Florida Press, Gainesville

Mayor F (2003) Preface. In: Sued-Badillo J (ed) General history of the Caribbean, vol 1, Autochthonous societies. UNESCO Publishing/Macmillan, Paris, pp 2–5

McClure S (1982) Parallel usage of medicinal plants by Africans and their Caribbean descendants. Econ Bot 36(3):291–301

Moreno Rodríguez E, Bermúdez A, Herrera Oliver PP (1995) Las plantas y el espiritismo cruzado en Cuba oriental. Fontqueria 42:289–308

Moret ES (2007) Afro-Cuban religion, diaspora ethnobotany and environmental change in post-Soviet Cuba. Ph.D. thesis, University of Oxford

Moret ES (2008a) The Caribbean. In: Stearns PN (ed) Oxford encyclopedia of the Modern World: 1750 to the present. Oxford University Press, Oxford

Moret ES (2008b) Afro-Cuban religion, ethnobotany and healthcare in the context of global political and economic change. Bull Lat Am Res 27(3):333–350

Moret ES (in press) Hegemony, identity and trans-Atlantic modernity: Afro-Cuban religion (re) politicisation and (de)legitimisation in the post-Soviet Era

Morgan PD (1997) The cultural implications of the Atlantic slave trade: African regional origins, American destinations and New World developments. Slavery Abolit 18(1):122–145

Nesheim I, Dhillion SS, Stølen KA (2006) What happens to traditional knowledge and use of natural resources when people migrate? Hum Ecol 34(1):99–131

Neuwinger HD (2000) African traditional medicine. A dictionary of plant use and applications. Medpharm Scientific Publishers, Stuttgart

Ortiz F (1940) Contrapunteo cubano del tobaco y el azúcar: Advertencia de sus contgrastes agrarios, económicos, históricos y sociales, su etnografía y su trasculturación. Cátedra, Madrid

Ortiz F (1975) Los negros esclavos. Editorial de Ciencias Sociales, Havana

Palmié S (1995) Against syncretism: 'Africanizing' and 'Cubanizing' discourses in North American *Òrìṣà* worship. In: Fardon R (ed) Counterworks: managing the diversity of knowledge. Routledge, London, pp 73–104

Palmié S (2002) Wizards and scientists: explorations in Afro-Cuban modernity and tradition. Duke University Press, London

Palmié S (2005) *Ackee and Saltfish* vs. *Amalá con Quimbombó?* A note on Sidney Mintz' contribution to the historical anthropology of African American cultures. Journal de la Société des Américanistes 91(2):89–122

Peet R, Watts M (eds) (1996) Liberation ecologies: environment, development, social movements. Routledge, London

Pels P (1997) The anthropology of colonialism: culture, history, and the emergence of western governmentality. Annu Rev Anthropol 26:163–183

Pérez Medina T (1998) La santería cubana: El camino de Osha. Ceremonias, ritos y secretos. Biblioteca Nueva, Madrid

Rashford J (1987) The search for Africa's baobab tree in Jamaica. Jam J 20(2):2–11

Rashford J (1993) Arawak, Spanish and African contributions to Jamaica's settlement vegetation. Jam J 24(3):17–23

Reyes-García V, Vadez V, Tanner S, McDade T, Huanca T, Leonard WR (2006) Evaluating indices of traditional ecological knowledge: a methodological contribution. J Ethnobiol Ethnomed 2(21):1–9

Roig y Mesa JT (1959) Plantas medicinales, aromáticas, o venenosas de Cuba. Guerrero Casa Mayor y Cia, Havana

Roig y Mesa JT (1988) Diccionario botánico de nombres vulgares cubanos, vols 1 and 2. Editorial Científico-Técnica, Havana

Santiago-Valentin E, Olmstead RG (2004) Historical biogeography of Caribbean plants: introduction to current knowledge and possibilities from a phylogenetic perspective. Taxon 53(2):299–319

Schwartz SB (ed) (2004) Tropical Babylons: sugar and the making of the Atlantic world. 1450–1680. University of North Carolina Press, Chapel Hill

Seoane Gallo J (1988) El folclor médico de Cuba: Provincia de Camagüey. Editorial de Ciencias Sociales, Havana

Siegel PE (1991) On the Antilles as a potential corridor for cultigens into eastern North America. Curr Anthropol 32(3):332–334

Thomas H (1971) Cuba: the pursuit of freedom. Harper & Row, New York

Torres Jiménez IB, Quintana Cárdenes IJ (2004) Análisis comparativo sobre el empleo de plantas medicinales en la medicina tradicional de Cuba e Islas Canarias. Revista Cubana de Plantas Medicinales 9(1):1–14 (January–April)

Vasquez Davila MA (2001) Plantas útiles de Cuba. Instituto Tecnológico e Agropecuario de Oaxaca, Oaxaca

Vázquez FM, Suarez MA, Pérez A (1997) Medicinal plants used in the Barros area, Badajoz province (Spain). J Ethnopharmacol 55:81–85

Verger P (1995) Ewé: the use of plants in Yoruba society. Oldebrecht, Rio de Janeiro

Voeks RA (1997) Sacred leaves of candomblé: African magic, medicine, and religion in Brazil. University of Texas Press, Austin

Voeks RA (2004) Disturbance pharmacopoeias: medicine and myth from the humid tropics. Ann Assoc Am Geogr 94(4):868–888

Volpato G, Godínez D (2004) Ethnobotany of *Pru*, a traditional Cuban refreshment. Econ Bot 58(3):381–395

Volpato G, Godínez D, Beyra A, Barreto A (2009) Uses of medicinal plants by Haitian immigrants and their descendants in the Province of Camagüey, Cuba. J Ethnobiol Ethnomed 5:16

Watlington F (2003) The physical environment: biogeographical teleconnections in Caribbean prehistory. In: Sued-Badillo J (ed) General history of the Caribbean, vol 1, Autochthonous societies. UNESCO Publishing, Paris

Wezel A, Bender S (2003) Plant species diversity of Cuba and its significance for household food supply. Agrofor Syst 57:39–49

Williams JJ (1932) Voodoos and obeahs: phases of West Indian witchcraft. Lincoln MacVeagh, New York

Zimmerer KS, Bassett TJ (2003) Approaching political ecology: society, nature, and scale in human-environment studies. In: Zimmerer KS, Bassett TJ (eds) Political ecology: an integrative approach to geography and environment-development studies. Guilford Publications, New York, pp 1–25

Chapter 10
What Makes a Plant Magical? Symbolism and Sacred Herbs in Afro-Surinamese *Winti* Rituals

Tinde van Andel, Sofie Ruysschaert, Kobeke Van de Putte, and Sara Groenendijk

Abstract The Surinamese *Winti* religion is probably the least known of all Afro-Caribbean beliefs. Magical plants are essential ingredients in Afro-Surinamese rituals, but little research has been done on the plants associated with the various deities. Why are certain plants thought to have magical power? How did Surinamese plants, which must have been unknown to the Africans at first, attain magical status during the course of history? During ethnobotanical surveys in 2005 and 2006, we collected all magical species mentioned to us and interviewed 20 *Winti* priests and several traditional healers, vendors and collectors of magical plants. We recorded at least 411 species and 1,100 different recipes for magical baths, potions and rituals. Fabaceae, Rubiaceae, Piperaceae and Asteraceae were the most important families. Most plants were used to expel evil, as luck charms, and for spirits of the forest, the sky and the recently deceased. The bulk of the plants were native to Suriname. Several things could make a plant magical: a connection with ancestors; a remarkable growth form, shape, scent or color; its habitat; relation with animals; an associated medicinal use; or a sacred status among other ethnic groups. Plants that once helped escaped slaves to survive in the forest (e.g. water-containing herbs,

T. van Andel, Ph.D. (✉)
Netherlands Centre for Biodiversity Naturalis (Section National Herbarium of the Netherlands), Leiden University, P.O. Box 9415, 2300 RA Leiden, The Netherlands
e-mail: Tinde.vanAndel@naturalis.nl

S. Ruysschaert, M.Sc.
Laboratory of Tropical and Subtropical Agriculture and Ethnobotany, Ghent University, Coupure Links 653, 9000 Ghent, Belgium

K. Van de Putte, M.Sc.
Department of Biology, Research Group Mycology (Spermatophytes), Ghent University, K.L. Ledeganckstraat 35, 9000 Ghent, Belgium

S. Groenendijk, M.Sc.
Social Sciences group, Wageningen University,
P.O. Box 8130, 6700 EW Wageningen, The Netherlands

R. Voeks and J. Rashford (eds.), *African Ethnobotany in the Americas*,
DOI 10.1007/978-1-4614-0836-9_10,

inflammable resins or seeds for ammunition) are now considered sacred by their descendents. Searching for the reason behind a plant's magical power revealed a wealth of botanical, ecological, historical, linguistic and cultural knowledge. It is this specific ethnobotanical wisdom that makes a plant magical in the end.

Keywords Afro-American ethnobotany • Herbal baths • Magic plants • Maroons • Suriname • Traditional healers

Introduction

The Surinamese *Winti* religion is probably the least known of all Afro-American belief systems, although it shares many aspects with Brazilian *Candomblé*, Haitian *Vodou*, Cuban *Santería* and Jamaican *Obeah*. *Winti* literally means 'wind' but also refers to spirits, invisible energy and the belief itself. Among practitioners, the religion is also known by the terms *Komfo* (gods), *Kulturu* (culture) and *Obia*, which is used to define both healing spirits and supernatural medicines or objects (Wooding 1979; Price 1983; Thoden van Velzen and van Wetering 1988). *Winti* adherents recognize a wide range of supernatural beings that, by means of music, dance, costumes, prayers and magic plants, can possess human beings and bring them into a state of trance. Living in harmony with one's ancestors and guardian angels is thought to bring health and prosperity, but the same spirits can cause disease and ill fortune if they are disregarded or not provided with regular offerings (Herskovits and Herskovits 1934; Green 1978; Stephen 1998). Several authors regard the Afro-Surinamese *Winti* religion as the most traditional magical-religious system among blacks in the Western Hemisphere, as it preserves more West African religious elements than *Vodou*, *Santería* or *Candomblé* (Herskovits 1941; Wooding 1972, 1979; Green 1978; St.-Hilaire 2000; Glazier 2001). Mintz and Price (1992), however, argue that due to their ethnic, cultural and linguistic heterogeneity, the West African slaves could not transfer their way of life and accompanying beliefs and values intact from one region to another. Contact between the enslaved and their masters ignited the creation of an entirely new, creolized language, culture and religion. Nonetheless, the *Winti* as it is practised by Maroons, descendants of Africans who escaped the coastal plantations and sought refuge in the country's dense forests, has been little influenced by Christianity. Despite two centuries of Moravian missionary activity in tribal territory, relatively few Maroons have converted to Christianity (Thoden van Velzen and van Wetering 1988; Hoogbergen 1990; Price 1990). In contrast, most urban Creoles identify themselves as Christians, although many also adhere to the *Winti* belief (St.-Hilaire 2000). Their rituals contain much more syncretic elements due to their prolonged contact with Christians and Jews in coastal Suriname and, more recently, with East Indian (Hindu) and Javanese (Muslim) immigrants (Wooding 1972; Stephen 1998).

African magic, poison and sorcery were the few areas in which blacks could manipulate their white masters. Their widespread belief in *obia* (magical medicine), and the fear it caused among the whites, constituted perhaps the most potent weapon of the Africans (Davis 1988; Stedman 1988; Voeks 1993). Most successful rebel groups were led by religious leaders and skilled herbalists, who consulted the spirits for the best way to escape and provided the enslaved with *obia* that made them invulnerable for their prosecutors (Price 1983, 2008). Considered by the authorities as hotbeds of resistance, African dances and rituals were forbidden by the end of the seventeenth century (van Lier 1971). Until the early 1980s, *Winti* remained prohibited by Surinamese law and was strongly condemned by most churches in the country as *afkodrei* (idolatry). After independence from the Netherlands in 1975, however, a new nationalism emerged, which created a revaluation of cultural identity and traditions (Stephen 1998). *Winti* was also more openly practised by the growing community of Surinamese migrants in the Netherlands, who had much less to fear from the religious authorities. Even though many Surinamese still regard it as sorcery, the merchandise of magic paraphernalia as well as the multitude of *Winti* websites proves that this belief is still very much alive today.

Two major characteristics of *Winti* are its connection with the ancestors and its relationship with the natural environment (Stephen 1998). Detailed anthropological research has been carried out on Maroon oral history, the various *Winti* spirits, their origin and behaviour (e.g. Wooding 1972; Price 1983, 1990, 2008; Thoden van Velzen and van Wetering 1988). While most scholars agree that nature plays a major role in this belief, little scientific attention has been paid to the plants associated with the different deities. Several Surinamese traditional healers have published books on medicinal herbs and magic rituals (e.g. Sedoc 1992; Stephen 1998; Zaalman 2002), but they list only local plant names or unreliable scientific names and provide little information on *why* certain herbs are considered as essential ritual ingredients.

In Cuba and Brazil, the ceremonial function of magic plants is (among others) to cleanse, refresh and prepare individuals and objects for contact with the deities by whom they can become possessed. Each of the deities has specific herbs associated with them, which are appropriate for making their respective magic medicine (Brandon 1991; Voeks 1997). This seems also the case in Suriname, where every *Winti* is said to have its own favorite herbs, music, color, food, jewellery, costumes, behaviour and spiritual language (Wooding 1972; Stephen 1998). The question remains, however: Why spirits prefer some herbs above others? Do these magical plants represent historic symbols? Or do they possess specific characteristics? Plants with strong odours, striking colours or odd shapes have long attracted the attention of humans, leading to the widespread belief that the gods provided signs within plants that indicated their uses (Bennett 2007). Does the Doctrine of the Signatures, a theory by which the physical characteristics of plants reveal their therapeutic value, also play a role in ritual plant use in Suriname?

African plants that were introduced to the New World by slave ships play an important symbolic role in *Candomblé* and *Santería* ceremonies (McClure 1982; Voeks 1993). The same accounts for some Neotropical species that closely resemble Old World species of the same genus (Voeks 1997). A (perceived) African origin may be another reason for people to ascribe divine power to plants, something that has never been investigated in Suriname. By means of literature research, interviews, plant collection and market surveys, both in Suriname and the Netherlands, we investigate what makes a plant magic. In this chapter, we report which plants are used in *Winti* rituals and why. We discuss the role of symbols, Old World origin and signatures and speculate on how elements of the Surinamese flora, which must have been unknown to the Africans at first, became magical in the course of history.

Methods

Fieldwork

Fieldwork was conducted by the first and last author between January and August 2006 and took place in the capital Paramaribo and several Maroon communities around Rijsdijkweg (Pará District), Bigiston (Marowijne), Marchallkreek and Nieuw Lombé (Brokopondo). Fieldwork locations are indicated on the map (Fig. 10.1). After explaining the nature of our research to our informants and obtaining their prior informed consent, we collected botanical vouchers of all plants mentioned to us as being used in *Winti* rituals. For all plants collected, we recorded vernacular names, growth form, vegetation type, preparation methods, doses, perceived effects and associated beliefs. We spoke with several collectors and vendors of magic plants and conducted semi-structured interviews with Maroon and Creole *Winti* specialists in Suriname (eight persons) and the Netherlands (ten). In Bigiston, we worked for 2 months with the Ndyuka Maroon traditional healer Ruben Mawdo to document his recipes for herbal baths and protective *obias*. We witnessed several healing sessions performed by Mawdo and other healers and attended a number of ritual dances (*wintipré*) in honor of the snake, earth, sky and forest spirits. Additional data were drawn from a 4-ha study plot combined with in situ and ex situ interviews with 25 local plant specialists, two *Winti* experts and 108 women (generalists) from randomly selected households (using pictures of plants and free-listing) in Brownsweg (Brokopondo), carried out by the second and third author in 2005 and 2006. They obtained more specific information from Saramaccan ritual specialists Michel Alubutu and Blacky Finfin and attended several healing sessions as well as a *dungulali* ritual to protect people from danger. Duplicates of botanical vouchers have been deposited at the National Herbarium of Suriname (BBS), the Ghent University Herbarium (GENT) and the National Herbarium of the Netherlands (L). Plant and author names were updated using Kew's online Plant List (http://www.theplantlist.org).

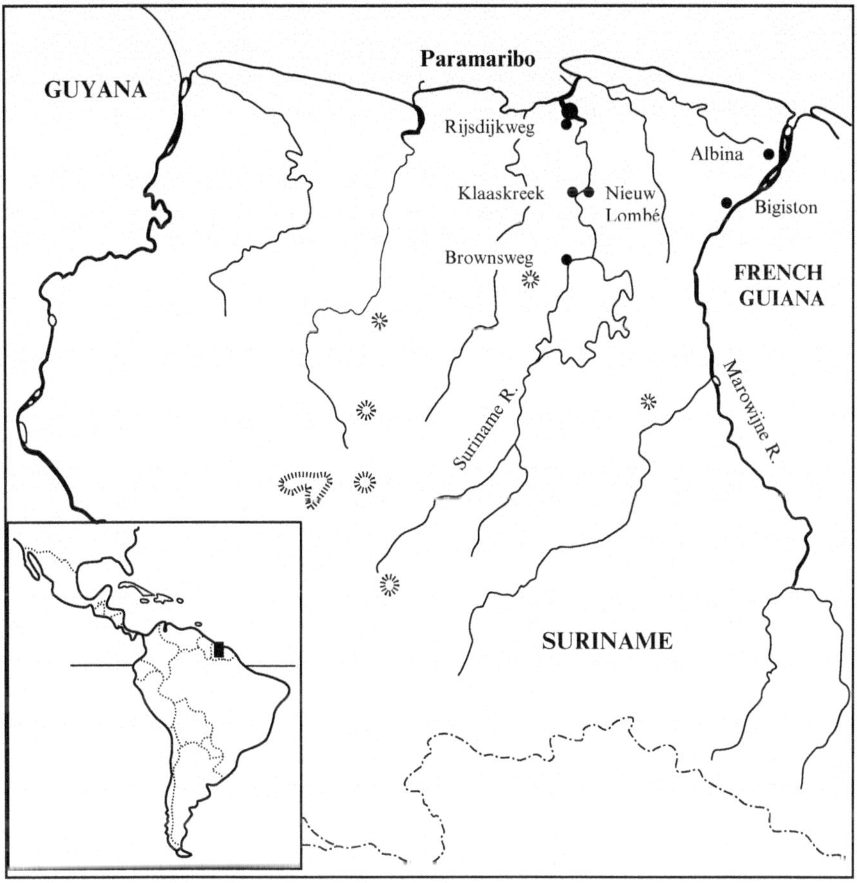

Fig. 10.1 Map of Suriname showing fieldwork locations (Based on a drawing by H. Rypkema)

Data Analysis

When asked why a specific plant had magical power, informants often referred to its appearance (e.g. strangler figs that killed their host, parasites), a strong smell, typical colour or sharp spines. When a plant grew deep in the forest or attracted specific animals, this was also stated as a reason to be associated with certain deities. To see whether these variables (growth form, habitat, ecology, scent, colour or the presence of spines or stinging hairs) were consistently associated with certain magical uses or particular spirits, we used the Pearson χ^2 test for comparison. Statistical analyses were conducted using the statistical program SPSS version 17.0. To see whether today's magical plants were used differently in the past, we compared our results with the scanty eighteenth-century ethnobotanical records from the diaries of Daniel Rolander (2008), John Gabriel Stedman (1988) and some Moravian missionaries quoted in Price (1990).

Results

Great Numbers and Recipes

Plants play a leading role in *Winti* rituals, in particular among the Maroons. We recorded at least 411 species (belonging to 114 families) that were employed in *Winti* rituals (see Appendix). Of these species, we could identify 404 to species level, 15 to genus level and 1 only to family level. The most important families were Fabaceae (47 species), followed by Rubiaceae (18 spp.), Piperaceae (16), Asteraceae (15), Poaceae (13), Araceae (14) and Lamiaceae (11). The most important genus was *Piper* with 13 species. Seven species could not be identified at all, because of incomplete voucher material or the fact that some of these vegetable products represented several species. *Babar' udu* (noisy wood), for example, consisted of the bark of two tree trunks that leaned against each other and made a scary sound when the wind moved their branches. *Draai tetey* was a piece of one or two lianas that were tightly twisted around a small tree trunk, used for the *prati* ritual, during which a person was separated from an evil spirit or the influence of a malicious person.

People not only recognized a large number of magical species but also had a multitude of ways to prepare them. We documented over 1,100 different recipes to prepare ritual baths and concoct a magical potion or methods to deal with trees inhabited by spiritual beings. Most species (77%), however, ended up as ingredients in herbal baths. These were prepared by crushing or pounding specific leaves, bark or wood and blending them with hot or cold water in a tub or a *prapi*, a richly decorated earthenware pot manufactured by Amerindians. Single plants were rarely used; some baths required no less than 15 different species of fragrant leaves, supplemented with perfume, alcohol and even pieces of fabric. An empty calabash (*Crescentia cujete* L.) was used to pour water over one's head and body (see Fig. 10.2).

Winti as a Lucrative Business

Winti is one of the driving forces behind the medicinal plant trade in Suriname. In 2006, more than 56% of the species (139 out of 247) sold at the medicinal plant markets in Paramaribo had one or more applications in ancestor rites, herbal baths or protective *obias*. Many traditional healers did their shopping at the herb market (van Andel et al. 2007). *Winti* is a lucrative business, in particular for those religious specialists who administer herbal purification baths to tourists (well-to-do Dutch Surinamers on holiday). In 2006, *Winti* specialists charged diaspora tourists around US$300 and local villagers US$150 for an elaborate herbal bath with more than ten different herbs, perfume, alcohol and accompanying prayers. Some traditional healers specialized in the ritual preparation of drug runners before they took on their flight, making them and their cargo untraceable for the custom officers by using

Fig. 10.2 Ndyuka *Winti* priest Ruben Mawdo giving a patient an herbal bath, Bigiston, 2006 (Picture by S. Groenendijk)

magical plants. Moreover, about 55,000 kg of herbal medicine is shipped annually to the Netherlands. Market surveys in the Netherlands revealed that 30% of the plants sold were used for *Winti* purposes (van Andel and van't Klooster 2007; Niekoop 2008). Apart from plants, shops and market stalls also sold a variety of non-vegetable magical ingredients, such as traditional costumes, coloured candles, essential oils, perfumes, Reckitt's blue cubes, *pemba* (sacred white clay), old coins, rusty nails, corks, bird feathers, animal bones and *papamoni* (cowrie shells).

Emic Categories of Magical Plant Use

Traditional healers gave their patients a *wasi* (ritual bath) to heal magical diseases, make their bodies accessible for good spirits, chase away bad ones, cleanse their souls, restore their balance with their *dyodyo* (guardian angels) or protect them against black magic and violence. In order to diagnose a supernatural illness, they tried to call the spirit bothering the patient by alternating different leaf baths. Once the person fell into trance and the spirit started to talk, the healer knew that he had chosen the right herbs and continued asking the spirit what was wrong and how the healing ceremony should proceed. Other techniques included lighting candles, holding a patient's hand and trying to diagnose the problem by singing and praying. In Maroon villages, it was common to see plastic tubs with leaves standing in front of people's houses for weeks or even months, used over and over again by simply adding fresh water and stirring the leaf mixture. Another familiar sight was the small leaf remains in people's hair, as herbal baths should be dried on the body and not rinsed off with clean water. Herbal decoctions were also sprinkled with broom-shaped shrubs around the house or the village to expel witchcraft or other negative spiritual influences. A calabash with a leaf decoction was placed on a wooden tripod in the garden to protect the crops from theft and enhance their growth. Magical objects maintained their power by regularly submerging them in leaf decoctions. Another way to use plants to cleanse a house or person from evil was the *smoko oso* ritual, during which strong-smelling bark or wood was burnt to produce a thick smoke. Magical potions were drunk to protect against snake bites or witchcraft or rubbed on the body as a love charm. Offerings for spirits were usually put on shrines, placed at the forest edge or between tree buttresses. Finally, there were plants that were considered sacred because spirits lived in them. This meant that they could never be cut down and their leaves or bark only collected when preceded by the proper prayers and libations. Our informants mentioned a total of 13 different spirits (or spirit groups), ten types of charms, two supernatural diseases and four other rituals for which magical plants were used (Table 10.1).

More than half of the species were grouped in the broadly defined categories 'bad spirits' or '*Winti* unspecified'. On one hand, this indicates that lots of species can be used for a broad range of magical purposes. Informants stated that 'they can do many good or bad things and can be used for anything related to *Winti*, as they are very powerful plants'. One the other hand, these vague categories also reveal that much specific information is still missing. In these cases, informants either did not know the exact use, did not want to divulge the details, or the researchers failed to ask the proper questions. Although our inventory may be far from complete, Table 10.1 does suggest a certain 'hierarchy' in spirits. If the number of species used for a particular spirit or charm is a measure for its importance in people's lives, the most prominent spirits addressed with herbs are those of the forest (*apuku*), the recently deceased (*yorka*), the air (*kromanti*) and the snakes (*fodu* and *papa*). Charms to attract luck or protect against black magic and baths for spiritual cleansing (*krinskin*) and fortifying one's soul (*swit' watra*) are also accompanied with a relatively high number of different herbs.

Species used to attract, pacify or exorcise spirits were often named after them. We recorded dozens of species that carried their deities' name, like *kromantiwiri* ('kromanti's herb', *Struchium sparganophorum* (L.) Kuntze), *apuku baasi* ('the boss of the *apukus*', *Crotalaria micans* Link), *pawintiwiri* ('papa winti's herb', *Justicia pectoralis* Jacq.) and *fodu kama* ('bed of the magical snake', *Coccocypselum guianense* (Aubl.) K.Schum.). When we asked our informants *why* specific plants were included in the different rituals, we did not always get a straightforward answer. People either had no idea or replied that plants had magical power 'as long as you believed in it' or that 'certain persons know how to give these herbs an assignment'. Several religious specialists said that it was their own *Winti* who told them which plants to use when they were in a state of trance. Others explained the reason for including a particular herb as follows: 'these leaves just make *fodu* feel

Table 10.1 Number and percentages of species used for the emic categories of magical plant use

Winti's (spirits)[a]	No. of species[b]	Description
Takru sani	114 (28%)	Bad spirits (general), haunting villages, houses, people and yards
Winti sani	109 (27%)	Spirits, magic (general, unspecified)
Apuku	99 (24%)	African forest spirits, jealous tricksters who love women
Yorka	84 (20%)	Spirit of (recently) deceased persons that harass family members
Kromanti	64 (16%)	Powerful spirits of the air, warriors and traditional healers
Fodu	60 (15%)	Various snake spirits, each one mounts a different species of snake
Papa	46 (11%)	Dangerous snake spirits associated with *Boa constrictor* snakes
Prati	42 (10%)	Ritual to separate a patient from evil spirits or negative energy
Bere winti, kabratafra	35 (9%)	Ancestor spirits, often honored with libations and food offerings
Ingi	35 (9%)	Amerindian spirits, speak native tongue, drink alcohol, smoke cigars
Wata wenu (incl. *tone*)	32 (8%)	Water spirits, *tone* are specific spirits of the rapids
Bakru	24 (6%)	Short, dark, evil demon, 'bought' to become rich and successful
Alakondre	19 (5%)	Combination of many different *winti's* at the same time
Akantasi	17 (4%)	Fierce spirit that lives in termite nests
Mama aisa	16 (4%)	Goddess of the earth, fertility and house yards
Bongo	8 (2%)	Specific family spirit in Brownsweg, Brokopondo
Leba	6 (1%)	Spirits of the streets, guardians of crossroads and villages, derived from *legba*, a major village deity in Benin and Nigeria
Charms		
Luck charm	84 (20%)	To get lucky or solve problems with money, police, court or school
Tapu fu wisi	57 (14%)	Protect against black magic (carried out by people to harm others)
Krinskin	53 (13%)	Lit. 'clean skin', herbal bath to wash off supernatural dirt
Swit' watra	53 (13%)	Lit. 'sweet water', herbal bath to fortify one's *kra* or *yeye* (soul)
Kroi	37 (9%)	Love charms, both to attract partners and to get rid of love spells
Feti obia	36 (9%)	Magic war medicine (protects against bullets, machetes, knives)
Ontiman obia	19 (5%)	Hunting charms
Meki pikin óbia	13 (3%)	Fertility charms, to avoid miscarriage or get pregnant
Kostgrond tapu	10 (2%)	Protect agricultural fields from theft and evil, enhance growth

(continued)

Table 10.1 (continued)

Winti's (spirits)[a]	No. of species[b]	Description
Puu mii a doo	6 (1%)	Protective bath for newborns when first shown around in village
Miscellaneous		
Ogri-ai	35 (9%)	Evil eye: magical children's disease caused by envy of others
Fyo fyo	27 (6%)	Magical children's disease caused by family quarrels and gossip
Ogri dren	17 (4%)	Bad dreams, nightmares
Wisi	11 (3%)	Black magic (to be carried out yourself to harm others)
Kankantri, katu, etc.	11 (3%)	Sacred trees or epiphytes harbouring spirits that cannot be felled
Sweli obia	5 (1%)	Ordeal poisons (used to point out witches or thieves)

[a]Local names are given in Creole (Sranantongo) or Maroon (Ndyuka or Saramaccan) languages
[b]Several species were employed for more than one category, so the total percentage exceeds 100%

good', or '*apuku* lives in this shrub', or '*akantasi* always reacts very strongly if you add these leaves in a bath'. In other cases, however, people had a clear idea on the efficacy of certain species: they had a remarkable growth form, obvious color or strong scent; they grew at a place where 'their' spirits also dwelled; or they had a specific relation with the ancestors or certain forest animals. In the following paragraphs, we will shed a light on the morphological, historical and ecological characteristics that were given as the cause of a plant's supernatural power.

Cultivated Species

Only 24% (98 spp.) of the magical plants were cultivated exotics, 19 of which were probably introduced to Suriname during the slave trade. Species used in ancestor rituals were significantly more often cultivated (Table 10.2). Ancestor meals (*ala mofo nyan* or *kabra tafra*) frequently contained food crops of African origin, such as yam, pigeon peas, sesame seeds, Bambara groundnut (*Vigna subterranea* (L.) Verdc., locally called *awoo pinda* 'peanut of the ancestors') or African black rice (*Oryza glaberrima* Steud.). The latter two crops are no longer cultivated along the coast, but they are still grown by Maroon farmers for food and to honor their ancient relatives. Ancestor offerings also included traditional dishes, like sugarcane juice, banana pudding and tea from the leaves of *Cecropia sciadophylla* Mart. or *Lippia alba* (Mill.) N.E. Br. ex Britton & P. Wilson but also old-time domestic items such as lime twigs as chewing sticks, Dutch clay pipes, tobacco leaves, traditional fabric, resin candles from Burseraceae bark and *Heliconia* leaves as plates.

Almost all plants used for *mama aisa* (defined by Surinamese *Winti* specialist Stephen (1998) as 'Mother Earth') were cultivated, which seems reasonable as she is the goddess of home gardens. Plants used for Amerindian spirits were also more often cultivated or weedy herbs than expected (Table 10.2). Many of them were

Table 10.2 Statistically significant features of magical plants used for the major emic categories[a]

Spirit type	Vegetation	Growth form	Color	Smell	Other features
Amerindian	Cultivated, secondary $\chi^2=11.504$	Herbs, shrubs $\chi^2=20.136$	Red $\chi^2=84.472$	–	Plants used by Amerindians
Air, war	–	Epiphytes, parasites $\chi^2=18.048$*	White $\chi^2=38.123$	–	War medicine $\chi^2=4.974$* Wound medicine $\chi^2=3.857$*
Forest	Primary $\chi^2=16.869$	Herbs, shrubs n.s.	–	Good $\chi^2=12.875$	*Piper*, *Costus*, Rubiaceae
Mama aisa	Cultivated $\chi^2=31.661$	–	Red, brown n.s.	Good $\chi^2=8.561$*	–
Evil eye	Cultivated $\chi^2=13.319$	–	Blue n.s.	Good $\chi^2=9.708$*	–
Ancestor	Cultivated $\chi^2=22.397$	–	–	–	Cultivated food plants
Deceased		Spines, itching hairs $\chi^2=13.134$	Blue n.s.	Bad n.s.	–
Black magic	–	Lianas n.s.	–	Bad n.s.	–
Snake	Secondary $\chi^2=8.547$*	Herbs n.s.	Red, brown $\chi^2=26.634$	–	–
Swit' watra	Cultivated $\chi^2=14.987$	Herbs n.s.	–	Good $\chi^2=45.946$	Asteraceae Piperaceae
Krinskin	Cultivated, secondary $\chi^2=10.528$	Herbs, shrubs n.s.	Red $\chi^2=14.931$*	Good $\chi^2=14.377$	–
Bad spirits (general)		–	–	Good and bad $\chi^2=14.025$	–
Fyo fyo	Cultivated $\chi^2=14.717$	Herbs n.s.	Blue n.s.	Bad $\chi^2=10.601$*	–
Hunting charm	Primary $\chi^2=13.947$	Epiphytes n.s.	–	–	–

n.s. trend visible, but not significant, – no trend visible
All p values <0.01, except *=0.01 $<p<$0.05
[a]Emic categories with no statistically significant features were omitted from the table

typical indigenous crops, like cassava, red pepper, tobacco, *Bixa orellana* L. and *Maranta arundinacea* L. Plants used in *swit' watra* and *krinskin* or to protect babies against the evil eye or *fyo fyo* were also significantly more often cultivated. These baths frequently contained fragrant herbs grown in home gardens (e.g. *Ocimum campechianum* Mill., *Lippia alba* and *Lantana camara* L.).

The Javanese and East Indian migrants that came to Suriname to work on the plantations after the abolition of slavery not only introduced several Asian food crops but also took some of their magical plants with them. Several of these Asian cultivars have now been incorporated in Afro-Surinamese religious ceremonies. *Cordyline fruticosa* (L.) A. Chev., a shrub commonly planted on Javanese cemeteries, has found its way to Maroon villages. During funerals, people waved its purple leaves while singing their songs for the deceased. According to *Winti* priest Ruben Mawdo, the shrub's slender stem allows the dead person's spirit to ascend easily into the sky, preventing it from staying on earth and troubling its family members. A total of 29 magical plant species (7%) had been taken over from other ethnic groups, most of which (12 spp.) were magical Amerindian species, followed by East Indian (9 spp.) and Javanese (5 spp.) plants. Most of the Asian ritual species (e.g. *Azadirachta indica* A. Juss., *Ocimum tenuiflorum* L. and *Piper betle* L.) were used by Afro-Surinamers living around Paramaribo.

Vegetation Type

Most magical plants we recorded (85%) were of Neotropical origin. More than 79% of the species (327 spp.) were native to Suriname and collected from the wild or from semi-domesticated sources. The majority of the wild species (40%, 166 spp.) originated from primary forest, while 154 species (38%) were found in secondary forest or occurred as weeds in open vegetation. Habitat played a role in several magical categories. Plants employed in hunting charms were more frequently growing in primary forest than expected (Table 10.2), which is explained by the higher abundance of wildlife in undisturbed forest. Plants associated with *apuku* forest spirits came also more often from primary forest. Once a respectable god from the Congo-Angola region (Wooding 1972), *apuku* transformed in Suriname into a lower-class forest spirit, a jealous trickster who should never be trusted. According to Thoden van Velzen and van Wetering (1988), *apuku* is a symbol of the uncultivated and untamed part of nature. The forest spirit is often pictured as short, dark figure with his feet pointed backwards, which hides between the buttresses of giant trees, and does not like to be disturbed. *Apuku* also bears a striking resemblance to the Ghanaian forest dwarfs (*mmotia* in Twi) that have similar features, live deep in the forest and, although mischievous, are quite proficient in the use of herbs (Grottanelli 1988). In Suriname, Piperaceae (11 spp.) and Rubiaceae (7 spp.) are its favourite families. People believe *apuku* lives inside these shrubs. We heard Ruben Mawdo sing a song for *apuku* before he picked the leaves of *Piper anonifolium* (Kunth) Steud. *Apuku* loves women, and he will become jealous and offensive when the woman he has chosen as his medium is approached by other men. Several recipes exist for herbal baths that should be taken by a man 'to tame the *apuku*' of the woman he likes, when his advances have not yet been successful.

The wild plants that were used for Amerindian spirits represented species that were (and sometimes still are) essential elements of the local indigenous cultures, such as *Ischnosiphon arouma* (Aubl.) Körn., the plaiting fibre used for cassava processing equipment, and *Brosimum acutifolium* Huber, a tree that yields hallucinogenic exudate that was used by indigenous shamans in the past. Plants to attract or to chase away snake spirits were more often found in successional vegetation, something that might be explained by a greater abundance of snakes in dense shrubland. For all other spirits and charms, the vegetation type where the plants grew or their state of domestication did not seem to matter.

Growth Form

Among all 411 magical plants, herbs were the most frequent growth form (158 spp., 38%), followed by trees (127 spp., 31%), shrubs (67 spp., 16%), lianas (45 spp., 11%), (hemi-) epiphytes (17 spp., 4%), parasites (7 spp., 1.7%) and water plants (2 spp., 0.48%). People generally believed that large trees with impressive buttresses were inhabited by forest spirits. Among the sacred forest giants were *Parkia* spp., *Pseudopiptadenia suaveolens* (Miq.) Grime and *Ceiba pentandra* (L.) Gaertn., the largest tree in Suriname and worshipped throughout its range in the Old and New World. Other sacred species that are seldom damaged out of fear for their inhabitants are strangler figs (e.g. *Ficus nymphaeifolia* Mill., *F. schumacheri* (Liebm.) Griseb.) and the large hemi-epiphytes *Clusia grandiflora* Splitg. and *Coussapoa angustifolia* Aubl. People carefully harvested the bark or aerial roots of these species and paid for this with offerings and libations to pacify the spirits that lived in them.

Kromanti is the general name of the gods of the sky, thunder, lightning, iron, war and medicinal knowledge. The name is derived from the Dutch Fort at Kormantse in Southern Ghana, from where large numbers of enslaved Africans were shipped towards Suriname. According to the legend, Coromantine blacks were fierce warriors and medicine men (Stedman 1988; Price 2008; Rolander 2008), who formed the backbone of armed resistance against the planters. In Suriname, *kromanti* are believed to be African flying spirits that travelled with the enslaved to the New World (Wooding 1972; Stephen 1998). Their various names and properties, however, suggest that they are a mixture of several African warrior and sky gods. Coromantine's knew how to prepare *obias* that granted them invulnerability to bullets, broken glass, fire and machetes. Special herbal baths were made 'to call *kromanti* to protect you against your enemies'. The *kromanti* plants we recorded were often epiphytes (*Philodendron* spp., *Ficus* spp., *Coussapoa angustifolia*) or parasites (Viscaceae, Loranthaceae) because they were regarded as 'powerful, as they come from the sky and can kill trees with their roots like *kromanti* itself'. The sky gods can change themselves into a bird of prey or a vulture and fly high in the sky, the reason why bird feathers are standard ingredients in *kromanti* rituals. The delicate fern *Trichomanes vittaria* DC. ex Poir., which shrinks when dry but unfolds open like a feather when placed in water, was therefore used in a ritual bath to call *kromanti*.

Herbs were favored for several magical categories, but only for Amerindian spirits was this preference significant (Table 10.2). Several water plants (e.g. *Eichhornia crassipes* (Mart.) Solms) or species commonly growing on river banks (*Guarea gomma* Pulle) were used in rituals for water spirits, but too many other plants without an association with water were used to make numbers significant. Although growth form was not significant for most categories, it was often mentioned as the reason for the magical power of individual plant species. *Philodendron scandens* K. Koch & Sello, for example, with its heart-shaped leaves that grow tightly pressed against a tree trunk, was called *abrasa* (embrace) and used in love charms.

Throwing Spines on the Road of Life

Growth form plays a key role in plants used to deal with *yorkas*, the angry spirits of the dead. Ghosts of recently deceased persons have the nasty habit of following and harassing their family members, especially when they have done bad things in life or when their relatives do not keep themselves to the strict mourning prescriptions (Polimé 1998; Stephen 1998). It is said that a *yorka* can 'blow spines on your path', making your life miserable with diseases as *yorka* fever, venereal diseases and AIDS. Plants with unpleasant spines (e.g. *Xylosma tessmannii* Sleumer or *Randia armata* (Swartz) DC.) or irritating hairs (e.g. *Mucuna sloanei* Fawc. & Rendle or *Cnidoscolus urens* (L.) Arthur) were put in an herbal bath to wash the person that was bothered by a *yorka*. Specific prayers, like 'yorka waai maka gwe' (lit. yorka, blow these spines away), helped to pacify the angry ghost so that it could peacefully ascend to heaven. Spiny plants were not exclusively used against *yorkas*. If someone is possessed by an *akantasi* spirit, he will be able to climb spiny *Astrocaryum* palms or run through thickets of *Machaerium lunatum* (L.f.) Ducke without being hurt. To diagnose whether a patient is bothered by this ferocious spirit, a *Winti* priest will prepare an herbal bath for him with macerated branches of these spiny plants. If the person in trance reacts strongly on this bath, *akantasi* is blamed for causing the trouble.

Color

We know from eighteenth-century eye witnesses that Surinamese Amerindians painted their bodies red with *Bixa orellana* when they travelled to the city or visited plantations (Stedman 1988; Rolander 2008). Plants used for Amerindian spirits had significantly more red flowers, stems or leaves. *Ingi tyen* (Indian cane), for instance, a sugarcane variety with dark red leaves, is specially cultivated for its use in medicine and *ingi winti* rituals. People possessed by indigenous spirits wear traditional red Carib costumes and drink red anise lemonade. The correlation of red plants and their use in *ingi* rituals was highly significant (Table 10.2). Herbal baths for ritual cleansing (*krinskin*) also contained more red plants than expected (Table 10.2), but this might be a coincidence, as none of our informants ever suggested a relation between the color red and ritual cleansing. In a rainy country like Suriname, the sky is often white. Herbs used in *kromanti* rituals therefore often have white leaves (e.g. *Cecropia* spp., *Parinari campestris* Aubl.), flowers (*Heliotropium indicum* L.) or latex (*Bagassa guianensis* Aubl., *Ficus* spp.). Local names as *kromanti pemba* (for

Fig. 10.3 Men possessed by *kromanti* spirits during a *Winti* dance in Elenakondre, Suriname (Picture by H. Rypkema)

the fern *Pityrogramma calomelanos* (L.) Link, which leaves are powdery white below) confirm this preference. People possessed by sky spirits paint themselves with *pemba* (white kaolin) and dress in white (see Fig. 10.3).

According to a widespread belief, blue is a color that can ward off evil. Plants to chase away spirits of the dead had a tendency to be blue, like *Indeed, suffruticosa* Mill., *Commelina diffusa* Burm. f. and *Stachytarpheta* spp. This trend, however, was not significant as there were many species used without a specific color (other than green). The same accounted for plants used to treat the symptoms of evil eye: a loss of appetite, vomiting and weakness of babies caused by the envious looks of adults. In the past, the blue dye extracted from indigo leaves was used more often in herbal baths to ward off evil (Wooding 1979), but indigo is no longer grown as a plantation crop, and the color in these rituals now comes largely from Reckitt's blue cubes. Babies are washed in blue baths to protect them against the evil eye and *fyo fyo*, and sometimes given small amounts of the liquid to drink as well (Ruysschaert et al. 2008). For the other magical categories, color did not play a significant role, but for some individual herbs, it was mentioned as the reason for their magical power. By far the most expensive magical species on the market was *Psychotria ulviformis* Steyerm., a small, creeping herb with purple-brown leaves very similar in color to rotten leaves lying on the forest floor. The herb, which was quite difficult to detect in between the litter, was locally known as *kibriwiwiri* (hiding herb) and believed to make people or objects invisible. This herb may have been the ingredient in the magic invisibility *obia* described by

Price (1983) that gave the rebel slaves the confidence to undertake the dangerous flight from the plantations. Today, the herb is popular among cocaine smugglers, who are convinced it makes themselves and their drugs invisible during their flight from Suriname to the Netherlands.

Scent

Most of our informants agreed that fragrant herbs had a positive influence on spirits, while stinking plants or smoke could exorcise evil. People generally distinguished between good and bad scent, without much further detail. There was, however, some variation in personal preference. The strong garlic scent of *Mansoa alliacea* (Lam.) A.H. Gentry, for example, was defined as sweet by some but disliked by others. Pungent-smelling herbs were frequently put in herbal baths to drive out demons, black magic, spirits of the deceased and *fyo fyo*. The latter disease, named after the foul-smelling bug *Nezara viridula* (Heteroptera), was thought to be caused by family strife and affect the health of small children or unborn babies.

Aromatic herbs were significantly more often used in herbal baths for ritual cleansing (*krinskin*), taken 'when things do not go well' and people feel miserable, a condition named *poi* by Saramaccans. Sweet-smelling plants (e.g. *Ocimum campechianum*, *Renealmia* spp., *Lantana* camara, *Lippia alba*) were also dominant ingredients in *swit' watra*, herbal baths to fortify one's soul. Attractive flowers (e.g. *Ixora coccinea* L. and *Jasminum multiflorum* (Burm.f.) Andrews) often floated on the surface of these baths. According to respondents in Brownsweg, beauty and sweet odour would antagonize the supernatural causes of bad luck. *Apuku* baths were also more often sweet-scented, which could be caused by the frequent inclusion of *Piper* and *Costus* species. Remarkably, sweet-scented herbs were also used in baths to expel the evil eye and other bad spirits (Table 10.2) and sometimes combined with bad-smelling ones. Moreover, many plants included in these baths did not have any particular odour (or we failed to notice this). Therefore, scent only appeared as a significant feature in a few magical categories. We feel, however, that scent is quite important in *Winti* rituals, as a myriad of synthetic fragrances is sold for *Winti* rituals, varying from essential oils, incense and perfumes to asafetida, camphor blocks and bull's urine.

Medicinal Use

The majority (298 spp., 73%) of the magical plants also had one or more medicinal uses, although the distinction between supernatural illnesses (*wintisiki*) and physical diseases (*datrasiki*) and their succeeding therapy was not a rigid one. Female infertility, for example, was treated both with herbs that cleaned the uterus and with ritual baths, as women generally thought that their barrenness was caused by *wisi* (black magic). The cause of this curse was jealousy and hatred of other women, which could only be alleviated by spiritual mixtures. Strengthening herbal baths used to increase one's magical power were seen as a general health promotion and disease prevention. Most informants agreed that annoyed spirits could cause disease, and in some cases, plants that were used to call or chase away certain spirits were also used to combat diseases caused by these spirits. A decoction of *Mansoa alliacea*, for example, was taken internally to 'expel bad spirits from the body when suffering from stomach ache', while the garlic-scented wood was also used in herbal baths to drive away evil. Plants that were said to protect against gunshots, machetes and knives (*feti obias*) were significantly more often used to disinfect wounds and cuts ($\chi^2=7.071, p=0.008$) than others. In the majority of cases, however, medicinal applications seemed unrelated to the magical uses of the plants. A total of 127 species (31%) did not have additional medicinal uses.

External Features or Ecological Symbolism?

We estimate that signatures (growth form, color, shape or scent) caused the perception of supernatural powers for more than half of the species (210 spp.). This means that morphological characters often make a plant magical, but there is more to the story. Some 25 species (6%) had a clear symbolic function, such as the ancestor foods and the calabash (*Crescentia cujete*), which served as a plate, spoon and cup in times of slavery. For a total of 152 species (37%), we were unable to pin down the reason why they were used for *Winti* rituals. Finally, some 35 plants (9%) were chosen as ritual ingredients for their 'ecological behaviour'. We have seen earlier that plants growing in specific habitats (e.g. deep in the forest, in creeks) were associated with spirits that dwelled in these places. The relationships between plants and animals, however, also called people's attention. Some plants that were used for *fodu* or *papawinti*, a snake cult that probably originated in Grand and Little Popo in present-day Togo (Price 1983), were chosen because snakes often slept on them. Additionally, *Guarea gomma* is a slanting riverbank tree that provides a resting place for sunbathing anacondas. Ichthyotoxic plants (e.g. *Lonchocarpus* spp., *Clibadium surinamense* L.) were used in luck charms: if they could bring you fish, they might bring you money and love as well.

On several occasions, people used their detailed ecological knowledge to explain spiritual presence. The moss *Octoblepharum albidum* Hedw., for example, is a *kromanti obia* because it catches dew from the air and thus collects the power from the sky god. The large *Ceiba pentandra* tree must have divine power because many animals seek refuge in its crown and numerous epiphytes cling to its branches. *Akantasi* is a malevolent spirit that lives in termite nests. When people disturb him or burn a field in his vicinity, he will get so angry that he starts to produce froth. In fact, these foaming termite nests are caused by spittle bugs (Delphacidae, Bart de Dijn, personal commmunication, 2006) that infest the tree roots that grow into the

Table 10.3 The 15 most widely used *Winti* plants and their morphological and ecological characteristics

Species	No. of categories	Growth form	Vegetation	Color	Scent	Reason for magic[a]
Piper marginatum	21	Shrub	Secondary	–	Good	Scent
Begonia glabra	18	Epiphyte	Primary	–	–	Vegetation, water storage
Lippia alba	18	Herb	Cultivated	–	Good	Scent
Costus spp.[b]	18	Herb	Secondary	Red, white	Good	Water storage, color
Lantana camara	17	Shrub	Secondary	Red	Good	Scent, color
Ocimum campechianum	16	Herb	Cultivated	–	Good	Scent
Crescentia cujete	15	Tree	Cultivated	–	Bad	Symbolic utensil
Justicia pectoralis	15	Herb	Secondary	–	Good	Scent, snake-shaped
Scoparia dulcis	13	Herb	Secondary	–	Bad	Broom-shaped
Euphorbia thymifolia	12	Herb	Secondary	Red, white	–	Color, flat-shaped
Peperomia pellucida	12	Herb	Secondary	–	–	Water storage
Aciotis purpurascens	11	Herb	Secondary	Red	–	Water storage, color
Lycopodiella cernua	11	Herb	Secondary	–	–	Unknown
Musa sp.	11	Herb	Cultivated	Red	–	Symbolic food

[a]Motives for the use of magical plants were all provided by our informants
[b]Data from four *Costus* species were merged as they were used similarly

termite hills in search of fertile soil. People greatly feared this phenomenon, which was said to have killed several people in the past. Pieces of termite nests were sold at the Paramaribo market for rituals to calm down an angry *akantasi*.

In Table 10.3, the 15 most widely used *Winti* plants and the motivation for their inclusion in rituals are listed. We can assume that these 'multipurpose' species, which turned up as ingredients in hundreds of recipes, were the most frequently used magical plants in Suriname. Eight of the 15 species also featured in the top ten of most sold herbs in Dutch *Winti* shops (van Andel et al. 2007; Behari-Ramdas 2008). In terms of frequency of use, herbs, secondary forest species, fragrant leaves and red plants were important. When asked why especially these 15 species were used so often, our respondents answered something like 'all *Winti's* love this herb' or '*sangrafu* (*Costus* spp.) is the master of all plants', which does not advance our quest to find out what makes a plant magical.

Plants That Helped People Survive

Some plants that lack obvious features, ecological associations or Amerindian magic simply 'earned' their magical power in the past centuries by being extremely useful. Several species used in *Winti* rituals today have played a crucial role in the most decisive periods in Maroon history: their escape from slavery, their survival in the unknown forests and their bitter fight for freedom. When anthropologist Richard Price (1983) asked Saramaccan historians how their ancestors managed to escape from the plantations and stay alive, they replied that their forefathers had extraordinary magical power. It was their *obia* that showed them their way through the forest. Lieutenant J. G. Stedman, hired in the 1770s by the Dutch authorities to capture escaped slaves, admired the ability of the runaways to cross the brackish coastal swamps and the hot, white sand savannahs without food, ammunition or drinking water. While his own soldiers were dying from fever and thirst, he pondered about the secret water source of the Maroons. Had he read the unpublished diary of Linnaeus' student Daniel Rolander, written some 15 years earlier, he might have gotten the clue. On 15 September 1755, Rolander (2008: 1372) observed that the inflorescences of various *Costus* species contained 'a great supply of water … that the American Indians and black slaves usually drink, when they lack other good, sweet water'. People also frequently chewed the juicy stems of these herbs when wandering in the forest. Rolander did not notice any magical connotation with *Costus*. Today, the plant is a key ingredient in many different herbal baths. *Costus* is cultivated around shrines in Suriname and grown as an indoor plant by Surinamers in the Netherlands. Leaves and stems are chewed by people in trance to pacify angry spirits, and the juice is drunk by people to calm down after being possessed. During traditional Maroon funerals, *Costus* flowers are enclosed in the hands of the dead. When a baby is born, Saramaccans bury the placenta with some *Costus* leaves. After closing the hole, they plant a living *Costus* shoot on the spot.

A total of 19 species were mentioned for their ability to store water, four of which are among the 15 most frequently used plants (Table 10.3). Rainwater harvested from the bracts of *Heliconia* flowers is sold for ca. US$25 per litre at the Paramaribo market, and water collected from the crown of a *Mauritia flexuosa* L.f. palm was even shipped to the Netherlands. According to *Winti* specialist Ro Faria, the moisture collected in epiphytic Bromeliads is spiritually pure and can be used for ritual cleansings. Lianas from the Dilleniaceae family that store clean water are the chief ingredients of the *dungulali* medicinal cult of present-day Saramaccan Maroons. Chewing the succulent stems of *Begonia glabra*, *Peperomia pellucida* (L.) Kunth and *Aciotis purpurascens* (Aubl.) Triana also provided people with water. These plants now bear the general name of *kowru ati* (lit. cold heart) and are said to work as a 'tranquillizer' on persons or spirits with *ati bron*, a hot temper, anger or stress. They also cause a god to depart from the body of its medium (Herskovits and Herskovits 1936). Cold water is also essential in libations given to ancestors and other spirits to honor and pacify them.

Forest products that were used in the past to make fire, like the tinder fungus *Ganoderma applanatum* (Pers.) Pat., Burseraceae resin and *fungu* (lit. 'sparkle', balls of hair from *Parinari campestris* leaves collected in ant's nests), are now burnt to chase away evil. Rolander (2008: 1383) also described how 'perfidious, hostile, fugitive blacks' industriously cultivated *Canna indica* L. at the foot of the mountains where they lived and used the seeds 'to shoot animals for food and for their brave resistance to the annual military expeditions of the whites. The seeds are usually imbued with a poisonous liquid, enabling them to inflict lethal wounds'. The hard, round seeds, known in Suriname as *weglopershagel* (runaway hail), are nowadays used in herbal baths to protect against bullets. The ancient function of ammunition to shoot the white enemy has now changed into a protective *obia* against general violence. Stedman (1988: 392) also reported that plantations were attacked with fake guns 'with only a crooked stick shaped something like a musket to supply it in appearance'. We encountered several species that referred to their resemblance to guns, such as *baba'i lopu* (noisy barrel) for the aerial roots of *Clusia grandiflora* and *Philodendron solimoesense* A.C. Sm. and *goni lopu* (gun barrel) for the flattened wood of the lianas *Coccoloba gymnorrachis* Sandwith and *Abuta* sp. that look like double-barrelled guns. These pieces of wood, which may have served in the past to manufacture these (fake) guns, are now used in various rituals to ward off evil, just like gunpowder itself. These examples prove that a good knowledge of their natural environment was fundamental for the successful escape of the Maroons. Those plants that helped the formerly enslaved to survive at critical moments are now considered sacred among their descendants.

Discussion

Suppressed and Secret Knowledge

The vast numbers of plants that play a role in magical rituals prove that the *Winti* religion is very much alive and is strongly connected to the Surinamese flora. Our results confirm that *Winti* specialists have a great knowledge of their natural environment. They use their careful observations on the morphology and ecology of plants and animals to describe and understand their spiritual world. We did not find any magical plant without a local name, but we found many with multiple names. However, knowledge of ritual plants was not evenly distributed and recipes varied greatly among respondents. We also are aware that our list of magical plants is far from complete. Many species, recipes and associated beliefs on magical plants would be added to our list if research would be continued in Maroon villages deeper in the interior and more *Winti* specialists were interviewed. As local names often provided us with clues on present or historical magical uses, further linguistic studies on vernacular names would yield additional information on the origin of useful plants.

For centuries, European scholars tended to ignore, deride and ridicule the ritual aspects prominent in slave medicine (Schiebinger 2004). As a result, very few records exist on the use of magical plants from the last three centuries, making it difficult to trace how and when plants 'earned' their magical power among the Afro-Caribbean population. In Suriname, there is still a reserved attitude towards the dissemination of magical expertise to outsiders. The knowledge of medicinal plants and the claim to possession by certain spirits is restricted to certain lineages or families (Price 1983; Stephen 1998; Groenendijk 2007). Too much openness would decrease people's exclusive magical power. Of all *Winti* rituals, the use of black magic and the ingredients for ordeal poison are probably the most secret. Given the widespread fear of being bewitched and extensive use of poison ordeals in the twentieth century (Thoden van Velzen and van Wetering 1988), the number of plants we recorded for these categories is probably an underestimation. As the knowledge of *Winti* plants and rituals provides an attractive income, many traditional healers and plant vendors we spoke with were concerned that outsiders would abuse or commercially exploit their information.

Doctrine of the Signatures

Although each *Winti* may have its favorite color, food, music and dance, they are somewhat less discriminating in regard to their herbs. Our results point out that although some species were strictly linked to certain deities, there were also many 'flexible' plants that catered to a large number of spirits and charms. The Doctrine of the Signatures, however, definitely plays an important role in the use of ritual plants. Growth form, leaf shape, scent, color or the presence of spines or stinging hairs is often cited as the reason for a plant's divine power. This phenomenon also prevails in spiritual pharmacopoeias elsewhere in the New World. Plants with prickles or stinging hairs, associated in Suriname with spirits of the dead, are the major ingredients in magical powders used in Afro-Brazilian sorcery (Brazeal and Bauml 2009). These powders, associated with the troublesome god Exu, are known in Brazil as *pemba*, a name that is reserved for sacred white clay in Suriname.

Throughout the Black Atlantic, fragrant herbs are said to have a positive influence on spirits, whereas foul-smelling plants are used to exorcise evil (McClure 1982; Grénand et al. 1987; Voeks 1997; van Andel 2000). In the Dutch Antilles, many of the magical plants taken orally appear to possess poisonous or mind-altering substances (Rutten 2003). Some toxic plants, like *Spigelia anthelmia* L. and *Cascabela thevetia* (L.) Lippold, are used in *Winti* rituals, but almost always externally in herbal baths. Yet no studies have been done so far on the physical and psychological effects of ritual herb baths. There is a general lack of scientific studies regarding the effect of aromatic herbs on human health, even though aromatherapy is one of the fastest growing complementary therapies in the world (Thomas 2002). Strong odours have led to the discovery of some healing plants, as they indicate the presence of bioactive compounds (Bennett 2007). According to Voeks (1997), the scent of certain leaf combinations alone can induce possession trance among trained *Candomblé* adherents.

According to Bennett (2007), the Doctrine of the Signatures is a post hoc attribution used to facilitate oral transmission of knowledge of plants with effective medicinal properties, rather than a priori clues to the utility of these species. In the case of some ritual plants, however, this might be the other way around. *Winti* plants are mostly employed externally or away from the human body and therefore depend less on their pharmaceutical properties. When Africans arrived in the New World, they had to find new plants for their ancient African gods, and they looked for species that matched the spiritual 'requirements' regarding color, scent, shape and growth form (Voeks 2009). Still, Bennett (2007) states that researchers were never present when the utility of medicinal plants was discovered, and, therefore, it is impossible to say whether signatures influenced plant selection. Plant use, however, is never a static process. When people migrate to new floristic regions, for instance, they have to reinvent their pharmacopoeia all over again. We gathered first-hand evidence of the discovery of new plants for old deities, for example, when we saw the Dutch herb *Oxalis corniculata* L. for sale in a Surinamese *Winti* shop in the Netherlands. According to the shopkeeper, the plant represented the 'Dutch variety' of the Surinamese weed *Euphorbia thymifolia* L. During the rainy season, when home gardens in Paramaribo were flooded, he had difficulties importing *E. thymifolia*. When he saw *O. corniculata* growing on the pavement outside his shop, he decided to use it as a substitute because it had similar dark brown leaves, a flat growth form and also occurred as a weed.

Disturbance Pharmacopoeia

The Maroons, regarded in Suriname as the masters of ritual plant knowledge, have lived for centuries surrounded by dense rainforest. This is reflected by the high proportion of trees and primary forest species in their magical flora. Almost 40% of the wild species were harvested from climax vegetation, and spirits of the deep woods like *apuku* or charms to encounter game animals frequently featured in these recipes. This differs considerably from the black diaspora in Brazil, who have a disturbance pharmacopoeia consisting of weeds, cultivated plants and secondary shrubland species, while trees are generally lacking (Voeks 2004). If we look at the most frequently used and commercial *Winti* plants, however, herbs and shrubs from disturbed habitats also play a prominent part. Just like in coastal Brazil (Voeks 2009), some of these pantropical weeds must have been known to the enslaved Africans when they debarked in Suriname. The assumption that the paucity of trees in black diaspora healing floras is a product of lack of time and experience in the South American forest (Voeks 2009) may hold for the Afro-Brazilian population around Bahia, but certainly not for the Suriname Maroons.

In the eyes of many African and Afro-Caribbean healers, it is not cultivated or domesticated but wild plants that possess real powers for healing. Even plants found growing in vacant lots or growing on the pavement are preferable to those bought in a store (Brandon 1991; Cunningham 1993; McMillen 2008). This belief has led to the scarcity of wild resources for medicinal and magical plants in many countries (Cunningham 1993). Suriname is an exception to this rule. Only 18% of the traditional healers interviewed by Behari-Ramdas (2008) preferred wild plants, believing that they were 'purer' than cultivated ones. Less than half of the 249 commercial species marketed at the Paramaribo market were harvested exclusively from the wild, and Maroons were actively cultivating wild species for medicinal and magical purposes (van Andel and Havinga 2008).

Plant Use from the Motherland

Pantropical weeds of American origin like *Peperomia pellucida*, *Mimosa pudica* L. and *Petiveria alliacea* L., were introduced during the early years of the slave trade in West Africa and are used there today for magical rituals as well (Voeks 2009). Did the enslaved recognize these herbs from their motherland when they arrived in Suriname? Or were these plants already used in Amerindian rituals? Certain local names of plants refer to their African origin, like *nengre-kondre pepre* (lit. negro country pepper) for *Aframomum melegueta* K. Schum. Such names can also point towards the occurrence of similar-looking plants in Africa, as in the case of *nengre-kondre adru* (negro country *adru*) for the strictly Neotropical *Cyperus prolixus* Kunth. More fieldwork is needed on magical plant use in those West African countries where the ancestors of the black Surinamese were enslaved, in order to make a similar comparison between African and Afro-American plant use, as was done for Nigeria and Brazil (Voeks 1997). The recent discovery of African black rice (*Oryza glaberrima*) in a Maroon garden (van Andel 2010) suggests that there might be more Old World crops cultivated in the Surinamese interior than have been to date documented.

Amerindian Influence

The importance of Amerindian spirits and their associated plant species in the *Winti* belief reflect the exchange of spiritual and ethnobotanical knowledge between Africans and indigenous peoples in seventeenth- and eighteenth-century Suriname. Similar references to indigenous spirits are known form Haitian *Vodou* (Deren 1953) and Brazilian *Candomblé* (Voeks 1997). Unfortunately, published information on magical plant use among Surinamese indigenous people is too scanty to compare their ritual flora with the Maroons. Moreover, the number of Amerindian magical species during the period of slavery was probably much higher than today, as a great deal of indigenous spiritual knowledge has been lost due to evangelization (Heemskerk et al. 2007; Herndon et al. 2009; Hoffman 2009). It is likely that some of the indigenous magical species of the past now live on in the *Winti* belief as herbs to pacify Amerindian spirits (van Andel et al. 2012).

Ritual Knowledge Under Threat

Their geographic isolation and a conspiracy of secrecy against outsiders have enabled Maroons to retain a good deal of their cultural autonomy (Thoden van Velzen and van Wetering 1988). In the last decade, however, access to Suriname's remote interior has increased substantially as a result of better roads, airstrips and growing tourism. Even though *Winti* is gaining popularity among higher educated urban Creoles (Stephen 1998; Zaalman 2002), and urban Maroons have created a flourishing market in magical herbs and objects (van Andel et al. 2007), traditional Maroon culture is disappearing due to the migration of the youth to the capital, globalization and evangelical pressure. Maroon elders complained to us that their grandchildren had moved to the city or migrated to the Netherlands and knew nothing about the forest. Most Surinamese shops in the Netherlands sell synthetic oils and perfumes for *Winti* baths rather than scented herbs. Protestant sects are increasing their influence in Maroon villages, and for education and primary health care, Maroons still depend on Moravian and Catholic organizations. Although some of these groups tolerate the use of medicinal plants for physical ailments, they all strongly condemn the worshipping of *Winti* spirits and ancestor shrines and generally ignore Maroon culture and language (Price 1990; Kambel 2006). Magical plant knowledge and ancient landraces cultivated for ancestor offerings, such as the African black rice and the Bambara groundnut, are the first to disappear when people are converted to Christianity. The recent plans for the construction of a road to Brazil straight through Maroon territory and the expansion of the Brokopondo Reservoir further endanger their traditional way of life (van Dijck 2003, 2010). Only a limited number of village elders have the specific ritual knowledge of how plants received their magical power, who gave them their name and why spirits prefer specific plants. Our study has pointed out that behind every sacred plant, there may be an extensive body of botanical, ecological, historical, linguistic and cultural knowledge. It is this specific oral history, largely undocumented and under great danger of being lost, that makes a plant magic in the end.

Conclusions

A total of 411 plant species were used in Afro-Surinamese *Winti* rituals. Our informants employed plants for 13 different spirits, ten charms, two magical diseases and four other ritual categories. Most species were used to expel evil or deal with spirits in general, followed by luck charms, and spirits of the forest, the sky and the recently deceased. Although in some cases it was a question of an exclusive god-leaf correspondence, there were also many 'multi-use' species. Most plants were used as ingredients in herbal baths. Some Old World species, introduced during the slave trade, became magical *because* of their ancestral origin. The bulk of the magical plants, however, were native to Surinamese forest. There were many things that could make a plant magical: a connection with the ancestors, either from Africa or Suriname; a remarkable growth form, shape, scent or color, a typical habitat; an intriguing ecological relationship with certain animals; an associated medicinal use; or simply a ritual status among other ethnic groups. Finally, several forest products (e.g. water-containing herbs, inflammable resins or seeds for ammunition) became magical because they once helped the escaped slaves to survive in the forest and win their struggle for freedom. Finding out what makes plants magical reveals some unknown history of this interesting case of African ethnobotany in the Americas.

Acknowledgements This research would not have been possible without the collaboration, trust and patience of our informants in Suriname and the Netherlands, in particular our principal *Winti* specialists Ruben Mawdo, Norbert Eersteling, Solé Ponica, Bibi van Essen, André Kendo, Michel Alubutu and Blacky Finfin. MSc students Karolien van Kerckhove, Reinout Havinga, Joelaika Behari-Ramdas and Lucienne Niekoop assisted in the fieldwork. We are grateful for the botanists of the National Herbarium of the Netherlands for their help with identifying our specimens. This research was funded by the Netherlands Organisation for Scientific Research (NWO), CWO Faculty of Bio-Engineering and BOF, Ghent University, Leopold III Fund and VLIR-IUS (Belgium).

Appendix A: Species Used in *Winti* Rituals, Suriname

Acanthaceae
Dicliptera sp. KVP0036
Justicia calycina (Nees) V.A.W. Graham
Justicia cayennensis (Nees) Lindau
Justicia pectoralis Jacq.
Lepidagathis alopecuroidea (Vahl) R. Br. ex Griseb.
Acoraceae
Acorus calamus L.
Adoxaceae
Sambucus canadensis L.
Agavaceae
Furcraea foetida (L.) Haw.
Amaranthaceae
Alternanthera brasiliana (L.) Kuntze
Amaranthus blitum L.
Amaranthus dubius Mart. ex. Thell.
Cyathula prostrata (L.) Blume
Pfaffia glomerata (Spreng.) Pedersen
Amaryllidaceae
Allium cepa L.
Allium sativum L.
Crinum erubescens L.f. ex Aiton
Hippeastrum sp. TvA5158
Hymenocallis tubiflora Salisb.
Anacardiaceae
Anacardium occidentale L.
Mangifera indica L.
Spondias mombin L.
Tapirira guianensis Aubl.
Annonaceae
Annona muricata L.
Duguetia pycnastera Sandwith
Guatteria anthracina Scharf & Maas
Guatteria punctata (Aubl.) R.A. Howard
Guatteria scandens Ducke
Xylopia discreta (L.f.) Sprague & Hutch.

(continued)

Appendix A (continued)

Xylopia frutescens Aubl.
Apiaceae
Apium graveolens L.
Eryngium foetidum L.
Foeniculum vulgare Mill.
Pimpinella anisum L.
Apocynaceae
Allamanda cathartica L.
Asclepias curassavica L.
Cascabela thevetia (L.) Lippold
Tabernaemontana heterophylla Vahl
Tabernaemontana siphilitica (L.f.) Leeuwenb.
Tabernaemontana undulata Vahl
Araceae
Alocasia plumbea van Houtte
Caladium bicolor (Aiton) Vent.
Colocasia esculenta (L.) Schott
Dieffenbachia seguine (Jacq.) Schott
Dracontium asperum K. Koch
Heteropsis flexuosa (Kunth) G.S. Bunting
Montrichardia arborescens (L.) Schott
Philodendron deflexum Poepp. ex Schott
Philodendron linnaei Kunth
Philodendron melinonii Brongn. ex Regel
Philodendron scandens K. Koch & Sello
Philodendron solimoesense A.C. Sm.
Araliaceae
Hydrocotyle umbellata L.
Schefflera morototoni (Aubl.) Maguire, Steyerm. & Frodin
Arecaceae
Astrocaryum sciophylum (Miq.) Pulle
Astrocaryum vulgare Mart.
Cocos nucifera L.
Desmoncus polyacanthos Mart.
Euterpe oleracea Mart.
Mauritia flexuosa L.f.
Attalea maripa (Aubl.) Mart.
Oenocarpus bacaba Mart.
Socratea exhorrhiza (Mart.) H. Wendl.
Aristolochiaceae
Aristolochia cf. consimilis Mast.
Asphodelaceae
Aloe vera L.
Asteraceae
Bidens cynapiifolia Kunth
Chromolaena odorata (L.) R. King & H. Rob.
Clibadium surinamense L.
Cyanthillium cinereum (L.) H. Rob.

(continued)

Appendix A (continued)

Eclipta prostrata (L.) L.
Elephantopus mollis Kunth
Erechtites hieracifolia (L.) Raf. ex DC.
Eupatorium triplinerve Vahl
Mikania guaco Bonpl.
Mikania micrantha Kunth
Rolandra fruticosa (L.) Kuntze
Struchium sparganophorum (L.) Kuntze
Tagetes spp.
Tilesia baccata (L.f.) Pruski
Unxia camphorata L.f.
Begoniaceae
Begonia glabra Aubl.
Bignoniaceae
Bignonia nocturna (Barb.Rodr.) L.G. Lohmann
Crescentia cujete L.
Mansoa alliacea (Lam.) A.H. Gentry
Bixaceae
Bixa orellana L.
Boraginaceae
Cordia curassavica (Jacq.) Roem. & Schult.
Cordia sagotii I.M. Johnst.
Cordia schomburgkii DC.
Heliotropium indicum L.
Bromeliaceae
Bromelia alta L.B.Smith
Guzmnia pleiosticha (Griseb.) Mez
Burseraceae
Protium cf. aracouchini Marchand
Protium guianense (Aubl.) March.
Protium heptaphyllum (Aubl.) Marchand
Tetragastris altissima (Aubl.) Swart
Tetragastris panamensis (Engl.) Kuntze
Cannaceae
Canna indica L.
Caricaceae
Carica papaya L.
Cecropiaceae
Cecropia obtusa Trécul
Cecropia sciadophylla Mart.
Coussapoa angustifolia Aubl
Chrysobalanaceae
Licania incana Aubl.
Licania leptostachya Benth.
Licania membranacea Sagot ex Laness.
Parinari campestris Aubl.

(continued)

Appendix A (continued)

Clusiaceae
Clusia grandiflora Splitg.
Garcinia madruno (Kunth) Hammel
Symphonia globulifera L.f.
Combretaceae
Terminalia catappa L.
Commelinaceae
Commelina diffusa Burm. f.
Commelina erecta L.
Commelina rufipes Seub.
Tripogandra serrulata (Vahl) Handlos
Convolvulaceae
Cuscuta americana L.
Ipomoea batatas L.
Ipomoea sp.
Costaceae
Cheilocostus speciosus (J. König) C. Specht
Costus arabicus L.
Costus scaber Ruiz & Pav.
Costus spiralis (Jacq.) Roscoe var. spiralis
Costus sp.
Cucurbitaceae
Lagenaria siceraria (Molina) Standl.
Luffa aegyptiaca Mill..
Momordica charantia L.
Cyperaceae
Cyperus odoratus L.
Cyperus prolixus Kunth
Scleria secans (L.) Urb.
Scleria stipularis Nees
Dilleniaceae
Davilla kunthii A.St.-Hil.
Doliocarpus brevipedicellatus Garcke
Dioscoreaceae
Dioscorea sp.
Ebenacee
Diospyros guianensis (Aubl.) Gürke
Elaeocarpaceae
Sloanea grandiflora Sm.
Euphorbiaceae
Cnidoscolus urens (L.) Arthur
Croton trinitatis Millsp.
Codiaeum variegatum (L.) Rumph. ex A. Juss.
Euphorbia hirta L.
Euphorbia thymifolia L.
Jatropha curcas L.
Jatropha gossipyfolia L.
Manihot esculenta Crantz

(continued)

Appendix A (continued)

Maprounea guianensis Aubl.
Ricinus communis L.
Fabaceae
Abarema jupunba (Willd.) Britton & Killip
Abrus precatorius L.
Adenanthera pavonina L.
Arachis hypogaea L.
Balizia pedicellaris (DC.) Barneby & J.W. Grimes
Bauhinia guianensis Aubl.
Bauhinia surinamensis Amsh.
Bocoa prouacensis Aubl.
Caesalpinia pulcherrima (L.) Sw.
Cajanus cajan (L.) Millsp.
Crotalaria incana L.
Crotalaria micans Link
Crotalaria pallida Aiton
Desmodium adscendens (Sw.) DC.
Desmodium barbatum (L.) Benth.
Desmodium triflorum (L.) DC.
Dicorynia guianensis Amshoff
Dipteryx odorata (Aubl.) Willd.
Eperua falcata Aubl.
Hymenaea courbaril L.
Indigofera suffructicosa Mill.
Inga alba (Sw.) Willd.
Inga heterophylla Willd.
Inga stipularis DC.
Inga virgultosa Desv.
Lonchocarpus chrysophyllus Kleinhoonte
Lonchocarpus negrensis Benth.
Lonchocarpus heptaphyllus (Poir.) DC.
Machaerium lunatum (L.f.) Ducke
Machaerium myrianthum Spruce ex Benth.
Mimosa pudica L.
Mora excelsa Benth.
Mucuna sloanei Fawc. & Rendle
Ormosia coutinhoi Ducke
Parkia nitida Miq.
Parkia pendula (Willd.) Walp.
Parkia ulei (Harms) Kuhlm. var. surinamensis Kleinh.
Pseudopiptadenia suaveolens (Miq.) Grimes
Sclerolobium melinonii Harms
Senna alata (L.) Roxb.
Senna occidentalis (L.) Link
Senna quinquangulata (Rich.) Irwin & Barneby
Tephrosia sinapou (Buchholz) A.Chev.
Vigna unguiculata (L.) Walp. ssp. unguiculata

(continued)

Appendix A (continued)

Vigna subterranea (L.) Verdc.
Zygia inaequalis (Humb. & Bonpl. ex Willd.) Pittier
Zygia latifolia (L.) Fawc. & Rendle var. latifolia
Gentianaceae
Chelonanthus alatus (Aubl.) Pulle
Gesneriaceae
Codonanthe crassifolia (H. Focke) C.V. Morton
Haemodoraceae
Xiphidium caeruleum Aubl.
Heliconiaceae
Heliconia acuminata A. Rich.
Heliconia bihai (L.) L.
Heliconia psittacorum L.f.
Heliconia richardiana Miq.
Humiriaceae
Sacoglottis cydonioides Cuatrec.
Hypericaceae
Vismia cayennensis (Jacq.) Pers.
Vismia guianensis (Aubl.) Pers.
Vismia macrophylla Kunth
Iridaceae
Eleutherine bulbosa (Mill.) Urb.
Lamiaceae
Amasonia campestris (Aubl.) Moldenke
Hyptis atrorubens Poit.
Hyptis lanceolata Poir.
Hyptis mutabilis (Rich.) Briq.
Hyptis recurvata Poit.
Leonotis nepetifolia (L.) R. Br.
Ocimum campechianum Mill.
Ocimum gratissimum L.
Ocimum tenuiflorum L.
Plectranthus amboinicus (Lour.) Spreng.
Thymus vulgaris L.
Lauraceae
Cassytha filiformis L.
Cinnamomum verum J. Presl.
Laurus nobilis L.
Licaria guianensis Aubl.
Lauraceae sp. TVA 4791
Laxmanniaceae
Cordyline fruticosa (L.) A. Chev.
Lecythidaceae
Couratari guianensis Aubl.
Loganiaceae
Spigelia anthelmia L.
Spigelia hamelioides Kunth

(continued)

Appendix A (continued)

Loranthaceae
Oryctanthus florulentus (Rich.) Tiegh.
Phthirusa stelis (L.) Kuijt
Struthanthus syringifolius (Mart.) Mart.
Lythraceae
Lawsonia inermis L.
Malpighiaceae
Byrsonima spicata (Cav.) Rich. ex Kunth
Hiraea faginea (Sw.) Nied.
Malvaceae
Abelmoschus esculentus (L.) Moench
Abelmoschus moschatus Medik.
Ceiba pentandra (L.) Gaertn.
Gossypium barbadense L.
Hibiscus bifurcatus Cav.
Lueheopsis rugosa (Pulle) Burret
Pachira nervosa (Uittien) Fern. Alonso
Quararibea guianensis Aubl.
Sida cordifolia L.
Sterculia excelsa Mart.
Sterculia pruriens (Aubl.) K. Schum.
Theobroma cacao L.
Waltheria indica L.
Marantaceae
Ischnosiphon arouma (Aubl.) Körn.
Ischnosiphon gracilis (Rudge) Körn.
Ischnosiphon petiolatus (Rudge) L. Andersson
Ischnosiphon puberulus Loes.
Maranta arundinacea L.
Mayacaceae
Mayaca sp. TvA 5444
Melastomataceae
Aciotis purpurascens (Aubl.) Triana
Clidemia hirta (L.) D.Don
Miconia ciliata (Rich.) DC.
Miconia gratissima Benth. ex Triana
Miconia lateriflora Cogn.
Miconia racemosa (Aubl.) DC.
Miconia tomentosa (Rich.) D. Don ex DC.
Miconia sp. TvA5237
Nepsera aquatica (Aubl.) Naudin
Meliaceae
Azadirachta indica A. Juss.
Carapa guianensis Aubl.
Guarea gomma Pulle
Menispermaceae
Abuta sp. TvA5488

(continued)

Appendix A (continued)

Moraceae
Bagassa guianensis Aubl.
Brosimum acutifolium Huber
Brosimum rubescens Taub.
Ficus nymphaeifolia Mill.
Ficus schumacheri (Liebm.) Griseb.
Ficus sp. GBA0162
Musaceae
Musa sp. (several cultivars)
Myristicaceae
Myristica fragrans Houtt.
Myrtaceae
Campomanesia aromatica (Aubl.) Griseb.
Campomanesia grandiflora (Aubl.) Sagot
Eugenia florida DC.
Pimenta dioica (L.) Merr.
Psidium guajava L.
Psidium guineense Sw.
Syzygium aromaticum (L.) Merr. & L.M. Perry
Syzygium cumini (L.) Skeels
Nyctaginaceae
Neea ovalifolia Spruce ex J.A. Schmidt
Nymphaeaceae
Nymphaea amazonum Mart. & Zucc.
Olacaceae
Heisteria cauliflora J.E. Smith
Oleaceae
Jasminum multiflorum (Burm.f.) Andrews
Onagraceae
Ludwigia erecta (L.) H. Hara
Ludwigia octovalvis (Jacq.) P.H. Raven
Passifloraceae
Passiflora coccinea Aubl.
Passiflora foetida L.
Passiflora glandulosa Cav.
Pedaliaceae
Sesamum indicum L.
Phyllanthaceae
Phyllanthus amarus Schumach. & Thonn.
Phytolaccaceae
Petiveria alliacea L.
Pinaceae
Pinus caribaea Morelet
Piperaceae
Peperomia pellucida (L.) Kunth
Peperomia quadrangularis (J.V. Thomps.) A. Dietr.
Peperomia rotundifolia (L.) Kunth
Piper aduncum L.

(continued)

Appendix A (continued)

Piper anonifolium (Kunth) Steud.
Piper arboreum Aubl.
Piper cf. avellanum (Miq.) C. DC.
Piper bartlingianum (Miq.) C. DC.
Piper betle L.
Piper brownsbergense Yunck.
Piper demeraranum (Miq.) C. DC.
Piper divaricatum G. Mey.
Piper hispidum Sw.
Piper marginatum Jacq.
Piper peltatum L.
Piper pulleanum Yunck.
Poaceae
Andropogon bicornis L.
Bambusa vulgaris Schrad.
Cynodon dactylon (L.) Pers.
Eleusine indica (L.) Gaertn.
Ichnanthus breviscrobs Döll
Imperata brasiliensis Trin.
Imperata contracta (Humb., Bonpl. & Kunth) Hitchc.
Oryza glaberrima Steud.
Oryza sativa L.
Olyra micrantha Kunth
Paspalum conjugatum P.J. Bergius
Saccharum officinarum L.
Zea mays L.
Podostemmaceae
Mourera fluviatilis Aubl.
Polygonaceae
Coccoloba gymnorrhachis Sandwith
Coccoloba mollis Casar.
Coccoloba sp. SRU0926
Pontederiaceae
Eichhornia crassipes (Mart.) Solms
Portulacaceae
Portulaca oleracea L.
Putranjivaceae
Drypetes variabilis Uittien
Rhizophoraceae
Rhizophora racemosa G. Mey.
Rosaceae
Rosa sp.
Rubiaceae
Capirona decorticans Spruce
Coccocypselum guianense (Aubl.) K.Schum.
Coffea liberica Hiern.
Duroia aquatica (Aubl.) Bremek.

(continued)

Appendix A (continued)

Ixora coccinea L.
Oldenlandia corymbosa L.
Psychotria apoda Steyerm.
Psychotria bracteocardia (DC.) Müll. Arg.
Psychotria capitata Ruiz & Pav.
Psychotria cuspidata Bredem. ex Schult.
Psychotria poeppigiana Müll. Arg.
Psychotria rosea (Benth.) Müll. Arg.
Psychotria sphaerocephala Müll. Arg.
Psychotria ulviformis Steyerm.
Randia armata (Swartz) DC.
Spermacoce latifolia Aubl.
Spermacoce verticillata L.
Uncaria guianensis (Aubl.) J.F. Gmel.
Rutaceae
Citrus aurantifolia (Christm.) Swingle
Citrus aurantium L.
Citrus limon (L.) Burm.f.
Citrus sinensis (L.) Osbeck
Ertela trifolia (L.) Kuntze
Ruta graveolens L.
Zanthoxylum pentandrum (Aubl.) R.A. Howard
Salicaceae
Banara guianensis Aubl.
Casearia javitensis Kunth
Xylosma tessmannii Sleumer
Santalaceae
Phoradendron cf. bathyoryctum Eichler
Phoradendron perrottettii Nutt.
Sapindaceae
Cupania hirsuta Radlk.
Paullinia sp. TvA5014
Vouarana guianensis Aubl.
Sapotaceae
Micropholis guyanensis (A. DC.) Pierre
Pouteria cf. cuspidata (A. DC.) Baehni
Pouteria engleri Eyma
Scrophulariaceae
Scoparia dulcis L.
Simaroubaceae
Quassia amara L.
Siparunaceae
Siparuna guianensis Aubl.
Smilacaceae
Smilax schomburgkiana Kunth
Solanaceae
Capsicum frutescens L.
Nicotiana tabacum L.

(continued)

Appendix A (continued)

Physalis angulata L.
Solanum americanum Mill.
Solanum leucocarpon Dunal
Solanum stramoniifolium Jacq.
Solanum subinerme Jacq.
Trigoniaceae
Trigonia laevis Aubl.
Ulmaceae
Trema micrantha (L.) Blume
Urticaceae
Laportea aestuans (L.) Chew
Verbenaceae
Lantana camara L.
Lavandula angustifolia Mill.
Lippia alba (Mill.) N.E. Br. ex Britton & P. Wilson
Priva lappulacea (L.) Pers.
Stachytarpheta cayennensis (Rich.) Vahl
Stachytarpheta jamaicensis (L.) Vahl
Vitaceae
Cissus erosa Rich.
Cissus sp.
Vochysiaceae
Vochysia guianensis Aubl.
Zingiberaceae
Aframomum melegueta K. Schum.
Alpinia purpurata (Vieill.) K.Schum.
Renealmia alpinia (Rottb.) Maas
Renealmia floribunda K. Schum.
Renealmia guianensis Maas
Renealmia monosperma Miq.
Renealmia sp.
Zingiber montanum (J.Konig) Link ex A.Dietr.
Zingiber zerumbet (L.) Roscoe ex Sm.

Appendix B: Lower Plants

Adiantaceae
Adiantum latifolium Lam.
Adiantum leprieurii Hook.
Adiantum serratodentatum Humb. & Bonpl. ex Willd.
Pityrogramma calomelanos (L.) Link
Aspleniaceae
Asplenium serratum L.

(continued)

Appendix B (continued)

Dennstaedtiaceae
Pteridium caudatum (L.) Maxon.
Dryopteridaceae
Tectaria sp. KVP92
Gleicheniaceae
Dicranopteris flexuosa (Schrad.) Underw.
Hymenophyllaceae
Trichomanes vittaria DC. ex Poir.
Polypodiaceae
Campyloneurum repens (Aubl.) C. Presl.
Schizaeaceae
Lygodium volubile Swartz
Thelypteridaceae
Thelypteris leprieurii (Hook.) R.M. Tryon
Selaginellaceae
Selaginella parkeri (Hook & Grev.) Spring.
Selaginella novae-hollandiae (Sw.) Spring
Lycopodiaceae
Lycopodiella cernua (L.) Pic. Serm.
Leucobryaceae
Octoblepharum albidum Hedw.
Fungi
Ganoderma applanatum (Pers.) Pat.
Unidentified higher plants
onalibano' SR3657
yuwairi' SR3655
babar' udu' SR2443
bakúlu fulu' SR3653
switi sei tombo' SR2201
ogii ay dyingo' SR2178
draai tete' TVA5095

References

Behari-Ramdas JA (2008) Evaluating ecological impact of commercial trade on Surinamese medicinal plants sold in the Netherlands. MSc thesis, Utrecht University, Utrecht

Bennett BC (2007) Doctrine of the Signatures: an explanation of medicinal plant discovery or dissemination of knowledge? Econ Bot 61:246–255

Brandon G (1991) The uses of plants in healing in an Afro-Cuban religion, Santeria. J Black Stud 22:55–76

Brazeal B, Bauml J (2009) Evil nature: the malicious use of plants in Afro-Brazilian sorcery. Paper presented at the 50th annual meeting of the society of economic botany, College of Charleston, Charleston, 30 May–8 June 2009

Cunningham AB (1993) African medicinal plants: setting priorities at the interface between conservation and primary healthcare. UNESCO People and Plants Working Paper, Paris

Davis W (1988) Passage of darkness: the ethnobiology of the Haitian Zombie. The University of North Carolina Press, Chapel Hill

Deren M (1953) Divine Horsemen: the living Gods of Haiti. McPherson and Company, New York
Glazier SD (2001) Encyclopedia of African and African-American religions. Routledge, New York
Grénand P, Moretti C, Jacquémin H (1987) Pharmacopées Traditionnelles en Guyane: Créoles, Palikur, Wayapi. ORSTOM, Paris
Green EC (1978) *Winti* and Christianity: a study in religious change. Ethnohistory 25:251–276
Groenendijk S (2007) *Winti* practices in Bigiston, Suriname. MSc thesis, Utrecht University, Utrecht
Grottanelli VI (1988) The python killer: stories of Nzema life. The Chicago University Press, Chicago
Heemskerk M, Delvoye K, Noordam D, Teunissen P (2007) Wayana baseline study. Amazon Conservation Team, Paramaribo
Herndon CN, Uiterloo M, Uremaru A, Plotkin MJ, Emanuels-Smith G, Jitan J (2009) Disease concepts and treatment by tribal healers of an Amazonian forest culture. J Ethnobiol Ethnomed 5:27
Herskovits MJ (1941) The myth of the Negro past. Beacon, Boston
Herskovits MJ, Herskovits FS (1934) Rebel destiny: among the Bush Negroes of Dutch Guiana. S. Emmering, Amsterdam
Herskovits MJ, Herskovits FS (1936) Suriname folk-lore. Columbia University Press, New York
Hoffman B (2009) Drums and arrows: ethnobotanical classification and forest use by a Maroon and Amerindian community in Suriname, with implications for biocultural conservation. PhD thesis, University of Hawaii, Manoa
Hoogbergen W (1990) The history of the Surinamese Maroons. In: Brana-Shute G (ed) Resistance and Rebellion in Suriname: old and new. College of William and Mary, Williamsburg, pp 65–102
Kambel ER (2006) Policy note on indigenous peoples and Maroons in Suriname. Inter-American Development Bank, Washington, DC
McClure S (1982) Parallel usage of medicinal plants by Africans and their Caribbean descendants. Econ Bot 36:291–301
McMillen HL (2008) Conserving the roots of trade: local ecological knowledge of ethnomedicines from Tanga, Tanzania markets. PhD thesis, University of Hawaii, Manoa
Mintz SW, Price R (1992) The birth of African-American culture: an anthropological approach. Beacon, Boston
Niekoop L (2008) Ferberdesiki no abi dresi. from genital steam baths to laxatives: plant use amongst the Surinamese community in the Netherlands. MSc thesis, Utrecht University, Utrecht
Polimé T (1998) Dood- en begrafenisrituelen bij de Ndjuka. Oso 17:71–73
Price R (1983) First-time: the historical vision of an Afro-American people. The Johns Hopkins University Press, Baltimore
Price R (1990) Alabi's world. The Johns Hopkins University Press, Baltimore
Price R (2008) Travels with Tooy: history, memory, and the African American imagination. University of Chicago Press, Chicago
Rolander D (2008) Daniel Rolander's Journal (1754–1756). The Suriname Journal: composed during an exotic journey (trans: from Latin by Dobreff J, Dahlman C, Morgan D, Tipton J). In: Hansen L (ed) The Linnaeus apostles: global science & adventure, vol 3(3). IK Foundation & Company, London, pp 1215–1576
Rutten AMG (2003) Magische kruiden in de Antilliaanse folklore: etnofarmacologie van het Caribisch gebied. Erasmus Publishing, Rotterdam
Ruysschaert S, van Andel TR, van de Putte K, van Damme P (2008) Bathe the baby to make it strong, healthy: plant use and child care among Saramaccan Maroons in Suriname. J Ethnopharmacol 121:148–170
Schiebinger L (2004) Plants and empire: colonial bioprospecting in the Atlantic world. Harvard University Press, Cambridge
Sedoc N (1992) Afro-Surinaamse natuurgeneeswijzen. Vaco Press, Paramaribo
St.-Hilaire A (2000) Global incorporation and cultural survival: the Surinamese Maroons at the margins of the world-system. J World Syst Res VI:101–131
Stedman JG (1988) In: Price R, Price S (eds) Narrative of a five years' expedition, against the Revolted Negroes of Surinam (1790). The Johns Hopkins University Press, Baltimore

Stephen HJM (1998) *Winti* culture: mysteries, voodoo and realities of an Afro-Caribbean religion in Suriname and the Netherlands. Karnak, Amsterdam
Thoden van Velzen HUE, van Wetering W (1988) The great father and the danger. Floris Publications, Dordrecht
Thomas DV (2002) Aromatherapy: mythical, magical, or medicinal? Holist Nurs Pract 17:8–16
van Andel TR (2000) Non-timber forest products of the North-West district of Guyana. PhD thesis, Utrecht University. Tropenbos-Guyana Series 8B, Utrecht
van Andel TR (2010) African rice (*Oryza glaberrima* Steud.): lost crop of the enslaved Africans discovered in Suriname. Econ Bot 64(1):1–10
van Andel TR (2012) Etnobotanical notes from Daniel Rolander's Diarium Surinamicum (1754-1756): Are these plants still used in Suriname today? Taxon 61(4).
van Andel TR, Havinga RM (2008) Sustainability aspects of commercial medicinal plant harvesting in Suriname. For Ecol Manage 256:1540–1545
van Andel TR, van't Klooster CIEA (2007) Medicinal plant use by Surinamese immigrants in Amsterdam, the Netherlands. In: Pieroni A, Vandebroek I (eds) Traveling cultures, plants and medicines: the ethnobiology and ethnopharmacy of human migrations. Berghahn, New York, pp 122–144
van Andel TR, Havinga RM, Groenendijk S, Behari-Ramdas JA (2007) The medicinal plant trade in Suriname. Ethnobot Res Appl 5:351–373
van Dijck PFFM (2003) The impact of Arco Norte on Northern Amazonia and the Guiana Shield: methodological reflections. Eur Rev Lat Am Caribb Stud 75:101–108
van Dijck PFFM (2010) The IIRSA Guyana Shield Hub: the case of Suriname. Cedla, University of Amsterdam, Amsterdam
van Lier RAJ (1971) Frontier society: a social analysis of the history of Surinam. Marinus Nijhoff, The Hague
Voeks RA (1993) African medicine and magic in the Americas. Geogr Rev 83:66–78
Voeks RA (1997) Sacred leaves of Candomblé: African magic, medicine and religion in Brazil. University of Texas Press, Austin
Voeks RA (2004) Disturbance pharmacopoeias: medicine and myth from the humid tropics. Ann Assoc Am Geogr 94:868–888
Voeks RA (2009) Traditions in transition: African diaspora ethnobotany in lowland South America. In: Alexiades M (ed) Mobility and migration in indigenous Amazonia: contemporary ethnoecological perspectives. Berghahn, London, pp 275–294
Wooding CJ (1972) *Winti*: een Afro- Amerikaanse godsdienst in Suriname. Krips Repro B.V, Meppel
Wooding CJ (1979) Traditional healing an medicine in *Winti*: a sociological Interpretation. Issue 9:35–40
Zaalman JA (2002) August, een bonoeman. Ralicon, Paramaribo

Chapter 11
Medicinal and Cooling Teas of Barbados

Sonia Peter

Abstract Barbados is the most easterly island of the Caribbean archipelago. The tradition of plant use for medicinal remedies has its origins in the self-help practices of the Africans, Amerindians, and Europeans whose lives were intertwined in the early history of the island. A survey conducted in 2007 to assess the current knowledge base in the rural communities on the island unearthed 93 plant species that are still employed in traditional medicine. The survey encompassed 8 of 11 parishes, 35 rural communities, and over 400 participants. Survey results showed that males and females with some knowledge of the use of plants for medicinal purposes represented 28.5 and 35.9%, respectively, of the sample population. The principal custodians of the knowledge base proved to be the females in the communities aged 45 and older. Families sought for therapeutic agents included Apiaceae, Asteraceae, Boraginaceae, Euphorbiaceae, Liliaceae, Papaveraceae, and Piperaceae. The leaf proved to be the best medicine matrix, and extraction invariably involved boiling or steeping in hot water. In the practice of brewing teas for medicine, hypertension and diabetes prove to be two conditions for which a variety of plants are sourced for natural products. In addition, the therapeutic regimen involves a strategy for good health revolving around the minimization of strain on the body by the elimination of toxins and heat stress. This was evident in the number of plants applied in medicinal teas for detoxification (33) and specifically in cooling teas (37). Denuding of the land by the years of sugarcane cultivation and subsequent development has resulted in a shift in medicinal sources from forested areas to roadsides and pastures and from trees to herbs. Phytochemical investigation of a selection of the plants used for cooling teas revealed a high antioxidant profile that is known to offer protection from degenerative diseases. The plants selected for analysis were shown to be high in polyphenolic compounds that possess bioactive properties,

S. Peter, B.Sc., Ph.D. (✉)
Departments of Chemistry and Environmental Science,
Division of Natural Sciences, Barbados Community College,
Howell's Rd., St. Michael, Bridgetown, Barbados
e-mail: speter@caribsurf.com

R. Voeks and J. Rashford (eds.), *African Ethnobotany in the Americas*,
DOI 10.1007/978-1-4614-0836-9_11,

including antioxidative, antihypertensive, anti-inflammatory, antiproliferative, and anti-thrombogenic. This protective profile, silently embodied in ethnomedical tradition, may have contributed to Barbados having one of the highest populations of centenarians per capita.

Keywords Traditional knowledge • Ethnomedicine • Natural products • Longevity

Introduction

Barbados is located at the coordinates 13° 10′ N, 59° 32′ W, lying most easterly in the archipelago between North and South America. The census report for 2000 gave a population of 268,792 with ethnicity made up of African descent 93.00%, European descent 3.19%, mixed 2.60%, and East Indian descent 1.03% (National Census Report 2000 (2009). The island area is 166 square miles and is inhabited by approximately 650 species of indigenous flowering plants (Carrington 2007). The Caribbean region is well documented as a biodiversity hotspot. In a recent report by the Critical Ecosystem Partnership Fund (CEPF), it was estimated that there were approximately 11,000 plant species in the region with 72% endemism among the species (CEPF Report 2009). Barbados, however, does not share this high level of island endemism found in the neighboring territories due to the extensive clearing that occurred during the period of sugarcane-based economic activity spanning the seventeenth and eighteenth centuries. Currently, there are only two endemic species cited by Carrington (2007, *Wild Plants of Barbados*)—*Metastelma barbadense* Schltr. and *Phyllanthus andersonii* Muell. There are limited forests remaining on the island, with one of the few old-growth areas being Turner's Hall Wood, the most species rich site on the island. Access to medicinal species has become increasingly restricted with the most extensive area of natural vegetation found in the deep gully systems that run the length of the limestone based landscape (Fig. 11.1). Practitioners of the tradition invariably source their plants by foraging pastures and roadsides or by development of herb gardens.

The British established the first documented community on the island around 1625 (Schomburgk 1848). Interestingly, the island is thought to have been named after the Barbados fig tree *Ficus citrifolia* Mill., which has an extensive aerial root system (Fig. 11.2). It is noted that the Portuguese, on an Atlantic journey from Brazil, named the island Las Barbadas for the bearded appearance of the trees that were abundant on the island. The Portuguese, however, did not claim the island at that time. The population rapidly increased after the occupation by the British with the demand for a labor force thought more suited to the rigors of a tropical landscape than the European settlers. The enslavement of Africans, many from Guinea, and a smaller percentage of Amerindians from neighboring islands provided the human capital (Handler 1969; Schomburgk 1848). Between June 24, 1698, and December 25, 1707, approximately 35,000 Africans were brought to the island to live under the laws of enslavement. Handler and Jacoby (1993) state that the

Fig. 11.1 Gully system in the parish of St. Thomas. In this image, the breadfruit tree, *Atrocarpus altilis* (Parkinson) Fosberg, features prominently in the foreground

Fig. 11.2 *Ficus citrifolia* Mill. The bearded fig tree features as a prominent species in the gully system and on the island of Barbados in general

thousands of Africans and their descendants in Barbados suffered from a wide array of ailments and diseases. Based on the descriptions, these included gastrointestinal and respiratory disorders, joint deterioration, contact dermatitis, and scurvy, among others, resulting from the harsh living and working conditions. European medical practitioners were considered ineffective at delivery of relief from these medical challenges, and the early healing practices of the Barbadian enslaved is believed to have been based mainly on a traditional West African system that incorporated spiritual and supernatural elements (Handler 2000). It was known, but not systematically documented, that knowledge of plants and roots played a significant role in the healing traditions of inhabitants on the island. A pharmacopoeia was said to have been established by the first few decades of the eighteenth century. It can be assumed that plant use traditions changed over time. However, Bayley (1949) reported that in the 1940s, some plants had to be treated magically for efficacy, harvesting of plant material could only take place at certain times, and some medicinal extracts had to be left in the dew to attain maximum potency. This suggests that some aspects of the West African belief system were still of value to the healing traditions of the time.

This chapter examines the level of medicinal plant knowledge within the rural communities of Barbados as indicated by the results of a survey conducted in 2007. Rural communities were selected under the assumption that cultural erosion due to acculturation would have been lower in these settings (Srithi et al. 2009). This strategic position is reinforced further by Gazzaneo et al. (2005) and Prance (1991), who agreed that rural communities are the most neglected in terms of ethnobotanical studies. Some examination of the phytochemistry of the plant list generated was considered essential to establish a chemical profile of value in supporting the tradition by providing a basis for efficacy. Medicinal and cooling teas typically possess an antioxidant profile that imparts physiological properties that are responsible for the therapeutic effect (Miller 1996; Nijveldt et al. 2001). This examination of the use of plants for medicinal and cooling teas is a subunit of a project being funded by the National Council for Science and Technology with the broad objective of preserving the heritage knowledge associated with the use of plants on the island via documentation, validation through scientific evaluation, education, and promotion of sustainable utilization of species (Report by Peter et al. 2007).

Methods

Survey

A national survey was conducted in 35 rural communities throughout 8 of the 11 parishes on the island. Snowball sampling was done in order to capture the most knowledgeable participants in each community. Four hundred and forty participants were interviewed using a structured questionnaire to record the heritage knowledge according to age and gender. Participants were asked if they had any knowledge regarding the use of plants for medicinal applications. In each cited use, the part of

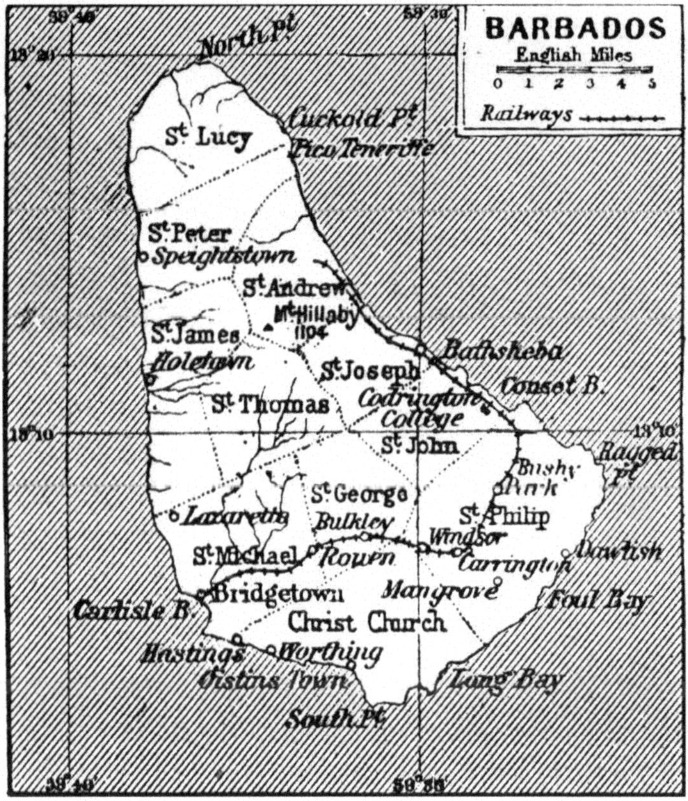

Fig. 11.3 Map of Barbados from 1922. The parishes in which communities participated in the survey included St. Lucy, St. Andrew, St. Peter, St. James, St. John, St. George, St. Thomas, and St. Joseph

the plant was recorded as well as the method of usage and frequency of usage, where possible and applicable. Figure 11.3 highlights the parish boundaries and the sites of the survey. Generally, these parishes are subjected to less industrial development and tourism activity and possess greater plant biodiversity. Herbarium samples were collected for accessible species and vouchers lodged in the collection of the National Council for Science and Technology. Some species were unavailable due to the season of collection or over harvesting by practitioners. This collection is being extended as additional species are collected (Fig. 11.4).

Qualitative Analysis

Fresh plant material was located in traditional areas including gullies, roadsides, and pastures. Plant material was identified and voucher specimen lodged in the

Fig. 11.4 (**a**) Herbarium sample of mosquito bush, *Ocimum Campechianum* Mill. This herb finds varied applications as a medicinal including detoxification and cleansing, sedation, and to relieve symptoms of colds and flu. (**b**) Herbarium sample of Black sage, *Cordia Currasavica* (Jacq.) Roemer & Schultes. A shrub, or small tree, found useful for laryngitis, rash, acne, postnatal care, detoxification, and as a tonic. (**c**) Herbarium sample of Christmas bush, *Chromolaena Odorata* L. King & H. E. Robins. A shrub of value during the cold and flu season and also found efficacious for regeneration of body and spirit

herbarium at the National Council for Science and Technology. Leaves and flowers of 11 plants used for making cooling teas were air dried for 4 days and subdivided using an industrial blender to a fine powder. For each analysis, 1 g of dried plant material was macerated with 5 cm^3 methanol and the extract used to prepare Merck Silica Gel F_{254} Thin Layer Chromatography plates for separation and identification of natural products of the aromatic and polyphenol classes. The TLC plates were developed in a solvent system of hexane/acetone 3:1 and observed under ultraviolet light at 254 and 365 nm using a Spectroline E-series UV lamp. Developed chromatograms were also sprayed with Folin-Ciocalteu reagent (10%) solution for identification of polyphenols.

Quantitative Analysis

A total phenol assay was carried out on each plant extract according to McDonald et al. (2001). Gallic acid was used as the reference phenol, and a standard curve was obtained using a series of solutions of concentrations 0, 50, 100, 150, 200, and 250 mg dm^{-3} of gallic acid in methanol: water (50:50 v/v). The plant extracts were made by maceration of 1 g of dried plant material in 10 cm^3 of methanol. Plant extracts and standard solutions, 0.5 cm^3, were treated with 4 cm^3 of anhydrous Na_2CO_3 in aqueous solution (1 Molar) and 5 cm^3 of 10% Folin-Ciocalteu reagent.

Color development was allowed for 15 min and absorbance at 765 nm measured spectrophotometrically. Duplicate estimates were made, and total phenol content was expressed as gallic acid equivalents in mg/100 g of dry plant material.

Results

Survey

The survey sample was composed of 59% females and 41% males. This closely matches the current population composition. A total of 440 respondents were interviewed (Fig. 11.5). Both male and female respondents had some knowledge of the use of medicinal plants. The average percentage of females in the sample with knowledge of the traditional practices was 35.9% while the average percentage for the males was 28.5%. In many instances, the "medicine women" in the village were deceased, leaving a void of key information regarding plant usage. Women in the communities with a greater repository of information were 45 years and older. Figure 11.6 displays the variation in knowledge with respect to gender for each parish surveyed. Respondents readily identified the plants in relation to their medicinal applications and were clear on the parts used, but dosage and frequency were often variable. Ninety-three plants were identified as useful for a variety of medical problems. A preference for herbs was evident (36%), closely followed by trees

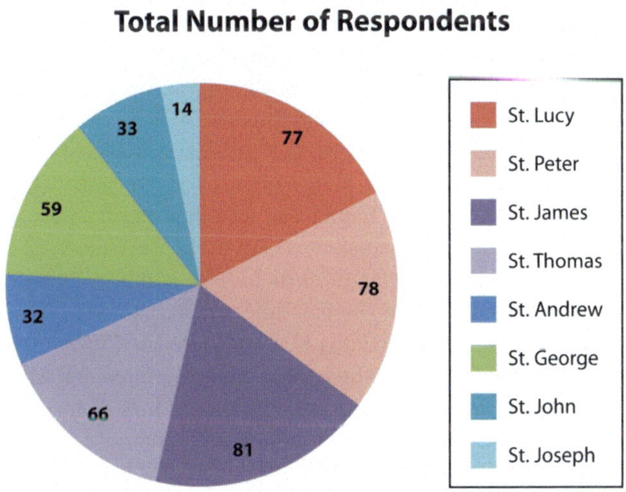

Fig. 11.5 Chart showing the number of respondents per parish

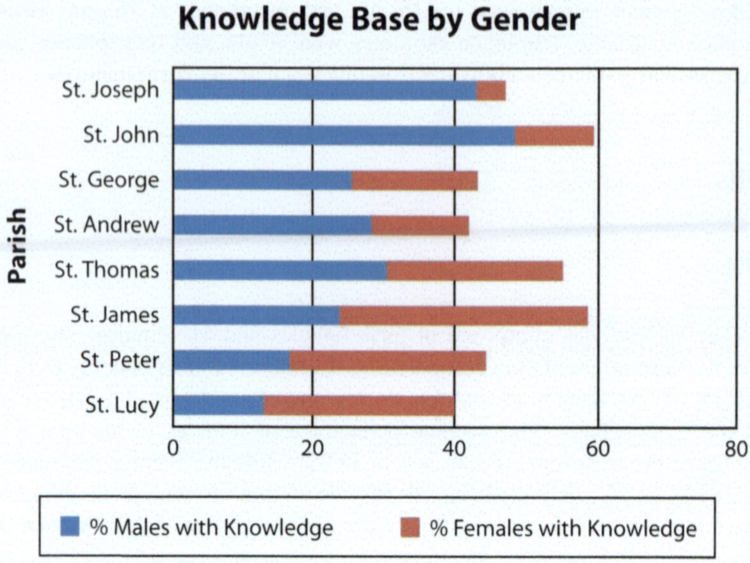

Fig. 11.6 Chart displaying the relationship between gender and knowledge of medicinal plant usage in the sample population. Overall, 39.5% of the females in the sample had knowledge of the traditional usage of indigenous plant species while 29.5% of the males interviewed demonstrated knowledge of the heritage

(30%), shrubs (22%), and grasses (7%). Most of the herbs were located in pastures, roadsides, or gully verges, but some were cultivated in gardens. Sixty-three medicinal uses for these plants were documented for species belonging to 43 families. Two herbs and one vine were identified as effective for a myriad of problems in all 8 parishes and 35 communities by multiple respondents. These three plants are aloes (herb), cerasee bush (vine), and wonder-of-the-world (herb). Treatment of "heat" in the body (37 species), the common cold and the flu (24), detoxification (33), hypertension (21), and diabetes (16) showed the greatest variety of plants utilized. Figure 11.7 displays the number of plant sources for the more frequently cited medical or health conditions. The Lamiaceae and Poaceae families provided the most treatments (Fig. 11.8). Medicinal or cooling teas were made as infusions or decoctions. In some instances, preparations included a number of plants to produce a brew thought more effective than singular component teas. Some researchers are investigating the efficacy of this modality (Ulrich-Merzenich et al. 2010). The leaf is considered most reliable in providing relief for symptoms, but all parts of the plants were on occasion found to be useful. Table 11.1 lists plants applied for the varied medicinal purposes as captured from the questionnaires.

All of the communities interviewed acknowledged the use of plants for the formulation of "cooling teas." In practice, cooling teas were administered as hot teas, but there was no standardization evident in terms of the frequency of usage. In addition, the plant material could be applied as harvested or dried to improve potency. Great variability was also noted in response to the question on the mass of plant material used in formulations. In most instances, the leaves were used, but

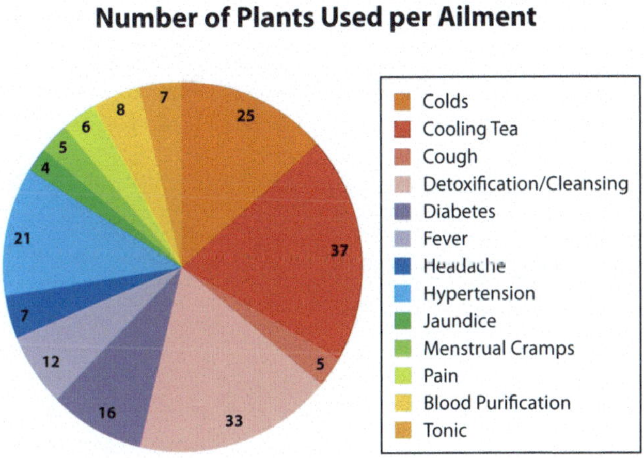

Fig. 11.7 Chart highlighting the number of plants used per medicinal application. This data shows that a significant variety of plant sources were sought for colds, cooling teas, and detoxification

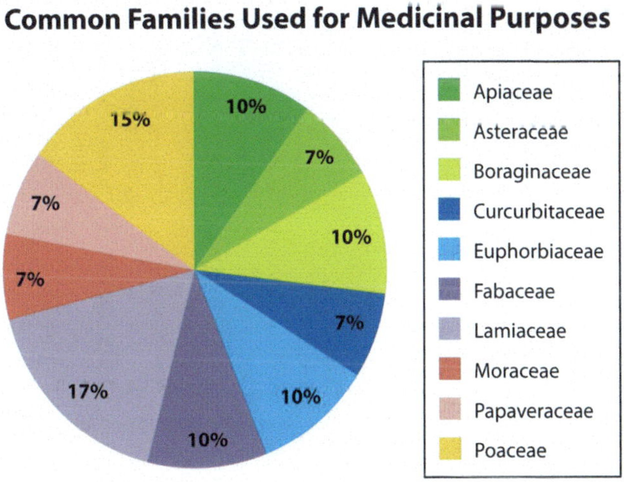

Fig. 11.8 Chart showing frequency of citation of selected plant families. Lamiaceae and Poaceae species feature prominently in the pharmacopoeia

Table 11.1 Plants used for medicinal teas in rural communities in Barbados

Family	Common name	Specific name	Type	Part used	Usage
Acanthaceae	Minnie root	*Ruellia tuberosa* L.	Herb	WP	Aphrodisiac, arthritis
	Garden balsam	*Justicia pectoralis* Jacq.	Tree	L	Colds, flu
	Blood root	*Justicia secunda* M. Vahl.	Shrub	L	Blood purifier, colds, flu, earache worms
Alliaceae	Garlic	*Allium sativum* L.	Herb	P	Cholesterol, detoxification/cleansing, diabetes, gas, hypertension
Amaranthaceae	Spinach	*Spinacia oleracea* L.	Vine	WP	Anemia
Anacardiaceae	Golden apple	*Spondias cytherea* Sonn.	Tree	L	Cough, detoxification, colds, flu
	Mango	*Mangifera indica* L.	Tree	L	Fever
	Cashew	*Anacardium occidentale* L.	Tree	L	Hypertension
Annonaceae	Soursop	*Annona muricata* L.	Tree	L	Cancer, appetite, fever, gas, hypertension, kidney stones, urinary tract infection
	Sugar apple	*Annona squamosa* L.	Tree	L	Worms
Apiaceae	Celery	*Apium graveolens* L.	Herb	L	Back pain, detoxification/cleansing
	Fennel	*Foeniculum vulgare* Mill.	Herb	L	Eye wash, gas, stomach inflammation
	Parsley	*Petroselinum crispum* L.	Herb	L	Detoxification/cleansing, diabetes
	Fitweed	*Eryngium foetidum* L.	Herb	L	Detoxification, post natal care
Apocynaceae	Periwinkle	*Catharanthus roseus* L.	Herb	L	Cancer
	Frangipani	*Plumeria rubra* L.	Tree	L	Acne, rash
Aracaceae	Wild eddoes	*Colocasia esculenta* (L.) Schott.	Herb	L	Asthma
Arecaceae	Coconut	*Cocos nucifera* L.	Tree	N	Hypertension
Asteraceae	Christmas bush	*Chromolaena odorata* L. King & H. E. Robins	Shrub	L	Colds, flu, cough, detoxification, sore throat, rejuvenation
	Cure-for-all	*Pluchea carolinensis* (Jacq.) G. Don.	Shrub	L	Colds, flu, appetite, jaundice, menopause, children's tonic
Boraginaceae	Black sage	*Cordia curassavica* (Jacq.) Roemer & Schultes	Shrub/small tree	L	Laryngitis, postnatal care, tonic, acne, rash, detoxification
	Clammy cherry	*Cordia obliqua* Willd.	Tree	L	Colds, flu, detoxification, fever
Brassicaceae	Wild mustard	*Brassica juncea* (L.) Czernj.	Herb	L	Blood circulation, hypertension

Cactaceae	Cochineel	*Opuntia cochinellifera* (L.) Miller	Shrub	L	Pain
Caesalpiniaceae	Tamarind	*Tamarindus indica* L.	Tree	L	Detoxification/cleansing
	Pride of Barbados	*Caesalpinia pulcherrima* (L.) Sw.	Shrub	L	Detoxification/cleansing
Caricaceae	Paw paw	*Carica papaya* L.	Tree	L	Detoxification
				GF	Hypertension
Capparaceae	White willow	*Capparis indica* (L.) Druce	Tree	L	Urinary tract infection
Chenopodiaceae	Wormwood	*Chenopodium ambrosioides* L.	Herb	L	Worms
Cecropiaceae	Trumpet tree (pop-a-gun)	*Cecropia schreberiana* Miq.	Tree	L	Diabetes, hypertension, colds, flu
Commelinaceae	Pond grass	*Commelina* spp.	Herb	L	Diabetes
Crassulaceae	Wonder-of-the-world	*Kalanchoe pinnata* (Lam.) Pers.	Herb	L	Detoxification/cleansing, fever, diabetes, sore throat (colds/flu), wounds, worms
Cucurbitaceae	Cerasee bush	*Momordica charantia* L.	Vine	WP	Bronchitis, colds, flu, coughs, detoxification (blood purifier), diabetes, fatigue, fever, hypertension, tonic
	Cucumber	*Cucumis sativus* L.	Vine	F	Hypertension
	Pumpkin	*Cucurbita* spp.	Vine	S	Prostate health
Euphorbiaceae	Cassava	*Manihot esculenta* Crantz.	Shrub	RT	Cancer, cholesterol, colds, flu, pain, headache
	Castor oil	*Ricinus communis* L.	Shrub	L	Colds, flu, hypertension, laxative, purging, cleansing after birth, fever
	Physic nut	*Jatropha gossypiifolia* L.	Shrub	L	Revitalizer
	Seed-under-leaf	*Phyllanthus amarus* Schum.	Herb	L	Prostate health
Fabaceae	Garfield bush	*Stylosanthes hamata* (L.) Taubert	Herb	L	Colds, flu, detoxification/cleansing

(continued)

Table 11.1 (continued)

Family	Common name	Specific name	Type	Part used	Usage
Lamiaceae	Mosquito bush (duppy basil)	*Ocimum campechianum* Mill.	Herb	L	Colds, flu, detoxification/cleansing, sedative
	Thyme	*Thymus vulgaris* L.	Herb	L	Detoxification/cleansing
	Peppermint	*Mentha piperita* L.	Herb	L	Detoxification/cleansing
	Rosemary	*Rosmarinus officinalis* L.	Herb	L	Gas, memory
	Broad-leaved-thyme	*Plectranthus amboinicus* (Lour.) Spreng.	Herb	L	Hypertension
	Ball bush (Governor's balls)	*Leonotis nepetifolia* (L.) Ait. f.	Herb	FH	Colds, flu worms
	Horehound	*Marrubium vulgare* L.	Shrub	L	Aphrodisiac, asthma, colds, flu detoxification, tonic
Lauraceae	Avocado (pear)	*Persea americana* Mill.	Tree	L	Colds, flu, cough, detoxification, Diabetes, fever, headache, hypertension, jaundice, rejuvenation, tonic
Liliaceae	Aloes	*Aloe vera* (L.) Burm. f.	Herb	G	Blood purifier, bronchitis, cataract, Colds, flu, constipation/laxative, purging, detoxification, diabetes, glaucoma, hypertension, inflammation, menstrual cramps
Malvaceae	Hibiscus	*Hibiscus spp.*	Shrub	L	Abortion, headache
	Okra	*Abelmoschus esculentus* (L.) Moench	Shrub	L	Detoxification
	White broomweed	*Sida spinosa* L.	Herb	L	Detoxification
Meliaceae	Neem	*Azadirachta indica* A. Juss.	Tree	L	Aphrodisiac, blood purifier, diabetes, fatigue, hypertension infections, pain, worms
Moraceae	Breadfruit	*Artocarpus altilis* (Parkinson) Fosberg	Tree	L	Diabetes, hypertension, colds, flu inflammation

Musaceae	Banana	*Musa acuminata* Colla	Tree	L	Diarrhea
				FR	Menstrual cramps
Myoporaceae	Olive	*Bontia daphnoides* L.	Tree	L	Colds, flu, headache, Menstrual cramps, post natal care
Myrtaceae	Bay leaf	*Pimenta racemosa* (Mill.) J. W. Moore	Tree	L	Blood purifier, colds, flu, diabetes diarrhea, fever, gas, hypertension menopause, nausea, sedative, tonic
	Eucalyptus	*Eucalyptus* spp.	Herb	L	Colds, flu, diabetes
	Guava	*Psidium guajava* L.	Tree	L	Colds, flu, detoxification/cleansing hypertension
Papaveraceae	Holly hock (yellow hock)	*Argemone mexicana* L.	Herb	L	Arthritis, diabetes, fever, gas, jaundice
				B	Earache
				S	Glaucoma
Phytolaccaceae	Gully root	*Petiveria alliaceae* L.	Herb	R	Blood purifier, bronchitis, immune, system, tonic
				R/L	Colds, flu, fractures
				L	Aphrodisiac, laxative, purging
Piperaceae	Rock balsam	*Peperomia magnoliifolia* (Jacq.) A. Dietr.	Herb	L	Appetite, blood purifier, detoxification/ cleansing
				L/ST	Colds, flu, fractures
	Shine bush	*Peperomia pellucida* (L.) Kunth	Herb	L	Neuralgia, children's tonic, detoxification/ cleansing
	Elderbush	*Piper dilatatum* Rich.	Shrub	L	Colds, flu
Plantaginaceae	English plantain	*Plantago major* L.	Herb	L	Colds, flu

(continued)

Table 11.1 (continued)

Family	Common name	Specific name	Type	Part used	Usage
Poaceae	Bamboo	*Bambusa vulgaris* Schrad. ex Wendl	Tree like	L/S	Colds, flu, diabetes, fever
	Bird plimpler	*Cenchrus echinatus* L.	Grass	L	Colds, flu
	Dutch grass	*Eleucine indica* (L.) Gaertn.	Grass	WP	Detoxification/cleansing
	Lemon grass	*Cymbopogon citratus* (DC. ex Nees) Stapf	Grass	L	Arthritis, blood purifier, colds, flu detoxification/cleansing, fever, menstrual cramps, rejuvenation, worms
	Sour grass	*Digitaria insularis* (L.) Mez ex Ekman	Grass	L	Colds, flu, detoxification
Rubiaceae	Dog dumpling (Noni)	*Morinda citrifolia* L.	Tree	L	Sedative, inflammation, headache, fever, arthritis
				L/FR	Hypertension, pain
				FR	Detoxification/cleansing, rejuvenation
	St. John's bush	*Psychotria nervosa* Sw.	Shrub	L	Colds, flu, laxative, purging, hypertension, appetite detoxification/cleansing
Rutaceae	Fingle-me-go (finger grow)	*Zanthoxylum spinifex* (Jacq.) DC.	Shrub	L	Aphrodisiac, arthritis, ring worms, detoxification/cleansing
		Citrus aurantiifolia (Christm.) Swingle	Shrub	FR	Appetite, colds, flu
				FR/PT	Stomach cramps
Sterculiaceae	Buffcoat	*Waltheria indica* L.	Herb	L	Colds, flu, rejuvenation detoxification/cleansing
Urticaceae	Nettle	*Laportea aestuans* (L.) Chew	Herb	L	Pain, kidney stones, diabetes, Urinary tract infection, prostate health
Verbenaceae	Coffee fence	*Clerodendrum aculeatum* (L.) Schlecht.	Shrub	L	Bronchitis, sore throat
	Blue Vervain	*Stachytarpheta jamaicensis* (L.) Vahl.	Herb	L	Depression, diabetes, nerve tonic
Zingiberaceae	Ginger	*Zingiber officinale* Roscoe	Herb	RT	Colds, flu, diarrhea, gas inflammation, nausea, stimulant
	Turmeric	*Curcuma longa* L.	Herb	L	Blood purifier, colon cancer, hypertension, detoxification/cleansing
Zygophyllaceae	Lignum vitae	*Guaiacum officinale* L.	Tree	L	Arthritis

L leaf, *RT* root, *WP* whole plant, *S* seed, *N* nut, *P* pod, *GF* green fruit, *F* fruit, *PT* pith, *FH* flower head, *G* gel, *B* bud

Table 11.2 Plants used for cooling teas in rural communities in Barbados

Family	Common name	Specific name	Type	Part used
Acanthaceae	Minnie root	*Ruella tuberosa* L.	Herb	WP
	Garden balsam	*Justica pectoralis* Jacq.	Herb	L
Amaranthaceae	Spinach	*Spinacia oleraceae* L.	Vine	L
Anarcardiaceae	Golden apple	*Spondias cytherea* Sonn.	Tree	L
	Mango	*Mangifera indica* L.	Tree	L
Annonaceae	Soursop	*Annona muricata* L.	Tree	L
	Sugar apple	*Annona squamosa* L.	Tree	L
Asteraceae	Christmas bush	*Chromolaena odorata* L. King & H. E. Robins	Shrub	L
	Cure-for-all	*Pluchea carolinensis* (Jacq.) G. Don.	Shrub	L
	Daisy	(not identified)	Herb	L
Boraginaceae	Soldier bush	*Tournefortia volubilis* L.	Shrub	L
	Clammy cherry	*Cordia obliqua* Willd.	Tree	L
Caesalpiniaceae	Pride of Barbados	*Caesalpinia pulcherrima* (L.) Sw.	Tree/shrub	L
Crassulaceae	Wonder-of-the-world	*Kalanchoe pinnata* (Lam.) Pers.	Herb	L
Cucurbitaceae	Cerasee bush	*Momordica charantia* L.	Vine	WP
Malpighiaceae	Barbadian cherry	*Malpighia emarginata*	Tree	L
Fabaceae	Iron weed	*Desmodium procumbens* (Mill.) A. Hitchc.	Herb	L
Lamiaceae	Ball bush	*Leonotis nepetifolia* (L.) Ait. f.	Herb	L
	Mosquito bush (duppy basil)	*Ocimum micranthum* Mill.	Herb	L
	Peppermint	*Mentha piperita* L.	Herb	L
Lauraceae	Pear (avocado)	*Persea Americana* Mill.	Tree	L
Malvaceae	Hibiscus	*Hibiscus* spp.	Shrub	F
Myrtaceae	Bayleaf	*Pimenta racemosa* (Mill.) J. W. Moore	Tree	L
Phytolaccaceae	Gully root	*Petiveria alliaceae* L.	Herb	L
Plantaginaceae	English plantain	*Plantago major* L.	Herb	L
Piperaceae	Rock Balsam	*Peperomia magnolifolia* (Jacq.) A. Dietr	Herb	L
	Shine bush	*Peperomia pellucida* (L.) Kunth	Herb	L
Poaceae	Lemon grass	*Cymbopogan citratus* (DC ex Nees) Stapf	Grass	L
Rubiaceae	St. John's bush	*Psychotria nervosa* Sw.	Shrub	L
Urticaceae	Nettle	*Laportea aestuans* (L.) Chew	Herb	L
Zingerberaceae	Ginger	*Zingiber officinale* Roscoe	Herb	RT

L leaf, *F* flower, *RT* root, *WP* whole plant

flowers or the whole plant was also found to be efficacious. All communities expressed the use of cooling teas as a regimen for better health or as a preventative health measure. The survey unearthed 31 plant species belonging to 22 families involved in the traditional cooling teas. Two families had the highest frequency of application—Lamiaceae and Asteraceae. The leaves were chosen as the vehicle for "cooling" phytochemicals in 88% of the species documented. Table 11.2 shows the families, plants, and parts used in the cooling teas.

Qualitative Analysis

The developed chromatograms were observed for complexity of natural products as detected by quenching or fluorescing bands at the two analytical wavelengths. Polyphenols are aromatic compounds with conjugated multiple bonds and such complexity of structure is indicated by dark zones against a green background at 254 nm. It should be reinforced, however, that not all aromatic natural products detected would be in the polyphenol class. Detection of fluorescent bands at 365 nm, dark yellow, blue, or green zones, can indicate the presence of the protective flavonoid group of phenols (Harborne and Baxter 1999). Blue bands that develop on spraying with Folin-Ciocalteu reagent confirm the presence of the aromatic phenol class of natural products. All of the extracts gave positive results for the phenol class of compounds with the Folin-Ciocalteu reagent. These blue zones were of typical low R_f values and concentrated near the point of application. Correspondingly, this region also fluoresced with colors characteristic of the flavonoid group (cf. Fig. 11.9).

Quantitative Analysis

The total phenol assay gave gallic acid equivalents (GAE) ranging from 60 mg/100 g to 1,835 mg/100 g of dry plant material. The green tea brand gave an estimate that was close to the average GAE equivalents for the selected series of plants, 653 mg/100 g. Pear leaf gave the highest reading while clammy cherry gave the lowest. Table 11.3 lists the results of the total phenol assay.

Discussion

Medicinal Teas

The pharmacopoeia revealed by this study bears some resemblance to that established during the period of enslavement on the island, the seventeenth and early eighteenth centuries. Handler and Jacoby (1993), in their examination of medicine and plant use during this period, highlighted 61 plant species that were used for varied medicinal purposes. Approximately 22% of those listed species were

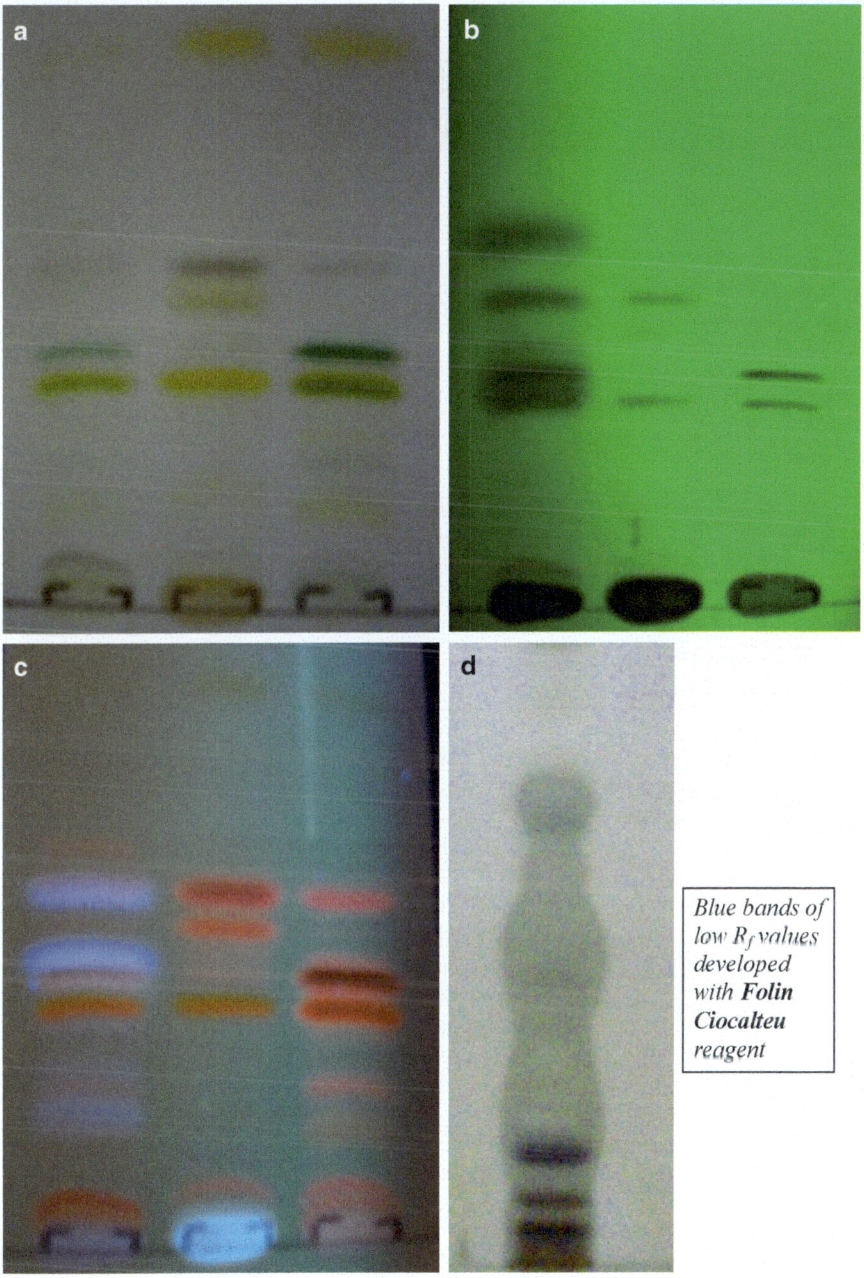

Fig. 11.9 (**a**) Thin layer chromatography of extracts from "cooling tea" species. Natural product bands are shown for (l-r) leaves of Lime, *Citrus aurantifolia* (Christm.) Swingle, Golden apple, *Spondias cytherea* Sonn., and Barbados cherry, *Malphigia emarginata* Sesse & Moc. (**b**) Thin layer chromatography visualization of leaf extracts from "cooling tea" species. Analysis under ultraviolet of wavelength 254 nm shows natural products as *dark bands* in the separated extracts. (**c**) Thin layer chromatography visualization of "cooling tea" extracts under UV 365 nm. The fluorescing *blue/violet bands* are typical of phenolic natural products, which typically exhibit antioxidant properties. (**d**) Thin layer chromatography visualization of soursop, *Annona muricata* L., leaf extract using Folin-Ciocalteu spray reagent. The *dark blue bands* also confirm the presence of phenolic natural products in the separated extract (Color figure online)

Table 11.3 Total phenol content of selected "cooling tea" plants

Plant assayed (common name/part)	Total phenol content (mg/100 g dry mass GA equivalents)
Pear (l)	1,835
Bay leaf (l)	1,150
Mango (l)	1,115
Pride of Barbados (l)	940
Ball bush (l)	850
Pride of Barbados (f)	700
Green tea brand[a] (l)	660
Hibiscus (f)	385
Golden apple (l)	300
Barbados cherry (l)	255
Lime (l)	140
Soursop (l)	103
Clammy cherry (l)	60

[a]Reference tea; *l* leaf, *f* flower

Table 11.4 Pre-1834 use of medicinal plants common to survey list

Common name	Traditional use
Aloe	Burns, purgative
Black sage	Sudorific (cooling tea)
Christmas bush	Vulnerary, detergent (topical cleanser)
Fingrigo (fingle-me-go)	*Not clearly defined*
Fitweed	"Fits" (convulsions) in children
Gully root	Obstructions (bowel)
Holly hock	Purgative, ulcers
Lignum vitae	Blood purifier
Paw paw	Fever, ring worms and skin disorders
Physic nut	Fresh wounds, swellings, purgative
Rock balsam	Vulnerary
Soldier's bush	Vulnerary, inflammations, swellings
Vervain	Menstrual flow
Wild basil (mosquito bush)	Vulnerary (wound healing)

Source: Handler and Jacoby (1993)

common to the list generated from this survey (Table 11.4). However, there was a notable shift in the applications of these plants, with treatment for wounds, swellings, infections, diarrhea and dysentery, poison antidotes, skin eruptions, and venereal disease featuring more prominently in the pre-1834 pharmacopoeia. Notable also is the importance of aloe, which to this day features as the most reliable and relied upon medicinal species. According to Handler and Jacoby (1993) and Schomburgk (1848), free Amerindians and Europeans as well as enslaved Africans and Amerindians were involved in growing aloe for commerce. Aloe is known for its bioactive gel, which is rich in myriad healing natural products, including mucopolysaccharides (Pugh et al. 2001). The value of this species in Barbados has been passed on and expanded with experimentation by successive generations. One respondent in the parish of St. Lucy claims to have developed a cure for asthma from the application of aloe with a specific additive. The clarity with which the cure was described, in terms of the quantity of active material, dosage, time of repeat dosage, and expected outcome, suggests that this may be a convergence point for the centuries of knowledge of application regarding this specific species.

Many pharmacopoeias feature Lamiaceae and Asteraceae species in prominence (Gazzaneo et al. 2005). This is corroborated by this study, as members of the Lamiaceae family are frequently employed to mitigate the symptoms associated with a range of health issues. Most notable is its use in the management of common colds and flu and as detoxification agents. The leaves of these species are highly aromatic, a property easily detected organoleptically (Abdelshafeek et al. 2009). The aromatic natural products stored in these leaves are potent and useful for respiratory relief and possess many antibacterial and antiviral properties (Mimica-Dukic and Bozin 2008). It should be noted that some of these species are not native to the island and are cultivated based on the knowledge of the healing value of the family. Of the seven Lamiaceae species listed in the medicinal teas table, only two grow naturally on the island—ball bush (*Leonotis nepetifolia* (L.) Ait. F) and mosquito bush (*Ocimum campechianum* Mill.) (Carrington 2007). This prevalence for including introduced species in the pharmacopoeia has been demonstrated in other studies, such as that by Bennett and Prance (2000), who looked at the importance of introduced species to the indigenous pharmacopoeia of northern South America, and Voeks (2009), who examined the role of exotic species in the continuity of African-derived healing traditions in Amazonia (see also Chap. 9 by Moret in this volume).

The Poaceae family is also a sustained source of phytomedicines in the rural communities in Barbados. The five species listed in this family in Table 11.1 grow freely in Barbados (Rogers and Holder 1999). Their role as sources of medicine can be explained in terms of their genetic ability to withstand harsh conditions being imbued with natural protective agents. These allelopathic natural agents that protect the grasses from competition with other plant species and organisms are the bioactive compounds sourced by practitioners of the tradition. Chemical and ecological factors underpin the selection of species in this pharmacopoeia. Most of the species fall into the category of herbs and shrubs and are typically sourced from disturbed areas due to limited accessibility (but see Stepp and Moerman 2001; Voeks 2004). The persistence of development with a shift from an agricultural economy to one based on tourism has impacted the indigenous plant stock and hence availability. On occasion, individuals are observed competing with traffic to garner the herbs and shrubs required to prepare their formulated teas. Roadside plants exist in a stressful ecological environment and this promotes a more vigorous and diverse biosynthesis of natural products for survival.

The management of diabetes and hypertension is currently a significant health challenge on the island. This is of more importance to Barbadians of African descent (Hennis et al. 2002). The prevalence of type 2 diabetes among the Barbadian participants of African descent was approximately 18%, while among the participants of European and other descent, this figure dropped to approximately 6%. In addition, there was a positive association with hypertension. Barbadians therefore actively seek redress from the symptoms of these two maladies from a variety of plants that are sourced as antidiabetic and antihypertensive modulators. Participants often admitted to foregoing the therapy provided by their medical practitioner because of a greater confidence in the traditional medicine. Popular species include cerasee bush (*Momordica charantia* L.) and the leaves of breadfruit (*Artocarpus altilis* (Parkinson) Fosberg), which are used as decoctions at a dosage based on the knowledge of individual blood sugar patterns. The breadfruit leaf may be used green, when yellow, or dried, on steeping or boiling in water to make the tea. The number of plants reflects the level of trial and error experimentation that ensued through the ages.

Cooling Teas

The rural population considered the tradition of using plants for "cooling teas" as a key component in the health-care regimen. Participants were adamant that the regular administration of cooling teas relieved heat stress, which was beneficial for fortification of the body against maladies such as the common cold and flu as well as chronic diseases. This therapeutic system is seen in many cultures, including Chinese, Indian, and African. One aspect of this system is based on the humoral theory of disease etiology which implies that a diseased state occurs when the balance between "hot" and "cold" in the body has been disrupted (Logan 1975). This concept is not based on a literal meaning of heat but rather on opposing forces that can dictate the state of health. Curing of disease or illness, therefore, requires the application of a system to work in opposition to that which led to the imbalance. This may involve the use of cold, hot, wet, or dry modalities (Currier 1966; Queiroz 1984; Coo 1989). Lans (2006) reports that in the Caribbean island of Trinidad, traditional systems surrounding heat loss in health-care management are not based on the humoral principles, and that the cooling effect is related to the removal of heat stress in the body. Lans (2006) argued that Trinidad's hot/cold system developed through syncretism of Amerindian, African, European, and East Indian traditions. In Barbados, it is possible that early applications of the cooling belief system did have some link to humoral principles, but there is no evidence one way or the other. Participants could not explain or account for the origin of cooling teas; they simply believed that a "cooled state" is a healthy state.

Within this Barbadian context, there is evidence to support the idea that heat stress relief represents a good health-care strategy. Heat stress has been linked to generation of reactive oxygen species (ROS) (Zuo et al. 2000), free radicals, and studies reveal that aging is accompanied by a decrease in small antioxidant molecules in the vascular system as well as decrease in effectiveness of antioxidant enzymes (Wei et al. 2001). Free radicals are very reactive oxidizing species implicated in degeneration of tissues, aging, and diseases such as cardiovascular and

Alzheimer's. Antioxidants mitigate free radicals by hindering their oxidative capacity. Studies have shown that there is a natural defense enzymatic system in the body that is effective at neutralizing oxidative species and the related stress (Scapagnini et al. 2002, 2006). Scholarly examination of the role of antioxidants in human health is revealing their broad spectrum benefits (Miller 1996; Rietveld and Wiseman 2003). It is now acknowledged that the reinforcement of antioxidants in the diet is valuable as a component of a preventative health-care strategy (Peng and Kuo 2003; Nijveldt et al. 2001). Validation of the health value of teas in relation to the phytochemistry of the extracts is being rigorously sought in many scientific studies (Friedman et al. 2005; Phan et al. 2001; Akinmoladun et al. 2007; Velayutham et al. 2008). It is quite likely, therefore, that this oral tradition of health management, rooted in home-based scientific observation and discovery, and passed down through generations, possesses at least some medical validity.

Further reinforcement for the value of the cooling tea system was obtained from the results of the qualitative and quantitative analyses conducted. The thin layer chromatography results highlighted the presence of natural products with conjugation such as is present in aromatic natural products. Furthermore, qualitative confirmation of the presence of polyphenols was obtained. Quantitative estimation of polyphenol content validated the selected species as significant sources of these agents, which also function as antioxidants (Huang et al. 2005). Polyphenols exhibit antioxidant capacity by scavenging for free radicals. They have also been shown to possess significant pharmacological properties including anti-inflammatory, antibacterial, antifungal, antiulcer, vascular protective, and anticancer (Di Carlo et al. 1999; Fresco et al. 2006; Kandaswami et al. 2005; Loa et al. 2009).

A regimen of fortifying the diet with antioxidant cooling teas may well act to enhance health status and slow the aging process. In Barbados, life expectancy for males is estimated as 73 years and for females 77 years (Barbados Statistical Services 2008). Barbados is rated second only to Japan in centenarians per capita. Global figures give a centenarian population ratio for the island of 1 per 2,800, which places Barbados in the top percentile (United Nations 2007). This population of African or mixed descent, culturally steeped in ethnomedical traditions, may have benefited from the recognition of the value of cooling teas to human health and hence longevity.

Conclusion

Traditional medicine still carries considerable cultural significance in rural Barbados. Casual conversation with individuals regarding the practice invariably yields anecdotal evidence. Stories of bitter aloes being forced on children as protective medicine are abundant, and "bush tea" brewed from cerasee vine is a staple during the cold and flu season. Home brews of cooling teas can be found bottled and refrigerated, including by the author, and many need that cup of soursop tea as a night cap. This ethnomedical legacy, although impacted by acculturation, is alive and well. The objective of this study was to seek systematic documentation, analysis, and ultimately preservation of this knowledge domain. The wealth of data obtained will be further subjected to statistical and phytochemical analysis to inform strategies for heritage protection, conservation, and potential natural product development. A quantitative study of the link between medicinal and cooling tea use and longevity will be conducted. It is also hoped that collaboration with ethnobotanists in the distant African territories from which some of this heritage was transported across the Atlantic will be established for a comparative study.

Acknowledgments This study was conducted in collaboration with the National Council for Science and Technology, NCST. I am expressing gratitude to the former Director of the NCST, Mr. Lennox Chandler, who appropriated funding for the survey, and the graduate students who assisted with the field work.

References

Abdelshafeek KA, Abdelrahem F, Elwahsh MA, Abdelkhalek IA (2009) Investigation of the flavonoid constituents and insecticidal activity of *Teucrium zanoii*. Pharmacogn Res 1(6):410–416

Akinmoladun AC, Ibukun EO, Afor E, Obuotor EM, Farombi EO (2007a) Phytochemical constituent and antioxidant activity of extract from the leaves of *Ocimum gratissimum*. Sci Res Essay 2(5):163–166

Akinmoladun AC, Ibukun EO, Dan-Ologe IA (2007b) Phytochemical constituents and antioxidant properties of extracts from the leaves of *Chromolaena odorata*. Sci Res Essay 2(6):191–194

Barbados Statistical Services (2008). www.barstats.gov.bb/census. Last accessed in 2011

Bayley I (1949) Bush teas of Barbados. J Barbados Mus Hist Soc 16:103–112

Bennett BC, Prance GT (2000) Introduced plants in the indigenous pharmacopoeia of Northern South America. Econ Bot 54(1):90–102

Carrington S (2007) Wild plants of Barbados. Macmillan, Oxford

Coo LC (1989) Ethnonutrition in Hong Kong traditional dietary methods of treating and preventing disease. Ethnonutrition 11(5):221–231

Critical Ecosystems Partnership Fund Report (2009) Ecosystem profiles: the Caribbean islands biodiversity hotspot. http://www.cepf.net/Documents/Finaldraft_Caribbean_EP.pdf. Last accessed in 2011

Currier R (1966) The hot-cold syndrome and symbolic balance in Mexican and Spanish-American folk medicine. Ethnology 5(3):25–263

Di Carlo G, Mascolo N, Izzo AA, Capasso F (1999) Flavonoids: old and new aspects of a class of natural therapeutic drugs. Life Sci 65(4):337–353

Fresco P, Borges F, Diniz C, Marques MP (2006) New insights on the anticancer properties of dietary polyphenols. Med Res Rev 26(6):747–766

Friedman M, Kim SY, Lee SJ, Han GP, Han JS, Lee KR, Kozuhue N (2005) Distribution of catechins, theaflavins, caffeine, and theobromine in 77 teas consumed in the United States. J Food Sci 70(9):C550–C559

Gazzaneo LRS, Paiva de Lucena R, Paulina de Albuquerque U (2005) Knowledge and use of medicinal plants by local specialists in a region of Atlantic forest in the state of Pernambuco (Northern Brazil). J Ethnobiol Ethnomed 1:9

Handler JS (1969) The Amerindian slave population of Barbados in the 17th and early 18th centuries. Caribbean Stud 8(4):38–64

Handler JS (2000) Medicine and Obeah in Barbados circa 1650–1834. New West Indian Guide 74(1 and 2):57–90

Handler JS, Jacoby JA (1993) Slave medicine and plant use in Barbados. J Barbados Mus Hist Soc XLI:75–98

Harborne JB, Baxter H (1999) The handbook of natural flavonoids, vol 2. Wiley, Chichester

Hennis A, Wu SY, Nemesure B, Li X, Leske CN (2002) Diabetes in a Caribbean population: epidemiological profile and implications. Int J Epidemiol 31:234–239

Huang D, Ou B, Prior RL (2005) The chemistry behind antioxidant capacity assays. J Agric Food Chem 53:1841–1856

Kandaswami CL, Lee T, Lee PP, Hwang JJ, Ke FC, Huang YT, Lee MT (2005) The antitumor activities of flavonoids. In Vivo 19(5):895–909

Lans C (2006) Ethnomedicines used in Trinidad and Tobago for urinary problems and diabetes mellitus. J Ethnobiol Ethnomed 2:45

Loa J, Chow P, Zhang K (2009) Studies of structure-activity relationship on plant polyphenol-induced suppression of human liver cancer cells. Cancer Chemother Pharmacol 63(6):1007–1016

Logan M (1975) Hot-cold theory of disease. Med Anthropol Newsl 6:8–14

McDonald S, Prenzler PD, Antolovich M, Robards K (2001) Phenolic content and antioxidant capacity of olive extracts. Food Chem 73(1):73–84

Miller ML (1996) Antioxidant flavonoids: structure, function and clinical usage. Alt Med Rev 1(2):103–111

Mimica-Dukic N, Bozin B (2008) Mentha L. species (Lamiaceae) as promising sources of bioactive secondary metabolites. Curr Pharm Des 14:3141–3150

National Census Report for 2000 (2009) 2000 round of population and housing census sub-project. National Census Report Barbados

Nijveldt RJ, van Nood E, van Hoorn D, Boelens PG, van Norren K, van Leeuwen P (2001) Flavonoids: a review of probable mechanisms of action and potential applications. Am J Clin Nutr 74:418–425

Peng I-W, Kuo S-M (2003) Biochemical and molecular actions of nutrients research communication. J Nutr 133:2184–2187

Peter S, Hill A, Chandler L (2007) Report: a survey of the applications of medicinal plant species in rural Barbados. National Council for Science and Technology, Barbados

Phan TT, Wang L, See P, Graayer RJ, Chan SY, Lee ST (2001) Phenolic compounds of *Chromolaena odorata* protect cultured skin cells from oxidative damage: implication for cutaneous wound healing. Biol Pharm Bull 24(12):1373–1379

Prance GT (1991) What is ethnobotany today? J Ethnopharmacol 32:209–216

Pugh N, Ross SA, Mahmoud AS, Pasco DS (2001) Aloeride, a new high molecular weight polysaccharide from *Aloe vera* with potent immunostimulatory activity. J Agric Food Chem 49:1030–1034

Queiroz MS (1984) Hot and cold classification in traditional Iguape medicine. Ethnology 23(1): 63–72

Rietveld A, Wiseman S (2003) Antioxidant effects of tea: evidence from human clinical trials. J Nutr 133:3285S–3292S

Rogers G, Holder A (1999) The grasses of Barbados – Poaceae. Botanical Research Institute of Texas/The University of the West Indies, Fort Worth

Scapagnini G, Foresti R, Calabrese V, Stella Giuffrida AM, Green CJ, Motterlini R (2002) Caffeic acid phenethwyl ester and curcumin: a novel class of heme oxygenase-1 inducers. Mol Pharmacol 61:554–561

Scapagnini G, Colombrita C, Amadio M, D'Agata V, Arcelli E, Sapienza M, Quattrone A, Calabrese V (2006) Curcumin activates defensive genes and protects neurons against oxidative stress. Antioxid Redox Signal 8:395–403

Schomburgk RH (1848) History of Barbados. Longman, Brown, Green and Longmans, London

Srithi K, Balsla H, Srisanqa P, Trisonthi C (2009) Medicinal plant knowledge and its erosion among the Mien (Yao) in Northern Thailand. J Ethnopharmacol 123(2):335–342

Stepp JR, Moerman DE (2001) The importance of weeds in ethnopharmacology. J Ethnopharmacol 75:19–23

Ulrich-Merzenich G, Panek D, Zeitler H, Vetter H, Wagner H (2010) Drug development from natural products: exploiting synergistic effects. Indian J Exp Biol 48:208–219

United Nations (2007) World population ageing. http://www.un.org/esa/population/publications/WPA2007/Full_text.pdf. Last accessed in 2011

Velayutham P, Babu A, Liu D (2008) Green tea catechins and cardiovascular health: an update. Curr Med Chem 15(18):1840–1850

Voeks RA (2004) Disturbance pharmacopoeias: medicine and myth from the humid tropics. Ann Assoc Am Geogr 94(4):868–888

Voeks RA (2009) Traditions in transition: African diaspora ethnobotany in lowland South America. In: Alexiades M (ed) Mobility and migration in indigenous Amazonia: contemporary ethnoecological perspectives. Berghahn, London, pp 275–294

Wei Y-H, Lu C-Y, Wei C-Y, Ma Y-S, Lee H-C (2001) Oxidative stress in human aging and mitochondrial disease: consequences of defective mitochondrial respiration and impaired antioxidant enzyme system. Chin J Physiol 44(1):1–11

Zuo L, Christofi FL, Wright VP, Liu CY, Merola AJ, Berliner LJ, Clanton TL (2000) Intra- and extracellular measurement of reactive oxygen species produced during heat stress in diaphragm muscle. Am J Physiol Cell Physiol 279:C1058–C1066

Part IV
Ethnobotanical Continuity and Change

Chapter 12
Candomblé's Cosmic Tree and Brazil's *Ficus* Species

John Rashford

Abstract Candomblé identifies largely orally transmitted religious traditions in Brazil tracing back to various parts of Africa. This research identifies the species of *Ficus* that serve as Candomblé's cosmic tree. Nineteen religious centers (terreiros) were surveyed and 17 had fig trees. Contrary to the general assumption of a single species, five native figs were identified, including *Ficus elliotiana* (S. Moore), *F. clusiifolia* (Schott), *F. gomelleira* (Kunth), *F. cyclophylla* (Miquel), and *F. tomentella* ((Miq.) Miq.) The most common was *F. elliotiana*, followed by *F. gomelleira* and *F. clusiifolia*. These results suggest that Candomblé has a complex relation with a variety of *Ficus* species, both native and exotic, and a wider survey that includes more rural communities and other urban areas will probably yield many surprises.

Keywords Sacred tree • World tree • Tree of life • Iroko • African religions • Tempo • Loko

Candomblé identifies largely orally transmitted religious traditions of Brazil tracing back to various parts of Africa. Its manifestations are everywhere in Salvador, Bahia, especially in association with elaborate calendrical rites and festivities. With the country's rapidly expanding national economy, including the growing importance of domestic and international tourism, Candomblé imagery has become emblematic, not only of Salvador but of Brazil (Walker 1990; Shirey 2009). Celebrated spiritual beings called orixás are the focus of devotees, and through music, dance, and possession, they establish contact with them to access the "energy" that ensures the success of their most important concerns. Approximately 20 orixás are widely recognized, and in addition to their individual drum rhythms, dance steps, and

J. Rashford, M.A., Ph.D. (✉)
Department of Sociology and Anthropology, College of Charleston,
66 George Street, Charleston, SC 29424, USA
e-mail: RashfordJ@cofc.edu

R. Voeks and J. Rashford (eds.), *African Ethnobotany in the Americas*,
DOI 10.1007/978-1-4614-0836-9_12, © Springer Science+Business Media New York 2013

Fig. 12.1 The *Ficus cyclophylla* at the Instituto Médico Legal Nina Rodrigues, the landscape of which was once a terreiro

familiar songs, all have their place in religious narratives that account for their distinctive personalities, associations with nature and special powers, as well as their characteristic colors, ceremonial attires, emblematic instruments, ritual salutations, and preferred offerings. Their relative popularity is based, in part, on the number of their *children*, as devotees are called in Candomblé, and on their representation in popular culture. Iroko is one of the most overlooked orixás, judging from the number of his devotees, scholarly publications, popular representations, and public art. So too is the cosmic iroko tree which is his domain (e.g., Murphy 1992; Barnet 1999; Verger 2002; Gondim 2004; Falola and Childs 2004; Prandi 2000; Murrell 2010). It is generally assumed that the iroko tree at the center of Candomblé is represented by a single species of *Ficus*, but this is not the case. This chapter presents the results of a survey of 19 religious centers called terreiros where several species of figs were identified (Fig. 12.1).

Some version of the cosmic tree motif, often involving figs, appears in religious cosmologies worldwide (Anonymous 1890; Philpot 1897; Beech 1913; Maguire 1931; Collis 1954; Gautier 1996; Condit 1969; Cook 1974; Reno 1977; Simoons 1998; Altman 2000; Kunwar and Bussmann 2006; Kislev et al. 2006), and the classic view of the cosmic tree owes much to Mircea Eliade, one of the most influential historians of religion, who emphasized its essential nature as the center of the

Fig. 12.2 Spontaneous examples of the cosmic tree motif on walls in Salvador. (**a**) *Ficus clusiifolia*. (**b**) *Ficus gomelleira*

universe connecting the sky and earth. Summarizing his view of the cosmic tree and its relationship to the religious importance of trees generally, Eliade (1991:44) wrote: "The most widely distributed variant of the symbolism of the Center is the Cosmic Tree, situated in the middle of the Universe, and upholding the three worlds as upon one axis. It may be said, in general, that the majority of the sacred and ritual trees that we meet within the history of religions are only replicas, imperfect copies of this exemplary archetype, the Cosmic Tree. Thus, all these sacred trees are thought of as situated in the Center of the World." This conception has extensive ethnographic support and is widely represented in scholarly publications, especially in encyclopedias and other reference works, but it is limited. If the cosmic tree is truly cosmic, it must be both the center and the periphery, that is, it must be the connector as well as the sky and the earth that it connects.

An alternative perspective, not as widely discussed as Eliade's but cited by researchers for more than a 100 years, is the cosmic tree as cosmic whole. With its branches above and roots below, the tree has been widely chosen to model the universe, most likely because it fits well the human experience of the sky above and earth below. The cosmic whole is the grounding motif of the cosmic tree, and ideas of the cosmic tree as cosmic center or cosmic connector, or as cosmic center connector, are only its varying expressions (Fig. 12.2a, b). For a more coherent understanding of understanding Candomblé, as well as other African-derived religious traditions like Vodou of Haiti and Santeria of Cuba, the conception of a tree-like universe provides a more adequate framework than Eliade's emphasis on the cosmic tree as preeminently the cosmic center (Deren 1970:36; Desmangles 1977:15; Rey 2005:1659; Prussin 1999:430).

Fig. 12.3 (**a**) A popular postcard of the orixá iroko from a set of orisha paintings by Gil Abelha. (**b**) The *Ficus gomelleira* tree of the terreiro Ilê Alabaxé in Maragogipe

In Candomblé, Iroko is the orixá of the treelike universe that is his domain, and in the spiritual landscape of terreiros, the iroko tree is set apart from other trees by a broad white ribbon and bow around its trunk called an Oja. It is often surrounded by a low wall that is painted white, and sometimes there are white streamers under its canopy (Fig. 12.3a, b). The intersection of the base of the tree and the earth's surface forms a vertical/horizontal axis that is the center of the universe – the cosmic crossroad – where Iroko's seat and altar are ritually established. Here spirits gather, especially the orixás and ancestors, and cosmic spiritual energy, called Axé, is most concentrated. According to Walker (1990:123–124), "to increase human participation in the *ashe*, the spiritual force of the universe," is the "raison d'être of … Candomblé," and the "concept of *ase*," writes Abiodun (1994:71), which "has intrigued many scholars of Yoruba culture both in Africa and the African diaspora … remains foundational for religio-aesthetic discourse in Brazil, the Caribbean Islands, and the United States… [and] is fast becoming a Pan-Africanist term." Voeks (1997:184–185) rightly notes that the iroko tree is at the "the center of [Candomblé] ceremonies" and this is understandable. As a representation of the universe, the cosmic iroko tree shelters all orixás, whose domains, residences, seats, altars, and offerings are associated with its various parts. The branching trunk, which forms the sky, is the realm of Olorum the Creator, and it is also associated with the orixás Oxalá, linked to high places and the creation of humanity; Xangô, to lightning and thunder; Iansã, to wind and storms; and Oshumarê, to rainbows and snakes. The horizontal axis where the sky and earth meet to form the surface is associated with Nanã, linked to swamps and ruins; Oxum, to rivers and lakes; Obá, to waterfalls; Iemanjá, to the ocean; Ossaim, to leaves indispensable for healing and the conduct of rites; Oxóssi, to forests and hunting; Ogun, to metals and opening the way; and Exu, to roads and entrances. The branching root, which forms the earth, is the domain of Omulu, the guardian of cemeteries and guide to spirits of the dead.

Identifying the Iroko Fig

Accepting the assumption that the iroko tree is a single species of fig, Martins and Marinho (2002), members of the well-known terreiro, Ilê Axé Opô Afonjá, incorrectly identified it as *Ficus religiosa* L. (2002:34), an error that also appears in the work of Lody (1975) and Wafer (1991:172), and is found on blogs and other kinds of Internet sites, including Wikipedia. Prandi (2001) identified the iroko fig as *Ficus maxima* Mill., which he said was "worshiped as an orixá in the old [terreiros] … of Bahia and Pernambuco" (2001:566). *Ficus maxima* is found in Central America, the Caribbean, and several South American countries, and while Carauta and Diaz (2002:566) identified it as occurring in many parts of Brazil, they did not report it as present in Bahia. The single-species assumption also underlies the identification of Candomblé's fig by the Brazilian common names gameleira (Omari 1989:57; Magalhães 2003:109; Coffey 2005:262) and gameleira branca (Menezes 1949:107; Cacciatore 1977:130; Prandi 2001:566). These names are ambiguous, since some authors treat them as synonyms, while others regard them as referring to different species (Camargos et al. 2001), and the name gameleira is also used as synonymous with the genus *Ficus*, or a subset of the genus *Ficus*, especially strangler figs. From the 1950s to the 1990s, *Ficus doliaria* was widely regarded as the iroko tree (e.g., Bastide 1978; Cacciatore 1977:130, Voeks 1990; Murphy 1994). A specimen of *Ficus doliaria* in the herbarium of the Federal University of Bahia is correctly identified as *Ficus gomelleira* (Kunth & C. D. Bouché), and the species is described by the collector, G. C. Pereira Pinto, as the spiritual tree of Candomblé. Though accurate, this is still a limited description; it assumes *Ficus gomelleira* is the only *Ficus* species that serves as Candomblé's cosmic tree (Fig. 12.4).

Method

In the summer of 2005, 2007, and 2010, I surveyed 19 terreiros in order to identify the species of *Ficus* used as their iroko tree. The results of this study also incorporate observations of the cultivated and wild figs that are present in Salvador. Eighteen terreiros occurred in Salvador and surrounding areas, and one was located in Recife, Pernambuco. Permission to visit the terreiro for the express purpose of looking at its iroko tree was arranged in advance. Informal interviews were conducted with members of the terreiros, and the primary aim of the interviews was to identify the

Fig. 12.4 The *Ficus gomelleira* tree of the terreiro Oshumarê

orixás with which the dedicated figs were associated. This was essential, as there were often two or more figs of ritual importance belonging to the same or different species. Given the sensitivities surrounding the religious status of the iroko tree, no voucher specimens were collected. However, where permitted, each tree was photographed and leaf samples were collected from the ground. Dique do Tororó is a green space of historic importance in the heart of Salvador, and to provide a framework for understanding why a number of *Ficus* species and varieties were commonly found in the landscape of large terreiros, a survey was made of all the figs that inhabited the Dique. The scientific identities of the figs discussed here are correlated with the botanical names, descriptions, and illustrations provided by Carauta and Diaz in their excellent book *Figueiras no Brasil* (2002).

Results and Discussion

Table 12.1 below presents the results of the 19 terreiros surveyed, and it includes a reference to "nation," a word that is generally used in Candomblé to identify different traditions with reference to their geographical and cultural origins in Africa (Cacciatore 1977:150). Of the 1,163 terreiros included in the book *Mapeamento dos terreiros de Salvador* (Santos 2007), 57.8% self-identified as belonging to Ketu nation, 24.2% as Angola, 2.1% as Jêje, and 1.3% as Ijexá. The other 14.6% of the

Table 12.1 The *Ficus* species of 19 terreiros

	Terreiro	Nation	*Ficus* species
1	Afonjá (Ilê Axé Opô Afonjá)	Ketu	*Ficus elliotiana* S. Moore
2	Aganju (Ilê Axé Opô Aganju)	Ketu	*Ficus elliotiana* S. Moore
3	Bate Folha (Manso Banduquenque)	Angola	*F. clusiifolia* Schott
4	Bogum (Zoogodô Bogum Malë Rundó)	Jêje	*F. clusiifolia* Schott
5	Boiadeiro (Ilê Asé Omim J'Obá)	Ketu	*Ficus elliotiana* S. Moore
6	Casa Branca (Ilê Axé Iyá Nassô Oká)	Ketu	*F. gomelleira* Kunth & C. D. Bouché
7	Gantois (Ilê Axé Iyá Omi Axé Iyamassê)	Ketu	*Ficus elliotiana* S. Moore
8	Instituto Médico Legal Nina Rodrigues	Unknown	*F. cyclophylla* Miquel
9	Manso (Manso Dandalungua Concuazenza)	Angola/Ketu	*Ficus tomentella* (Miq.) Miq.
10	Mutalombo (Mutalombo Yê Kaiongo)	Angola	none
11	Obakan (Ilê Axé Obakan)	Ketu	*F. clusiifolia* Schott
12	Obanã (Ilê Axé Obanã)	Ketu	*Ficus elliotiana* S. Moore
13	Oshumarê (Ilê Axé Oxumarê)	Ketu	*F. gomelleira* Kunth & C. D. Bouché
14	Plataforma (Ilê Axé Kalé Bokun)	Ijexá	*Ficus elliotiana* S. Moore
15	Portão (Ilê Omoródè Axé Orixá N'Lá)	Ketu	*F. gomelleira* Kunth & C. D. Bouché
16	Sítio de Pai Adão (Ilê Axé Obá Oguntó)	Nago/Ketu/Egba	*F. clusiifolia* Schott
17	Tanuri (*Tanurí Junçara*)	Angola	None
18	Terreiro de Pai Edinho (Ilê Alabaxé)	Ketu	*F. gomelleira* Kunth & C. D. Bouché
19	Terreiro de Valdete (Terreiro Ilê Ewa Olodumare)	Nago/Vodum	*Ficus elliotiana* S. Moore

terreiros were a combination of nations that identified themselves, for example, as Angola/Ketu, Nago/Vodum, Jêje/Ketu, and so on. In the small sample of terreiros this chapter presents, ten (52.6%) were Ketu, three (15.8%) were Angola, and the remaining five were Angola/Ketu, Ijexá, Jêje, Nago/Ketu/Egba, and Nago/Vodum (Parés 2004; Reis and Mamigonian 2004).

Table 12.1 indicates that 17 of the 19 terreiros had fig trees representing five species (Fig. 12.5a–d). The most common, *Ficus elliotiana*, was found in seven terreiros; both *Ficus gomelleira* and *Ficus clusiifolia* occurred in four terreiros; and *F. cyclophylla* and *Ficus tomentella* were each found in only one terreiro. It is noteworthy that *Ficus elliotiana* is the iroko tree of Gantois and Afonja; these are prominent terreiros that value their historic links to Casa Branca, renowned as the oldest continually functioning terreiro in Salvador. All three are proud of their fidelity to African traditions, yet unlike Gantois and Afonja, the iroko tree of Casa Branca

Fig. 12.5 (**a**) The *Ficus elliotiana* of Terreiro Ilê Ewa Olodumare. (**b**) The *F. elliotiana* of Ilê Axé Kalé Bokun. (**c**) The *Ficus cyclophylla* at the Instituto Médico Legal Nina Rodrigues. (**d**) The *Ficus clusiifolia* at Oshumarê

is *Ficus gomelleira*. With the exception of *Ficus tomentella*, these are among the most common native figs of Salvador, often found growing on other trees, walls, buildings, and bridges. Seedlings and trees of all sizes can be seen along roadsides and in the city's many green spaces ranging in size from small squares to large areas like Campo Grande, Dique do Tororó, Parque da Cidade, and Zoobotánico (Fig. 12.6a–d).

Figs That Grow from the Sky to the Earth

That at least five species of figs have been adopted in Candomblé reveals what has long been recognized. Candomblé communities are not subject to any overarching ritual authority. Nevertheless, these figs have one thing in common. Candomblé's cosmic fig grows from the sky to the earth, and so do these native figs of Salvador. According to Valdina Pinto de Oliveira, a spiritual leader of the Angolan terreiro Tanurí Junçara, widely respected for her knowledge of Candomblé's relation to nature: "Generally people don't plant the iroko tree. It only appears by itself on trees and walls. Old people," she continued, "believe that such a tree has a strong spirit" (see also Cassidy 1961:380; Cassidy and Le Page 1980:200; Allsopp 1996:260; Coffey 2005:262). Her Angolan terreiro, Tanurí Junçara, did not have an iroko tree, and neither did the Angolan terreiro, Mutalombo. Bate Folha, another well-known Angolan terreiro, had a *Ficus clusiifolia* dedicated to Tempo. Most interesting, however, was the Angolan terreiro, Manso, where the figs "chose" the terreiro and not the terreiro the figs. A tropical almond tree (*Terminalia catappa* L.) dedicated to Tempo was colonized by a *Ficus tomentella* and a jackfruit tree (*Artocarpus heterophyllus* Lam.) by a *Ficus clusiifolia* (Fig 12.7a–c).

A situation similar to Manso also occurred at Oshumarê, which had three species of figs. The iroko tree of this well-known Ketu terreiro, as earlier noted, was *Ficus gomelleira*, and the two other species were *Ficus clusiifolia* and *Ficus elliotiana*. Where the *Ficus clusiifolia* now grows once stood a jackfruit tree (*Artocarpus heterophyllus*) dedicated to Ossaim, the orixá responsible for the healing and ceremonial leaves of Candomblé. All attempts to remove the colonizing *clusiifolia* and restore the jackfruit failed. The members of Oshumarê decided to dedicate the *Ficus clusiifolia* to Ossaim, since this was obviously the tree Ossaim himself had chosen. The situations at the terreiros Manso and Oshumarê reveal that the relationship between a particular orixá and a particular species is not only general; it is also specific. Figs that spontaneously colonize particular sites in the terreiro can become associated with specific orixás by virtue of where they grow.

The Number of Ficus Species in Terreiros

The relationship between Candomblé and various species of *Ficus* is far more complex than suggested by the common assumption that the iroko tree is represented by

Fig. 12.6 Some examples of young wild figs in Salvador. (**a**) *Ficus gomelleira* growing at the entrance to Palácio de Aclamação. (**b**) *Ficus elliotiana* along a sidewalk in Salvador. (**c**) *Ficus clusiifolia* on a gate in the Graça neighborhood. (**d**) *Ficus clusiifolia* and *Ficus microcarpa* var. *nitida* growing together on a building in the Piedade

Fig. 12.7 (**a**) The *Ficus tomentella* of the terreiro Manso Dandalungua Concuazenza. (**b**) Leaves. (**c**) The base of the tree

a single species of fig. It is noteworthy that large terreiros had up to four species of figs in their landscape, and in some cases, there were several specimens of the same species present. Bate Folha, for example, had *Ficus gomelleira* and *Ficus clusiifolia*, and both species were of ritual importance. Of the several specimens of *Ficus clusiifolia*, two were ritually dedicated, one of which was the iroko tree of the terreiro. Afonja had three species that, in addition to *Ficus elliotiana* as its iroko tree, included several specimens of *Ficus clusiifolia* and an impressive *Ficus elastica* (Roxb. ex Hornem) (Fig. 12.8a, b). Bogum had five figs representing four species, and two specimens of *Ficus clusiifolia* were associated with different orixás. One was dedicated to Tempo and the other to Exu, and in the latter case, it is noteworthy

Fig. 12.8 (**a**) The *Ficus elliotiana* of Afonjá (Ilê Axé Opô Afonjá) photographed on the evening the terreiro celebrated (**b**) 100 years of Candomblé

that Cacciatore (1977:126) defined "figueira" as a tree "of which there are diverse species" that "pertains to the Exus." The other two native figs of Bogum were a small *Ficus nymphaeifolia* (Miller) growing on the dead trunk of a caja tree (*Spondias mombin* L.) in the middle of the terreiro and several seedlings of the Asian fig *Ficus microcarpa* var. *nitida* (see Barrett 1949) growing on the inside and outside of the terreiro's boundary wall. The 12 figs of the terreiro Aganju represented three species comprised of three *Ficus clusiifolia* trees, including a mature specimen dedicated to ancestral spirits and two young specimens in African oil palm trees (*Elaeis guineensis* Jacq.); eight *Ficus elliotiana* trees, of which one of two mature specimens was dedicated to Iroko (Fig. 12.9a–c), and six young trees grew on buildings and in African oil palm trees; and a young *Ficus benjamina* L. used as a Christmas tree. Oshumarê was unique in having three species of figs, each associated with a different orixá. The iroko tree of the terreiro was *Ficus gomelleira*, and one of two *Ficus clusiifolia* trees was dedicated to Ossaim, and a *Ficus elliotiana* to Exu.

Looking at the distribution of figs at Dique do Tororó, a historic green space in the heart of Salvador, provides insights into why large terreiros often have a variety of fig species in their landscape (Fig. 12.10). Dique occupies a special place not only in Candomblé but in the cultural and recreational activities of Salvadorians. It has long been an important site where offerings are made to Oxum in her own right and to Oxum as a prelude to New Year offerings to Yemanja. Shirey (2009:72) notes that for many years the lake was surrounded by "overgrown trees, shrubs, and weeds" and was generally associated with crime. In 1993, however, it was com-a b

c

Fig. 12.9 (**a**) The *Ficus elliotiana* of the terreiros Aganju (Ilê Axé Opô aganju). (**b**) Leaves of *F. elliotiana*. (**c**) The *F. elliotiana* of Terreiro Ilê Ewa Olodumare

Fig. 12.10 Map of Dique do Tororó

pletely renovated. The lake area and roadsides around the lake were landscaped with a variety of native and exotic plants, and in addition to the orixá statues and fountain featured in the lake and the popular shaded path around it, the renewed Dique also offered public parking, restaurants, event venues, playgrounds, outdoor gym, lakeside seating, boating pier, and fishing decks (Fig. 12.11).

All figs found growing at Dique do Tororó in July 2010 were recorded, and for the purpose of this survey, Dique was divided into three areas: the lakeside, and in the northeastern area of the lake, the palm grove, and the roadside wall bordering Dique. Dique had six species of figs represented by 117 individuals of which 52% were *Ficus cyclophylla*, 27% were *Ficus clusiifolia*, 9% were *Ficus elastica*, 8% were *Ficus gomelleira*, 5% were *Ficus microcarpa* var. *nitida*, and 1% was *Ficus benjamina*. *F. cyclophylla*, *F. clusiifolia*, and *F. gomelleira* were present in all three

Fig. 12.11 The monumental orisha statues in the lake of Dique do Tororó, the work of Bahian artist Tati Moreno, framed by one of several silk cotton trees (*Ceiba pentandra* L. [Kuntze]) around the lake

areas, while the palm grove and roadside wall also had *F. microcarpa* var. *nitida* which was not found in the lakeside area, and the lakeside area had *F. elastica* and *F. benjamina*, which were not found in the palm grove or on the roadside wall. Of the 117 figs, 90% were seedlings. All 12 mature trees were around the lake, including the ten *F. elastica* trees, a *F. benjamina*, and a large *F. cyclophylla* growing on the northeastern side of the lake (Fig. 12.12). This *F. cyclophylla*, which grows across from the palm grove and roadside wall, probably accounts for the fact that more than half of the 117 figs were represented by *F. cyclophylla* – 35 in the palm grove, 14 on the roadside wall, and 12 in the lakeside area. This suggests large terreiros are likely to have many seedlings growing in the vicinity of their iroko tree. It also suggests why we are likely to encounter more than one species of fig in large terreiros when we find that of the 40 palm trees of Dique, 16 were without figs, but 24 were colonized by strangler figs. Of the 24 with figs, 12 had only one species, 11 had two, and 1 had three – half of the 24 palms with fig associates had been colonized by two or more species (Fig. 12.13a, b).

Falling Fig Trees

Another overlooked aspect of the dynamic relationship between Candomblé adherents and their *Ficus* species is the instability of large fig trees in the confines of terreiros, a concern expressed in many conversations. Mae Mininha of Gantois, for

Fig. 12.12 The mature *Ficus cyclophylla* on the northeastern side of Dique do Tororó

Fig. 12.13 The palm grove of Dique do Tororó. (**a**) *Ficus clusiifolia* and *Ficus cyclophylla* in a palm tree. (**b**) *Ficus microcarpa* var. *nitida* at the base of a coconut tree

example, feared the terreiro's large iroko tree could fall, but she said there were many *F. elliotiana* seedlings on the grounds of the terreiro waiting to take its place. The *F. clusiifolia* at Bate Folha fell in 2002 and was replaced in 2005. The same thing occurred at Afonja; the *F. elliotiana* observed in 2002 growing next to the house of Sango was replaced by a young tree in the center of the large open area just beyond the entrance to the terreiro. As previously noted, the iroko fig of Casa Branca is *F. gomelleira*, of which there were two large trees. The one growing on the steep hillside behind the main building of the terreiro was cut down because it posed a serious threat of toppling on the building. The other growing on the hillside in front of the terreiro was cut down in 2005. It had begun to drop large branches and they feared it too would fall. Nevertheless, at the time of this research, many seedlings of *F. gomelleira* could be seen growing in the landscape of Casa Branca.

Ficus gomelleira in Public Spaces

Ficus gomelleira is a particularly interesting fig with a place of distinction in Candomblé. No other species of fig seems to be as widely represented as a ritual tree in public places. *F. gomelleira* occurs widely in South America and is widespread in Brazil, where according to Carauta and Diaz (2002:59), it is found in the Brazilian states of "Pará, Amapá, Maranhão, Piauí, Minas Gerais, Rio de Janeiro, São Paulo, Paraná, [and] … Mato Grosso." Curiously, they did not mention its presence in Bahia. Yet, as earlier noted, it is common in Salvador. It is significant that this species is the cosmic fig of the venerable *Casa Branca*, and although *F. clusiifolia* and *F. gomelleira* were each found in four terreiros, neither *F. clusiifolia* nor any of the other *Ficus* species of Candomblé seem to be as widely represented as a ritual tree in public places as *F. gomelleira*.

The most significant *Ficus gomelleira* tree outside of terreiros that was frequently mentioned in conversations was the one at the crossroads leading to São Lazaro Church in Ondina, and both the church and the fig were especially associated with the orixá Omulu. Being an orixá who is a source of both life and death, Omulu, paradoxically, is one of the most celebrated and yet most feared. A guide to the spirit of the dead whose Yoruba name means Lord or King of the earth, Omulu's domain ranges from the earth's surface to its depth, and his seat, residence, and altar site are associated with the "crosses of cemeteries." Omulu, correlated with São Lazaro, patron saint of the poor, is the orixá of smallpox and other diseases, especially epidemics, and in his capacity as a life source, he is popularly regarded as the Physician of the Poor who heals as well as offers protection from diseases. But Omulu, as the Master of Death (Magalhães 2003:150), also inflicts individuals with disease as a form of punishment.

The literal and figurative association of roads, especially crossroads, with travel and communication, makes them profoundly important in Candomblé, and the orixás of roads play an important role as mediators in the relationship between humans and orixás. Iroko, as previously noted, is guardian of the cosmic crossroad,

Fig. 12.14 (a) The dead trunk of the *Ficus gomelleira* that grew on the bank of Dique do Tororó photographed in 2010. (b) A seedling of *Ficus gomelleira* growing on this trunk in 2005 when the photograph was taken

dwelling place of spirits, and a powerful source of Axé. Ogun, the vanguard orixá of war whose element is iron, is traditionally the patron of blacksmiths, and, by extension, all who depend on iron for implements and material, including warriors, farmers, and artisans; today his domain also embraces railroads and metallurgy. In the resistance to slavery and the racial oppression that continued in its aftermath, Ogun's embodiment of the fighting spirit has made him a popular orixá for African communities of the New World where he is regarded as "the opener of the way." Exu is the trickster orixá of communication, indispensable in all Candomblé rites because he guards thresholds, determining the positive or negative outcome of all exchange. And Omulu is guardian of the crossroad of life and death, the destiny of all human beings. While thresholds are generally understood to include crossroads, entrances, and doorways, at a deeper level they include all literal and figurative crossings that make the interaction between humans and orixás possible. The abundant offerings at the São Lazaro fig highlight the importance of these orixás, and they also highlight the importance of the orixá Xangô. Although not generally regarded as an orixá of roads, he is specifically associated with *F. gomelleira*.

The second significant *Ficus gomelleira* in a public area, which also came to light in conversations with members of Candomblé, was described as a "spiritually powerful" tree that grew on the banks of Dique do Tororó. It was cut down by the city, but the stump was still there, and the stump had a *Ficus gomelleira* seedling growing on it when it was photographed in 2005 (Fig. 12.14). According to Dique administrators, it was cut because a large branch fell, killing a pedestrian and crushing a passing car. The newspaper report of the incident said the tree was carefully examined by those responsible, who concluded it was a continued risk to the public and could not be saved. Dique administrators blamed the use of candles associated with offerings for weakening the tree. According to some members of Candomblé, the tree was killed because it was a popular place of offerings, and some city officials objected to this because of its African connection and because the offerings, they said, created excessive litter. Whatever reason, it is puzzling that the *F. gomelleira* trees of São Lazaro and Dique do Tororó were both damaged by fire.

Pai Anderson of Recife reported that in large terreiros with sufficient outdoor space for plants, people greet Ogun and Xangô at the iroko tree when they arrive. The epitome of extraordinary power and great humanity, Xangô is the just warrior orixá of lightning and thunder, who is arguably the most celebrated orixá in the Americas. His domain and altar site are quarries, and he is principally represented by stones, the double-winged ax, and fire; the latter accounts for candles being an important part of the offerings presented to him. According to Martins and Marinho (2002:39): "For many … [the iroko tree] is inhabited by different divinities, as for example, the orixá Ocô as well as some members of the families of Ogun and of Xangô, the 'Senhor' of fire." Is it the offerings to Xangô at the foot of the iroko tree that were responsible for the damage by fire that had affected the *Ficus gomelleira* trees of São Lazaro and Dique?

Conclusion

Candomblé is remarkable for its adaptability as evidenced by the intricate relationships that have been established with the native figs of Brazil. Five native species were found in 16 terreiros in Salvador and one in Pernambuco. Judging from the small sample presented, from the fact that figs sometimes "choose" terreiros, and from the literature, this research suggests we should expect to find more than five species of figs representing Iroko and other orixás in the landscapes of Candomblé. There are over a thousand terreiros in Salvador alone (Santos 2007), and *F. microcarpa* var. *nitida*, for example, is a good candidate for adoption (Fig. 12.15). This native of Asia is now naturalized not only in Brazil but in many parts of the tropics and subtropics, including Florida, Hawaii, and some territories of the Caribbean. It is widespread in Salvador, and if the salient point for Candomblé centers on trees that grow from the sky to the earth, one would expect *F. microcarpa* var. *nitida* to be among the *Ficus* species selected for ritual dedication in terreiros even though it is an exotic. The most likely explanation why this has not happened, as far as we know, is that *F. microcarpa* var. *nitida* only became naturalized after its pollinating wasp was introduced sometime in the decade between 1970 and 1980 (Carauta and Diaz 2002:155). Nevertheless, one can see mature specimens of this species in Salvador, such as the one colonizing a *Pachira aquatica* (Aubl.) tree in Nazaré Park (Fig. 12.16). *Ficus glabra* (Vell.) and *F. maxima* are also good candidates for being found in terreiros even though *F. maxima* was not in some of the oldest terreiros of Salvador as Prandi (2001:566) indicated.

Fig. 12.15 The naturalized *Ficus microcarpa* var. *nitida* growing in Nazaré Park, Salvador, colonizing a *Pachira aquatica* tree

Fig. 12.16 *Ficus glabra* in Salvador in the Vitoria neighborhood of Salvador

Fig. 12.17 One of several large *Ficus elastica* trees growing on the bank of Dique do Tororó

Published accounts also suggest other species of figs are to be expected in the landscape of terreiros. For example, Cacciatore (1977:130) reports that in the Batuque tradition of Belem, Pará, *Ficus citrifolia* (Mill.) was the spiritual tree that served as the residence of the Caboclo spirit, *Velho Jupetequara*. This species, emblematic of Barbados as the species from which it is said the island took its name (Gooding 1974; Fraser et al. 1990), is common throughout the Caribbean (Little et al. 1974) where it is widely associated with inspirational life. Cacciatore (1977:165) also reports that different species of figs were used to represent the vodun, Loko, of the Jêje nation, who is popularly viewed as correlated with the orixá Iroko in the Ketu tradition, and the inkice Tempo, in the Angola tradition (Cacciatore 1977:235; Wafer 1991:166; Sodré 1995; Coffey 2005:263). In Bahia, she reports Loco was represented by "gameleira branca" (which elsewhere she identified as *F. doliaria*, i.e., *F. gomelleira*); in Maranhão and Rio de Janeiro, by the cajazeira tree (*Spondias mombin* L.); and in the "jejenagô" tradition of Rio de Janeiro, by both the cajueiro tree (*Anacardium occidentale* L.) and "ficus italiano," the latter being the common name for *Ficus elastica*. It is interesting that Cacciatore says *F. elastica* is found in terreiros of Rio de Janeiro, since this species is an Asian exotic that is not naturalized in Brazil. In Salvador, as earlier noted, a large *F. elastica* was part of the landscape of Afonja, and magnificent specimens can be seen around the city, such as along major thoroughfares, on the campus of the Federal University of Bahia, and especially around Dique do Tororó (Fig. 12.17). None, however, were identified as ritually dedicated or associated with offerings. Nevertheless, this study has shown that Candomblé has a complex relation with a variety of *Ficus* species, both native and exotic, and a wider survey that includes more rural communities and other urban areas will probably yield many surprises.

Acknowledgments I thank all the communities of Candomblé whose support made this study possible. I am also grateful for the support of Dr. Ubiratan Castro de Araújo and Dr. Jocélio Teles dos Santos who served at different times as the Director of the Center for Afro-Asian Studies of the Federal University of Bahia. A special thanks to Valdina Pinto de Oliveira for sharing her love of Candomblé and nature and to my colleagues in Salvador and Recife, especially Cláudia Regina Muniz Barreto, Fernando Batista, and Hernandes Santos Souza, without whose assistance this project would not have been possible. This chapter has benefited from the helpful comments of Moore Quinn, Dee Dee Joyce, Henry Drewal, and Ted and Dale Rosengarten.

References

Abiodun R (1994) Understanding Yoruba art and aesthetics: the concept of Ase. Afr Arts 27(3):68–78, 102–103

Allsopp R (1996) Dictionary of Caribbean English usage. University of the West Indies Press, Jamaica/Barbados/Trinidad & Tobago

Altman N (2000) Sacred trees: spirituality, wisdom and well-being. Sterling, New York

Anonymous (1890) Cultus arborum. Privately printed. Available online through Kessinger Publishing

Barnet M (1999) La Regla de Ocha: the religious system of Santería. In: Olmos M, Paravisini-Gebert L (eds) Sacred possessions: Vodou, Santería, Obeah, and the Caribbean. Rutgers University Press, New Brunswick, pp 79–100

Barrett MF (1949) *Ficus retusa* L. Bull Torrey Bot Club 76(1):53–64

Bastide R (1978) The African religions of Brazil: toward a sociology of the interpenetration of civilizations. Johns Hopkins University Press, Baltimore

Beech WH (1913) The sacred fig-tree of the A-Kikuyu of East Africa. Man 13(3):4–6

Cacciatore OG (1977) Dicionário de cultos Afro-Brasileiros. Forense Universitária, Rio de Janeiro

Camargos JAA, Coradin VTR, Czarneski CM, Daniela de Oliveira D, Meuerditchian I (2001) Catálogo de árvores do Brasil. IBAMA, Brasilia

Carauta JPP, Diaz BE (2002) Figueiras no Brasil. Editora UFRJ, Rio de Janeiro

Cassidy FG (1961) Jamaica talk, three hundred years of the English language in Jamaica. Macmillan, London

Cassidy FG, Le Page RB (1980) Dictionary of Jamaican English. Cambridge University Press, Cambridge

Coffey J (2005) Orixá Iroko. In: Taylor B (ed) Encyclopedia of religion and nature. Continuum, London/New York, pp 262–265

Collis JS (1954) The triumph of the tree. The Viking Press, New York

Condit IJ (1969) Ficus: the exotic species. Division of Agricultural Sciences, University of California, Berkeley

Cook R (1974) The tree of life: images for the cosmos. Thames and Hudson, New York

Deren M (1970) Divine horsemen: Voodoo gods of Haiti. A Delta Book, New York

Desmangles LG (1977) African interpretations of the Christian cross in Vodun. Sociol Anal 38(1):13–24

Eliade M (1991) Images and symbols: studies in religious symbolism. Princeton University Press, Princeton

Falola T, Childs MD (2004) The Yoruba diaspora in the Atlantic world. Indiana University Press, Bloomington

Fraser H, Carrington S, Forde A, Gilmore J (1990) A–Z of Barbadian heritage. Heinemann Publishers (Caribbean) Ltd, Kingston
Gautier D (1996) *Ficus* (Moraceae) as part of agrarian systems in the Bamileke region (Cameroon). Econ Bot 50(3):318–326
Gondim AB (2004) Seu guia no Candomblé. Self-published, Salvador
Gooding EGB (1974) The plant communities of Barbados. Ministry of Education, Barbados.
Kislev ME, Hartmann A, Bar-Yosef O (2006) Early domesticated Fig in the Hordan valley. Science 312:1372–1734
Kunwar RM, Bussmann RW (2006) *Ficus* (Fig) species in Nepal: a review of diversity and indigenous uses. Lyonia 11(1):85–97
Little EL, Woodbury RO, Wadsworth FH (1974) Trees of Puerto Rico and the Virgin Islands. U.S. Department of Agriculture, Washington, DC
Lody RGM (1975) Ao som do Adjá. Departamento de Cultura da SMEC and Prefeitura Municipal de Salvador, Salvador
Magalhães EG (2003) Orixás da Bahia. Self-published, Salvador
Maguire RAJ (1931) Masai water-trees. Man 31(July):142–143
Martins C, Marinho R (2002) Iroko. Pallas, Rio de Janeiro
Menezes A (1949) Flóra da Bahia. Companhia Editora Nacional, São Paulo
Murphy JM (1992) Santería: African spirits in America. Beacon, Boston
Murphy JM (1994) Working the spirit: ceremonies of the African diaspora. Beacon, Boston
Murrell NS (2010) Afro-Caribbean religions. Temple University Press, Philadelphia
Omari MS (1989) The role of the gods in Afro-Brazilian ancestral ritual. Afr Arts 23(1):54–61, 103–104
Parés LN (2004) The "Nagôization" process in Bahian Candomblé. In: Falola T, Childs MD (eds) The Yoruba diaspora in the Atlantic world. Indiana University Press, Bloomington/Indianapolis, pp 185–208
Philpot JH (1897) The sacred tree in religion and myth. Macmillan and Co. Limited, London
Prandi R (2000) African gods in contemporary Brazil: a sociological introduction to Candomblé today. Int Sociol 15(4):641–663
Prandi R (2001) Mitologia dos Orixás. Companhia das Letras, São Paulo
Prussin L (1999) Non-Western sacred sites: African models. J Soc Archit Hist 58(3):424–433
Reis JJ, Mamigonian BG (2004) Nagô and Mina: the Yoruba diaspora in Brazil. In: Falola T, Childs MD (eds) The Yoruba diaspora in the Atlantic world. Indiana University Press, Bloomington/Indianapolis, pp 77–110
Reno SJ (1977) Religious symbolism: a plea for a comparative approach. Folklore 88(1):76–85
Rey T (2005) Trees in Haitian Vodou. In: Taylor BR, Kaplan J (eds) The encyclopedia of religion, vol 2. Thoemmes Continuum, New York, pp 1658–1659
Santos JT (2007) Mapeamento dos terreiros de Salvador. Centro de Estudos Afro Orientais, Salvador
Shirey H (2009) Transforming the Orixás: Candomblé in sacred and secular spaces in Salvador da Bahia, Brazil. Afr Arts 2009(Winter):62–79
Simoons FJ (1998) Plants of life, plants of death. The University of Wisconsin Press, Madison
Sodré J (1995) As histórias de Lokoirokotempo: a árvore sagrada o Candomblé. Self-published, Salvador
Verger PF (2002) Orixás: deuses Iorubás na África e no Novo Mundo. Corrupio, Salvador
Voeks R (1990) Sacred leaves of Brazilian Candomblé. Geogr Rev 80(2):118–131
Voeks R (1997) Sacred leaves of Brazilian Candomblé: African magic, medicine, and religion in Brazil. University of Texas Press, Austin
Wafer J (1991) The taste of blood: spirit possession in Brazilian Candomblé. University of Pennsylvania Press, Philadelphia
Walker SS (1990) Everyday and esoteric reality in the Afro-Brazilian Candomblé. Hist Relig 30(2):103–128

Chapter 13
Exploring Biocultural Contexts: Comparative Woody Plant Knowledge of an Indigenous and Afro-American Maroon Community in Suriname, South America

Bruce Hoffman

Abstract Despite the increasing use of quantitative methods in ethnobotany, there has been relatively little advance in understanding of cross-cultural patterns. Within three tropical forest vegetation zones, I documented and compared local plant knowledge, categorical use, and resource selection for a short-resident (±300 years) Afro-American Saramacca "Maroon" community and an indigenous Cariban-speaking Trio community. For 3–4 male specialists at each site, the "recognition" and use of stems ≥10 cm dbh in forest plots (0.5–1 ha) was recorded. Comparative methods included quantitative *use value* indices and categorized uses (construction, edible, medicine, technology, and trade). The Trio emphasized medicinal uses regardless of vegetation zone. Saramacca *use value* was greatest within fallow forest, exceeded Trio knowledge for "construction" and "trade," and emphasized timber and carving species. For both groups, the *use value* of most taxa (family and species) was correlated with "apparency" (abundance, species richness), with the exception of palms and major cultural species. As hypothesized, Trio indigenous knowledge was greater – with more biological species named (97.3% vs. 83.9%) and utilized (87.7% vs. 66.9%) and more uses cited per species. However, the shorter-resident Saramacca Maroon participants still revealed a robust knowledge and use of woody plant diversity.

Keywords Cross-cultural research • Quantitative ethnobotany • *Use value* indices • Traditional knowledge • Trio • Saramacca

---

B. Hoffman, Ph.D. (✉)
National Herbarium of the Netherlands, Naturalis Biodiversity Center,
P.O. Box 9514, RA 2300, Leiden, The Netherlands
e-mail: bruce.hoffman@naturalis.nl; bruce.hoffman.phd@gmail.com

R. Voeks and J. Rashford (eds.), *African Ethnobotany in the Americas*,
DOI 10.1007/978-1-4614-0836-9_13,

Introduction

Global-scale analyses have shown that the humid tropics possess an exceptional diversity of both plant species and biodiversity-dependent forest cultures (Collard and Foley 2002; Harmon 1996; Nettle 1998; Oviedo et al. 2000; Stepp et al. 2004; Sutherland 2003). The cultural diversity of neotropical forests includes thousands of indigenous groups, mixed ethnicity "folk" cultures (e.g., caboclos, ribereños, mestizos, creoles), and, of special interest in this volume, traditional Afro-American "Maroon" cultures formed by enslaved Africans and their descendents. The concept of "biocultural diversity" was coined in the 1990s to examine linkages between cultural and biological diversity and to emphasize the importance of both in conservation and community development initiatives (Etkin 2002; Harmon 2000; Hladik 1993; Maffi 2001, 2004; Posey 1999, 2001; Stepp et al. 2002).

A fundamental aim of ethnobotany is to examine and explain cultural similarities and differences in plant knowledge and use (Balée 1994b). Predictive knowledge of human-nature relationships contributes to both theory and conservation applications in ethnobotany. In recent decades, the capacity for pattern analysis has improved significantly due to the development of quantitative ethnobotanical methods (Castaneda and Stepp 2008; Hoffman and Gallaher 2007; Höft et al. 1999; Phillips 1996). Forest plot samples have been used widely to assess *knowledge or use diversity* (importance indices) for a given set of local *biotic diversity* (Balée 1987; Boom 1987, 1990; Byg and Balslev 2001; de Albuquerque et al. 2005; Galeano 2000; Grenand 1992; Johnston and Colquhon 1996; Kristensen and Lykke 2003; Kvist et al. 1995; Lykke et al. 2004; Milliken et al. 1992b; Phillips and Gentry 1993a, b; Phillips et al. 1994; Pinedo-Vasquez et al. 1990; Prance 1972; Prance et al. 1987; Reyes-García et al. 2006; Torre-Cuadros and Islebe 2003).

While the emergence of quantitative methods in ethnobotany has substantially increased scientific rigor, there has been relatively little gain in understanding of cross-cultural patterns (Moerman 2005; Zent 2001). One challenge is a long-standing emphasis upon indigenous groups, often viewed as more "colorful," more threatened, and in possession of "better" ethnobotanical knowledge than other forest peoples (Alcorn 1993; Brokenshaw et al. 1978; Clay 1988; Sillitoe 1999). From the late 1980s, the need to broaden the scope of ethnobotany was recognized and increasingly acted upon (Anderson 1990; Cocks 2006; Galeano 2000; Halme and Bodmer 2007; Parker 1989; Phillips et al. 1994; Silva et al. 2007; Soler Alarcón and Luna Peixoto 2008; Young 2005). Campos and Ehringhaus (2003) provided powerful evidence of the value of nonindigenous knowledge in a comparison of two indigenous groups and two folk societies in the Brazilian Amazon. They found that more than a third of the uses cited by indigenous informants for 17 palm species were learned from folk societies.

In the present study, I examine cross-cultural patterns in ethnobotany through quantitative and qualitative comparison of two distinct forest cultures in Suriname, an indigenous group, the Trio, and an Afro-American Maroon group descended from enslaved Africans, the Saramacca. I consider the Saramacca "nonindigenous" due to their relatively brief residency (±300 years) and connection with the African diaspora (Arends 2002; Price 1976; van der Elst 1971). Among the handful of ethnobotanical accounts of Maroons, this is the only published study I am aware of that has made a quantitative, empirical comparison between Maroon and indigenous plant use knowledge (see Grenand et al. 1987; Hurault 1965; van Andel et al. 2008a, b; van Andel and Havinga 2008; Voeks 1995, 1997).

Comparisons of universal and particular ethnobotanical attributes of the two cultures are presented within the following interrelated categories:

Extent of ethnobotanical knowledge: Who "knows more" about local biodiversity? It seems logical that a longer time frame (more experience) would support the accumulation of biodiversity knowledge through increased discovery and experimentation, observation of natural cycles and ecology, adoption of outside knowledge, adaptive management, and knowledge transfer. For pooled data and by vegetation zone, I addressed a "regional residency" null hypothesis that the body of ethnobotanical knowledge of a long-residing indigenous agrarian culture is no greater than that of a short-resident nonindigenous agrarian culture.

Categorical use patterns: How are forest types and forest species used? Use patterns often reflect cultural, socioeconomic, and spiritual contexts. I addressed a null hypothesis that the two forest cultures exhibit no major differences in the categorical use of resources (i.e., the relative importance of five use categories).

Resource selection patterns: What are the drivers of resource selection and what resources are most culturally important? I documented and compared the two cultures for (a) the ecological "apparency" hypothesis that more conspicuous taxa are more likely to be known or used (see de Albuquerque and de Lucena 2005) and (b) "inordinately important" plant resources, including the taxonomic affinities and use of these resources.

Research Location, Environment, and Biocultural Context

Research Country and Field Sites

The research was conducted in Suriname, a small tropical nation of 470,000 inhabitants in the remote *Guianan Shield* region of northeastern South America (Hammond 2005a) (see Fig. 13.1). Suriname has the second highest area of forest cover per inhabitant (34 ha) among tropical nations (FAO 2006). More than 80% of Suriname's inhabitants live in the northern extreme of the country, in the capital city of Paramaribo and surrounding former plantation lands (Census 2007). Approximately 90% of Suriname's land is relatively undisturbed tropical forest with communities of traditional forest peoples and small-scale gold-mining or timber operations.

I resided and conducted research in Suriname between 2004 and 2006. The research was conducted at two cultural field sites located approximately 200 km apart (see Fig. 13.1). The Saramacca Maroon (SA) research site was located in

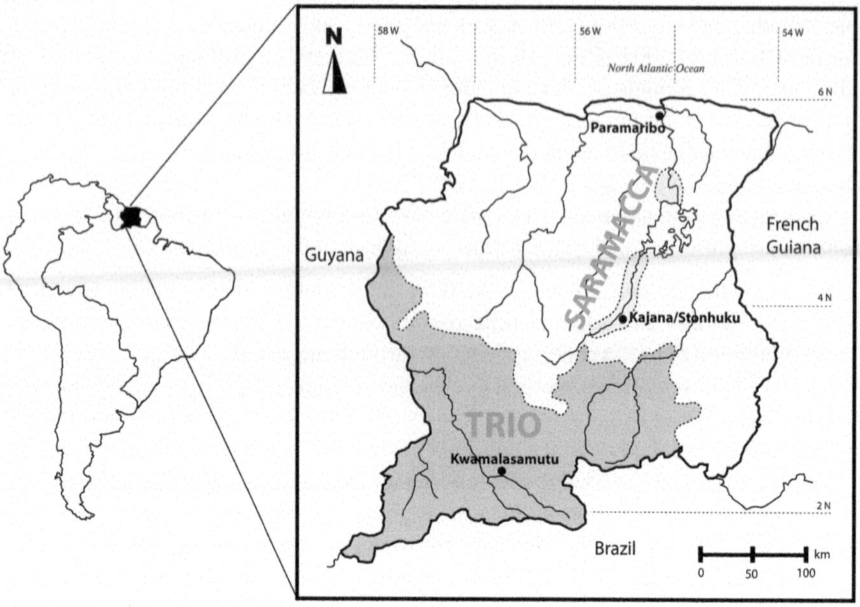

Fig. 13.1 Map of Suriname: lands and research villages of the Saramacca Maroons (Kajana/Stonhuku) and Trio Amerindians (Kwamalasamutu)

central Suriname, at Stonhuku village and adjoining lands on the upper Pikin Lio (3° 53′ N latitude, 55° 34′ W longitude, 100 m elevation). The Trio (TR) indigenous research site was located in far southern Suriname at Kwamalasamutu village and adjoining lands (2° 21′ N latitude, 56° 47′ W longitude, 200 m elevation).

Biophysical Context

The Köppen climate classification for the Saramacca site in central Suriname is *Af* (tropical rainforest), with 2,500 mm average yearly rainfall and a bimodal seasonal rainfall pattern (long wet→long dry→short wet→short dry). At the more southerly Trio site, the *Af* class merges with a drier, continental *Aw* class, resulting in somewhat reduced annual precipitation (ca. 2,000 mm) and a unimodal seasonal rainfall pattern (long wet – long dry) (Hammond 2005b; Köppen 1923; Nurmohamed and Naipal 2006; Teunnisen et al. 2003).

At a regional scale, both field sites occurred within low-elevation humid tropical forest on rolling hills of brown sand and clay (Alfisols and Ferralsols). In contrast to high-diversity forests in northwestern Amazonia, lowland tropical forests of the Guianan Shield are low to moderate in α-diversity and uniquely abundant in Lecythidaceae and caesalpinioid Fabaceae species (de Granville 1988; Johnston and Gillman 1995; Schulz 1960; Steege et al. 1993; van Andel 2001). Quantitative floristic accounts of central or southern Suriname forests were not available prior to this study. The tree diversity of southern Suriname forests was expected to be greater than central Suriname forests based upon large-scale diversity pattern analyses (Steege et al. 2001, 2003).

Cultural Context

The "Maroon" cultures of Suriname were established by groups of enslaved Africans that escaped into forests and swamps adjoining the plantations of Dutch colonial Suriname in the seventeenth and eighteenth centuries. The Saramacca were among the first of these groups to become established as a distinct culture with permanent settlements. Today, Saramacca villages occur along the middle and upper Suriname River, with an estimated population of 23,000–25,000 (Price 2002) (see Fig. 13.2a). Villagers speak mostly Saramaccan – a hybrid mix of English Creole (50%), Portuguese (35%), and "African" (10%, various origins) – or Sranantongo, the national English-based *lingua franca* (Smith 2002). I label the Saramacca "nonindigenous" due to their relatively brief residency (±300 years) and sociocultural connection with the African diaspora (Arends 2002; Price 1976; van der Elst 1971).

The Cariban-speaking *Trio* culture is indigenous to southern Suriname and adjoining Brazilian lands. It is uncertain when their specific group formed, but Cariban-speaking peoples have existed in the region for thousands of years (Basso 1977). Trio live today in five widely dispersed villages, with 800–900 Trio speakers in the research site village of Kwamalasamutu (Carlin 2004) (see Fig. 13.2b). Some Kwamalasamutu residents are non Trio minorities, including Okojana, Sikijana, Wai-wai (Tukujana), Mawajana, and Akurio peoples. Trio and Sranantongo are the most commonly spoken languages, but Dutch is used in formal communications and is taught in the government school.

Subsistence

The Saramacca and Trio are agrarian societies and the non-flooded forestlands around both research villages included mosaics of multiage swidden plots. Iconic Saramacca crops include African upland rice (likely varieties of *Oryza glaberrima* Steud, *O. latifolia* Desv., *O. rufipogon* Griff.), Asian rice (*O. sativa* L.), and tubers (*Colocasia* spp., *Manihot esculenta* Crantz, *Xanthosoma* spp.). In addition, many Saramacca families at Stonhuku supplemented their diet with wheat and rice from the city. The iconic staple for the Trio at Kwamalasamutu was cassava (*Manihot esculenta* Crantz), with many "bitter" and "sweet" varieties. The Trio did not grow rice but consumed coastally grown white rice from the city when possible. Both cultures are highly dependent upon local wild and semiwild plant and animal resources.

Fig. 13.2 (**a**) Kajana-Stonhuku villages and Saramacca research participants. (**b**) Kwamalasamutu village and Trio research participants

Economics

From the eighteenth century onward, the Saramacca have maintained economic ties with coastal society in Suriname through trade or wage labor. Many Saramacca individuals travel regularly by river and road between interior villages and Paramaribo to conduct business. The Trio, on the other hand, are geographically and socially isolated, and economic links are inconsistent and poorly developed. In Kwamalasamutu, only a few forest products such as Brazil nuts (*Bertholletia excelsa* O. Berg) and "maramara" seed necklaces (*Schefflera morototoni* L.) generated cash income. Opportunities for wage labor were few, and most individuals were not highly skilled in monetary transactions.

Forest and Resource Protection

The Saramacca have an elaborate land tenure system and traditional spiritual beliefs that result in strict controls over forest and resource use. During this research, old-growth forests in the village vicinity were off-limits to locals and researchers for two or more days a week to appease forest and ancestor spirits. Some plant taxa were taboo (e.g., *Ficus* spp., Saramacca=*taku*) and were avoided when cutting forest plots. Based upon Saramacca traditions, visiting the forest on a forbidden day, cutting down a taboo plant, or using a forest resource on someone's land without permission could result in untold suffering or death for an entire lineage. By contrast, the Kwamalasamutu Trio practiced communal land use and, due to missionary influence, the influence of traditional spiritual beliefs upon forest visitation and resource use was minimal. I was able to conduct research on any day and in any location in the forests adjoining Kwamalasamutu village. It appears that the Saramacca sociocultural system, as currently practiced, affords greater ecological resilience and protection of biodiversity in local forests.

Methods

Research Initiation

The two culture sites were chosen because the overall context – an indigenous and nonindigenous group living within an ecologically similar forested landscape – facilitated research questions. The Pikin Lio area was chosen for Saramacca Maroon research due to the presence of more remote and "traditional" communities (Frits van Troon, a Saramaccan botanist and forester, personal communication 2003). The research was facilitated by the Amazon Conservation Team (ACT), a US-based NGO with an active, locally managed biocultural conservation program in Suriname. I held pre-research meetings with village heads and villagers in both communities to ensure understanding of the research and prior-informed consent at the local level.

Forest Plots

Local assistants at each village participated in the selection of three forest zone sites to establish plots and conduct ethnobotanical interviews, including examples of non-flooded "upland forest" (UP), "seasonally flooded or lowland forest" (LO), and fallow forest of greater than 15 years of age (FA). These broad ecological categories were salient to both local peoples and myself and are common in regional accounts of neotropical vegetation (Berry et al. 1995; Daly and Mitchell 2000; de Granville 1988; Huber 1995; Mori 1991; Steege et al. 1993). Plots were all within 1.5 h travel from village centers, and, therefore, prior and ongoing resource use was expected. The chosen old-growth sites had not been cleared or otherwise majorly disturbed within living local memory.

A total of 2.5 ha were delineated per research village, including 1.0 ha in UP vegetation, 1.0 ha in LO vegetation, and 0.5 ha in FA vegetation. 1.0-ha plot shapes varied from square (100 m × 100 m) to rectangular (20 m × 500 m) due to differences in topography and vegetation extent. Each plot was divided into 0.1-ha subplots with all tree and liana stems ≥10 cm dbh (diameter at breast height, 1.37 m) taxonomically assessed, numbered, and tagged by myself and local assistants. To analyze the floristic similarity of morphospecies and abundances between plots, I applied a nonmetric multidimensional scaling (NMDS) ordination in PC–ORD 5.0. I chose to use 1.0 (0.5)-ha plots rather than 0.1-ha sample plots, first, to allow comparison with the 1.0-ha plots of Phillips and Gentry (1993a, b) in Peru and additional 1.0-ha plots established in the Guianas (van Andel 2000b) and, second, because the first forest zone (LO) selected by Trio participants was too small for the establishment of long transects.

Botanical Collections

Specimens were collected for each new or questionable folk and/or biological taxon encountered within plots, and out-of-plot specimens were collected when relevant to the research. Pooling data from both research sites, I collected 1,435 botanical voucher numbers including more than 95% of the folk taxa named by research participants. Vouchers were sorted into morphospecies and, after preliminary identifications, were distributed to the National Herbarium of Suriname (BBS), US National Herbarium (US), Netherlands National Herbarium (L), and international specialists. Unicate specimens were identified and returned to the BBS in Suriname.

Forest Interviews

Due to the large number of plant stems and species in plots, only 3–4 local people were able to participate in specialist knowledge interviews at each site. The participants were chosen nonrandomly through a two-step selection process. A larger group of male forest specialists (10–12 people) were first chosen based upon local reputation and availability. These people were assessed for their plant knowledge and interview interest in 20-m pilot transects and trail walks. After this initial assessment, a smaller group was selected for the forest plot research. The final participants included three older Trio indigenous men (aged 50s–60s) and four older Saramacca men (aged 50s–70s). Initially, there were four Trio participants, but one of them left the village. For every numbered plant, participants were queried on plant recognition ("Do you know this plant?"), local name ("If known, what is it called?"), and uses ("Is this plant useful or has no use?" and [if used] "What is done with it?"). In some cases, additional ethnobotanical data was collected in open and semistructured village and "walk in the woods" interviews (Martin 2004). Sranantongo (English-based Creole) was spoken initially, and local languages (Saramacca and Trio) were used increasingly as my language skills improved. Unfortunately, female specialists could not participate in plot interviews due to cultural restrictions. Additional nonplot interviews to assess generalist plant knowledge were conducted with more than 30 individuals per site, including men and women. This additional data will be presented elsewhere.

Quantitative Ethnobotanical Analysis

Quantitative methods were used to compare ethnobotanical data between the two study groups. Categorical variables included cultural affiliation, taxonomic identity (kinds of plants), use (kinds of uses), and vegetation zones (kinds of places). Quantitative (interval) data included the number of plot species and stems named and used, the number of use citations, and *use value* indices (described below). Comparisons were based largely on averages and percentiles due to between-site differences in floristic composition and the unequal number of participants (four Saramacca participants, three Trio participants).

The consensus "use value" method for collecting ethnobotanical data was introduced in Phillips and Gentry (1993a, b) and Phillips (1996) (Table 13.1). As originally presented, every interview citation (one person, one plant, one use) is considered to be a statistical "event." Initial data collection is a tally of citations without weighting by informant or researcher. The *use value* index or UV_s of a species is based upon the number of independent citation events divided by the number of participants. Family use value (UVF_s) is the sum of species UV_s within a given family and can be standardized through division by the number of species (Galeano 2000; Phillips and Gentry 1993a). My methods differed from the original Phillips and Gentry method as follows: (1) participants were not asked about the same plant more than once, (2) participants were chosen nonrandomly, (3) biological species were not lumped into folk taxa, and (4) to avoid inflated values for firewoods, only the "best" firewoods on a three-point participant-defined (emic) scale were included.

I sorted use citations into five broad categories: "construction," "edible," "medicine," "technology," and "trade/commerce" (Cook and Prendergast 1995; Prance et al. 1987). The "construction" category included both thatch material for roofs and wood material for poles, beams, and planks. The "edible" category included

Table 13.1 Formulas and description of use value methodology

Species use value	$UV_{is} = \left(\sum U_{is}\right) / \left(n_{is}\right)$	U_{is} = number of uses mentioned for species *s* by informant *i* and n_{is} = the number of "events" in which informant *i* cites a use for species *s* Tally the number of plant uses mentioned for a given species (all uses equal) and divide by the number of "events" (all use citations over time of the study for a species by one informant)
Species use value (one species across all informants)	$UV_s = \left(\sum UV_{is}\right) / \left(n_i\right)$	n_i = total number of informants interviewed for species *s* Sum the informant use values for a species and divide by the total number of informants
Family use value (one plant family across all informants)	$UVF_s = \sum UV_s$	Sum the use values for all the species within a given family Relative $UVF_s = \Sigma UV_s$/no. of species

wild-harvested edible plants (e.g., *Spondias mombin* L., Anacardiaceae), semidomesticated plants, and animal hunting attractants. The "technology" category encompassed a variety of uses and products such as canoes, household tools, hunting and fishing tools, resins, and subsistence fibers. As with Milliken et al. (1992b), the "technology" category included uses placed in an "other" category in some studies (van Andel 2000b). The "trade/commerce" category included craft items for casual tourist markets (tourist seed jewelry, calabash, carvings) and timber and non-timber forest products for more established markets.

The "medicine" category included all plants that are used for therapeutic, bioactive, and ritual purposes, including poisons, charms, and repellants. Medicinal and ritual uses were combined into one category to address a major difference in use citations between the Saramacca and Trio. In forest interviews, Saramacca participants often ascribed general spiritual or ritual purposes to plant species and were less concerned with addressing specific physical ailments. In contrast, Trio participants very commonly linked specific physical ailments with specific plants and made a clear distinction between these plants and spiritual or ritual plants.

The *use value* index was used to compare use knowledge patterns and quantify "importance" for plant families and species. To test the hypothesis that "importance" is a function of "apparency" – based upon stem density (family and species) or species richness (family) – I used the "regression-residuals" method introduced by Moerman for medicinal plants (1979, 1991) and adapted by Phillips and Gentry (1993b). This method also allows the detection of outliers – species with either higher or lower than expected *use value* for their ecological "apparency" (see Galeano 2000; Phillips and Gentry 1993a). For each taxon of interest, a regression of UV_s (or FUV_s) (*y*-axis) versus abundance or species richness (*x*-axis) was run in Minitab 15. Following this, a scatterplot was made of UV_s (or FUV_s) versus the *standardized* residuals of UV_s (or FUV_s). Data points occurring beyond ±1.96S.E.

(α=0.05) revealed outlier species or families. For a review on "apparency" theory in ethnobotany and its origins in ecological theory, see de Albuquerque and de Lucena (2005).

Limitations of the *use value* methodology have been observed (Kvist et al. 1995; Phillips et al. 1994). Of concern is that statistically equal use citations ("events") and researcher-defined (etic) use categories ignore cultural complexity and context. "Importance" may be exaggerated for a taxon with multiple minor uses and understated for a taxon with a single, yet unreplaceable use. I chose to apply *use values* and researcher-defined categories, despite their limitations, because I required strong data standardization for cross-cultural comparisons.

Results

Ecological Data

Species richness and stem density for the Saramacca (SA) and Trio (TR) forest plots (stems ≥10 cm dbh) are summarized in Tables 13.2 and 13.3. Comparing pooled-vegetation data sets between sites (2.5 ha per site), TR-site species richness was much greater than SA-site species richness (Fisher's α index: SA 43.3, TR 78). Stem abundance was similar between the two sites. At both sites, there was a similar hierarchical relationship across vegetation zones in Fisher's α. Tree diversity was greatest in non-flooded (UP) old-growth (31.67, SA; 59.96, TR), slightly less in seasonally flooded (LO) old-growth (29.22, SA; 47.85, TR), and much less in fallow (FA) forest (10.17, SA; 15.28, TR). Comparing specific vegetation zones, TR diversity was conspicuously greater in the two old-growth plots and moderately greater in the fallow plots.

Upland forest (UP) at both SA and TR sites was characterized by gently undulating ridges of brown sand, 25–30 m canopy height, and a relatively open understory. Caesalpinioid Fabaceae were abundant, comprising 21% (SA) to 35% (TR) of stems. A 5–8-m-tall palm, *Astrocaryum sciophilum* (Miq.) Pulle (Sranantongo: *bugrumaka*), was common at both sites (33% of stems, SA; 10% of stems, TR).

Seasonally flooded forest (LO) at both sites was characterized by alluvial soils on flat terrain, 25–40 m canopy height, and a relatively open understory. Caesalpinioid Fabaceae were abundant, comprising 48% (SA) to 30% (TR) of stems. Canopy emergents

Table 13.2 Woody plant diversity and abundance for plots pooled at the Saramacca (*SA*) and Trio (*TR*) research sites (stems ≥10 cm dbh)

	SA	TR
	2.5 ha	2.5 ha
Number of families	39	44
Number of genera	104	160
Number of species	153	234
Number of individuals	1,436	1,487
Fisher's α	43.3	78.0

Table 13.3 Woody plant diversity and abundance by vegetation zone at the Saramacca (*SA*) and Trio (*TR*) research sites (stems ≥10 cm dbh)

	Upland (*UP*)		Lowland (*LO*)		Fallow (*FA*)	
	SA 1.0 ha	TR 1.0 ha	SA 1.0 ha	TR 1.0 ha	SA 0.5 ha	TR 0.5 ha
Number of species	94	138	81	117	38	52
Number of stems	581	539	438	504	417	444
Fisher's α	31.67	59.96	29.22	47.85	10.17	15.28

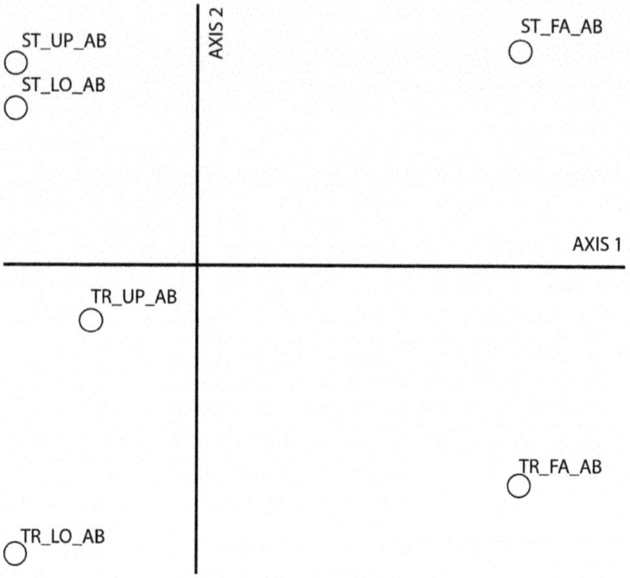

Fig. 13.3 Ordination of 335 morphospecies abundances (*AB*) at two cultural research sites

>100 cm dbh, such as *Ceiba pentandra* L. (TR) and *Pterocarpus* sp. (SA), were present at both sites. The TR site was distinguished by a high density of mid- to lower-canopy tree species, including *Inga* spp. (mimosoid Fabaceae), *Quararibea guianensis* Aubl. (Malvaceae), and *Sagotia racemosa* Baill. (Euphorbiaceae). Unique to the SA site were *Qualea* and *Vochysia* species (Vochysiaceae) in the upper canopy, and a high density of *Paypayrola guianensis* Aubl. (Violaceae) in the understory.

Fallow forest (FA) at both sites were characterized by gently undulating ridges of brown sand, 10–15 m canopy height, high light levels, and moderately dense shrub and herb layers. Common neotropical secondary forest genera dominated both sites, notably Urticaceae (*Cecropia*, *Pourouma*), *Vismia*, *Guatteria*, and *Casearia* at the TR site and *Henriettia*, *Jacaranda*, and *Isertia* at the SA site.

In the NMDS ordination (Fig. 13.3), a two-dimensional analysis explained 92% of variation (335 species, $r^2 = .92$, rel. Sorenson). In old-growth forest, diagram clustering (similarity) was greater for geographically proximal, ecologically distinct plots (SA-UP vs. SA-LO, TR-UP vs. TR-LO) than for distal, ecologically similar plots (SA-UP vs. TR-UP, SA-LO vs. TR-LO). The closest similarity was observed between the two old-growth plots at the SA site. Regional proximity of old-growth plots was thus a stronger predictor of species composition than moisture regime. The two fallow plots were floristically distinct from one another and, as expected, distinct from all old-growth plots.

Because the comparative analysis of resource use between matching vegetation zones at each site was based upon proportional differences and the *use value* index, dissimilarity in species diversity and composition was unproblematic. A limitation to analysis, however, was the strong similarity in species composition of the two SA old-growth sites. As a result, SA intersite UP and LO comparisons were of less value than expected in addressing research questions.

Comparative Extent of Knowledge on Biodiversity

Data Pooled

Overall, both Trio (TR) and Saramacca (SA) participants exhibited considerable knowledge in the recognition and use of plant diversity (see Fig. 13.4). Knowledge of forest plot resources ranged from 83.9 to 98.7% of species "named," 61.2–93.5% of species with one or more uses, 94.1–100% of stems "named," and 89.7–100% of stems with one or more uses. *Use value* (UV_s – per species) data included medians of 1.25 (SA) and 1.67 (TR), means of 1.67 ± 1.13 S.D. (SA) and 1.98 ± 1.06 S.D. (TR), and maximums of 6 (SA) and 7 (TR).

On a percentage basis, the indigenous TR participants possessed a greater body of knowledge on local species than the Maroon SA participants. The TR group named 13.4% more of total species (83.9% SA <97.3% TR) and "used" 20.8% more of total species (66.9% SA <87.7% TR) (Fig. 13.9). The number of uses cited per species was 1.32 for the SA group and 1.94 for the TR group. Trio UV_s was statistically significantly greater than SA UV_s for all species ($t = 5.4$, $p < 0.001$) and for a reduced data set with "unnamed" (unknown) species excluded ($t = 2.5$, $p < 0.01$) (Table 13.4).

Per stem, the number of uses cited was 0.14 for the SA participants and 0.31 for the TR participants. By percentage, TR knowledge was a minor 3% greater for stems "named" (96.9% SA <99.9% TR) and a moderate 7% greater for stems "used" (93% SA <99% TR). *Per-stem* percent comparisons were complicated by the lower diversity and thus greater species dominance (many stems of few species) in all SA plots. This ecological dissimilarity equalized *per-stem* knowledge differences between the SA and TR participants. Although cross-cultural ethnobotanical research would ideally be based upon the same forest plots, this was impractical due to the localized context of traditional knowledge.

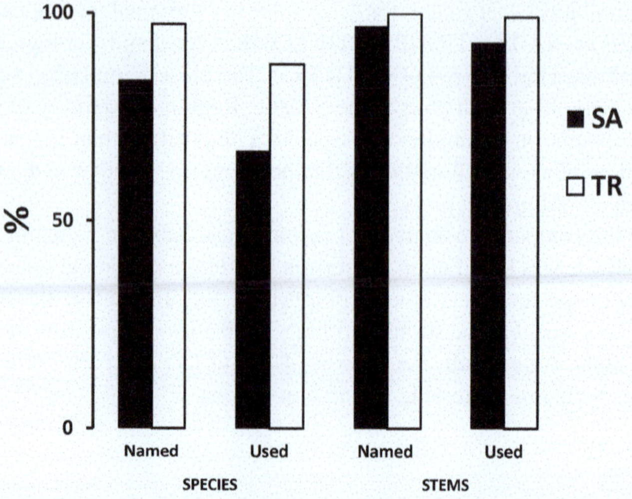

Fig. 13.4 Plots pooled knowledge-base comparison: percent of total plot species and stems named and/or utilized (≥1 use) by Saramacca Maroon (*SA*) and Trio Indigenous (*TR*) research participants (Species: [Trio, 234 species, 3 participants, 97.3% named, 87.7% used; Saramacca, 153 species, 4 participants, 83.9% named, 66.9% used]. Stems: [Trio, 1,487 stems, 99.9% named, 99.0% used; Saramacca, 1,436 stems, 96.9% named, 93.0% used]. A "used" species has ≥1 cited uses, only citations on the participant-determined "very good" firewood resources included)

Table 13.4 Plots pooled knowledge-base comparison: total and per-species *use value* (UV_s) for Saramacca (*SA*) and Trio (*TR*) participants

	SA	TR	
Total UV_s	1.27 ± 1.13 SD	1.91 ± 1.02 SD	
Per species UV_s named + unnamed	104	160	$t=5.4^{***}$
Per species UV_s named	153	234	$t=2.5^{**}$

SA (n = four participants), TR (n = three participants)
$^{**}p<0.01$; $^{***}p<0.001$

Vegetation Zones

On a per-species percentage basis, TR knowledge exceeded SA knowledge within UP and LO old-growth zones – including species "named" (UP – 10.1% difference, LO – 8.5% difference) and "used" (UP – 27.9% difference, LO – 10.9% difference) (Table 13.5). On a per-stem percentage basis, TR knowledge also exceeded SA knowledge in old-growth plots, but to a lesser degree – including stems "named" (UP – 2.5% difference, LO – 5.9% difference) and "used" (UP – 7.9% difference, LO – 9.7% difference). Within FA plots, Saramacca plant knowledge more closely matched Trio plant knowledge. TR participants had a 6–7% edge

Table 13.5 Knowledge-base comparison by vegetation zone: % plot species and stems "named" and "used" by Saramacca (*SA*) and Trio (*TR*) research participants (stems ≥10 cm dbh)

	Upland (*UP*)		Lowland (*LO*)		Fallow (*FA*)	
	SA 1.0 ha	TR 1.0 ha	SA 1.0 ha	TR 1.0 ha	SA 0.5 ha	TR 0.5 ha
Number of species	94	138	81	117	38	52
% Species named	87.2	97.3	90	98.5	92	98.7
% Species used	61.2	89.1	77.7	88.6	92	93.5
Number of stems	581	539	438	504	417	444
% Stems named	97.4	99.9	94.1	99.0	99.3	100
% Stems used	91.0	98.6	89.7	99.4	99.3	100

Saramacca (*SA*), four participants; Trio (*TR*), three participants; a "used" species has ≥1 cited uses, only citations on the participant-determined "very good" firewood resources included

over SA participants in the percent species "named" (98.7% TR, 92% SA), but percentage differences were minor for species "used," stems "named," and stems "used." Use value (UV_s) index culture comparisons across vegetation zones were similar to per-species percentage results. TR UV_s was significantly greater than SA UV_s in old-growth plot comparisons and insignificantly different between fallow plots (ANOVA, $F=8.64$, $p<0.0001$) (Fig. 13.5). SA knowledge was more extensive within fallow forest, and TR knowledge was even across all three vegetation zones.

Comparative Categorical Use of Biodiversity

Data Pooled

Analyzing the distribution of UV_s by *use category*, I documented dissimilar patterns for the two study cultures (Table 13.6). The hypothesis that cultural differences in categorical use were insignificant was thus rejected. Trio participants cited *medicine* category uses to an inordinate degree – 56.1% of total TR UV_s. Only two categories, *medicine* and *technology*, comprised 80% of Trio UV_s. *Construction* (10.2%) and *edible* (9%) category uses were less frequently cited and *trade* comprised only 2.6% of UV_s. Saramacca UV_s was distributed more evenly across the five categories than Trio UV_s ($t=.20$, $p=0.84$, unequal variances, percentages arcsine transformed). Three categories, *construction*, *medicine*, and *technology*, comprised 80% of total SA *use value*. *Trade* (9.8%) and *edible* (9%) category uses were less frequently cited.

Trio *medicine* mean UV_s was significantly greater than Saramacca *medicine* mean UV_s (1.11 TR, 0.35 SA, MWU, $p<0.001$). There was no significant difference in mean *edible* UV_a (MWU, NS, $p=0.26$) or mean *technology* UV_s (MWU, NS, $p=0.26$) between study groups. Saramacca mean UV_s exceeded Trio mean UV_s, however, in the *construction* (0.30, MWU, $p<0.05$) and *trade* (0.13, MWU, $p<0.001$) categories.

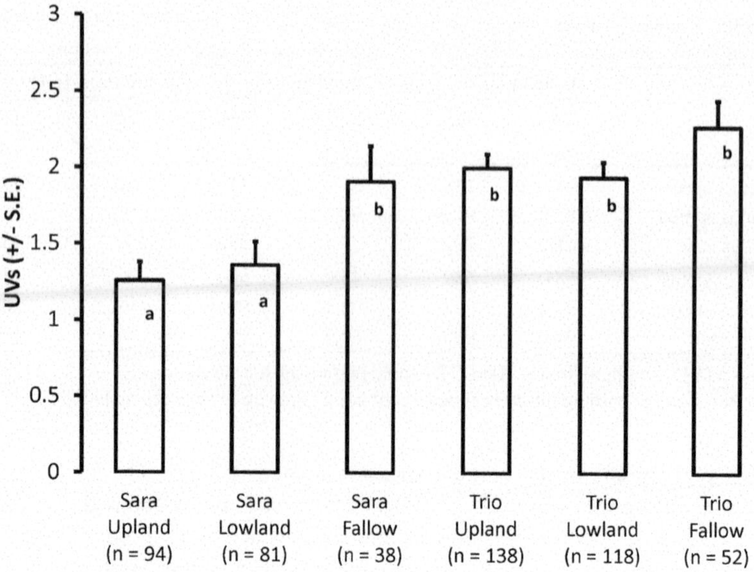

Fig. 13.5 Knowledge-base comparison by vegetation zone: mean per-species *use value* (UV_s) for Saramacca (*SA*) and Trio (*TR*) participants (ANOVA, $F=8.64$, $p<0.0001$) (n=number of species, total UV_s per vegetation zone for Saramacca (Sara) was 118 (upland), 110 (lowland), and 77 (fallow) – and for Trio was 273 (upland), 229 (lowland), and 116 (fallow))

Table 13.6 Resource use patterns: *use value* (UV_s) by use category for Saramacca (*SA*) and Trio (*TR*) participants

	SA site % of UV_s	TR site % of UV_s	SA site mean UV_s (SE)	TR site mean UV_s (SE)	Mann-Whitney
Construction ($_cUV$)	23.3	10.2	**0.30** (0.03)	0.20 (0.02)	*
Trade ($_\$UV$)	9.8	2.6	**0.13** (0.02)	0.053 (0.01)	**
Edible ($_eUV$)	9.1	9.0	0.12 (0.02)	0.18 (0.02)	NS $p=0.28$
Medicine ($_mUV$)	27.3	56.1	0.35 (0.04)	**1.11** (0.05)	**
Technology ($_tUV$)	31.1	22.5	0.48 (0.04)	0.44 (0.03)	NS $p=0.26$

SA $n=117$ sp., TR $n=229$ sp.; numbers in bold are significantly higher
*$p<0.05$; **$p<0.001$

Vegetation Zones

For the Trio, approximately the same distribution of categorical UV_s knowledge was observed for pooled data and for all three vegetation zones (Fig. 13.6). In other words, the Trio stuck to their preferential use pattern regardless of ecological or

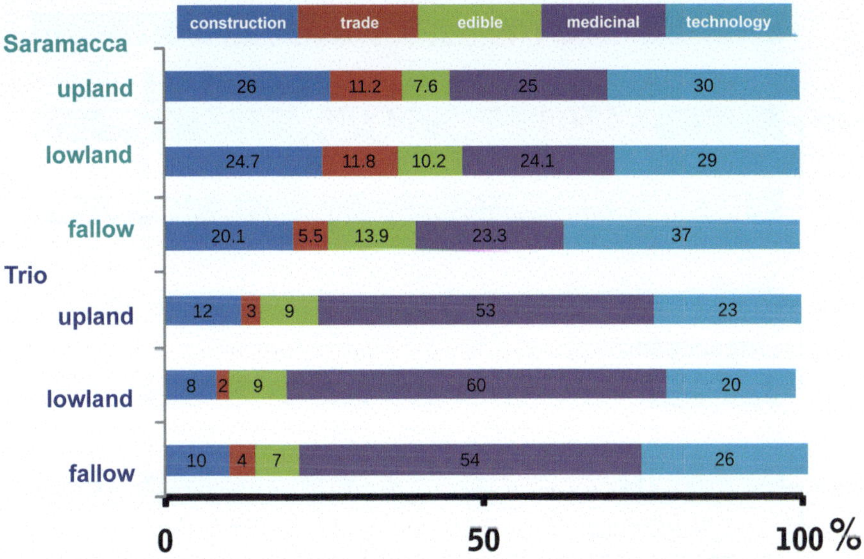

Fig. 13.6 Resource use patterns per vegetation zone: percent of total *use value* (UV_s) per use category for Saramacca (*SA*) and Trio (*TR*) participants (Upland = terra firme unflooded forest sites; lowland = seasonally flooded forest sites; fallow = older fallow (>15 years); total UV_s per vegetation zone for SA was 118 (upland), 110 (lowland), and 77 (fallow) – and for TR was 273 (upland), 229 (lowland), and 116 (fallow))

taxonomic variation. For the Saramacca, categorical UV_s distribution was similar for pooled and old-growth plots. Within the fallow (FA) zone, however, SA knowledge was more evenly distributed across use categories, due to a proportional increase in *technology* and *edible* uses. In the *medicine* category, TR knowledge exceeded SA knowledge in all three forest zones. In the *construction* and *trade* categories, SA knowledge exceeded TR knowledge only in the old-growth plots.

Patterns in Resource Selection and Use

Resource Use Patterns for Four Plant Families

In Fig. 13.7, I compare the distribution of family use value (UVF_s) across use categories for four plant families important at both research sites. For both cultures, the Arecaceae family was used for a wide range of purposes with a significant component of *edible* UV_s (32% SA, 24% TR). The *medicine* category predilection of Trio participants was still apparent, but less extreme. For the three non-palm families in Fig. 13.10, the previously described dissimilar cultural patterns were observed. *Medicine* category uses were cited more by the Trio (e.g., Lauraceae: 0% SA, 40% TR) and the Saramacca emphasized *construction* (e.g., Lauraceae: 52% SA, 27% TR) and *trade* (e.g., Lauraceae: 31%, SA 0% TR).

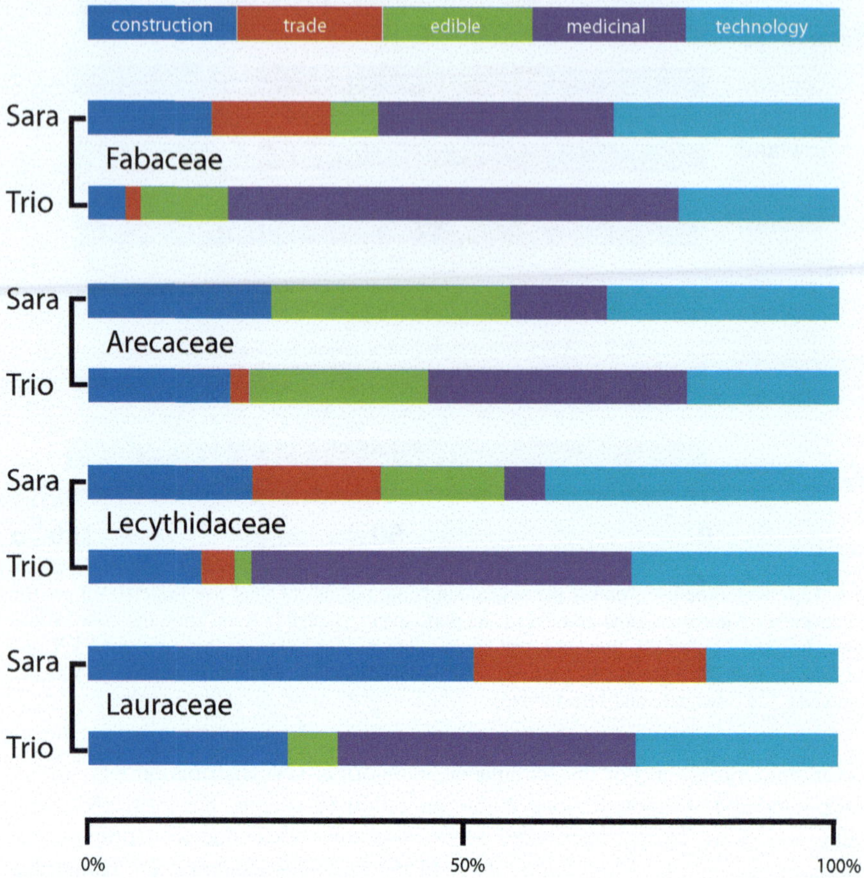

Fig. 13.7 Resource use patterns for four plant families: proportions of total *family use value* (UVF_s) per use category by Saramacca (*SA*) and Trio (*TR*) participants (Total UVF_s by family: Fabaceae (47, ST; 97.6, KW); Arecaceae (24.5, ST; 36.3, KW); Lecythidaceae (9, ST; 26, KW); Lauraceae (7.3, ST; 5, KW))

Ecological "Apparency" and Use Knowledge

If ecological "apparency" is a driver of ethnobotanical selection, then the most abundant or speciose taxa should also be the most useful. For large woody forest plants, species richness (*S*) and abundance (AB) are useful indicators of "apparency" (Phillips and Gentry 1993b). Sixteen plant families comprising the highest percentages of total *family use value* (UVF_s) in this study are shown in Fig. 13.8, arranged in descending order (green bars). The percent contribution of these families to total *S* (no. sp., blue bars) and total AB (no. indiv., red bars) is shown alongside % UVF_s.

At both sites, the Fabaceae (sensu lato, Password 2003) family had the highest % UVF_s and comprised a higher percentage of *S* and AB than any other family. For most

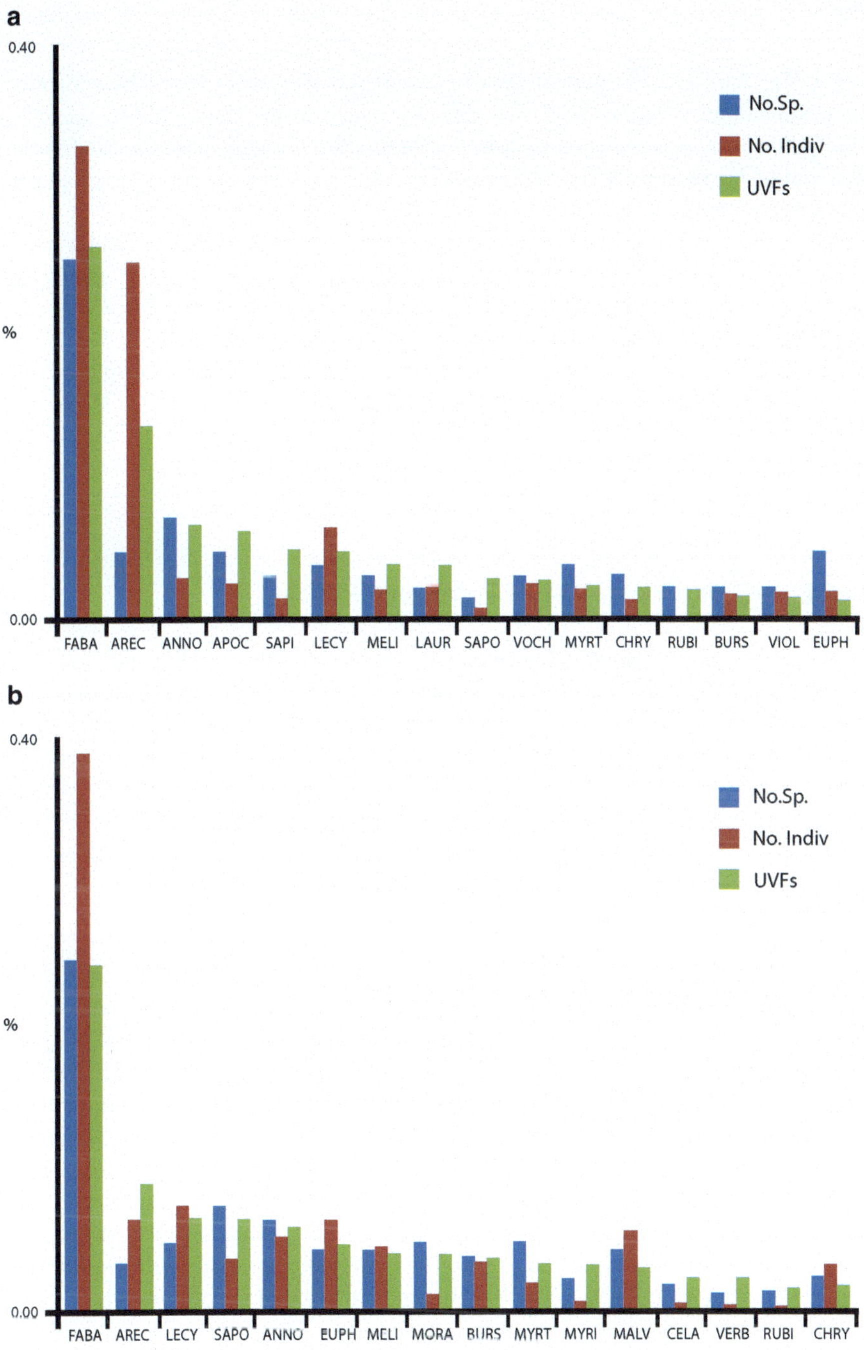

Fig. 13.8 (**a**), (**b**) Sixteen plant families compared by percent contribution to *family use value* (UVF_s), species richness (*S*), and abundance (*AB*) at the Saramacca (*SA*) and Trio (*TR*) research sites (arranged by high to low UVF_s) (*FABA* Fabaceae, *AREC* Arecaceae, *LECY* Lecythidaceae, *SAPO* Sapotaceae, *ANNO* Annonaceae, *EUPH* Euphorbiaceae, *SAPI* Sapindaceae, *MELI* Meliaceae, *LAUR* Lauraceae, *MORA* Moraceae, *BURS* Burseraceae, *MYRT* Myrtaceae, *MYRI* Myristicaceae, *MALV* Malvaceae, *CELA* Celastraceae, *VERB* Verbenaceae, *RUBI* Rubiaceae, *CHRY* Chrysobalanaceae, *VOCH* Vochysiaceae, *VIOL* Violaceae)

low UVF_s families, the percentages of *S* and AB were also low. These results suggest ecological "apparency" as a selection mechanism. The high Arecaceae % UVF_s, however, was matched by low *S* at both sites – suggesting that this family was valued for reasons beyond ecological abundance. This case is discussed in the following section.

Regression and residual analyses supported the "apparency" hypothesis for most families and species. UVF_s was positively correlated at the SA and TR sites with both *S* and AB (# species SA: $F=81.9$, $p<0.001$, $r^2=0.78$; # species TR: $F=337.0$, $p<0.001$, $r^2=0.92$; # stems SA: $F=93.4$, $p<0.001$, $r^2=0.80$; # stems TR: $F=120.1$, $p<0.001$, $r^2=0.80$). For species, I found a weak but positive relationship between UV_s and stem abundance at both sites (TR: $F=5.09$, $p<0.05$, $r^2=0.01$; SA: $F=10.48$, $p<0.01$, $r^2=0.11$). The regression-residual charts in Figs. 13.9, 13.10, and 13.11 show that the vast majority of taxa at both sites had no greater (or lesser) *use value* than expected for *S* and AB values.

Inordinately Important Plant Resources and Uses

Analysis of pooled citation data revealed an uneven distribution of use knowledge across taxa for both the Saramacca (SA) and Trio (TR) groups. Participants cited an inordinate number of uses (>5) for a minority of taxa and few uses (<2) for the great majority of taxa (Fig. 13.12). For some of the highest *use value* taxa, the positive correlation between ecological "apparency" and use knowledge did not apply. In Figs. 13.9 and 13.10, family outliers (to the right of the +1.96 S.D. line) were those "more-useful-than-expected."

Palm Resources

In regression-residual analysis of species richness, the Arecaceae family was inordinately "important" for both the Saramacca and Trio (Fig. 13.9). Based upon abundance, the Arecaceae was not an outlier due to the high density of one species, *Astrocaryum sciophilum*, in SA and TR old-growth zones. The 14 highest UV_s species outliers at both research sites are shown in Fig. 13.13. By species, palms dominated in straight UV_s tallies and as "more-useful-than-expected" resources in regression-residual analysis. The Saramacca top 14 UV_s list included four Arecaceae species (the top three highest UV_s for any study species) and half of the Trio top 14 UV_s list were Arecaceae. The palm species in study plots included *Attalea maripa* (Aubl.) Mart (code: ATMA, SA and TR sites); *Astrocaryum sciophilum* (Miq.) Pulle (ASSC, TR site only); *Euterpe oleracea* Mart. (EUOL, SA site only); *Euterpe precatoria* Mart. (EUPR, TR site only); *Oenocarpus bacaba* Mart. (OEBA, TR site only); and *Socratea exorrhiza* (Mart.) H. Wendl. (SOEX, SA site only). Some additional Arecaceae species of local importance were not present or occurred outside of plots.

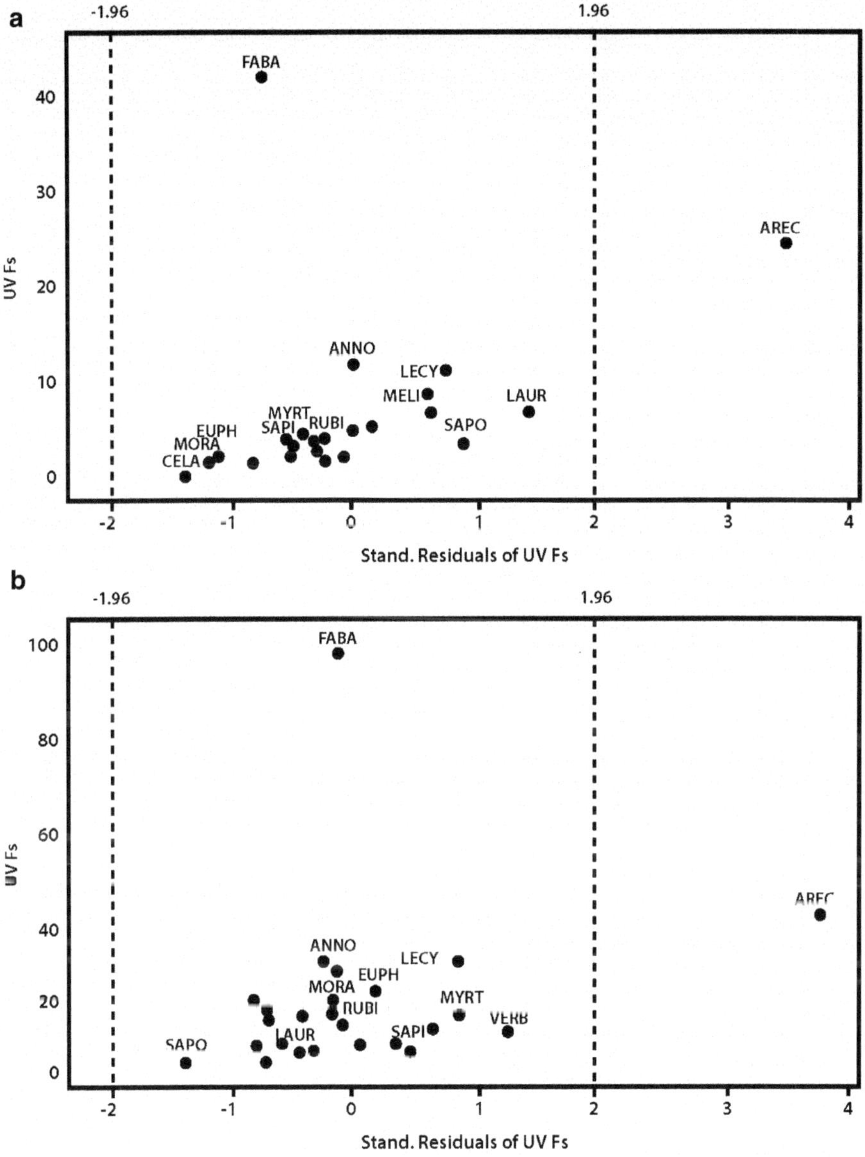

Fig. 13.9 (**a**) (SA), (**b**) (TR) Relationship of family "importance" (UVF_s) to species richness per family and detection of outliers (*x*-axis values are the standardized residuals of UVF_s, derived from the regression of species richness and rel. UVF_s) (Any family with standard residuals >1.96 or <−1.96 S.E. ($p<0.05$) has, respectively, greater or lesser use value (UVF_s) than expected from its relative diversity. Families with less than two species or seven individuals were excluded (SA 24 families, TR 25 families))

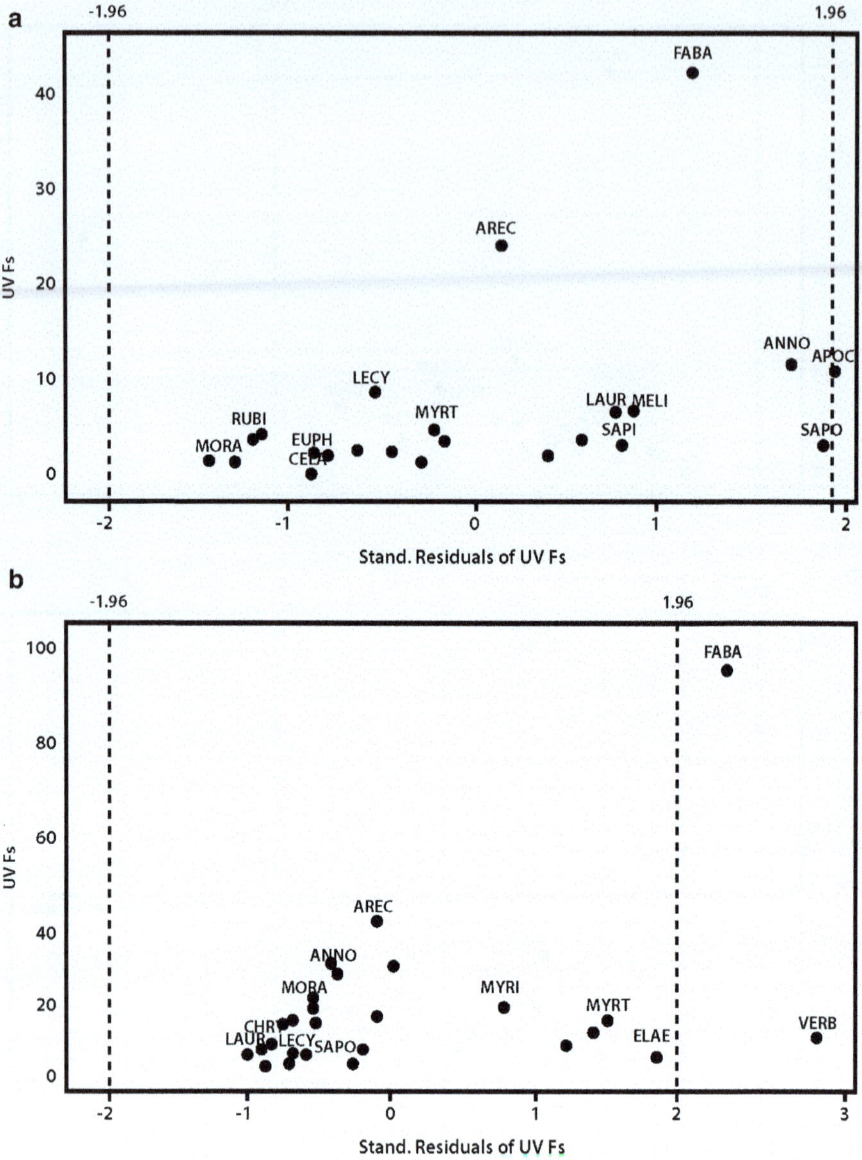

Fig. 13.10 (**a**) (SA), (**b**) (TR) Relationship of family "importance" (UVF_s) to stem abundance per family and detection of outliers (*x*-axis values are the standardized residuals of UVF_s, derived from the regression of stem abundance and rel. UVF_s) (Any family with standard residuals >1.96 or<−1.96 S.E. ($p<0.05$) has, respectively, greater or lesser use value (UVF_s) than expected from its relative abundance. Families with less than two species or seven individuals were excluded (SA 24 families, TR 25 families))

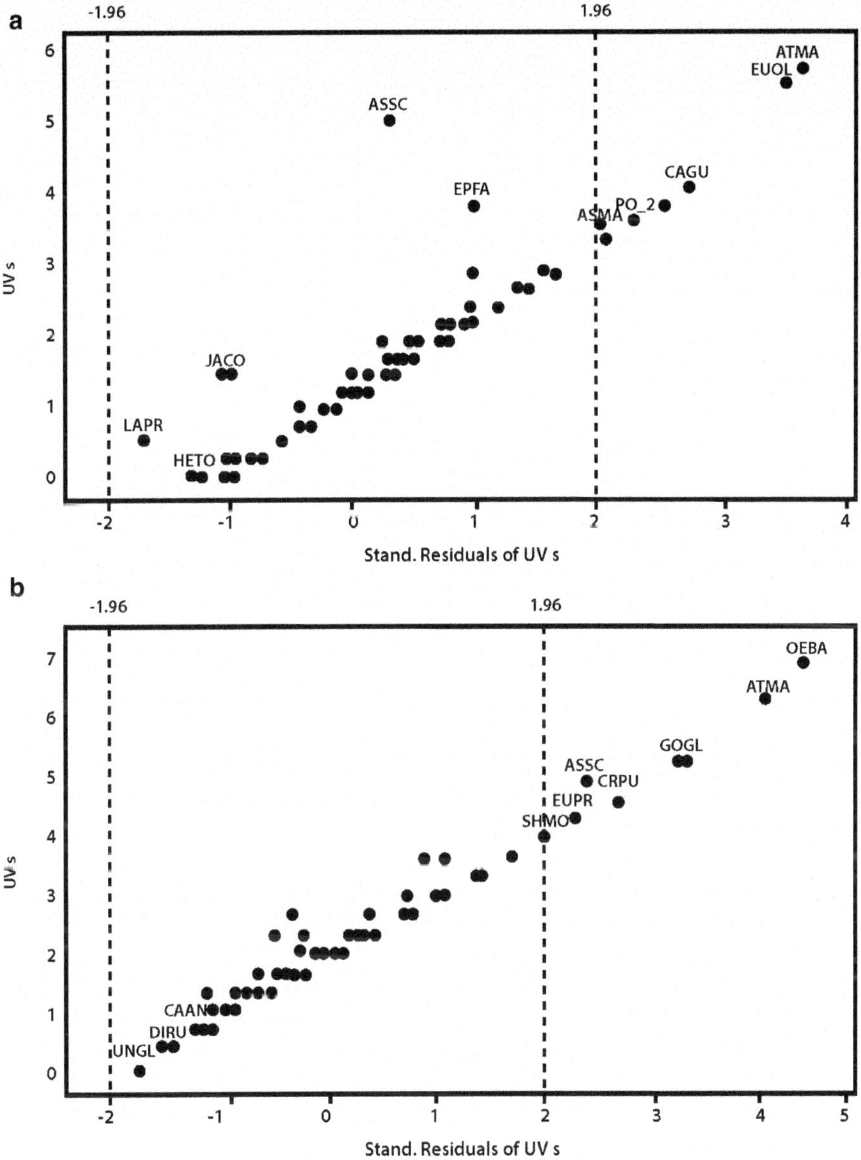

Fig. 13.11 (**a**) (SA), (**b**) (TR) Relationship of species "importance" (UV_s) to stem abundance per species and detection of outliers (*x*-axis values are the standardized residuals of UV_s, derived from the regression of stem abundance and rel. UV_s) (SA site, $N=153$; TR site, $N=235$; any species with standard residual >1.96 or <−1.96 S.E. ($p<0.05$) has, respectively, greater or lesser UV_s than expected for its abundance)

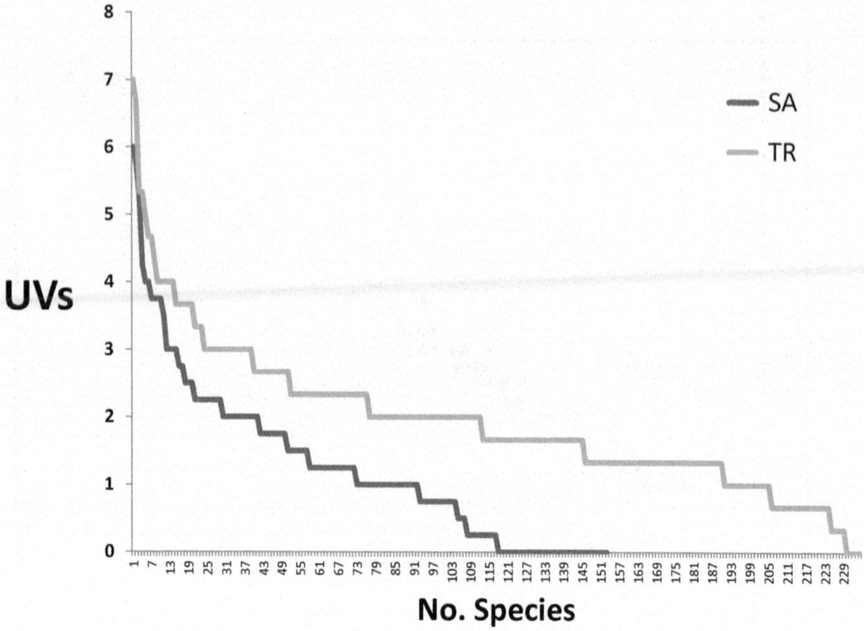

Fig. 13.12 Distribution of per-species UV_s data at the Saramacca (*SA*) and Trio (*TR*) sites

The categorical and specific uses of three high UV*s* palm species are listed below:

(a) *Attalea maripa* (Aubl.) Mart. (species code: ATMA) – uses cited by both groups: *construction* (leaves, thatch), *edible* (fruit mesocarp, oil), *technology* (seed, cosmetic oil, game attractants), and *medicine* (stem sap, healing wounds; leaves, ritual). Some Saramaccans additionally mentioned the production of salt (technology) from burnt fronds.

(b) *Euterpe oleracea* Mart. (EUOL) – uses cited by both groups: *construction* (leaves, thatch; stem, planks), *edible* (fruit mesocarp, oil; apical meristem, palm beetle grubs [(*Rhynchophorus palmarum* L.]), *technology* (seed, cosmetic oil), and *medicine* (stem sap, healing wounds; leaves, ritual). Saramacca participants again uniquely cited salt production. Unique Trio uses included a fruit beverage, craft and household tools woven from leaves, and medicinal remedies for malaria/fever, hepatitis, general sores, and leishmaniasis.

(c) *Astrocaryum sciophilum* (Miq.) Pulle. (ASSC) – uses cited by both groups: *construction* (leaves, roof thatch), *edible* (fruit mesocarp, oil), *technology* (seed, cosmetic oil), *medicine* (stem sap, healing wounds; leaves, rituals). Trio use was distinguished by crafts and household tools woven from leaves, game attractants, and certain medicinal uses, including general sores and leishmaniasis.

There was a close correspondence in SA and TR use knowledge for these palms. The main differences were the greater number of TR medicinal uses and the SA production of salt.

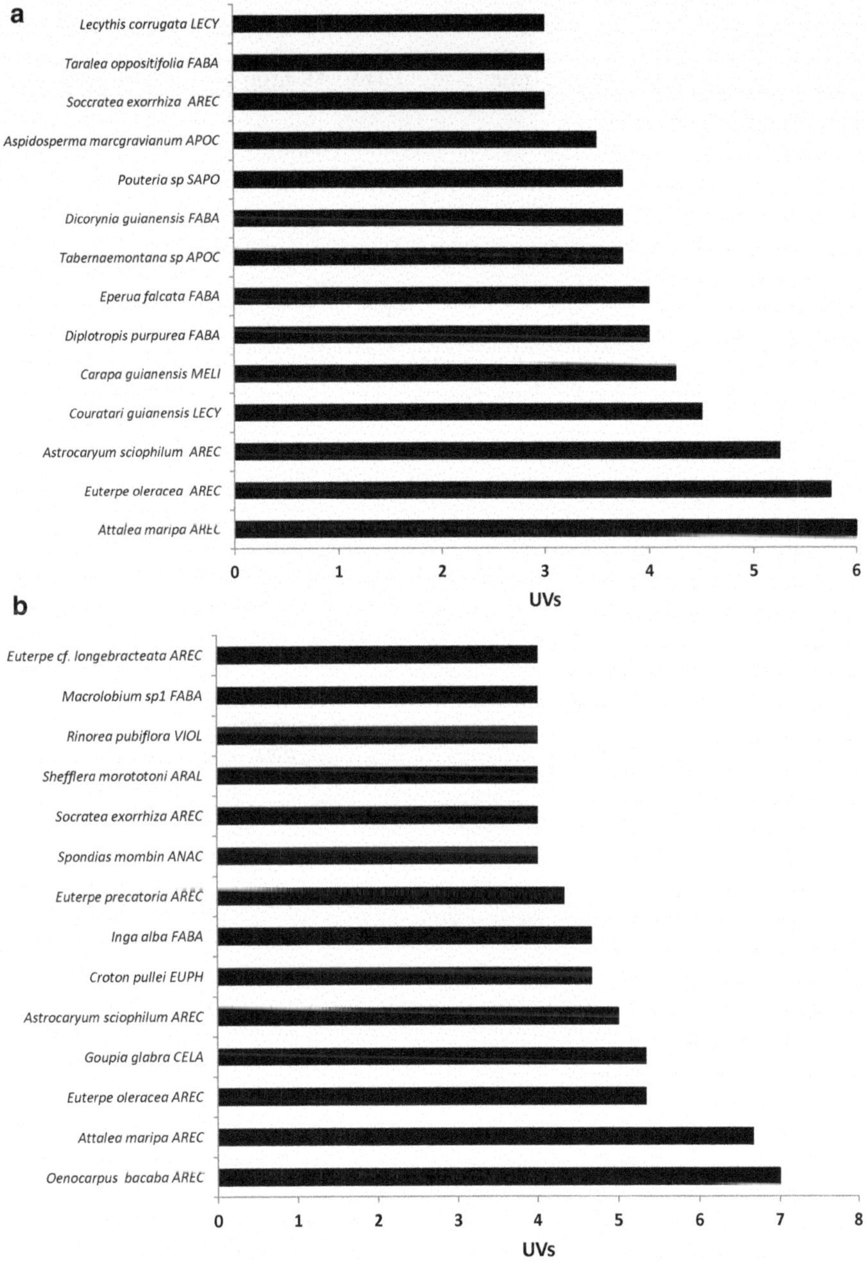

Fig. 13.13 (**a**), (**b**) Fourteen species with highest UV_s at the Saramacca (*SA*) and Trio (*TR*) sites

Table 13.7 Categorical and specific uses of the top five highest UV_s non-palm species at the Saramacca and Trio sites

	Construction	Trade	Edible	Medicine	Technology
Trio non-palm resources					
Croton pullei Lanj. (CRPU)	–	–	–	Sap: leishmaniasis	–
Goupia glabra Aubl. (GOGL)	Wood: house poles, planks	–	–	Sap: wounds	Wood: canoe, mortar
Inga alba (Sw.) *Willd.* (INAL)	–	–	Fruit	Sap: wounds	Sap: resin, paint; wood: firewood
Schefflera morototoni (Aubl.) Maguire, Steyerm. & Frodin (SCMO)	–	Seeds: tourist necklaces	–	Leaves/sap: ritual	Seeds: beads; wood: drums
Spondias mombin L. (SPMO)	–	–	Fruit	Sap: general health, ritual	–
Saramacca non-palm resources					
Carapa guianensis Aubl. (CAGU)	Wood: timber	Wood: timber	–	Seed oil: vermifuge, ritual	–
Couratari guianensis Aubl. (COGU)	Wood: timber	Wood: timber	–	Fruit pods: ritual	Wood: carved tools
Diplotropis purpurea (Rich.) Amshoff (DIPU)	Wood: timber	Wood: timber	–	–	Wood: carved tools
Eperua falcata Aubl. (EPFA)	Wood: timber, shingles	Wood: timber, shingles	–	–	Wood: carved tools
Pouteria sp. (PO__)	Wood: timber	Wood: timber	Fruit	–	Wood: carved tools

Non-palm Resources

In contrast to the Arecaceae, the top UV_s *non-palm* species and uses for each research site were largely dissimilar (Table 13.7). The Trio top five UV_s non-palms included *Croton pullei* Lanj. (CRPU); *Goupia glabra* Aubl. (GOGL); *Inga alba*

(Sw.) Willd. (INAL); *Schefflera morototoni* (Aubl.) Maguire, Steyerm. & Frodin (SCMO); and *Spondias mombin* L. The Saramacca top five UV_s non-palms included *Carapa guianensis* Aubl. (CAGU); *Couratari guianensis* Aubl. (COGU); *Diplotropis purpurea* (Rich.) Amshoff (DIPU); *Eperua falcata* Aubl. (EPFA); and *Pouteria* sp. (PO__).

With the exception of *Croton pullei*, the high UV_s TR non-palm species were iconic, well-documented resources for many regional indigenous groups (Balée 1994a; Grenand et al. 1987; Milliken et al. 1992a; van Andel 2000a). In accordance with their "cultural syndrome," the Trio selected taxa largely of medicinal and technology use. The only trade use was for tourist necklaces. *Goupia glabra*, the only well-known commercial timber species, was occasionally used by the Trio to make planks. By contrast, all five of the high UV_s SA species were commercial timber trees and, with the exception of *Carapa guianensis*, were not well-known regional indigenous resources (Lindeman and Mennega 1963). In accordance with their "cultural syndrome," the Saramacca emphasized construction, trade, and carving resources.

Discussion

Extent of Knowledge

An extensive knowledge of plant diversity has been documented for many indigenous forest cultures (Balée 1994a; Berlin 1992; Conklin 1967; Plotkin 1986; Prance et al. 1987; Schultes and Raffauf 1990), but this knowledge has rarely been directly compared with nonindigenous cultures. On a basic quantitative level, this study supported the hypothesis that longer resident, indigenous forest peoples "know more" about local biota than shorter-resident, nonindigenous forest peoples. Campos and Ehringhaus (2003) also documented a more extensive knowledge base on palm species for two indigenous groups compared to a ribereño and caboclo community in Brazil. These findings suggest that long-resident indigenous groups might have more to contribute to biodiversity-related projects (e.g., parataxonomy) when a choice must be made between two groups or two areas.

This study also supported a growing body of empirical evidence that short-resident, nonindigenous forest peoples are capable of developing robust utilitarian knowledge of local plant diversity (Da Cunha and De Albuquerque 2006; de Albuquerque et al. 2005; Halme and Bodmer 2007; Lira et al. 2009; Soler Alarcón and Luna Peixoto 2008; Voeks 1996; Young 2005). The 66.9% of upland old-growth woody species (≥10 cm dbh) used by Saramacca participants in this study was comparable to similar *terra firme* plot inventories among some indigenous (Prance et al. 1987, 61–78%; van Andel 2000b, 58%) and nonindigenous groups (Galeano 2000, 62.8%; Pinedo-Vasquez et al. 1990, 60%). I expect that Saramacca use knowledge would be quantitatively equal or greater than that of some indigenous groups, particularly considering the widespread decline of indigenous languages, traditions, and knowledge transfer.

Ethnobotanical Divergences (Vegetation, Categorical Use, Resources)

Previous research has highlighted the cultural importance of *terra firme* old-growth forest for various indigenous groups (Prance et al. 1987), old growth and edges for the indigenous Guaymi (Castaneda and Stepp 2008), floodplain old growth for mestizos (Phillips et al. 1994), and secondary forests for indigenous and nonindigenous medicinal uses (Voeks 1996). However, there has been insufficient empirical research to reveal principles of habitat preference among indigenous, folk, and tribal Afro-American groups. Recognition of greater or lesser "importance" does not make any particular forest zone useless. Local peoples depend upon a mosaic of different vegetation and land use zones to meet their needs, and some zones may harbor essential resources that are rarely used (Grenand 1992; Young 2005).

As was expected for the long-resident culture, the Trio exhibited extensive use knowledge across all three vegetation zones and use categories. Trio knowledge was consistently skewed toward *medicinal* and *technology* uses. This consistency reflected the adaptive capacity of TR participants to satisfy their main needs or interests regardless of forest zone. The lack of commercial uses was not surprising given the lack of access to the coastal economy. The basis of Trio ethnobotany as a product of long-term experience and a largely subsistence economy was reflected in the prevalence of widespread, iconic indigenous resources in the "top UV_s" Trio species lists.

In contrast to the Trio, Saramacca knowledge was not consistent across all vegetation zones. In old-growth forests, the SA emphasis upon *construction* and *trade* uses follows a pattern documented in previous nonindigenous ethnobotanical studies (Galeano 2000; Phillips and Gentry 1993a; Soler Alarcón and Luna Peixoto 2008). The prevalence of commercial uses reflected the greater Saramacca experience and access to regional markets. Modern economic influences upon the Saramacca were reflected in the status of most "top UV_s" non-palm resources as commercial timber resources.

In comparison with the Trio, the Saramacca appeared, both quantitatively and qualitatively, to have a special relationship with fallow forest. The high SA *use value* index in fallow forest and greater emphasis upon *technology* uses reflected combined cultural, economic, and biological influences. The SA fallow plot contained softwood species amenable to wood carving – a very significant cultural and economic activity for the Saramacca. Research participants made a major cultural distinction between fallow and old growth that highlights the importance of understanding local contexts. The previously mentioned taboos upon old-growth forest visitation and resource use were rarely applied to farms, trails, and fallow forests.

Ethnobotanical Convergences

The positive correlation of ecological "apparency" (species richness or stem abundance) and resource selection, documented for both the Trio and Saramacca, is a widespread phenomenon. Similar patterns have been observed in Brazil (Da Cunha and De Albuquerque 2006) and Peru (Phillips and Gentry 1993a), for an Afro-American nontribal community in the Colombian Chocó (Galeano 2000), and among indigenous peoples in Ecuador (Paz y Mino et al. 1995), Belize (Amiguet et al. 2005), and Indonesia (Caniago and Siebert 1998). Working with a Mayan community in the Yucatan, Torre-Cuadros and Isbele (2003) found a weak but positive relationship between availability (ecological Importance Value Index, IVI) and *use value* (UV_s). While the idea that the conspicuous species are more likely to be used in some way is well supported, it cannot be assumed that such "apparent" species are highly significant cultural resources.

The multicategory use and cultural significance of the Arecaceae family and many species can be reliably predicted for traditional cultures everywhere in the tropics (Balick 1988; Byg and Balslev 2001; Kvist et al. 1995; Reyes-García et al. 2006; Soler Alarcón and Luna Peixoto 2008). It was no surprise that Saramacca and Trio ethnobotany converged upon Arecaceae species as extraordinarily significant resources. Documenting the knowledge and use of Arecaceae resources is always a good starting point in ethnobotanical pattern analysis.

Limitations and Future Directions

Limitations of *use value* indices and researcher-defined use categories were discussed in the methods section. Hybrid methods that incorporate more cultural relativity, such as participant-defined (emic) use categories and citation weighting, have begun to play a greater role in quantitative ethnobotanical research (McClatchey et al. 2006; Reyes-García et al. 2006; Torre-Cuadros and Ross 2003; Young 2005).

Galeano (2000) showed how regression-residual analysis can be put to practical use in identifying critical resources for management. Documentation of a higher *use value* than expected may indicate that a species is naturally rare or, more likely, that it was once common and has been impacted by use. Galeano identified Lauraceae species as potentially threatened timber resources.

Bennett and Husby (2008) criticized the regression-residual analysis method. First, they observed that the method is prone to subjective hypothesis tests and residual analysis. Second, they found that the assumption of homogeneity of variation is violated because *use value* variation increases with family size. These authors presented contingency table and binomial analysis. It would be useful in future research to test and compare this new approach.

Based upon unpublished free list interviews with the Saramacca, the contribution of medicinal knowledge to Saramacca *use value* would likely have increased substantially if smaller plants, female informants, and anthropogenic zones had been included. A future study should document the knowledge of women, conduct interviews of plot-less generalist knowledge, include a greater variety of anthropogenic and forest zones, and sample a wider range of growth forms.

Conclusion

This study is unique as one of the relatively few direct, quantitative comparisons of indigenous and traditional nonindigenous people's knowledge and use of plants, and the only known such published comparison between an indigenous and Maroon culture. The overall aim of the research was not to assess the ethnobotanical "superiority" of any given culture but to explore dynamics and cross-cultural patterns in traditional plant knowledge. Documentation and comparison of multiple ethnobotanical contexts helps to build a predictive framework in ethnobotany and has conservation value, for example, in predicting and accommodating biocultural resource preferences. Etkin (2002) observed that the resource use knowledge of distinctive cultures is synergistic – understanding the forest resource use of more than one culture enhances the overall pool of adaptive knowledge and strategies for traditional peoples in the tropics. Further research is needed to document and elucidate use patterns in ethnobotany across biocultural contexts and scales. It is hoped that such research and dissemination of results will encourage greater exchange of ethnobotanical knowledge and/or materials between traditional peoples – particularly as the need for effective adaptive responses to global climate change increases.

Acknowledgements The research presented here was conducted in support of my doctoral candidacy at the Department of Botany, University of Hawai`i - Manoa, Honolulu, USA. I would like to especially thank the local communities of Kwamalasamutu and Stonhuku/Kajana for their friendship and participation in the fieldwork. In addition, I am grateful to the following organizations for their generous logistical and financial support: Amazon Conservation Team, Arlington, Virginia, USA http://www.amazonteam.org/; Anton de Kom University of Suriname, Paramaribo, Suriname; Botany in Action Fellowship Program (BIA), Phipps Conservatory and Botanical Gardens, Pittsburgh, Pennsylvania, USA http://phipps.conservatory.org/project-green-heart/botany-in-action/.

Appendices

Appendix 1 Kwamalasamutu (Trio) Plant List: Alphabetical by Scientific Name

Woody families and species in plots at the Kwamalasamutu (Trio) field site (≥10 cm dbh) – including habit (*T* tree, *L* liana), vernacular names, collection numbers (all B. Hoffman), species abundance within three vegetation zones (*UP* non-flooded, *LO* seasonally flooded, *FA* fallow), total *use value* (UV_s), and *use value* per use category (*CON* construction, *$$$* cash/trade, *EDI* edible, *MED* medicinal/ritual, *TEC* technology)

Appendix 1

Trio list

Family	Species	Habit	Vern. name	Coll #	# UP	# LO	# FA	UV_s	CON	$$$	EDI	MED	TEC
Anacardiaceae	*Spondias mombin* L.	T	mope	2,712	0	1	0	4	0	0	1	3	0
Anacardiaceae	*Tapirira* sp. *1*	T	awaimë	5,623	1	0	0	2	0	0	0.33	0.67	1
Annonaceae	*Anaxagorea* sp. *1*	T	turiimë	6,126	1	0	0	1.67	0.67	0	0	0.67	0.33
Annonaceae	*Annona haematantha* Miq.	L	ariminaimë (siminate)	6,370	0	1	0	1.67	0	0	0	1.33	0.33
Annonaceae	*Bocageopsis multiflora* (Mart.) R.E. Fr.	T	rasai	6,131	1	0	0	1.67	1	0.5	0	0	0.33
Annonaceae	*Duguetia cauliflora* R.E. Fr.	T	kapai ejamï (sikiman)	5,871	1	0	0	2.33	1	0	0	1	0.33
Annonaceae	*Duguetia sp1*	T	kapai ejamï (tikorijan)	6,035	1	0	0	2.33	1	0	0	1	0.33
Annonaceae	*Ephedranthus guianensis* R.E. Fr.	T	mekoro wewe	6,267	2	0	0	2.67	1	0	0	0.67	1
Annonaceae	*Fusaea longifolia (Aubl.)* Safford	T	kurija uru	6,103; 6,124	3	1	0	3	0.67	0	0.33	1.33	0.67
Annonaceae	*Guatteria punctata* (Aubl.) R.A. Howard	T	warumë	5,844	0	0	1	1.33	0.33	0	0	0.33	0.67
Annonaceae	*Guatteria 1*	T	warumë2	6,184	2	0	0	1.33	0.33	0	0	0.33	0.67
Annonaceae	*Rollinia exsucca* (DC. ex Dunal) A. DC.	T	karosiwa	6,200; 6,025	2	0	30	2	0	0	0	1.33	0.67
Annonaceae	*Unonopsis glaucopetala* R.E. Fr.	T	warumëimë 1	6,181	1	0	0	0	0	0	0	0	0
Annonaceae	*Xylopia nitida* Dunal	T	turi (sikiman)	6,201	2	0	5	2	1	0	0	0.67	0.33
Annonaceae	*Xylopia pulcherrima* Sandwith	T	turi (tikorijan/ tamiring)	5,943; 5,942	0	0	6	2	1	0	0	0.67	0.33

Apocynaceae	*Ambelania acida* Aubl.	T	kamaki	5,629; 5,894; 5,957; 6,346	1	0	0	1.33	0	0	1	0.33	0
Apocynaceae	*Geissospermum argenteum* Woodson	T	wataki	5,569	6	0	1	2.33	0	0	0	2.33	0
Apocynaceae	*Lacmellea aculeata* (Ducke) Monach.	T	ariwepu	6,120	1	1	0	2.33	0	0	0.33	2	0
Araliaceae	*Schefflera morototoni* (Aubl.) Maguire, Steyerm. and Frodin	T	maramara	6,149	1	0	1	4	0	1	0	2	1
Arecaceae	*Astrocaryum sciophilum* (Miq.) Pulle	T	murumuru	5,798	49	5	1	5	1	0	1	2	1
Arecaceae	*Attalea maripa* (Aubl.) Mart.	T	maripa	No coll	0	1	4	6.67	1	0	1	1.67	3
Arecaceae	*Euterpe longebracteata* Barb. Rodr.	T	wapuimë/ wajime	5,935	1	0	0	4	1	0	1	2	0
Arecaceae	*Euterpe oleracea* Mart.	T	wapu	6,308	0	1	0	5.33	1	0	2	2	0.33
Arecaceae	*Euterpe precatoria* Mart.	T	mikiri	6,688	6	0	0	7	1	0.33	1.67	1.67	2.33
Arecaceae	*Oenocarpus bacaba* Mart.	T	kumu	No coll	2	0	0	4.27	1	0.5	0	2	0.67
Arecaceae	*Socratea exorrhiza* (Mart.) H. Wendl.	T	pïeura	6,916	2	0	0	4.33	1	0	2	1.33	0
Bignoniaceae	*Adenocalymna inundatum* Mart. ex DC.	L	arawata kariwa	5,682; 5,791	1	1	0	1.33	0.33	0	0	1	0
Bignoniaceae	*Jacaranda copaia* (Aubl.) D. Don	T	marimari	6,226	0	0	35	1.67	0.33	0	0	0.33	1
Bignoniaceae	*Memora schomburgkii* (DC.) Miers	L	kuraiweimë	5,709; 6,304	0	0	1	2	0	0	0	2	0
Bignoniaceae	*Tabebuia* 2	T	arani 2	5,862; 5,993	1	0	0	1	0	0	0	1	0
Bignoniaceae	*Tabebuia serratifolia* (Vahl) G. Nicholson	T	arani 1	5,853	0	1	0	1.33	0	0	0	1.33	0
Bixaceae	*Bixa orellana* L.	T	wïseimë 1	6,148	1	0	0	2	0	0	0	0.67	1.33

(continued)

Appendix 1 (continued)

Trio list

Family	Species	Habit	Vern. name	Coll #	# UP	# LO	# FA	UV_s	CON	$$$	EDI	MED	TEC
Boraginaceae	*Cordia cf. tentrandra* Aubl.	T	mamï (itu)	6,298	1	0	2	1.33	0	0	0	0.33	1
Boraginaceae	*Cordia 1*	T	mamï (apun)	5,770	0	1	0	1.33	0	0	0	0.33	1
Burseraceae	*Crepidospermum goudotianum* (Tul.) Triana and Planch.	T	sirisiri 3	6,024	11	2	0	2.33	0.33	0.5	0	0.67	1
Burseraceae	*Protium apiculatum* Swart	T	sirisiri 1 (sikiman)	5,653	8	11	0	1.67	0.33	0.5	0	0.33	0.67
Burseraceae	*Protium decandrum* (Aubl.) Marchand	T	sirisiri 2 (tikorijan)	6,180	0	1	0	1.33	0.33	0.5	0	0	0.67
Burseraceae	*Protium tenuifolium* (Engl.) Engl.	T	kupapi	6,273	2	0	0	1.67	0	0.33	0	0.67	0.67
Burseraceae	*Protium decandrum* (Aubl.) Marchand	T	sawarëimë	6,690	1	0	0	2	0	0.33	0	0.33	1.33
Burseraceae	*Tetragastris panamensis* (Engl.) Kuntze	T	arita	6,157; 6,280	4	0	0	2.33	0.33	0	0.33	0.67	1
Burseraceae	*Trattinikia rhoifolia* Willd.	T	awa, sororo	6,222; 6,333	0	0	2	2	0.33	0	0	0.67	1
Burseraceae	*Burs 1*	T	pïtïre	6,142	1	0	0	1.33	0.33	0	0.67	0	0.33
Celastraceae	*Cheiloclinium hippocrateoides* (Peyr.) A.C. Sm.	L	sipun	6,150; 6,171	1	0	0	5.33	1	1	0	1.67	2
Celastraceae	*Goupia glabra* Aubl.	T	pasisi	5,530	3	0	1	2	0	0	0	2	0
Celastraceae	*Prionostemma asperum* (Lam.) Miers	L	kapi	6,038	1	0	0	1.33	0	0	0	1.33	0
Celastraceae	*Salacia/Tontalea sp. 1*	L	sipunïimë	5,673; 5,858; 6,057	0	1	0	2	0	0	0	1.33	0.67
Chrysobalanaceae	*Hirtella* sp. *1*	T	rapahkë ihpotï	6,069	4	0	1	1	0.67	0	0	0	0.33
Chrysobalanaceae	*Licania cf. micrantha* Miq.	T	paripo	5,730	7	4	0	1.67	0.67	0	0	0.67	0.33

Chrysobalanaceae	*Licania* sp. *2*	T	paripo, pokoro apëri	6,140	17	0	0	1.33	0	0	0	1.33	0
Chrysobalanaceae	*Licania* sp. *3*	T	paripo, tutuman	6,195; 6,128	2	0	0	0.33	0	0	0	0.33	0
Chrysobalanaceae	*Parinari rodolphii* Huber	T	paripoimë 1	6,001	2	0	0	0.67	0	0	0	0.67	0
Clusiaceae	*Caraipa cf. densifolia* Mart.	T	opë	5,713	0	5	0	1	0	0	0	1	0
Clusiaceae	*Clusia palmicida* Rich. ex Planch. and Triana	L	wetu etaku	5,854; 6,078	0	1	0	1.67	0	0	1	0.67	0
Clusiaceae	*Rheedia macrophylla* (Mart.) Planch. and Triana	T	aimara ere toto	5,883	0	1	0	2.67	0	0	1	1.33	0.33
Clusiaceae	*Tovomita* sp. *1*	T	kurepu	6,036; 6,818	4	0	0	2.67	0	0.33	0	1.33	1
Clusiaceae	*Tovomita* sp. *3*	T	pïeuraimë	5,906; 6,659	0	1	1	1.33	0.33	0	0	1	0
Clusiaceae	*Vismia guianensis* (Aubl.) Pers.	T	weri sepï (tikorijan)	5,599; 6,199	0	0	48	1.33	0.33	0	0	1	0
Clusiaceae	*Vismia* sp. *2*	T	weri sepïïme (sikiman)	6,015	0	1	23	1.33	0.33	0	0	1	0
Clusiaceae	*Vismia* sp. *3*	T	weri sepïïme (tamiring)	6,100; 6,197	0	0	3	0.67	0	0	0	0.67	0
Combretaceae	*Combretum rotundifolium* Rich.	L	jekara	6,282	0	1	0	2	0	0	0.33	1.33	0.33
Combretaceae	*Terminalia* sp. *1*	T	mapirijaimë	6,211	0	10	0	1.67	0	0	0.67	1	0
Convolvulaceae	*Maripa cf. glabra* Choisy	L	kuturiimë	6,278	3	0	0	0.33	0	0	0	0.33	0
Dichapetalaceae	*Dichapetalum rugosum* (Vahl) Prance	L	awarimë pitëkënë	6,026; 5,877; 6,094	1	0	0	0.67	0	0	0	0.67	0
Dilleniaceae	*Davilla kunthii* A. St.-Hil. Doliocarpus spp.	L	ënkomikomi	5,220;5,593; 6,182; 6,681; 6,379	1	0	0	1.67	0	0	0	1.67	0
Dilleniaceae	*Davilla nitida* (Vahl) Kubitzki	L	sakataetu	6,271	4	0	0	0.67	0	0	0	0.67	0
Elaeocarpaceae	*Sloanea* sp. *3*	T	iiwa	6,313	1	0	0	3	0	0.33	0.33	1	1.33

(continued)

Appendix 1 (continued)

Trio list

Family	Species	Habit	Vern. name	Coll #	# UP	# LO	# FA	UV_s	CON	$$$	EDI	MED	TEC
Elaeocarpaceae	*Sloanea* sp. 2	T	tëpaimë/ wiseimë 2	6,118	1	0	0	1.33	0	0	0	0.67	0.67
Euphorbiaceae	*Conceveiba guianensis* Aubl.	T	okomokë	5,539; 5,605; 5,996; 6,054	3	1	0	1.67	0	0	0	1.67	0
Euphorbiaceae	*Croton pullei* Lanj.	L	katamiïmë	6,053	2	1	0	4.67	0	0	0	4.67	0
Euphorbiaceae	*Drypetes variabilis* Uittien	T	pakoko	6,139; 6,190	2	0	0	2	0	0	0	0.67	1.33
Euphorbiaceae	*Hevea guianensis* Aubl.	T	ëwee	5,893	0	1	0	0.67	0	0	0	0.67	0
Euphorbiaceae	*Mabea cf. taquari* Aubl.	L	mëree asoso	5,835	0	1	0	1.67	0	0	0.67	1	0
Euphorbiaceae	*Omphalea diandra* L.	L	warikë	5,748	0	2	0	1.33	0	0	0.33	1	0
Euphorbiaceae	*cf. Pera decipiens* (Tul.) A. DC.	T	nono atï	6,252	1	0	0	1.67	0.67	0	0	0.67	0.33
Euphorbiaceae	*Sagotia racemosa* Baill.	T	akoha/manari repï	5,764; 5,843; 5,988	24	32	0	3.67	0	0.33	0	2.67	0.67
Fabaceae	*Alexa imperatricis* (R.H. chomb.) Baill.	T	kïnotokë	5,723; 5,913	24	60	2	2.67	0	0	0	1.67	1
Fabaceae	*Bauhinia cf. guianensis* Aubl.	L	mïrokoko ehe (tikorijan)	6,063	0	1	0	2	0	0	0	2	0
Fabaceae	*Bauhinia cupreonitens* Ducke	L	mïrokoko ehe (tamirin)	6,064; 6,239	7	2	0	2	0	0	0	2	0
Fabaceae	*Candolleodendron brachystachyum* (DC.) R.S. Cowan	T	kutari	5,952; 6,205	9	2	0	2.33	0.33	0	0	1.33	0.67
Fabaceae	*cf. Diplotropis* sp. *1*	T	ërukëu	6,269; 6,144	1	0	0	1.67	1	0	0	0.67	0
Fabaceae	*Copaifera guyanensis* Desf.	T	mapataru	6,283	1	0	0	2.67	0	0	0.33	1.67	0.67
Fabaceae	*Crudia aromatica* (Aubl.) Willd.	T	wapa	6,002; 6,111	16	0	0	1	0.33	0	0	0.33	0.33
Fabaceae	*Derris amazonica* Killip	L	inekuipë	6,644	0	1	0	0.67	0	0	0	0.67	0
Fabaceae	*Dialium guianense* (Aubl.) Sandwith	T	tokiriman	6,061	0	4	1	3	0	0	0.67	2	0.33

Fabaceae	*Dinizia excelsa* Ducke	T	awareimë	6,385; 6,691	1	0	1	2	0.33	0	0	1.67	0
Fabaceae	*Dioclea virgata* (Rich.) Amshoff	L	tamoko enu	5,502; 5,466	0	1	0	1.67	0	0	0.33	1.33	0
Fabaceae	*Dipteryx odorata* (Aubl.) Willd.	T	ooto	6,169	1	0	0	3.67	0.33	0	0	3.33	0
Fabaceae	*Elizabetha princeps* Schomburgk ex Benth.	T	kakaimë	6,119	24	0	0	1.67	0	0	0	1	0.67
Fabaceae	*Eperua jenmanii* Oliv.	T	pararan	5,538; 6,146	42	13	0	2.33	0.33	0	0	1.67	0.33
Fabaceae	*Inga alba* (Sw.) Willd.	T	karau, apurikui	5,532	0	0	3	4.67	0	0.33	0.67	1.33	2.33
Fabaceae	*Inga* sp. *13*	T	karau, arawata	5,665; 6,029	1	0	2	1.67	0	0	0.67	0.67	0.33
Fabaceae	*Inga cf. rubiginosa* (Rich.) DC.	T	karau, makui wahiriri	5,680; 5,998; 6,037; 6,236; 6,351	6	14	0	2	0	0	1	0.33	0.67
Fabaceae	*Inga cf. thibaudiana* DC.	T	karau, awiki	6,177; 6;324	3	0	10	1.67	0	0	1	0.33	0.33
Fabaceae	*Inga* sp. *11*	T	karau, parawa apëri	5,678; 5,694; 6,314; 6,336	0	18	1	1.67	0	0	1	0.33	0.33
Fabaceae	*Inga* sp. *10*	T	karau, pai aotï	5,776; 5,953; 6,143	2	0	0	1.67	0	0	1	0.33	0.33
Fabaceae	*Inga* sp. *12*	T	karau, pohtï	5,753; 5,837	0	1	0	2.33	0	0	1	0.67	0.67
Fabaceae	*Inga* sp. *3*	T	karau, arimi arokï	5,666; 5,778; 5,874; 6,310	0	17	0	2	0	0	0.67	1	0.33
Fabaceae	*Inga* sp. *7*	T	karau, kurami	5,679; 5,934; 6,676	0	1	0	2	0	0	1	0.33	0.67
Fabaceae	*Inga* sp. *8*	T	karau, kurija tëpu	6,312; 6,382	3	2	1	1.33	0	0	1	0	0.33
Fabaceae	*Inga* sp. *9*	T	karau, norou	6,074	1	0	0	2.33	0	0	1	1	0.33

(continued)

Appendix 1 (continued)

Trio list Family	Species	Habit	Vern. name	Coll #	# UP	# LO	# FA	UV_s	CON	$$$	EDI	MED	TEC
Fabaceae	*Inga* sp. *1*	T	karau, anakara	5,616; 5,720; 5,789	3	8	0	1.33	0	0	0	1.33	0
Fabaceae	*Machaerium* sp. *3*	L	mĭkakëjee 2	5,992; 6,391	4	0	0	1	0	0	0	1	0
Fabaceae	*Machaerium* sp. *4*	L	mĭkakëjee 1	5,833; 5,901	0	0	1	1	0	0	0	1	0
Fabaceae	*Macrolobium angustifolium* Benth. R.S. Cowan	T	paree	6,317	0	1	0	2.33	0	0	0	1.67	0.67
Fabaceae	*Macrolobium* sp. *1*	T	tamara	5,657	0	2	0	4	0.33	1	0.33	1.33	1.33
Fabaceae	*Mimosa* sp. *1*	L	pai tĭkĭtĭkĭkai	5,508; 5,876	0	1	0	0	0	0	0	0	0
Fabaceae	*Ormosia cf. coarctata* Jacks.	T	wëtëu	5,523	1	0	0	3.67	0	0.67	0	1.67	1.33
Fabaceae	*Parkia nitida* Miq.	T	siri apëri	6,160; 6,386	2	0	2	0.67	0	0	0	0.67	0
Fabaceae	*Parkia pendula* (Willd.) Benth. ex Walp.	T	kunkui iputu	6,187	1	0	0	0.67	0	0	0	0.67	0
Fabaceae	*Pseudopiptadenia psilostachya* (DC.) G. P. Lewis and M. P. Lima	T	arimi putu	6,189; 6,332	1	0	0	1.67	0	0	0	1	0.67
Fabaceae	*Pterocarpus santalinoides* L'Hér. ex DC.	T	wakapurimë	5,811	0	30	0	2.67	0.33	0	0	2	0.33
Fabaceae	*Senna quinquangulata* (Rich.) H.S. Irwin and Barneby	L	pokopoko uru	5,500; 6,685	0	1	0	1.33	0	0	0	1.33	0
Fabaceae	*Swartzia benthamiana* Miq.	T	kuijari empata	5,632; 5,945; 6,178	6	3	0	2.33	0	0	0	2	0.33
Fabaceae	*Swartzia panacoco (Aubl.) R.S. Cowan*	T	kurimao popeta	5,695; 6,210	0	2	0	3.33	0	0	0	2.67	0.67
Fabaceae	*Swartzia* sp. *3*	T	kinoirawa	5,644; 5,714	1	1	0	2	0	0	0	1.33	0.67
Fabaceae	*Tachigali paniculata* Aubl.	T	sireinje	6,268; 6,335	2	6	0	1	0.33	0	0	0.67	0
Fabaceae	*Tachigali* sp. *1*	T	arapani	6,136	1	0	0	1.33	0	0	0	1	0.33
Fabaceae	*Taralea oppositifolia* Aubl.	T	arimai	5,621; 6,130	0	9	0	1.33	0	0	0	1	0.33

Fabaceae	Unkn Fab 2	T	kanai iputu	No coll	0	0	1	0.67	0	0	0	0.67	0
Fabaceae	Unkn Fab 3	T	makurereu	6,217	0	1	2	1.33	0	0	0	1.33	0
Fabaceae	Unkn Fab 4	T	taari jaran	5,692	0	3	0	2.67	0.67	0	0	1.33	0.67
Fabaceae	Unkn Fab 5	T	puura	6,156	1	0	0	2.67	0	0	0	1.67	1
Fabaceae	Unkn Fab 1	T	mëree	6,300	0	1	0	2	0	0	0	1	1
Fabaceae	*Vouacapoua americana* Aubl.	T	wakapu	6,270; 6,274	22	0	0	2	0.33	0	0	0.67	1
Fabaceae	*Zygia cataractae* (Kunth) L. Rico	T	karau, jaran	5,719; 5,743; 5,796; 5,925	0	12	0	2	0	0	0	1.67	0.33
Humiriaceae	*Sacoglottis guianensis* Benth.	T	makararan, makara-imë	6,079; 6,145	2	1	0	1.33	0	0	0.67	0.67	0
Lamiaceae (Verbenaceae)	*Vitex compressa* Turcz.	T	manakaimë 3	6,006	0	1	0	3	0	0	0	2	1
Lamiaceae (Verbenaceae)	*Vitex stahelii* Moldenke	T	manakaimë 1	5,732	0	2	0	3	0	0	0	2	1
Lamiaceae (Verbenaceae)	*Vitex triflora* Vahl	T	manakaimë 2	5,620	0	1	0	3	0	0	0	2	1
Lauraceae	*Licaria cf. chrysophylla*	T	wai 1 (mono)	5,672; 5,779	0	3	0	2	0	0	0.33	1.33	0.33
Lauraceae	*Licaria* sp. *2*	T	wai 2 (pijasa)	6,076; 6,049	7	0	2	1.33	0.67	0	0	0.33	0.33
Lauraceae	*Licaria* sp. *4*	T	waija	5,830	1	2	0	1.67	0.67	0	0	0.33	0.67
Lecythidaceae	*Couratari guianensis* Aubl.	T	pono	No coll	1	0	1	3.67	0.33	0	0	2.33	1
Lecythidaceae	*Eschweilera coriacea* (DC.) S.A. Mori	T	meripono 2	6,164	0	2	0	2.33	0.33	0	0	1.33	0.67
Lecythidaceae	*Eschweilera corrugata* Poit.	T	aimara ewa (tamiren)	6,229	4	4	1	3	0.33	0.33	0	1	1.33
Lecythidaceae	*Eschweilera. corrugata Poit. var. 2*	T	aimara ewa (tikorijan)	6,363	1	27	0	3	0.33	0.33	0	1	1.33
Lecythidaceae	*Eschweilera decolorans* Sandwith	T	meripono 1	6,080	7	2	0	2.33	0.33	0	0	1.33	0.67

(continued)

Appendix 1 (continued)

Trio list

Family	Species	Habit	Vern. name	Coll #	# UP	# LO	# FA	UV_s	CON	$$$	EDI	MED	TEC
Lecythidaceae	*Eschweilera pedicellata* (Rich.) S.A. Mori	T	tuhaimë 1	5,774	0	8	0	3	0.67	0.5	0.33	1	0.67
Lecythidaceae	*Eschweilera sagotiana* Miers	T	watara	6,176	8	0	0	2.67	1	0	0	1.33	0.33
Lecythidaceae	*Gustavia hexapetala* (Aubl.) Sm.	T	arekore antura/ pakira antura	5,792	7	0	0	1	0.33	0	0	0.67	0
Lecythidaceae	*Lecythis poiteaui* O. Berg	T	ariwera	6,133	8	0	0	3	0.33	0	0.33	1.67	0.67
Lecythidaceae	*Lecythis zabucajo* Aubl.	T	turaran	5,512	1	1	0	2.67	0	0	0	2	0.67
Malpighiaceae	*Byrsonima stipulacea* A. Juss.	T	urutuma	5,944; 6,221	0	0	5	1	0	0	0	0.67	0.33
Malpighiaceae	*Diplopterys lucida* (Rich.) W.R. Anderson and C. Cav. Davis	L	karinama 3	5,852; 6,151	0	1	0	0	0	0	0	0	0
Malpighiaceae	*Stigmaphyllon cf. convolvulifolium* (Cav.) A. Juss.	L	matukru	5,205	1	0	0	1.33	0	0	0	1.33	0
Malvaceae	*Apeiba albiflora* Ducke	T	wïkapau waku	5,654	0	1	3	2	0	0.33	0	0.33	1.33
Malvaceae	*Apeiba petoumo* Aubl.	T	mïkëtë	6,167	2	0	12	1.33	0	0	0	0.67	0.67
Malvaceae	*Ceiba pentandra* (L.) Gaertn.	T	kumaka	6,320	0	1	0	1.33	0	0	0	1.33	0
Malvaceae	*cf. Herrenia* sp. *1*	T	kajakë ipana	5,851	0	1	0	1	0	0	0	0.67	0.33
Malvaceae	*Luhea alternifolia* (Mill.) Mabb.	T	pompokoi	5,741	1	0	0	1.67	0	0	0	1	0.67
Malvaceae	*Pachira aquatica* Aubl.	T	mekumpë	5,731	0	5	0	1	0	0	0	1	0
Malvaceae	*Quararibea guianensis* Aubl.	T	paaraimë	5,661; 5,823	0	29	0	1.33	0	0	0	1.33	0
Malvaceae	*Sterculia pruriens* (Aubl.) K. Schum.	T	pamahta	5,781; 6,265	3	4	0	0.67	0	0	0	0.67	0
Malvaceae	*Theobroma subincanum Mart.*	T	arikanama	5,757; 6,266	1	0	0	2	0	0	1	0.67	0.33
Melastomataceae	*Mouriri* sp. *2*	T	pasisiïmë	6,218	0	0	8	3	0	0	0	2	1

Meliaceae	*Carapa guianensis* Aubl.	T	karapa	6,264	4	1	0	3.33	1	0	0	2	0.33
Meliaceae	*Cedrela odorata* L.	T	simajae	6,279	0	1	1	3	1	0.33	0	1.33	0.33
Meliaceae	*Guarea cf. kunthiana* A. Juss.	T	karapaimë 1	5,721	10	2	0	1.33	0.67	0	0	0.67	0
Meliaceae	*Guarea guidonia* (L.) Sleumer	T	karapaimë 2	5,829; 6,070	0	4	0	1.33	0.67	0	0	0.67	0
Meliaceae	*Guarea* sp. *1*	T	mene	No coll	0	1	0	1.67	0	0	0	1.33	0.33
Meliaceae	*Trichilia cipo* (A. Juss.) C. DC.	T	aritaimë 1	5,795; 5,880; 6,674	1	0	0	1.33	0	0	0	0.67	0.67
Meliaceae	*Trichilia quadrijuga* Kunth	T	kuikëpën	5,550; 6,664	0	7	0	1.67	0.33	0	0	1	0.33
Meliaceae	*Trichilia* sp. 2	T	aritaimë 2	5,663; 5,826	0	4	0	1.33	0	0	0	0.67	0.67
Meliaceae	*Trichilia surinamensis* (Miq.) C. DC.	T	aritaimë 3	5,828; 5,855	0	14	0	1.33	0	0	0	0.67	0.67
Moraceae	*Bagassa guianensis* Aubl.	T	pakasa	6,244	0	0	2	2.33	0.33	0	0	0.67	1.33
Moraceae	*Brosimum lactescens* (S. Moore) C.C. Berg	T	ararawao	5,697	0	2	0	0.67	0	0	0	0.33	0.33
Moraceae	*Clarisia racemosa* Ruiz & Pav.	T	pii	6,135	1	0	0	2.67	0.33	0	0	1	1.33
Moraceae	*Ficus maxima* Mill.	T	ruwii	5,518	0	1	0	2	0	0	0	1.33	0.67
Moraceae	*Helicostylis* sp. *1*	T	pauraran	5,866; 6,247	1	0	0	2.33	0.33	0	0	0.67	1.33
Moraceae	*Unkn Mora 1*	T	paara	5,728; 6,067	1	1	0	0.67	0	0	0	0.67	0
Moraceae	*Brosimum rubescens* Taub.	T	sijë	6,121	1	1	0	2.67	0	0	0	1.67	1
Moraceae	*Unkn Mora 3*	T	mapanu	5,785; 6,254	1	0	0	2.33	0.67	0	0.67	0.67	0.33
Moraceae	*Helicostylis tomentosa (Poepp. & Endl.) Rusby*	T	pijapijaman	6,284; 6,296; 6,299	1	0	0	0	0	0	0	0	0
Moraceae	*Unkn Mora 5*	T	ararama	6,039	1	0	1	0.67	0.33	0	0	0.33	0
Myristicaceae	*cf. Iryanthera* sp. *1*	T	ponïkïimë	5,637; 5,991	2	0	0	2	0	0	0	1.67	0.33
Myristicaceae	*cf. Compsoneura* sp. *1*	T	waroro	6,186; 6,192	2	0	0	3.33	0.67	0.5	0	2	0.33
Myristicaceae	*Virola 2*	T	ponïkï	5,759; 6,107; 6,295	3	0	0	2	0	0	0	2	0
Myristicaceae	*Virola* sp. *1*	T	wiri	5,737	0	1	0	3.67	0.67	0.5	0	2.33	0.33

(continued)

Appendix 1 (continued)

Trio list

Family	Species	Habit	Vern. name	Coll #	# UP	# LO	# FA	UV_s	CON	$$$	EDI	MED	TEC
Myristicaceae	*Virola surinamensis* (Rol. ex ottb.) Warb.	T	wërëtë	6,159; 6,048	1	0	0	2.33	0.67	0	0	1.67	0
Myrtaceae	*Calycorectes bergii* Sandwith	T	mïroko enu	5,711; 6,005	5	0	0	2	0.33	0	0.33	0.67	0.67
Myrtaceae	*Campomanesia aromatica* (Aubl.) Griseb.	T	sokui	5,987; 6,082	2	0	0	1.33	0	0	0	0.67	0.67
Myrtaceae	*Eugenia coffeifolia* DC.	T	mokoko enu	5,710; 5,693	1	2	0	0.67	0	0	0	0.67	0
Myrtaceae	*Eugenia florida* DC.	T	ëkëimë	6,009	0	1	0	0.67	0	0	0	0.33	0.33
Myrtaceae	*Myrcia decorticans* DC.	T	paijan epu 2	6,046	1	0	1	1.67	0	0	0	1.33	0.33
Myrtaceae	*Myrcia minutifolia*	T	mokoko enu	5,930	0	1	0	0.67	0	0	0	0.67	0
Myrtaceae	*Myrciaria floribunda* (H. West ex Willd.) O. Berg	T	soroisoroi irepa	5,990	1	0	0	2.33	0.33	0	0	1	1
Myrtaceae	*Unkn Myrt 3*	T	saha	5,793	1	2	0	1.67	0.33	0	0	1	0.33
Myrtaceae	*Unkn Myrt 1*	T	koi jaran	6,315	0	1	0	0.67	0	0	0	0.67	0
Myrtaceae	*Unkn Myrt 2*	T	mami enu	5,696; 5,734	0	3	0	2	0.33	0	0	0.67	1
Nyctaginaceae	*cf. Neea* sp. *1*	T	kariwaimë	5,716	1	1	16	1.33	0	0	0	1.33	0
Ochnaceae	*Ouratea* sp. *1*	T	paripoimë 2	6,122	7	0	0	0.33	0	0	0	0.33	0
Olacaceae	*Minquartia guianensis* Aubl.	T	otopïmï	6,185	1	2	0	1.33	0.33	0	0	1	0
Olacaceae	*Ptychopetalum olacoides* Benth.	T	japoko	5,609	1	0	0	2	0	0	0	2	0
Phytolaccaceae	*Seigueria aculeata* Jacq.	L	kurumuri 1	5,525	0	4	0	1	0	0	0	1	0
Picramniaceae	*Picramnia guianensis* (Aubl.) Jans.-Jac.	T	taari	5,504; 5,815	0	1	0	3	0	0.33	0	1.67	1
Rubiaceae	*Psychotria* sp. *1*	T	pai empaha (sikiman)	5,825	0	2	0	1.33	0	0	0	1.33	0
Rubiaceae	*Uncaria guianensis* (Aubl.) J.F. Gmel.	L	pijana iroi	6,355	0	0	1	2.33	0	0	0	2.33	0
Rubiaceae	*Unkn 2*	T	mirimirimë	5,658	0	2	0	3	0.67	0	0	1	1.33
Rutaceae	*Rauia subtruncata* Steyerm.	T	erï pakoro	5,515	0	10	0	2.33	0.33	0	0	1.67	0.33

Rutaceae	*Zanthoxylum rhoifolium* Lam.	T	werekiï	5,926	0	0	2	2.33	0	0	0	1.33	1
Salicaceae	*Casearia grandiflora* Cambess.	T	maruihpë	6,381	0	0	21	1.67	0	0	0	1	0.67
Salicaceae	*Casearia pitumba* Sleumer	T	akuri ampiri	5,725; 6,311	0	2	0	1.33	0	0	0.33	1	0
Salicaceae	*Laetia procera* (Poepp.) Eichler	T	maru	5,960	2	0	21	1.67	0	0	0	1	0.67
Sapindaceae	*Cupania hirsute* Radlk.	T	jaranoimë	5,602; 6,042	0	0	1	0.67	0	0	0	0.33	0.33
Sapindaceae	*Cupania* sp. *1*	T	anaije	6,322	0	1	0	0.33	0	0	0	0	0.33
Sapindaceae	*Unkn Sapi 1*	T	tïkëkin	5,819; 6,679	0	2	0	1.33	0	0	0	1	0.33
Sapotaceae	*Chrysophyllum cuneifolium* (Rudge) A. DC.	T	ëmori	5,947; 6,275	2	1	0	2.33	0	0	0	2	0.33
Sapotaceae	*Manilkara cf. huberi* (Ducke) Chevalier	T	awari parata	5,674	0	5	0	2	0.33	0	0.67	0.33	0.67
Sapotaceae	*Micropholis cf. cayennensis* T.D. Penn.	T	tarairuimë	6,183	1	0	0	2	0.67	0.5	0	0.67	0.33
Sapotaceae	*Micropholis* sp. *1*	T	kuusa	6,129	2	0	0	0.33	0	0	0.33	0	0
Sapotaceae	*Micropholis* sp. *2*	T	wapuri jaran	5,824	0	3	0	2.33	0.67	0	1	0.67	0
Sapotaceae	*Pouteria sagotiana* (Baill.) Eyma	T	tumuri	6,161	1	0	0	3	1	0	0	0.33	1.67
Sapotaceae	*Pouteria* sp. *10*	T	rapopimë	5,537	2	0	0	2	0	0	0.67	1.33	0
Sapotaceae	*Pouteria* sp. *2*	T	wapuri	6,056; 6,307	0	3	0	1.67	0.33	0	0.33	0.67	0.33
Sapotaceae	*Pouteria* sp. *4*	T	kunumïimë	5,783; 5,905	2	0	0	1.67	0.33	0	0.67	0.67	0
Sapotaceae	*Pouteria* sp. *5*	T	moweimë	5,817	0	1	0	3	0	0	0.67	1.33	1
Sapotaceae	*Pouteria* sp. *6*	T	waki	5,738; 6,191	1	1	0	1.33	0	0	0	0.67	0.67
Sapotaceae	*Pouteria* sp. *8*	T	mompë	5,733; 6,215	4	2	0	2	0	0	0.33	1	0.67
Sapotaceae	*Pouteria* sp. *9*	T	tamareimë	5,907; 6,116	1	0	0	1	0	0	0	1	0
Sapotaceae	*Unkn Sapo 1*	T	pakira auku	6,175	8	0	0	1	0	0	0	0.67	0.33
Sapotaceae	*Unkn Sapo 2*	T	ëmorijatëimë	6,163	1	0	0	0.67	0	0	0.67	0	0
Simaroubaceae	*Quassia cedron* L.	T	pai emu	5,810	0	1	0	1.33	0.33	0	0	1	0
Siparunaceae	*Siparuna* sp. *1*	T	irakë epu	5,939; 5,941; 6,232	1	1	1	2	0	0	0	2	0

(continued)

Appendix 1 (continued)

Trio list

Family	Species	Habit	Vern. name	Coll #	# UP	# LO	# FA	UV_s	CON	$$$	EDI	MED	TEC
Indet 1	Indet 1	T	aritaimë 4	5,848	0	1	0	1.33	0	0	0	0.67	0.67
Indet 2	Indet 2	T	nere uru	5,787; 6,032	1	2	0	1.33	0	0	0	1	0.33
Indet 3	Indet 3	T	–	6,039	1	0	0	0	0	0	0	0	0
Urticaceae	*Cecropia obtusa* Trécul	T	kurere	6,219	2	0	69	3.67	0	0	0	3.67	0
Urticaceae	*Cecropia sciadophylla* Mart.	T	kurere, anatara	6,224; 5,950	2	0	72	2.33	0	0	0	2.33	0
Urticaceae	*Cecropia* sp. *3*	T	kurere, ijuana	5,951	0	0	9	1.67	0	0	0	1.67	0
Urticaceae	*Pourouma* sp. *1*	T	puruma, arawata	6,683	1	0	0	1	0	0	1	0	0
Urticaceae	*Pourouma* sp. *2*	T	puruma, moi emu	6,041; 6,170	9	0	0	1.33	0.33	0	1	0	0
Urticaceae	*Pourouma* sp. *4*	T	puruma, wirinai enu	6,040; 6,250	1	0	1	1.33	0.33	0	1	0	0
Violaceae	*Rinorea pubiflora* (Benth.) Sprague and Sandwith	T	kurunje	5,503	0	1	0	4	0	0	0	3.33	0.67
Total					539	504	444	454	46.3	12.2	41	254	102

Appendix 2 Stonhuku (Saramacca Maroon) Plant List: Alphabetical by Scientific Name

Woody families and species in plots at the Stonhuku (Saramacca Maroon) field site (≥10 cm dbh) – including habit (*T* tree, *L* liana), vernacular names, collection numbers (all B. Hoffman), species abundance within three vegetation zones (*UP* non-flooded, *LO* seasonally flooded, *FA* fallow), total *use value* (UV_s), and *use value* per use category (*CON* construction, *$$$* cash/trade, *EDI* edible, *MED* medicinal/ritual, *TEC* technology)

Appendix 2

Saramacca list

Family	Species	Habit	Vern. name	Coll #	# UP	# LO	# FA	UV_s	CON	$$$	EDI	MED	TEC
Anacardiaceae	*Loxopterygium sagotii* Hook. f.	T	sángi udu	6,883	1	0	19	2	0.5	0.25	0	0	1.25
Annonaceae	*Anaxagorea acuminata* (Dunal) A.DC.	T	alukutú (matu)	6,533; 6,624; 6,721	1	1	0	0.75	0	0	0	0	0.75
Annonaceae	*Anaxagorea* sp. *1*	T	azau udu	6,566	1	0	0	1	1	0	0	0	0
Annonaceae	*Bocageopsis multiflora* (Mart.) R.E. Fr.	T	finu uwíi	6,438; 6,698; 6,873	5	0	0	1	0	0	0	0	1
Annonaceae	*Duguetia cf.* sp. *1*	T	watangóso	6,516	1	0	0	0.25	0	0	0	0	0.25
Annonaceae	*Duguetia paraensis* R.E. Fr.	T	akulí anza	6,506; 5,966	2	0	0	2.25	1	0	0	0.25	1
Annonaceae	*Ephedranthus guianensis* R.E. Fr.	T	SA- mömbë; TR- mekoro wewe	6,463	3	0	0	1.75	0	0	0	1.75	0
Annonaceae	*Fusaea cf. longifolia* (Aubl.) Saff.	T	kaká pau	6,613	0	1	0	1.25	1	0	0	0	0.25
Annonaceae	*Guatteria atra* Sandwith; Guatteria punctata (Aubl.) R.A. Howard	T	baáka pau	5,968; 6,477	3	0	12	1.75	1	0	0	0.25	0.5
Annonaceae	*Xylopia cayennensis* Maas	T	agámokáma1	6,347	0	0	3	2.25	0.25	0.25	0	1.25	0.5
Apocynaceae	*Ambelania acida* Aubl.	T	mambaái	6,396; 6,888	0	0	1	2	0	0	1	0	1
Apocynaceae	*Aspidosperma marcgravianum* Woodson	T	apokíta (wéti)	5,545	1	0	0	3.5	0	0.75	0	0.75	2
Apocynaceae	*cf. Couma guianensis* Aubl.	T	unkn	6,630	2	2	0	0	0	0	0	0	0
Apocynaceae	*Geisospermum argenteum* Woodson	T	leletí	5,591	13	0	1	1.25	0	0	0	1	0.25

Apocynaceae	*Apocynaceae 1*	T	tjɔntjutju	6,511; 6,550; 6,562	7	1	0	3.75	0.5	0.5	0	0.75	2
Apocynaceae	*Tabernaemontana undulata* Vahl	T	seki bita	6,519	1	0	0	1	0	0	0	1	0
Araliaceae	*Schefflera morototoni* (Aubl.) Maguire, Steyerm. and Frodin	T	atapíí	6,149	0	0	2	0.75	0	0	0	0.5	0.25
Arecaceae	*Astrocaryum aculeatum* G. Mey.	T	awaá	5,596	0	0	1	2.5	0	0	1.25	0	1.25
Arecaceae	*Astrocaryum sciophilum* (Miq.) Pulle	T	mumúu/maka	5,798	194	42	3	5.25	1	0	2	1	1.25
Arecaceae	*Attalea maripa* (Aubl.) Mart.	T	maipá	No coll	0	1	12	6	1	0	2	1	2
Arecaceae	*Euterpe oleracea* (Mart.)	T	pina	5,973	0	6	1	5.75	1.75	0	1.75	1	1.25
Arecaceae	*Oenocarpus bacaba* Mart.	T	tjangá	No coll	7	0	0	2	0.75	0	0.75	0	0.5
Arecaceae	*Socratea exorrhiza* (Mart.) H. Wendl.	T	pɛzáá	6,916	0	1	0	3	1.5	0	0	0.25	1.25
Bignoniaceae	*Adenocalymna inundatum* Mart. ex DC.	L	gɛán háti	6,717	0	4	0	0.25	0	0	0	0	0.25
Bignoniaceae	*Jacaranda copaia* (Aubl.) D. Don	T	já'fi	5,978	0	1	90	1.5	0.75	0	0	0.75	0
Bignoniaceae	*Memora flaviflora (Miq.) Pulle*	L	gaán háti unkn	6,493; 6,113	1	0	0	0	0	0	0	0	0
Bignoniaceae	*Tabebuia insignis* (Miq.) Sandwith	T	pantáá	5,775	0	1	0	0.75	0	0	0	0.75	0
Bignoniaceae	*Tabebuia serratifolia* (Vahl) G. Nicholson	T	gianti	6,724	0	1	0	2.5	0.5	0.25	0	1.75	0
Boraginaceae	*Cordia cf. tetandra*	T	tji-tji/túmba lobi	6,769	2	0	1	0.5	0	0	0	0	0.5
Burseraceae	*Protium morii* D.C. Daly	T	kandéa pau	5,740	2	0	0	0.75	0	0	0	0	0.75

(continued)

Appendix 2 (continued)

Saramacca list Family	Species	Habit	Vern. name	Coll #	# UP	# LO	# FA	UV_s	CON	$$$	EDI	MED	TEC
Burseraceae	*Protium* sp. *1*	T	unkn	6,546	1	0	0	0	0	0	0	0	0
Burseraceae	*Tetragastris altissima* (Aubl.) Swart	T	sáli	6,645	13	3	1	2.5	1	0.25	0.25	0.25	0.75
Cannabaceae	*Trema micrantha* (L.) Blume	T	kapeëë pau	6,429; 6,872	0	0	15	2	0	0	0	1	1
Celastraceae	*Cheiloclinium* sp. *1*	L	unkn	6,577	2	0	0	0	0	0	0	0	0
Celastraceae	*Goupia glabra* Aubl.	T	okó pepe	5,569	1	0	0	0	0	0	0	0	0
Celastraceae	*Salacia* sp. *1*	L	pongo tatái	6,620	1	2	0	0.75	0	0	0	0.75	0
Chrysobalanaceae	*cf. Licania* sp. *1*	T	kóóko 1	6,553	3	0	0	1	0.75	0	0	0.25	0
Chrysobalanaceae	*cf. Licania* sp. *2*	T	kóóko 2	6,554	3	0	0	1	0.75	0	0	0.25	0
Chrysobalanaceae	*Licania hypoleuca* Benth.	T	matu basöö	6,618	0	2	0	1.25	0	0	0	0.5	0.75
Chrysobalanaceae	*Licania micrantha* Miq.	T	tjawáá	6,486; 6,504	7	0	1	1	0	0	0	0	1
Clusiaceae	*Clusia grandiflora* Splitg. Clusia palmicida Rich.	L	katu	6,800; 6,812	4	1	0	0	0	0	0	0	0
Clusiaceae	*Tovomita* sp. *1*	T	mánni pau	6,490; 6,585	2	2	0	0.5	0.25	0	0	0	0.25
Clusiaceae	*Vismia* sp. *1*	T	píndjá páu (wéti)	6,017	0	0	13	2.25	0.75	0	0	1.5	0
Combretaceae	*Combretum* sp. 1	L	unkn	6,631	0	1	0	0	0	0	0	0	0
Combretaceae	*Terminalia* sp. *1*	T	anango súti	6,930	1	0	0	0.75	0	0	0	0.75	0
Dilleniaceae	*Doliocarpus dentatus* (Aubl.) Standl.	L	faja tatai	6,795	5	0	0	1	0	0	0.5	0.5	0
Euphorbiaceae	*Euph* sp. *1*	T	jóngo	6,601	0	0	2	1.25	1	0	0	0.25	0
Euphorbiaceae	*Conceveiba guianensis* Aubl.	T	agbagómatjáu pau	6,887	3	0	0	0.25	0	0	0	0	0.25
Euphorbiaceae	*Croton pullei* Lanj.	L	ateskoio tatái	6,622	0	1	0	0	0	0	0	0	0
Euphorbiaceae	*Croton* sp. *1*	T	ateskoio pau	6,814	0	2	12	0.25	0.25	0	0	0	0
Euphorbiaceae	*Euph* sp. *1*	T	unkn	6,569	2	0	0	0	0	0	0	0	0

Euphorbiaceae	*Mabea* sp. *1*	T	tjukunda	6,499	1	0	0	1	0	0	0	1	0
Fabaceae	*Alexa imperatricis* (R.H. Schomb.) Baill.	T	ndekú udu	5,751	0	23	0	1.5	0	0	0	1.5	0
Fabaceae	*Bauhinia outimouta* Aubl.	L	logososikáda (bë)	6,920	2	0	0	1	0	0	0	1	0
Fabaceae	*Bauhinia guianensis* Aubl.	L	logososikáda (wéti)	5,274; 5,597	0	1	0	1	0	0	0	1	0
Fabaceae	*Crudia aromatica* (Aubl.) Willd.	T	gawenti baka, unkn	6,611	0	18	0	0	0	0	0	0	0
Fabaceae	*Derris amazonica* Killip	L	mɨndeku (bë, mujeëë)	6,644	8	0	0	0.75	0	0	0	0.75	0
Fabaceae	*Dicorynia guianensis* Amshoff	T	síndjápéétu	5,556; 6,397	15	3	2	3.75	1	1	0	0.25	1.5
Fabaceae	*Diplotropis purpurea* (Rich.) Amshoff	T	tjɛbísi, baáka	6,575	1	0	0	4	1.25	1	0	0	1.75
Fabaceae	*Elizabetha cf. princeps* Schomburgk ex Benth.	T	líobandja	6,635; 6,794	0	23	0	1	0	0	0	0	1
Fabaceae	*Eperua falcata* Aubl.	T	bɨ́ɨ́du	6,522	31	80	0	4	1.5	1.25	0	0.5	0.75
Fabaceae	*Hymenolobium* sp. *1*	T	kɛdjú sii	6,709	1	1	0	0.25	0	0	0	0	0.25
Fabaceae	*Inga* sp. *1*	T	watjí, abonkíni	6,809; 6,880	0	0	5	1.5	0	0	0	1.25	0.25
Fabaceae	*Inga* sp. *2*	T	watjí, apukú	6,411; 6,439; 6,619	0	0	2	2	0	0	0	2	0
Fabaceae	*Inga* sp. *3*	T	watjí 1	6,345; 6,917	6	7	0	1.25	0	0	1	0	0.25
Fabaceae	*Inga* sp. *4*	T	watji 2	6,605; 6,632; 6,636	3	6	0	1.25	0	0	1	0	0.25
Fabaceae	*Inga* sp. *5*	T	watjí 3	5,522; 6,867	0	0	6	1.25	0	0	1	0	0.25
Fabaceae	*Inga* sp. *6*	T	watjí, séépi	6,580; 6,792	0	2	0	1	0	0	0	0.75	0.25
Fabaceae	*Inga cf. stipularis*	T	watjí, matu	6,528	1	0	0	0	0	0	0	0	0
Fabaceae	*Machaerium* sp. *1*	L	atɛ́ tatái (paata)	6,591	4	3	0	0.25	0	0	0	0.25	0

(continued)

Appendix 2 (continued)

Saramacca list

Family	Species	Habit	Vern. name	Coll #	# UP	# LO	# FA	UV_s	CON	$$$	EDI	MED	TEC
Fabaceae	*Macrolobium angustifolium* Benth. R.S. Cowan	T	watapánu	6,623	0	2	0	0	0	0	0	0	0
Fabaceae	*Mimosa* sp. *1*	L	akamáka tatái (bë)	6,655	2	1	0	1	0	0	0	1	0
Fabaceae	*Ormosia paraensis* Ducke	T	ógi pau (baáka)	6,523	3	0	0	2.75	0	0	0	1.75	1
Fabaceae	*Parkia cf. pendula*	T	asau (bë)/agrobigi	6,704	3	0	0	2.25	0.5	0	0	1.75	0
Fabaceae	*Pterocarpus* cf. *santalinoides* L'Hér. ex DC.	T	musansi	No coll	0	1	0	1	0	0.75	0	0	0.25
Fabaceae	*Pterocarpus* officinalis Jacq.	T	gwegwe	6,610	0	4	0	1.75	0.75	1	0	0	0
Fabaceae	*Swartzia benthamiana* Miq.	T	waiju (kaká búku)	6,539; 6,615	9	9	0	0.25	0	0	0	0	0.25
Fabaceae	*Swartzia schomburgkii* Benth.	T	búgubúgu	6,606	15	20	1	2	0	0	0	1	1
Fabaceae	*Tachigali albiflora* (Benoist) Zarucchi and Herend.	T	djedu	6,886	0	0	1	1	0	0	0	0	1
Fabaceae	*Tachigali guianensis* (Benth.) Zarucchi and Herend.	T	djedu (fini uwii)	6,649	0	2	0	1	0	0	0	0	1
Fabaceae	*Taralea* sp. 2	T	sándu pau (mujëë)	6,531	2	0	5	3	0.75	1	0	0	1.25
Fabaceae	*Taralea* sp. *1*	T	sándu pau (wómi)	6,617	0	1	0	3	0.75	1	0	0	1.25
Fabaceae	*Vouacapoua americana* Aubl.	T	boánti	6,642	16	3	0	2.25	1.25	0.25	0	0	0.75
Hugoniaceae	*Hebepetalum* sp. *1*	T	kwasíkáánu subí	6,571	1	2	0	0	0	0	0	0	0
Humiriaceae	*cf. Sacoglottis guianensis* Benth.	T	kóóko 3	6,480	4	0	0	1	0.75	0	0	0.25	0

Lauraceae	*Aniba taubertiana* Mez	T	apɛsí 2	6,513	5	0	0	2.75	1.25	1	0	0	0.5
Lauraceae	*Laur1*	T	apɛsí 1	6,491; 6,514; 6,530	5	0	0	2.25	1.25	0.75	0	0	0.25
Lauraceae	*Nectandra* sp. *1*	T	apɛsí 3	6,552	15	1	0	2.25	1.25	0.5	0	0	0.5
Lecythidaceae	*cf. Eschweilera* sp. *1*	T	unkn	6,657	1	0	0	0	0	0	0	0	0
Lecythidaceae	*Couratari stellata* A.C. Sm.	T	djuumú	6,452	5	0	3	3.75	1	0.75	0	0.5	1.5
Lecythidaceae	*Gustavia hexapetala* (Aubl.) Sm.	T	kwatíi	6,652	0	9	0	1.75	0.5	0	0	0.75	0.5
Lecythidaceae	*Lecythis poiteaui* O. Berg	T	gosó	6,567	10	0	0	1.25	0	0	0	0	1.25
Lecythidaceae	*Lecythis corrugata* Poit.; *L. coriacea* (DC.) S.A. Mori	T	baakaláka	6,196; 6,865	13	34	1	3	1	0.75	0.5	0	0.75
Lecythidaceae	*Lecythis zabucajo* Aubl.	T	zéntete	6,641	2	2	0	1	0	0	1	0	0
Loganiaceae	*Antonia ovata* Pohl	T	liká pau	6,555	1	1	0	0	0	0	0	0	0
Malvaceae	*Apeiba albiflora* Aubl.	T	kasába pau	6,893	1	0	3	1	1	0	0	0	0
Malvaceae	*Eriotheca macrophylla* (K. Schum.) A. Robyns	T	azɔnzé	6,637	0	1	0	0.75	0	0	0	0.25	0.5
Malvaceae	*Sterculia pruriens* (Aubl.) K. Schum.	T	kobé	6,529; 6,738	4	2	0	0.25	0	0.25	0	0	0
Marcgraviaceae	*Norantea guianensis* Aubl.	L	kau (tranga)	5,649	0	1	0	0	0	0	0	0	0
Melastomataceae	*Mela* sp. *1*	T	kódjítánda	6,579	0	1	0	0.75	0.25	0	0	0.25	0.25
Melastomataceae	*Henriettia* sp. 2	T	mispel (bë awe)	6,882	0	0	94	1.5	0.5	0	0.25	0	0.75
Meliaceae	*Carapa guianensis* Aubl.	T	kaapá	6,114	1	1	0	4.25	0.75	1	0	1.25	1.25
Meliaceae	*Guarea gomma* Pulle	T	kujaké fuúta	6,561	8	7	0	1.5	1	0	0	0.5	0
Meliaceae	*Guarea scabra* A. Juss.	T	jembuka	6,466	5	2	0	0	0	0	0	0	0
Meliaceae	*Trichilia septentrionalis* C. DC.	T	kujaké fuúta (wómi)	6,556	1	0	0	1.5	1	0	0	0.5	0

(continued)

Appendix 2 (continued)

Saramacca list

Family	Species	Habit	Vern. name	Coll #	# UP	# LO	# FA	UV_s	CON	$$$	EDI	MED	TEC
Moraceae	*Brosimum lactescens* (S. Moore) C.C. Berg	T	jijiji pau, leterhout	6,786	0	2	0	2	0	1	0	0	1
Moraceae	*Brosimum parinarioides* Ducke ssp. parinarinoides	T	sitopu oli	6,497	1	0	0	0	0	0	0	0	0
Moraceae	*Trymatococcus amazonicus Poepp. & Endl.*	T	jijiji pau, púu fínga	6,541	1	1	0	0	0	0	0	0	0
Moraceae	*Ficus* sp. *1*	T	unkn	No coll	1	0	0	0	0	0	0	0	0
Moraceae	*Pseudolmedia laevis* (Ruiz & Pav.) J.F. Macbr.	T	púu fínga	6,495	20	2	0	0	0	0	0	0	0
Myristicaceae	*Virola* sp. *1*	T	bómba	6,471	4	2	0	1.75	0	0.25	0	1.5	0
Myrtaceae	*Eugenia coffeifolia* DC.	T	nõunõu duumí	6,512	2	5	0	0	0	0	0	0	0
Myrtaceae	*Eugenia patrisii* Vahl	T	gulan tjango	6,503; 6,515; 6,540	5	0	0	1.75	0	0	0	1	0.75
Myrtaceae	*Eugenia* sp. *1*	T	logosofuúta	6,639	2	6	0	1.25	0.25	0	0	0.75	0.25
Myrtaceae	*Myrciaria floribunda* (H. West ex Willd.) O. Berg	T	gujába, matu	6,784	0	2	0	0.75	0.25	0	0	0	0.5
Myrtaceae	*Myrcia* sp. *1*	T	kwépi	6,729	2	0	0	0.75	0	0	0	0	0.75
Olacaceae	*Minquartia guianensis* Aubl.	T	bagibagi	6,526	0	3	0	2	1.25	0	0	0.75	0
Quiinaceae	*Lacunaria crenata* (Tul.) A.C. Sm.	T	unkn	6,494	1	0	0	0	0	0	0	0	0
Rubiaceae	*Duroia cf. eriopila* L. f.	T	maamaadósu	5,932	0	0	1	2	0	0	0.75	1.25	1.25
Rubiaceae	*Duroia micrantha* (Ladbr.) Zarucchi and J.H. Kirkbr.	T	siíndja udu	5,611	0	1	0	2	0	0.75	0	0	1.25

Rubiaceae	*Isertia coccinea* (Aubl.) Vahl	T	párɛgapánga	6,305	0	0	77	0.5	0	0	0	0	0.5
Rubiaceae	*Rubi* sp. *1*	T	unkn	6,824	0	1	0	0	0	0	0	0	0
Salicaceae	*Laetia procera* (Poepp.) Eichler	T	agɛmokáma1	5,600	2	1	2	1.75	0.75	0	0	0.25	0.75
Sapindaceae	*Cupania hirsuta* Radlk.	T	gaɔventí	6,876	0	4	0	1.5	0	0	0	1.5	0
Sapindaceae	*Cupania scrobiculata* Rich.	T	gaɔventí (baáka)	6,458; 6,478	7	2	0	1	0	0	0	1	0
Sapindaceae	*Talisia megaphylla* Sagot ex Radlk.	T	azɔbɛne	6,508	3	0	0	1.25	0	0	1	0.25	0
Sapindaceae	*Talisia* sp. *1*	T	gaɔventí (bë)	6,602	0	1	0	0	0	0	0	0	0
Sapotaceae	*Pouteria* sp. *1*	T	kwátabóbi	6,517	7	1	0	1.75	0.25	0	1	0.5	0
Sapotaceae	*Micropholis* sp. *4*	T	íntɔóbi	6,790	0	3	0	3.75	0.75	1	0.75	0	1.25
Simaroubaceae	*Simarouba amara* Aubl.	T	asɪmaípa	6,821	2	0	0	2.25	1	0.5	0	0	0.75
Siparunaceae	*Siparuna guianensis* Aubl.	T	fëbë pau	6,563	1	0	0	1.25	0	0	0	1.25	0
Unkn 1–4	Unkn 1–4	T	unkn	6,527	1	0	0	0	0	0	0	0	0
Unkn 2–2	Unkn 2–2	T	unkn	6,603	0	1	0	0	0	0	0	0	0
Unkn 2–3	Unkn 2–3	T	unkn	6,600	0	1	0	0	0	0	0	0	0
Unkn 2–5	Unkn 2–5	T	unkn	6,612	0	1	0	0	0	0	0	0	0
Uncollected	Unkns (7 taxa)	T/L	unkns	N/A	3	1	3	0	0	0	0	0	0
Urticaceae	*Cecropia peltata* L.	T	pɛnpantíi (mujëë)	5,652	1	4	10	1.25	0	0	0	0.75	0.5
Urticaceae	*Cecropia sciadophylla Mart.*	T	pɛnpantíi (wómi)	5,651	0	0	10	1.25	0	0	0	1	0.25
Urticaceae	*Pourouma minor* Benoist	T	akɛdam suta	5,911; 6,587	3	1	0	2	0.75	0.25	0	0	1
Urticaceae	*Pourouma mollis* Trécul or *Pourouma vellutina* Mart.	T	pɛnpantíi (bouwii)	5,911; 6,482	2	0	1	1.25	0.25	0	0	0	1
Violaceae	*Leonia* sp. *1*	T	awaá pau	6,521	1	3	0	1.25	0.75	0	0	0.5	0

(continued)

Appendix 2 (cɔntinued)

Saramacca list

Family	Species	Habit	Vern. name	Coll #	# UP	# LO	# FA	UV_s	CON	$$$	EDI	MED	TEC
Violaceae	*Paypayrola guianensis* Aubl.	T	físibía	6,564	0	12	0	1.75	0	0	0	0.5	1.25
Violaceae	*Rinorea* sp. *1*	T	wéti udu 2	6,524	7	0	0	0.25	0	0	0	0	0.25
Vochysiaceae	*Qualea cf. rosea* Aubl.	T	gúún foló	6,584	1	4	0	0.75	0.75	0	0	0	0
Vochysiaceae	*Qualea* sp. *1*	T	wéti udu 1	6,638	0	20	0	0.25	0	0	0	0	0.25
Vochysiaceae	*Vochysia guianensis* Aubl.	T	kwáíi	6,592	0	1	0	2	1	0	0	0.5	0.5
Vochysiaceae	*Vochysia* sp. *1*	T	alánkupi	6,906	1	2	0	2.25	1	1	0	0	0.25
Total					581	438	417	195	45.5	19.3	17.8	53.3	60.5

References

Alcorn JB (1993) Indigenous peoples and conservation. Conserv Biol 7:424–426

Amiguet VT, Arnason JT, Maquin P, Cal V, Sánchez-Vindas P, Alvarez LP (2005) A regression analysis of Q'eqchi' Maya medicinal plants from southern Belize. Econ Bot 60:24–38

Anderson AB (1990) Extraction and forest management by rural inhabitants in the Amazon estuary. In: Anderson AB (ed) Alternatives to deforestation: steps toward sustainable use of the Amazon rain forest. Colombia University Press, New York, pp 1–50

Arends J (2002) The history of the Surinamese creoles I – a sociohistorical survey. In: Carlin EB, Arends J (eds) Atlas of the languages of Suriname. KITLV Press, Leiden, pp 115–130

Balée W (1987) A etnobotanica quantitativa dos Índios Tembé (Rio Gurupi, Pará). Boletim do Museu Paraense Emilio Goeldi 3:29–50

Balée W (1994a) Footprints of the forest: Ka'apor ethnobotany – the historical ecology of plant utilization by an Amazonian people. Columbia University Press, New York

Balée W (1994b) Toward a comparative ethnobotany of lowland South America. In: Footprints of the forest: Ka'apor ethnobotany – the historical ecology of plant utilization by an Amazonian people. Columbia University Press, New York, pp 204–223

Balick MJ (ed) (1988) The palm, tree of life: biology, utilisation and conservation. The New York Botanical Garden Press, Bronx

Basso EB (ed) (1977) Cariban-speaking Indians: culture, society and language. University of Arizona Press, Tucson

Bennett BC, Husby CE (2008) Patterns of medicinal plant use: an examination of the Ecuadorian Shuar medicinal flora using contingency tables and binomial analyses. J Ethnopharmacol 116:422–430

Berlin B (1992) Ethnobiological classification: principles of categorization of plants and animals in traditional societies. Princeton University Press, Princeton

Berry PE, Huber O, Holst BK (1995) Floristic analysis and phytogeography. In: Berry PE, Holst BK, Yatskievych K (eds) Flora of the Venezuelan Guayana. Missouri Botanical Garden, St. Louis, pp 161–191

Boom BM (1987) Ethnobotany of the Chácobo Indians, Beni, Bolivia. In: Advances in economic botany, vol 4. The New York Botanical Garden Press, Bronx

Boom BM (1990) Useful plants of the Panare Indians of the Venezuelan Guayana. In: Prance GT, Balick MJ (eds) Advances in economic botany, vol 8. The New York Botanical Garden Press, Bronx, pp 57–65

Brokenshaw DW, Warren D, Werner O (1978) Indigenous knowledge systems and development. University Press of America, Washington, DC

Byg A, Balslev H (2001) Diversity and use of palms in Zahamena, eastern Madagascar. Biodivers Conserv 10:951–970

Campos MT, Ehringhaus C (2003) Plant virtues are in the eyes of the beholders: a comparison of known palm uses among indigenous and folk communities of southwestern Amazonia. Econ Bot 57:324–344

Caniago I, Siebert SF (1998) Medicinal plant ecology, knowledge and conservation in Kalimantan, Indonesia. Econ Bot 52:229–250

Carlin EB (2004) The Trio – Chapter 1. In: A grammar of Trio: a Cariban language of Suriname. Peter Lang, Frankfurt am Main, pp 1–43

Castaneda H, Stepp JR (2008) Ethnoecological importance value (EIV) methodology: assessing the cultural importance of ecosystems as sources of useful plants for the Guaymi people of Costa Rica. Ethnobot Res Appl 5:249

Census (2007) Suriname. The seventh population and housing census. Stichting Algemeen Bureau voor de Statistiek in Suriname. http://www.statistics-suriname.org. Accessed Sept 25 2007

Clay JW (1988) Indigenous peoples and tropical forests: models of land use and management from Latin America. Cultural survival report no. 27. Cultural Survival, Inc., Cambridge, MA

Cocks M (2006) Biocultural diversity: moving beyond the realm of 'indigenous' and 'local' people. Hum Ecol 34:185–200

Collard IF, Foley RA (2002) Latitudinal patterns and environmental determinants of recent human cultural diversity: do humans follow biogeographical rules? Evol Ecol Res 4:371–383

Conklin HC (1967) The relation of Hanunoo culture to the plant world. Ph.D. dissertation, University Microfilms, Ann Arbor, MI

Cook FEM, Prendergast HDV (1995) Economic botany data collection standard. Royal Botanic Gardens, Kew

Da Cunha L, De Albuquerque UP (2006) Quantitative ethnobotany in an Atlantic forest fragment of northeastern Brazil – implications to conservation. Environ Monit Assess 114:1–25

Daly DC, Mitchell JD (2000) Lowland vegetation of tropical South America – an overview. In: Lentz D (ed) Imperfect balance: landscape transformations in the pre-Colombian Americas. Columbia University Press, New York

de Albuquerque UP, de Lucena RFP (2005) Can apparency affect the use of plants by local people in tropical forests? Interciencia 30:506–511

de Albuquerque UP, Andrade LHC, de Silva ACO (2005) Use of plant resources in a seasonal dry forest (Northeastern Brazil). Acta Bot Bras 19:27–38

de Granville J (1988) Phytogeographical characteristics of the Guianan forests. Taxon 37:578–594

Etkin N (2002) Local knowledge of biotic diversity and its conservation in rural Hausaland, northern Nigeria. Econ Bot 56:73–88

FAO (2006) Global forest resources assessment 2005. FAO forestry paper 147. FAO, Rome

Galeano G (2000) Forest use at the Pacific Coast of Chocó, Colombia: a quantitative approach. Econ Bot 54:358–376

Grenand P (1992) The use and cultural significance of the secondary forest among the Wayapi Indians. In: Plotkin MJ, Famolare L (eds) Sustainable harvest and marketing of rain forest products. Island Press, Washington, DC, pp 27–40

Grenand P, Moretti C, Jacquemin H (1987) Pharmacopées traditionneles en Guyane: Créoles, Paliku, Wayapi. ORSTOM (Office de la Recherche Scientifique et Technique d'Outre-Mer), Paris

Halme KJ, Bodmer RE (2007) Correspondence between scientific and traditional ecological knowledge: rain forest classification by the non-indigenous Riberenos in Peruvian Amazonia. Biodivers Conserv 16:1785–1801

Hammond DS (2005a) Ancient land in a modern world. In: Hammond DS (ed) Tropical forests of the Guiana Shield: ancient forests of the modern world. CABI Publishing, Cambridge, MA, pp 1–14

Hammond DS (2005b) Biophysical features of the Guiana Shield. In: Hammond DS (ed) Tropical forests of the Guiana Shield: ancient forests of the modern world. CABI Publishing, Cambridge, MA, pp 15–194

Harmon D (1996) Losing species, losing languages: connections between biological and linguistic diversity. Southwest J Linguist 15:561–568

Harmon D (2000) Conserving biocultural vitality in protected landscapes. Discussion paper no. 15. Terralingua, Washington, DC

Hladik CM (1993) Tropical forests, people and food: biocultural interactions and applications to development. UNESCO, Parthenon, Paris

Höft M, Barik SK, Lykke AM (1999) Quantitative ethnobotany: applications of multivariate and statistical analyses in ethnobotany. People and Plants Working Paper 6, People and Plants Initiative, WWF, UK

Hoffman B, Gallaher T (2007) Importance indices in ethnobotany. Ethnobot Res Appl 5:201–218

Huber O (1995) Vegetation. In: Berry PE, Holst BK, Yatskievych K (eds) Flora of the Venezuelan Guayana. Timberland Press, Portland, pp 97–160

Hurault J (1965) La vie matérielle des Noirs Refugiés Boni et de Indiens Wayana du Haut-Maroni. ORSTOM (Office de la Recherche Scientifique et Technique d'Outre-Mer), Paris

Johnston M, Colquhon A (1996) Preliminary ethnobotanical survey of Kurupukari: an Amerindian settlement of Central Guyana. Econ Bot 50:182–194

Johnston M, Gillman M (1995) Tree population studies in low-diversity forests, Guyana. I. Floristic composition and stand structure. Biodivers Conserv 4:339–362

Köppen W (1923) Die climate der erde: grundriss der klimakunde. De Greyter, Berlin

Kristensen M, Lykke AM (2003) Informant-based valuation of use and conservation preferences of savannah trees in Burkina Faso. Econ Bot 57:203–217

Kvist LP, Andersen MK, Hesselsoe M, Vanclay JK (1995) Estimating use-values and relative importance of Amazonian flood plain trees and forests to local inhabitants. Commonw For Rev 74:293–300

Lindeman JC, Mennega AMW (1963) Bomenboek voor Suriname. Dienst's – Landsbosbeheer (LBB; Suriname Forest Service), Paramaribo

Lira R, Casas A, Rosas-López R, Paredes-Flores M, Pérez-Negrón E, Rangel-Landa S, Solís L, Torres I, Dávila P (2009) Traditional knowledge and useful plant richness in the Tehuacán-Cuicatlán valley, Mexico. Econ Bot 63:271–287

Lykke AM, Kristensen MK, Ganaba S (2004) Valuation of local use and dynamics of 56 woody species in the Sahel. Biodivers Conserv 13:1961–1990

Maffi L (ed) (2001) On biocultural diversity: linking language, knowledge, and the environment. Smithsonian Institution Press, Washington, DC

Maffi L (2004) Maintaining and restoring biocultural diversity: the evolution of a role for ethnobiology. In: Carlson TJS, Maffi L (eds) Ethnobotany and conservation of biocultural diversity. The New York Botanical Garden Press, Bronx, pp 9–36

Martin GJ (2004) Ethnobotany: a methods manual. Earthscan Publications, London

McClatchey W, Sirikolo MQB, Kaleveke L, Pitanapi C (2006) Differential conservation of two species of *Canarium* (Burseraceae) among the Babatana and Ririo of Lauru (Choiseul), Solomon Islands. Econ Bot 60:212–226

Milliken W, Miller RP, Pollard SR, Wandelli EV (1992a) Ethnobotany of the Waimiri-Atroari Indians of Brazil. Royal Botanic Gardens, Kew

Milliken W, Miller RP, Pollard SR, Wandelli EV (1992b) Waimiri-Atroari plant use. In: Ethnobotany of the Waimiri-Atroari Indians of Brazil. Royal Botanic Gardens, Kew, pp 19–45

Moerman DE (1979) Symbols and selectivity: a statistical analysis of native American medical ethnobotany. J Ethnopharmacol 1:111

Moerman DE (1991) The medicinal flora of native North America: an analysis. J Ethnopharmacol 31:1–42

Moerman DE (2005) "Symbols and selectivity" revisited. J Ethnopharmacol 100:40–42

Mori SA (1991) The Guayana lowland floristic province. Compte Rendu des Séances de la Société de Biogéographie 67:67–75

Nettle D (1998) Explaining global patterns of language diversity. J Anthropol Archaeol 17:354–374

Nurmohamed RJ, Naipal S (2006) Variability of rainfall in Suriname and the relation with ENSO-SST and TA-SST. Adv Geosci 6:77–82

Oviedo G, Maffi L, Larsen PB (2000) Indigenous and traditional peoples of the world and ecoregion conservation: an integrated approach to conserving the world's biological and cultural diversity. WWF International/Terralingua, Gland

Parker E (1989) A neglected human resource in Amazonia: the Amazon caboclo. In: Posey D, Balée W (eds) Resource management in Amazonia: indigenous and folk strategies. The New York Botanical Garden, Bronx, pp 249–259

Password F (2003) An update of the Angiosperm Phylogeny Group classification for the orders and families of flowering plants: APG II. Bot J Linn Soc 141:399–436

Paz y Mino C, Balslev H, Valencia R (1995) Useful lianas of the Siona-Secoya Indians from Amazonian Ecuador. Econ Bot 49:269–275

Phillips O (1996) Some quantitative methods for analyzing ethnobotanical knowledge. In: Alexiades MN (ed) Selected guidelines for ethnobotanical research: a field manual. The New York Botanical Garden Press, Bronx, pp 171–197

Phillips O, Gentry A (1993a) The useful plants of Tambopata, Peru: I. Statistical hypotheses tests with a new quantitative technique. Econ Bot 47:15–32

Phillips O, Gentry A (1993b) The useful plants of Tambopata, Peru: II. Additional hypothesis testing in quantitative ethnobotany. Econ Bot 47:33–43

Phillips O, Gentry A, Reynel C, Wilkin P, Gálvez-Durand C (1994) Quantitative ethnobotany and Amazonian conservation. Conserv Biol 8:225–248

Pinedo-Vasquez M, Zarin D, Jipp P, Chota-Inuma J (1990) Use values of tree species in a communal forest reserve in northeastern Peru. Conserv Biol 4:405–416

Plotkin MJ (1986) Ethnobotany and the conservation of the tropical forest with special reference to the Indians of southern Suriname. Unpublished Ph.D. dissertation, Department of Biology, Tufts University, Medford, MA

Posey DA (2001) Biological and cultural diversity: the inextricable, linked by language and politics. In: Maffi L (ed) On biocultural diversity: linking language, knowledge, and the environment. Smithsonian Institution Press, Washington, DC, pp 379–396

Posey DA (ed) (1999) Cultural and spiritual values of biodiversity. UNEP, London/Nairobi

Prance GT (1972) An ethnobotanical comparison of four tribes of Amazonian Indians. Acta Amazônica 2:7–27

Prance GT, Balee W, Boom BM, Carneiro RL (1987) Quantitative ethnobotany and the case for conservation in Amazonia. Conserv Biol 1:296–310

Price R (1976) The Guiana Maroons. The Johns Hopkins University Press, Baltimore

Price R (2002) Maroons in Suriname and Guyane: how many and where? New West Indian Guide 76:81–88

Reyes-García V, Huanca T, Vadez V, Leonard W, Wilkie D (2006) Cultural, practical, and economic value of wild plants: a quantitative study in the Bolivian Amazon. Econ Bot 60:62–74

Schultes RE, Raffauf RF (1990) The healing forest: medicinal and toxic plants of the Northwest Amazonia. Dioscorides Press, Portland

Schulz JP (1960) The vegetation of Suriname, vol II: Ecological studies on rain forest in northern Suriname. van Eedenfonds, Amsterdam

Sillitoe P (1999) The development of indigenous knowledge: a new applied anthropology. Curr Anthropol 39:223–252

Silva AL, Tamashiro J, Begossi A (2007) Ethnobotany of riverine populations from the Rio Negro, Amazonia (Brazil). J Ethnobiol 27:46–72

Smith N (2002) The history of the Surinamese creoles II – origin and differentiation. In: Carlin EB, Arends J (eds) Atlas of the languages of Suriname. KITLV Press, Leiden, pp 131–151

Soler Alarcón JG, Luna Peixoto A (2008) Use of terra firme forest by Caicubi Caboclos, middle Rio Negro, Amazonas, Brazil. A quantitative study. Econ Bot 62:60–73

Steege HT, Jetten VG, Polak AM, Weger MJA (1993) Tropical rain forest types and soil factors in a watershed area in Guyana. J Veg Sci 4:705–716

Steege HT, Sabatier D, Castellanos H, van Andel T, Duivenvoorden J, Adalardo De Oliveira A, Ek R, Lilwah R, Maas P, Mori S (2001) An analysis of the floristic composition and diversity of Amazonian forests including those of the Guiana Shield. J Trop Ecol 16:801–828

Steege HT, Pitman N, Sabatier D, Castellanos H, van Der Hout P, Daly DC, Duivenvoorden J, De Oliveira AA, Ek R, Lilwah R (2003) A spatial model of tree a-diversity and-density for the Amazon. Biodivers Conserv 12:2255–2277

Stepp JR, Wyndham F, Zarger R (eds) (2002) Ethnobiology and biocultural diversity. In: Proceedings of the 7th international congress of ethnobiology, Athens, Georgia, USA, October 2000

Stepp JR, Cervone S, Castaneda H, Lasseter A, Stocks G, Gichon Y (2004) Development of a GIS for global biocultural diversity. Policy Matters 13:267–270

Sutherland WJ (2003) Parallel extinction risk and global distribution of languages and species. Nature 423:276–279

Teunnisen P, Nordham D, Van Troon F (2003) Ethnoecological survey of the lands inhabited/used by the Trio people of Suriname. Part I. Ecological survey. Foundation Amazon Conservation Team Suriname, Paramaribo

Torre-Cuadros MA, Islebe GA (2003) Traditional ecological knowledge and use of vegetation in southeastern Mexico: a case study from Solferino, Quintana Roo. Biodivers Conserv 12:2455–2476

Torre-Cuadros MA, Ross N (2003) Secondary-biodiversity: local perceptions of forest habitats among the Maya of Solferino, Quintana Roo, México. J Ethnobiol 23:287–308

van Andel T (2000a) Non-timber forest products of the north-west district of Guyana. Part II. A field guide. Ph.D. dissertation, Utrecht University, Utrecht, the Netherlands. Tropenbos-Guyana Programme, Georgetown

van Andel T (2000b) Useful plant species in the seven forest hectare plots. In: Non-timber forest products of the north-west district of Guyana. Part I. Ph.D. dissertation, Utrecht University, Utrecht, the Netherlands. Tropenbos-Guyana Programme, Georgetown, pp 88–140

van Andel T (2001) Floristic composition and diversity of mixed primary and secondary forests in northwest Guyana. Biodivers Conserv 10:1645–1682

van Andel T, Havinga R (2008) Sustainability aspects of commercial medicinal plant harvesting in Suriname. For Ecol Manage 256:1540–1545

van Andel T, Behari-Ramdas J, Havinga R, Groenendijk S (2008a) The medicinal plant trade in Suriname. Ethnobot Res Appl 5:351–372

van Andel T, de Korte S, Koopmans D, Behari-Ramdas J, Ruysschaert S (2008b) Dry sex in Suriname. J Ethnopharmacol 116:84–88

van der Elst DH (1971) The Bush Negro tribes of Surinam, South America. Ph.D. dissertation, Northwestern University, Evanston, IL, p 377

Voeks RA (1995) Candomble ethnobotany: African medicinal plant classification in Brazil. J Ethnobiol 15:257–280

Voeks RA (1996) Tropical forest healers and habitat preference. Econ Bot 50:381–400

Voeks RA (1997) Sacred leaves of Candomble: African magic, medicine, and religion in Brazil. University of Texas Press, Austin

Young CA (2005) A comprehensive and quantitative assessment of Belizean Creole ethnobotany: implications for forest conservation. Department of Ecology and Evolutionary Biology, University of Connecticut, Storrs. Online: http://digitalcommons.uconn.edu/dissertations/AAI3187768

Zent S (2001) Acculturation and ethnobotanical knowledge loss among the Piaroa of Venezuela: demonstration of a quantitative method for the empirical study of traditional ecological knowledge change. In: Maffi L (ed) On biocultural diversity: linking language, knowledge, and the environment. Smithsonian Institution Press, Washington, DC, pp 190–211

Chapter 14
Ethnobotany of Brazil's African Diaspora: The Role of Floristic Homogenization

Robert Voeks

Abstract Nearly five million enslaved Africans were transported to the shores of Brazil over the course of the Atlantic slave trade. During the latter stages, from the 1780s to 1851, the majority hailed from the Bight of Benin, representing especially the Yoruba, Ewe, and Fon peoples. The belief systems introduced by these sub-Saharan peoples were reassembled in Brazil under the generic name of Candomblé. Among the noteworthy features of this religion is a profound spiritual association between a pantheon of deities (the orixás) and a host of edible and medicinal plant species. This chapter demonstrates that Brazil's African diaspora capitalized on a cornucopia of esculent and medicinal plants that had diffused back and forth across the Atlantic Ocean as part of the Columbian Exchange. Centuries before the kidnapping and transport of most African slaves, the anthropogenic habitats of South America and West Africa—the second-growth forests, swiddens, plantations, trails, and kitchen gardens—exhibited significant floristic similarity. This early transatlantic botanical homogenization greatly enhanced the ability of newly arrived Africans and their descendants to reassemble their ethnobotanical traditions in what was otherwise an alien floristic landscape.

Keywords Ethnobotany • Brazil • African diaspora • Slave trade • Candomblé • Spiritual • Traditional knowledge • Invasive species • Cultural diffusion

R. Voeks, Ph.D. (✉)
Department of Geography, California State University—Fullerton,
800 N. State College Blvd, Fullerton, CA 92834, USA
e-mail: rvoeks@fullerton.edu

R. Voeks and J. Rashford (eds.), *African Ethnobotany in the Americas*,
DOI 10.1007/978-1-4614-0836-9_14,

Introduction

From 1519 until 1867, nearly 11 million forced African immigrants were transported to the New World. Brazil alone witnessed the arrival of nearly five million Africans, more than any other colony or country in the Americas (Eltis and Richardson 2008). Because of the tremendous volume of men and women arriving, Brazil's coastal frontier was dominated numerically throughout most of its history by people of African descent. But given the brutality of the Middle Passage and the horrific living conditions endured by enslaved Africans, there was surely no diaspora in Earth's history more constrained in their ability to introduce their culture and lifeways to their new lands. Nevertheless, in spite of these impediments, Africans contributed formidably to what would emerge over time as Brazil's unique cultural identity—from language to foodways, music to dance, and healing traditions to belief systems.

Ethnobotanists investigate the relationship between plants and people. The rich and nuanced botanical traditions and uses for which indigenous rural peoples are so well known are assumed to be the outcome of long-term occupation and gradual experimental familiarity with the floristic environment (Voeks 2011). Received wisdom often depicts indigenous people as environmentally conscious stewards of ancient biological wisdom, transmitted vertically as ossified oral text from generation to generation. In contrast, the plant knowledge and skills sustained by nonnative peoples, whose familiarity with their floristic landscape is brief by comparison, has until recently seldom been considered (but see Alexiades 2009; Pieroni and Vandebroek 2008). Whether stated explicitly or simply implied, the biocultural relations between diaspora communities and the flora of their immigrant landscapes have been perceived as unworthy of academic investigation.

This presumed ignorance of nature is particularly striking in regard to the descendants of enslaved Africans in the Americas. Confronted with the challenges of living and working as chattel, the African diaspora indeed seem poor candidates to have either introduced their original plant-based traditions or learned the indigenous names and useful properties of these alien forests and fields. The belief systems were different, the languages were different, the cropping systems were different, and, most importantly in this context, the native flora was different. Two thousand miles of ocean and a hundred million years of species diversification separated sub-Saharan Africans from their familiar cultural and biological landscape. And unlike European and later Asian immigrants, Africans were profoundly constrained in their ability to import useful plants from their homeland (but see Carney and Rosomoff 2009: 123–125). Given these monumental barriers to biocultural diffusion and assimilation, it is not surprising that ethnobiologists have largely ignored African diaspora communities in favor of indigenous groups.

This chapter explores the apparent anomaly of African diaspora ethnobotanical knowledge and skills in tropical America. By means of a long-term case study among the descendants of enslaved Africans in Northeast Brazil, I suggest that the anthropogenic landscapes of tropical America were floristically similar to their sub-Saharan counterparts by the time the largest contingent of slaves were arriving. Several centuries of plant introductions, intentional and accidental, between tropical Africa and the Americas, created a common domesticated and disturbance flora in these distant regions. This process of botanical homogenization, termed the Columbian Exchange by environmental historian Alfred Crosby (1993), facilitated continuity and transculturation of African plant-based food and magico-medical traditions among their New World descendants and, in so doing, provided avenues of cultural resistance to Euro-American hegemony.

Study Area and Methods

Fieldwork has been carried out since 1990 in and around the cities of Salvador, Ilhéus, and Itabuna, Bahia (Fig. 14.1). Primary methods include open-ended interviews and participant observation, as well as field excursions with practitioners to spiritual gardens, disturbed habitats, and old-growth rainforest. I worked with priests and adherents representing the range of the Candomblé traditions—Ketu, Ijexá, Jeje

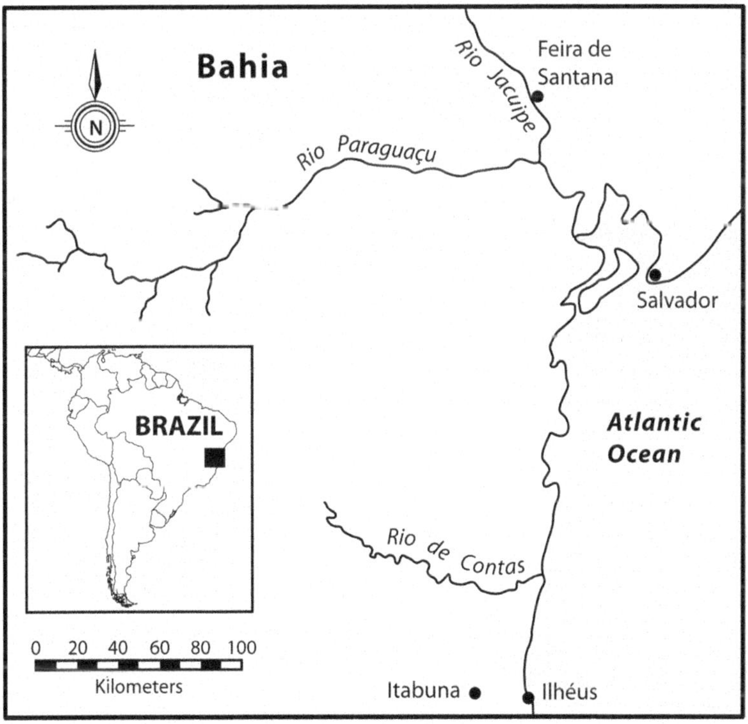

Fig. 14.1 Study areas in Salvador, Itabuna, and Ilhéus, Bahia

(Vodun), and Candomblé de Angola. Species were vouchered and stored at the Herbarium at the Centro de Pesquisas do Cacau, in Itabuna.

Extending from 12° to 18° south of the equator, the region's climate is classified as a Koppen Af, tropical rainforest type, with consistently warm temperatures and evenly distributed moisture. Mean annual rainfall ranges from 1,800 to 2,100 mm, and mean monthly temperatures vary from 21 to 27 °C (Silva 1984). Sandy-soiled beach communities inhabited by cosmopolitan herbaceous species and palms grade quickly into the tall, evergreen Atlantic Coastal rainforest. Dominated floristically by Myrtaceae and Fabaceae, these forests are highly endemic, extremely diverse, and critically threatened (Mori et al 1983; Ribeiro et al. 2009). They are, according to Myers et al. (2000), one of the hottest of the 25 global biodiversity hotspots.

Northeastern Brazil witnessed a series of economic cycles following the arrival of Portuguese merchants and settlers in the early sixteenth century. Extraction of the dye-wood pau brasil (*Caesalpinia echinata* Lam.) and piassava palm (*Attalea funifera* Mart.) was followed quickly by the exponential spread of sugarcane plantations and, later, gold-mining operations in the interior state of Minas Gerais (Voeks 1996). During the subsequent three centuries, over 10% of the total African slave population that would be unloaded in the Americas was transported to the captaincy and later state of Bahia, Brazil. Although these captive laborers arrived from various points in sub-Saharan Africa, the provenance of over 70% during the final decades of the Brazilian slave cycle was the Yoruba, Ewe, and Fon peoples of the Bight of Benin (Ribeiro 2008). Counted among these many thousands of late arrivals were captive priests and religious leaders, important potential sources of traditional West African medical and magical traditional knowledge, sold into servitude from their homelands during the Yoruba wars (Verger 1987: 10–11).

Candomblé Religion

The Candomblé religion constitutes a set of beliefs and practices introduced by Yoruba slaves and freedmen, negotiated and blended over time with Catholicism and with other West African-derived traditions. Engenho Velho, the first Candomblé terreiro (temple), was clearly present by 1830, and some accounts trace its existence to the mid-1700s. Candomblé is separated into several denominations, known as nacões (nations), including Candomblé de Ketu, Candomblé de Angola, Candomblé de Jeje, Candomblé de Congo, and others (Carneiro 1948). Each nation, to some degree, maintains its own unique lexicon, chants, ceremonies, deities, and offerings. Each also sustains, to a greater or lesser extent, its own medicinal, spiritual, and magical crops and wild plant species (Voeks 1997)

Candomblé represents an "exaltation turned toward life and its continuance" (Verger 1966: 35), as opposed to a religion of salvation directed toward the hereafter. Adherents recognize the existence of a supreme god, Olórun, the unknowable creator of all things, but he is perceived as distant and unapproachable by humans. It is the orixás of the Yoruba pantheon, serving as the earthly ambassadors of Olórun, who are directly linked to the everyday world of mortals. There is considerable variation among houses of worship, but roughly a dozen of these Yoruba-inspired gods and goddesses are well developed and find devotees in nearly all Candomblé temples. These deities include Xangô, Ogun, Oxalá, Oxóssi, Omolu, Ossâim, Iroko, Yemanjá, Oxum, Iansã, Nanã, and Oxumarê. Much of Candomblé worship is dedicated to cultivating and sustaining spiritual equilibrium between adherents and their "owner" deities, with the objective of maximizing prosperity, good health, fertility (at least in years past), and general good luck.

Candomblé temples are directed by the mãe- or pai-de-santo (also known by their Yoruba labels babalorixá or ialorixá, hereafter referred to as the mãe-de-santo), which translates literally to mother- or father-of-saints. She (or less commonly he) represents the principal line of communication between the material world of mortals and the realm of the orixás and other spiritual entities. In addition to taking responsibility for temple functions, administrative and spiritual, the mãe-de-santo serves as community healer (curandeiro), divining the source of medical, magical, and spiritual problems and prescribing culturally acceptable remedies. These remedies are often complemented by an array of plant-based recipes, drawn from a pharmacopoeia of sacred foods and medicinal leaves (Voeks 1993).

The mãe-de-santo treats everyday illness episodes, such as infections and body aches, but more often addresses spiritual and magically derived health problems. Spiritual illness is believed to result from disequilibrium between the patient and one or more of the orixás and is treated (in part) with a combination of plant species dedicated to one or more of the deities. Trained in sorcery, the mãe-de-santo can also neutralize the effects of black magic and, if called upon, employ the occult arts for her own ends or for those of her clients (Brazeal 2007). These latter activities, which represent one of the reasons that African-derived religions specifically have yet to achieve the legitimacy of monotheistic religions, Catholicism and Evangelical Protestantism, are usually cloaked in secrecy.

Sacred Leaves

Ossâim is the guardian of the sacred leaves and medicine. He is the orixá most intimately associated with health and healing, and his domain ranges from the forests to the fields, wherever curative plants grow. Originally, the exclusive guardian of the Yoruba ethnoflora, Ossâim's medicinal knowledge, according to legend, was coveted by other deities who sought to share in his secrets. The following oral text, related by pai-de-santo Ruy do Carmo Póvoas in Bahia (Voeks 1997) and also recorded in West Africa by Pierre Verger and in Cuba by Lydia Cabrera (Cabrera 1971: 100; Verger 1981: 122–124), describes how the orixás came to possess individualized plant pharmacopoeias:

> There is a legend of rivalry between Ossâim, the orixá of medicine and leaves, and Iansã, the orixá of stars, winds, and storms. Everything began as a result of jealousy. Iansã went to visit Ossâim. Ossâim is very reserved, quiet, silent. Iansã wanted to know what he

> was doing. When Ossâim has the opportunity, he explains things. But Iansã is always rushed, she wants everything done immediately. She is always asking questions, and she needs to know everything that's going on. When Iansã arrived at the house of Ossâim, he was busy working with his leaves. It happens that there are certain types of work with leaves that you can't talk about, you need to remain silent. Iansã started asking, "What are you doing? Why are you doing this? Why are you doing that?" And Ossâim remained silent. "Alright, if you don't want to tell me what you're doing, then I'll make you talk." That's when Iansã began to shake her skirt and make the wind blow. The house of Ossâim is full of leaves, with all of their healing properties, and when the wind began to blow, it carried the leaves in every direction. Ossâim began to shout "Ewe O, Ewe O." [Yoruba--my leaves, my leaves]. Ossâim then asked the help of the orixás to collect the leaves, and the orixás went about gathering them. And it happens that every leaf that an orixá collected, every species, he or she became the owner of that leaf.

Scattered by the winds of Iansã, the sacred leaves drifted into the habitats of the other deities. Oxum, goddess of freshwater, collected leaves near her rivers; Yemanjá, deity of the oceans, collected her leaves near the shoreline. Oxalá gathered white leaves, while Exu incorporated those that burned and pierced the skin. In this way, Ossâim retained the mysterious power of the plant kingdom, but each divinity came to possess his or her own personal ethnoflora.

The resultant leaf-deity correspondence represents a fundamental element of Candomblé ceremony and ritual. Each plant species, at least in principle, is an element in the personal pharmacopoeia of each individual orixá (Voeks 2000). For human devotees who belong to one or another divinity, healing is mediated through recourse to the spiritual energy (axé) of his or her guardian's leaves. A leaf bath (abô) for one of Oxum's followers, for example, will usually include three or seven of Oxum's species. A leaf whipping (sacudimento) intended to clean the negative fluids from a son or daughter (filho- or filha-do-santo) of Ogun will include some of Ogun's leaves. If a client suffers from an ailment associated with another deity, such as Xangô's notorious anxiety or Oxum's obsession with material wealth, then the mãe-de-santo will incorporate leaves from the appropriate orixá. In this way, the sacred leaves represent a crucial link between the world of mortals (aiê) and the world of the deities (orun).

Sacred Foods

In addition to the pharmacopoeia of healing species associated with the Yoruba deities, each orixá is complemented by plant and animal food preferences and prohibitions suited to his or her archetype. Oxalá, for example, is the masculine god of creation, peace, and love. He is the supreme god of the pantheon, is clothed in white from head to foot, and he assiduously avoids wearing black clothing. Oxalá requires the sacrifice of chickens, doves, and female goats, and his primary food preference is white corn cooked without salt. Iansã, the hot-tempered female orixá associated with wind, storms, and lightning, is especially fond of acarajé, black-eyed pea dumplings fried in dendê (African palm oil) and filled with shrimp. Omolu,

Fig. 14.2 Baiana selling Candomblé foods on the street in Salvador (Photo by Robert Voeks)

divinity of disease and infectious illness, prefers popcorn; Ogun, god of the forge and war, prefers yams cooked in dendê. Consecrated animals serve as sacrificial offerings for the divinities, whereas foods are served at public gatherings, particularly those that are dedicated to one or another of the orixás. Ritual food offerings are also placed in the altars of each orixá, the peji, on the day of the week dedicated to each deity.

Candomblé's sacred cuisine is also sold as street food in Bahian cities and elsewhere. Filhas-do-santo tend to their small food stalls attired in the traditional clothing of the temple, voluminous skirts and petticoats, white turbans, and necklaces dedicated to the orixás (Fig. 14.2). The ubiquitous street-corner presence of Baianas, as they are popularly known, represents the strongest visual signal of the endurance of African-inspired foodways among the Brazilian population. Besides homemade confections and fried fish, Baianas dispense a host of West African-inspired cuisine, including acarajé, vatapá (manioc paste cooked with dendê, shrimp, and hot peppers), and abará (steamed black-eyed pea meal wrapped in banana leaves). Long an important source of income and independence for women of faith, street food serves as a form of ritual obligation to the black gods. In the process of purchasing these exotic delicacies, locals and tourists unwittingly make offerings to one or another of the African deities.

The Candomblé Flora

In the original plant collection, Candomblé mães-de-santo discussed the medicinal, spiritual, and magical properties of roughly 200 wild and cultivated species. I collected and vouchered 140 of these taxa (Voeks 1997). The species list reveals the Candomblé pharmacopoeia to be floristically rich, with 117 genera distributed in 54 families. Ninety-six of the genera, or over 80%, are represented by only a single species. Although it is tempting to view this taxonomic diversity as a reflection of the region's protean species richness, such a conclusion misses the mark entirely. Although the

Atlantic Coastal rainforests are dominated, floristically and physiognomically, by trees and treelets, arborescence does not at all characterize the Candomblé pharmacopoeia. Trees and treelets make up less than 14 and 11%, respectively, of the taxa. Many of these species, 41%, are exotic rather than native, and several of the native trees are domesticated or otherwise managed. Less than 10% of the Candomblé flora was collected in old-growth rainforest, and several of these, including pau pombo (*Tapirira guianensis* Aubl.), aroeira (*Schinus terebinthifolia* Raddi), and canela-de-velho (*Miconia* sp.), are more abundant in secondary than in old-growth forests.

Roughly 43% (60 species) of the plants identified by Candomblé mães-de-santo are cultivated or otherwise spared or encouraged. Many of these, such as urucum (*Bixa orellana* L.), jaca-de-pobre (soursop—*Annona muricata* L.), papaya (*Carica papaya* L.), and sabugueiro (*Sambucus australis* Cham. & Schltdl.), are domesticated species that depend entirely on humans for their existence. Others exist naturally outside of cultivation, usually in secondary habitats, such as mamona (castor bean—*Ricinus communis* L.), transagem (English plantain—*Plantago major* L.), corneta (*Datura metel* L.), and aroeira, but are often purposely planted in home gardens as ornamentals or for medicinal use. Still others occur spontaneously in gardens as weeds but are spared because of their perceived value, such as bom dia (vinca—*Catharanthus roseus* (L.) G. Don), and fedegoso (*Senna occidentalis* L. Link), quebra-pedra (*Phyllanthus amarus* Schumach. & Thonn.).

Many of these plants are not native to Brazil. Of the 122 taxa for which origins are known or suspected, 25% are of Old World origin. Two of these that are widely known and used in West Africa arrived serendipitously via waif dispersal well after the continents of Africa and South America separated. Both the bottle gourd (*Lagenaria siceraria* (Molina) Standl.) and beach morning glory (*Ipomoea pes-caprae* Roth) have notoriously effective flotation devices that facilitated their transatlantic colonization. The bottle gourd, a symbolic representation of the parallel Yoruba worlds of the material and the spiritual, may well have suggested to immigrant Africans the presence of their deities in the New World. Beach morning glory, known in some Candomblé terreiros by the Yoruba term aboro aibá, is likely the first recognizable species that enslaved Africans would have recognized as they made landfall in Brazil (Fig. 14.3). Dedicated to Nana, the elderly female divinity of rain, soil, and mud, beach morning glory is employed in West Africa and in Bahia in spiritual baths (Verger 1995; Voeks 1997).

A few species, including obí (kola—*Cola acuminata* (P. Beauv.) Farw.) and akokô (*Newbouldia laevis (*P. Beauv.) Seem.), were purposely introduced from sub-Saharan Africa specifically for use in Candomblé worship. Others, such as orobô (*Garcinia kola* Heckel) and atarê (malagueta—*Aframomum melegueta* K. Schum.) failed to bear fruit in Brazil and continued, until quite recently, to be imported from Nigeria (Voeks 1993). Still others, such as the African oil palm (*Elaeis guineensis* Jacq.) and jambo branco (*Syzygium jambos* (L.) Alston), a fruit tree introduced from Asia, were brought for commercial purposes by the Portuguese colonists and incorporated by Afro-Brazilian healers. A host of Iberian medicinal species also found their way into the Afro-Brazilian pharmacopoeia, including erva doce

Fig. 14.3 Beach morning glory (*Ipomoea pes-caprae*). Name in West Africa and in some Brazilian Candomblé houses—aboro aibá. (**a**) Painting in Pernambuco, Brazil, late 1600s, by Dutch artist Albert Eckhout (Note morning glory growing behind the figure). (**b**) Photo of beach morning glory near Salvador, Bahia, by Robert Voeks

(anise—*Pimpinella anisum* L.), losna (wormwood—*Artemisia cf. absinthium* L.), and several mints, such as manjericão (*Ocimum americanum* L.), and poejo (pennyroyal—*Mentha pulegium* L.).

Plant life form is the other common feature of the Candomblé pharmacopoeia. The herbaceous habit predominates, with 55% of species, followed by shrubs with 15%. Few if any of these inhabit old-growth Atlantic Coastal rainforest. Rather, most are characterized by weedy life histories—they are small, they reproduce rapidly, and they are poor competitors. These plants owe their abundance and diversity to removal of the native forest rather than to its preservation. They occur most often in secondary forests, cattle pastures and plantations, along roadsides, and in kitchen gardens. And many are noxious weeds of pantropical distribution, including quitoco (*Pluchea sagittalis* (Lam.) Cabrera), malissa (*Mimosa pudica* L.), and guiné (*Petiveria alliacea* L.).

The primary forests of the Atlantic Coastal zone are dominated floristically by native trees, treelets, palms, lianas, and epiphytes, yet these habitats and life forms clearly fail to draw the medicinal and magical plant foraging attention of mães-de-santo. Rather than a natural outgrowth of the region's elevated tropical biodiversity, the Candomblé ethnoflora is highly representative of the profound human landscape modification this region has witnessed during the previous five centuries—it is cultivated, it is disturbed, it is exotic, and it is successional. These species exist within one of the most floristically diverse forests to date identified on Earth, but they are not of it.

These botanical features represent something of a puzzle. On the one hand, the primacy of disturbed habitats as the preferred collection sites for healers has been established in recent years by quantitative studies of the differing ethnobotanical value of old-growth and second-growth tropical forests (cf. Chazdon and Coe 1999; Gavin 2009; Voeks 2004). Familiar, accessible, and often rich in secondary compounds, herbs, cultivars, and nonnative species represent ideal candidates for plant pharmacopoeias (Stepp and Moerman 2001). In this regard, the plants employed in Candomblé magical and medical ceremonies are consistent with other similar ethnofloras. Nevertheless, the near total absence of native, old-growth trees stands in contrast to most indigenous forest pharmacopoeias. Native communities living in the lowland tropics not only possess biologically diverse medicinal and magical pharmacopoeias, but they also employ the full range of life forms, nearly always including the bark, roots, flowers, and fruit of native trees (cf. Bennett et al. 2002). Why is the Candomblé ethnoflora so different?

Black Atlantic Floristic Homogenization

By most standard measures of botanical similarity, Africa and South America have little in common. So taxonomically different are the two floras that Paul Richards termed Africa the botanical "odd man out" (Richards 1973: 21–26). The combined effects of continental drift, 100 million years of geographical isolation, pronounced climatic oscillations, and taxonomic divergence add up to a relatively minor shared floristic ancestry between these distant biomes. Indeed, according to this "nature minus people" view of Brazil's biogeographical legacy, newly arrived African bondsmen would have encountered few opportunities to reconstitute their plant-based food and healing traditions in the New World.

But this myopic view of plant geography disregards the profound botanical enrichment that occurred over the five centuries of the Columbian Exchange. Determined to recreate elements of their lost floral landscape, European traders, colonists, and missionaries went to considerable lengths to acclimate their useful Old World plant species and to introduce recently discovered foods and medicines and spices from Africa and Asia. The scope of this biotic homogenizing project was astonishingly rapid and global, linking the distant and barely known colonized shorelines of Southeast Asia, India, sub-Saharan Africa, the Mediterranean, and tropical America. In the New World, the arrival of European, African, and Asian cultivars doubled or even tripled the number of indigenous cultivated food crops

(Crosby 1993). So effective were these introduction and acclimation efforts that by the end of the sixteenth century, the abundance of introduced plants and animals of European origin in Bahia moved Padre Fernão Cardim to state that "This Brazil is already another Portugal" (Cardim 1939 [1584]).

The quest to learn about new and useful species, and the motivation to effect their near-global dispersion, was, according to Cañizares-Esguerra (2006), a product of scientific curiosity, patriotism, individual enterprise, and nationalistic campaigns. Missionaries were especially effective botanical vectors, their efforts to disseminate recently identified edible and medicinal species matched only by their proselytizing zeal. The Jesuits in particular, for whom studying and exploring nature was "one way of worshipping God," were committed to learning and spreading the world's *materia medicas* (Anagnostou 2007: 294). Stationed in missions throughout the world and in long-term intimate contact with indigenous people, the clergy was probably better placed to apprehend and distribute the world's useful species than any group on Earth (Harris 2005). At the same time, as infectious diseases navigated distant points of the tropical compass, European settlers—especially the Portuguese, French, Spanish, English, and Dutch—went to considerable lengths to globalize newly encountered indigenous pharmacopoeias for humanitarian relief and for profit (Rutten 2000). Concerted efforts to acclimate foods and medicines were initiated early by European medical schools and physic gardens. Beginning in the mid-sixteenth century, seeds and seedlings were dispatched to European centers and to distant tropical colonies, for example, the Cape of Good Hope (1652), Calcutta (1787), Ceylon (1821), Java (1811), and the Jardin Botanico in Rio de Janeiro (1808), guaranteeing that farmers and merchants had access to a cornucopia of novel agronomic opportunities (Gunn 2003). The great European botanical gardens, including Kew in London and the Jardin du Roi in Paris, were pivotal in this botanical homogenization endeavor, particularly when the intent was to breach lucrative botanical monopolies. Employing various means, from patronage to outright theft (Gramiccia 1988; Spary 2000), agents of the garden circuits orchestrated the transfer of clove, cinnamon, vanilla, cinchona, rubber, ginger, and many others. For British botanist Sir Joseph Banks, in a 1787 letter to Sir George Yonge (Chambers 2000: 90), these transfers of botanical resources were not so much "Filching from another country its commercial advantages" but rather fulfilled the lofty goal of "the production of nature usefull for the support of mankind that are present confined to one or the other of them."

Importantly, many of the most common exotic esculents from Asia, Africa, and tropical America had made their way around the globe well before these nationalistic efforts took place (Figs. 14.4 and 14.5). For example, according to Ho (1998), China witnessed the arrival of the American peanut (groundnut) by 1540 and the sweet potato by 1563. New World maize was being cultivated in the Chinese interior in 1555, suggesting its arrival several decades earlier. American guava was growing in India by 1673 and in Kampuchea by 1676. New World pineapple was cultivated in India by 1564, as was avocado by 1750 (Achaya 1998). French botanist Michel Adanson (1759: 165), working in present-day Senegal, noted in 1750 that on dry ground you can see "guavas, acajous [cashew], two sorts of paw paws [papaya],

Fig. 14.4 Early colonial diffusion of African crop plants to Latin America and the Caribbean (Sources: Acosta 1970 [1604]; Cardim 1939 [1584]; Carney and Rosomoff 2009; Crosby 1993; Edwards 1798; Labat 1724; Monardes 1580; Nóbrega 1886 [1554]; Piso 1948 [1648]; Rochefort 1681; Sloane 1707; Sousa 1971 [1587]; Stedman 1988 [1790]; Trapham 1694; Rolander 2008 [1754–1756]. Map drawn by Kelly Donovan)

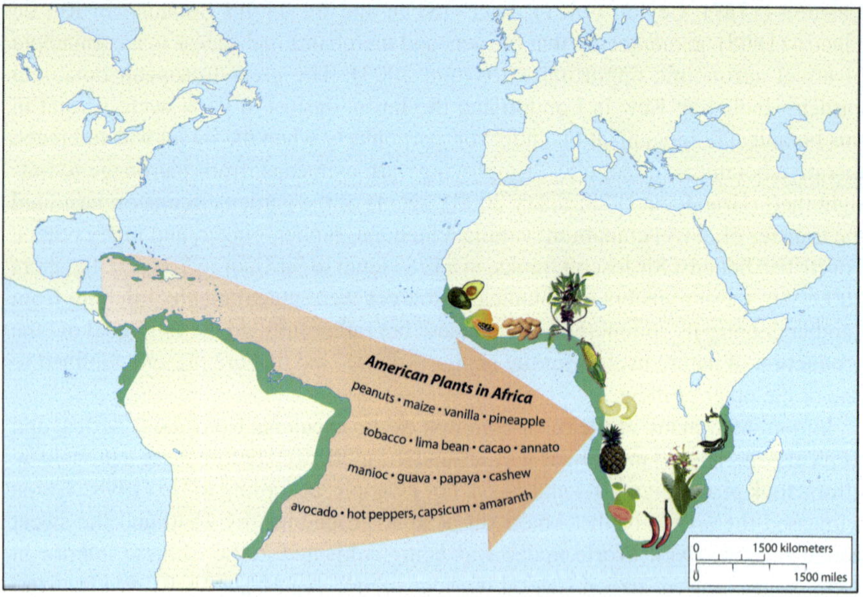

Fig. 14.5 Early diffusion of crop plants to sub-Saharan Africa (Sources: Achaya 1998; Adanson 1759; Alpern 2008; Atkins 1737; Bosman 1721; Dampier 2007 [1697]; Orta 1913 [1563]; Hair et al. 1992; Poivre 1770; Purseglove 1974; Santos 1609; Wadstrom 1795. Map drawn by Kelly Donovan)

with orange and citron trees of exquisite beauty … [and] the roots of manioc [cassava]," all of which are East Asian or American domesticates. In the early seventeenth century, Dominican missionary João dos Santos (1609: 9) observed from present-day Mozambique that "There are lots of pineapples, excellent as the ones in Brazil." Alpern (2008) reports that American prickly pear, papaya, guava, soursop, and cashew were all present in West Africa prior to 1650. Manioc from South America was cultivated in the Congo Basin by 1558, and maize was recorded as early as 1540 growing on the Cape Verde Islands (Camargo 2005; McCann 2005). In the Caribbean, fruits of Asian origin were so thoroughly acclimated by the late sixteenth century that Jesuit missionary José de Acosta (1970: 265) could report "whole woods and forests of orange trees." Likewise, Asian banana and plantain were reported in eighteenth-century Suriname, so common that Dutch officer John Stedman (1988: 296–297) described them as "the Bread of this Country." From present-day São Paulo, Brazil, Jesuit Padre Joseph de Anchieta noted in a 1585 letter that "from Africa there are lots of melons and beans that are better than those from Portugal" (Cartas Jesuíticas 1933). Clearly, by the end of the seventeenth century, wherever one traveled in the equatorial latitudes, he or she was likely encounter a wealth of familiar food plants.

Accompanying these intentional introductions were multiple, but largely unreported, accidental transfers of weedy herbs and shrubs. Arriving as stowaways on the thousands of ships plying the tropical sea lanes, opportunistic species, many invasive, silently colonized the increasingly disturbed floristic landscapes of Asia, America, and Africa. Although explorers and settlers failed to notice weedy taxa, many clearly arrived early in the colonial period, were tested by local communities, and were ultimately incorporated into local foodways and pharmacopoeias (Pfeiffer and Voeks 2008). Among the few species that observers noticed was the castor bean (*R. communis*), known to Europeans as Palma Christi, or the hand of Christ (Fig. 14.6). A readily identifiable shrub both cultivated and spontaneous, this plant has been known and used medicinally since ancient times. It was discovered in Pharaonic pots dating to 2,000 BCE and is recorded in the Bible (Scarpa and Guerci 1982). Although likely native to Egypt, castor bean had arrived and was being employed medicinally in India by at least the early sixteenth century (Pires 1967: 69) and was reported growing in West Africa by 1697 (Petiver 1697) as well as the Antilles and Brazil by 1681 and 1648, respectively (Rochefort 1681; Piso 1948: 384–385). Today, this globally invasive species is counted among the personal pharmacopoeia of orixá Omolu and is known throughout Brazil by its Bantu lexeme, mamona.

The effect of these intentional and accidental botanical transfers was a wholesale reorganization and dramatic enrichment of the humanized landscapes of the tropical latitudes. Primary habitats that were distant from centers of human activities, then as now, sustained high numbers of endemic trees, lianas, epiphytes, and shrubs known in most cases only by local indigenous peoples. But those vegetation associations occupied and created by people—swiddens, plantations, fallows, trails, and dooryard gardens—were occupied by many of the same crop plants and medicinal species, regardless of the continent in which they were domesticated. This increasing

Fig. 14.6 Historical reference to castor Bean (*Ricinus communis*). Known in Brazil as mamona, a Bantu term, this species is associated with the Candomblé god of smallpox and infectious disease—Omulu—perhaps due to the plant's explosive capsules and to the deity's renowned explosive temper (Map drawn by Kelly Donovan)

floristic similarity of the tropical latitudes was noted or at least suspected by colonial physicians and naturalists. From Goa, India, the Portuguese doctor Garcia da Orta concluded that "The medicines were never better known than at present, especially by the Portuguese … as a result of having transplanted them from one land to another" (Orta 1913 [1563]: 121). Sir Hans Sloan, who was both well traveled and possessed botanical collections from all over the world, also supported this view, reporting in 1707 that "I find a great many plants common to Spain, Portugal and Jamaica and the East Indies and most of all Jamaica and Guinea [West Africa]." (cited in Stearn 1988). Likewise, Carl Linnaeus, whose formidable familiarity with the world's tropical floras came from his apostles rather than personal experience, "tended to regard the tropical flora of the world," according to Stearn (1988), "as rather uniform" (Stearn 1988: 781).

Candomblé Spiritual Flora: Continuity and Change

The ebb and flow of crop plants, medicinal plants, and useful weeds, as well as the knowledge of their uses and cultural meanings, between Asia, Africa, the Mediterranean, and the Americas, facilitated cultural continuity and transculturation of foodways and religious ceremonies among Brazil's black diaspora. Some were direct transfers, such as the African oil palm (*E. guineensis*), first introduced by the Portuguese to Brazil for commercial purposes. The oil of this palm (dendê) plays a pivotal role in animal sacrifices and offerings (ebó) to orixá Exú and is used as cooking oil for many foods that are dedicated to one or another of the black deities. Other introduced African cultivars served as provisions on the slave ships and were transplanted, either by Africans or Europeans, for cultivation in slave or freedmen kitchen gardens (Carney and Rosomoff 2009). These included okra (*Abelmoschus esculentus* (L). Moench), an iconic staple of many African-American cuisines. In eighteenth-century Guyana, according to Bancroft (1769: 73), okra continued its dual West African role as food and abortifacient among enslaved Africans. Known in Brazil as quiabo, a Bantu term, okra is a sacred food for orixás Xangô, Iansa, and Ibeji. It is served freely to the public on December 5 as a stew (caruru) containing okra, dendê, and shrimp (Lordelo and Marques 2010). Caruru is also associated with the celebration of Ibeji, the mythical Yoruba twins, and in many cities is served freely as a religious obligation. But while the ingredients of caruru are mostly African in origin, and its preparation is associated with African-derived ceremonies, caruru was in fact originally an indigenous South American stew made with cultivated amaranth (*Amaranthus* sp.), a New World grain. Several of these edible amaranths, as well as the tradition of preparing them as caruru, were introduced to Africa early in the slave trade by the Portuguese. Over time, the amaranth component of this introduced South American cuisine was replaced by okra. Finally, having been adopted into West African culture, this Africanized caruru was reintroduced to Brazil by slaves and their descendants, who continue to prepare what has become a sacred food (Camargo 2005; Voeks 1997).

As in the case of okra, many of the plants and their uses that today represent spiritually relevant species to the Candomblé community arrived by circuitous means, and many were temporally contingent. Thus, whereas the majority of the species employed in Candomblé ceremonies are in fact American in origin, several are considered part of an African botanical legacy, and several maintain West African names in Brazil. This circumstance resulted from the asynchronous transatlantic movement of plants and people. As noted above, Europeans transferred many American species to Africa and Asia within a few short years of colonization, and most of these crops and (speculatively) weeds arrived during the first two centuries of exploration and colonization, that is, the sixteenth and seventeenth centuries. However, the bulk of slave traffic occurred later. Of the total of 4.8 million Africans brought to Brazil between 1550 and 1851, 80% (roughly four million) arrived after 1700 (Eltis and Richardson 2008: 49–50). Consequently, during the many decades after these exotic crops diffused to sub-Saharan Africa, local people must have come to perceive many as natives, which from a local perspective they were, and went on to incorporate them into cultivation regimes and ethnomedical practices. All of this would have occurred well before the peak of slave traffic in the eighteenth century. When Africans were enslaved and transported across the Atlantic generations after American species had been incorporated into their regional ethnofloras, they would have encountered food and medicinal species they considered to be their own. One elderly pai-de-santo (Vicente de Ogun) was clear on this point, noting that one or another spiritual species "é nossa," (is ours), meaning from Africa. Consequently, although many of the species used in Candomblé worship are New World in origin,

Fig. 14.7 Author holding folha-da-fortuna growing along a path in Bahia

it is quite likely that they had been assimilated into their ethnofloras centuries earlier in Africa. New World maize, for example, was assimilated into Yoruba recipes in present-day Nigeria for good luck, wealth, and healthy births (Verger 1995: 41–46). Today, it is the sacred food of several of the black divinities. And pantropical weeds of American origin, such as malissa (*M. pudica*), guiné (*P. alliacea*), and alfavaquinha-de-cobra (*Peperomia pellucida* (L.) Kunth), are now employed for similar medicinal and magical ends in West Africa, Brazil, and Cuba (Barros 1983; Voeks 1997).

The integration of the American peanut (*Arachis hypogaea* L.) into African diaspora ethnobotanical usage is instructive. Native to southern South America, peanuts were transplanted and naturalized by the Portuguese in West Africa by 1660 and possibly earlier (Alpern 2008). They spread rapidly from farmer to farmer, perhaps because of their similarity to the native Bambara nut (*Vigna subterranean* (L.) Verdc), a domesticated but less productive African tuber (Smith 2002: 9–13). The frequency of their appearance in contemporary African pharmacopoeias suggests that peanuts became both a food crop and a significant ethnomedical component (cf. Dalziel 1948; Verger 1995). Later, as a common provision on slave ships, peanuts were transferred over time to African-American gardens along the Atlantic coast. The Virginia variety, which had earlier journeyed to West Africa via Mexico and the Antilles, was thus (re)introduced in the seventeenth century to the West Indies and North America. Thus, following a journey to and from West Africa, peanuts over time came to be identified as a cultural marker and magical species for Africans and their American descendants (Voeks 2009).

The importance of weedy plant transfers is suggested by the presence and use of folha-da-fortuna (*Kalanchoe pinnata* (Lam.) Pers.), also known in some terreiros by its Yoruba lexeme, oju orô (Fig. 14.7). Although most of the species in the genus

Kalanchoe appear to be of African origin, their widespread dispersion during colonial times complicates pinpointing their original distribution. What is apparent is that newly arriving Africans in Brazil recognized this and other exotic Kalanchoes, applied its West African name, and continued traditional (or perhaps developed novel) meanings and uses. This weed and ornamental is today an element in orixá Oxum's pharmacopoeia. This deity's passion for wealth accounts for her association with this "leaf of fortune," which has the habit of sprouting roots and seedlings at its leaf margins (vivipary). This plant's perceived ability to create something from nothing (it is sometimes hung on the door to "attract money") underpins its connection with Oxum (Voeks 1997).

Conclusions

Brazil's African diaspora maintain a biologically diverse spiritual ethnoflora. Some species and uses represent a direct transfer of sub-Saharan traditions to the Americas, others suggest fusion and negotiation with Amerindian and European traditions, and still others likely constitute completely novel uses and meanings of native and exotic plants. Africans arriving as chattel slaves in Brazil encountered primary forests that were strange and mostly unknown to them, but the settled landscape into which they were forced to reside and toil was by the time of their arrival brimming with familiar esculents and medicinal plants. The grand transatlantic biotic exchange initiated in the sixteenth century created coastal poles of useful plant similarity facilitating a relatively seamless ethnobotanical transition for forced African immigrants. Many of these species were incorporated into ceremonies associated with the development of Candomblé, itself a reformulation of various Yoruba-Dahomey traditions and practices. Some species were perceived as especially reflective and symbolic of distant Africa and became over time iconic and even "sacred" in the sense that they retained their West African lexeme, became associated with one or another of the pantheon of black divinities, and were incorporated into ceremonies believed to be practiced by their ancestors (see Rashford 2012, this volume).

The notion that the plant knowledge sustained by African diaspora communities is lacking or somehow inferior is not supported by this project or others. The descendants of enslaved peoples throughout tropical America employ plants for food, construction, craft, medicine, magic, and the other use categories, just as do indigenous groups. But whether or not qualitative or quantitative differences exist between native and immigrant groups remains to be seen. The inchoate nature of immigrant ethnobotanical traditions, at least in the case of African descendants, is suggested by the numerical dominance of successional, domesticated, exotic, and herbaceous species. For example, in and around Recife, Brazil, Albuquerque (2001) reported that most of the 60 species used by Afro-Brazilian religious groups are either New World or European in origin, and nearly all are herbaceous. In the Afro-Brazilian community of Sao Benedito, Schardong and Cervi (2000) found 182 ethnospecies (122 identified to species rank) employed for medicinal or cosmetic purposes, and an additional 17 species used in spiritual healing ceremonies. Again, nearly all were herbs, the majority cultivated in kitchen gardens or maintained in otherwise weedy habitats. In another study carried out in Candomblé temples, the three most speciose families were Asteraceae, Lamiaceae, and Fabaceae, all herbaceous (Pires et al. 2009), and in a study of Afro-Cuban medicinal species, the majority were determined to be herbaceous, exotic, and weedy (Moret E, 2008, personal communication).

Does the dominance of herbaceous, cultivated, and weedy species in immigrant ethnofloras suggest a stage in the ethnobotanical acquisition and/or retention process? Recent research with Maroon communities in Suriname appears to support this view. Van Andel et al. (2012, in this volume) discovered among Afro-Surinamese practitioners of Wintu religion a high proportion of trees and primary forest species (40%) in their magical flora, a dramatic departure from previous findings. What is different about this group is their extreme isolation in the interior old-growth forests, their remoteness from coastal anthropogenic habitats, and their lengthy residence in the region, upward of three centuries. As van Andel et al. point out, "The assumption that the paucity of trees in black diaspora healing floras is a product of lack of time and experience in the South American forest (Voeks 2009) may hold for the Afro-Brazilian population, but certainly not for the Suriname Maroons." Further light on this question is shed by Hoffman (2012, this volume) who carried out the first cross-cultural comparison of ethnobotanical knowledge among indigenous (Trio Amerindians) and African diaspora (Saramaccan Maroons) groups. Although he reports that after 350 years the Saramacca have developed a "robust hybrid ethnobotany," Hoffman also discerned significant differences in plant knowledge between the groups. The indigenous Trio "know more" species overall, but the Saramacca have a particularly elevated knowledge of secondary as opposed to primary forest. If ethnobotanical traditions are indeed space and time contingent (Alexiades and Peluso 2009), the results of this and other work among America's African diaspora may provide a guidepost as to how the process progresses—from ubiquitous crops and weeds, to disturbed forests and fields, and ultimately to old-growth habitats. In order to understand the processes of continuity and change that shape foodways and pharmacopoeias, ethnobotanists are encouraged to pay greater attention to human diasporas.

Acknowledgments This project would not have been possible were it not for the kindness and generosity of the Candomblé community in Itabuna, Ilhéus, and Salvador. I thank Charlotte Greene for tracking down and translating primary material in the Muséum National d'Historie Naturelle, Paris, and Kelly Donovan for producing the maps and figures. I also thank Stanley Alpern and Tinde van Andel for their insightful reviews of an earlier version of this manuscript.

References

Achaya KT (1998) Bounty from the New World. In: MacLeod MJ, Rawski ES (eds) European intruders and changes in behaviour and customs in Africa, America and Asia before 1800, vol 30, An expanding world. Ashgate, Brookfield, pp 259–282

Adanson M (1759) A voyage to Senegal, the Isle of Goree, and the River Gambia (trans: From French). J. Nourse, London

Albuquerque U (2001) The use of medicinal plants by the cultural descendents of African people in Brazil. Acta Farm Bonaer 20:139–144
Alexiades M (ed) (2009) Mobility and migration in indigenous Amazonia: contemporary ethnoecological perspectives. Berghahn Books, London
Alexiades M, Peluso D (2009) Plants of the ancestors, plants of the outsiders: history, migration and medicinal plants among the Ese Eja (Peru, Bolivia). In: Alexiades M (ed) Mobility and migration in indigenous Amazonia: contemporary ethnoecological perspectives. Berghahn Books, London, pp 220–248
Alpern S (2008) Exotic plants of western Africa: where they came from and when. Hist Afr 35:63–102
Anagnostou S (2007) The international transfer of medicinal drugs by the Society of Jesus (sixteenth to eighteenth centuries) and connections with the work of Carolus Clusius. In: Egmond F, Hoftijzer P, Visser R (eds) Carolus Clusius. Royal Netherlands Academy of Arts and Sciences, Amsterdam, pp 293–313
Atkins J (1737) A voyage to Guinea, Brasil and the West Indies; in his majesty's ships the Swallow and Weymouth, 2nd edn. Ward and Chandler, London
Bancroft E (1769) Essay on the natural history of Guiana, in South America. T. Becket and P. R. DeHondt, London
Bennett B, Baker M, Andrade P (2002) Ethnobotany of the Shuar of eastern Ecuador, vol 14, Advances in economic botany. New York Botanical Garden Press, Bronx
Bosman W (1721) A new and accurate description of the coast of Guinea, 2nd edn. Knapton, London
Brazeal B (2007) Blood, money and fame: Nago magic in the Bahian backlands. Unpublished PhD dissertation, University of Chicago
Cabrera L (1971) El Monte. Np, Miami
Camargo M (2005) Estudo etnobotánico da mandioca (*Maniot esculenta* Crantz Euphorbiaceae) na Diaspora Africana. In: Anais do Seminário Gastronomia de Gilberto Freyre. Fundacão Gilberto Freyre, Recife, pp 22–30
Cañizares-Esguerra J (2006) Nature, empire, and nation: explorations of the history of science in the Iberian world. Stanford University Press, Stanford
Cardim F (1939 [1584]) Tratados da terra e gente do Brasil. Np, São Paulo
Carneiro E (1948) Candomblés da Bahia. Editora Tecnoprint. Np
Carney J, Rosomoff R (2009) In the shadow of slavery: Africa's botanical legacy in the Atlantic world. University of California Press, Berkeley
Cartas Jesuíticas (1933) Cartas, informacões, fragmentos históricas e sermões do Padre Joseph de Anchieta, S. J. (1554–1594). Civilização Brasileira, Rio de Janeiro
Chambers N (ed) (2000) The letters of Sir Joseph Banks: a selection, 1768–1820. Imperial College Press, London
Chazdon RL, Coe FG (1999) Ethnobotany of woody species in second-growth, old-growth, and selectively logged forests of northeastern Costa Rica. Conserv Biol 13:1312–1322
Crosby AW (1993) Ecological imperialism: the biological expansion of Europe, 900–1900. Cambridge University Press, Cambridge
Dalziel JM (1948) Useful plants of west tropical Africa. Crown Agents for the Colonies, London
Dampier W (2007 [1597]) A new voyage round the world. Warwick, New York, 1,500 Books
de Acosta J (1970 [1604]) The natural & moral history of the Indies (trans: Grimston E), vol 61 (2). The Hakluyt Society/Burt Franklin, New York
de Barros JFP (1983) Ewé o Osányin: Sistema de classificação de vegetais nas casas de santo Jêje-Nagô de Salvador, Bahia. Unpublished PhD dissertation, Universidade de São Paulo, Brazil
de Orta G (1913 [1563]) Colloquies on the simples and drugs of India (trans: Sir Markham C). Henry Sotheran and Co., London
de Rochefort C (1681) Histoire naturelle et morale des iles Antilles de l'Amérique. Reinier Leers, Rotterdam
de Sousa GS ([1587] 1971) Tratado descritivo do Brasil em 1587, 4th edn. Editora Nacional, São Paulo

dos Santos J (1609) Ethiopia oriental, e varia historia de coisas notauois de Oriente. Impressa no Conuento de S. Domongos, np
Edwards B (1798) The history, civil and commercial, of the British colonies in the West Indies, vol 1. Mundell & Son, London
Eltis D, Richardson D (2008) A new assessment of the transatlantic slave trade. In: Eltis D, Richardson D (eds) Extending the frontiers: essays on the new transatlantic slave trade database. Yale University Press, New Haven, pp 1–60
Gavin M (2009) Conservation implications of rainforest use patterns: mature forests provide more resources but secondary forests supply more medicine. J Appl Ecol 46:1275–1282
Gramiccia G (1988) The life of Charles Ledger (1818–1905): alpacas and quinine. Macmillan, London
Gunn G (2003) First globalization: the Eurasian exchange, 1500–1800. Rowman and Littlefield, Lanham
Hair P, Jones A, Law R (eds) (1992) Barbot on Guinea: the writings of Jean Barbot on West Africa, 1678–1712, vol 1. The Hakluyt Society, London
Harris SJ (2005) Jesuit scientific activity in the overseas missions, 1540–1733. Isis 96:71–79
Ho P (1998) The introduction of American food plants into China. In: MacLeod MJ, Rawski ES (eds) European intruders and changes in behaviour and customs in Africa, America and Asia before 1800, vol 30, An expanding world. Ashgate, Brookfield, pp 283–293
Hoffman B (2012) Exploring biocultural contexts: comparative woody plant knowledge of an indigenous and Afro-American Maroon community in Suriname, South America. In: Voeks R, Rashford J (eds) African ethnobotany in the Americas. Springer, Dordrecht
Labat JB (1724) Nouveau voyage aux isles de l'Amérique: contenant l'histoire naturelle de ces pays, l'origine, les moeurs, la religion & le gouvernement des habitants anciens & modernes. Alberts R, van Duren J, Gosse P, Husson P, Johnson T, & Le Vier C, Paris
Lordelo M, Marques J (2010) E o caruru de Iansã se acabou? Resiliência cultural de uma festa de comida em Salvador e Feira de Santana (BA). Paper presented at II Congresso Latinamericano de Etnobiologia, 10 November, Recife, Brazil
McCann J (2005) Maize and grace: Africa's encounter with a New World crop, 1500–2000. Harvard University Press, Cambridge
Monardes N (1580) Joyful newes out of the new found world, wherein are declared the rare and singular vertues of divers and sundrie herbs, trees, dyes, plants, and stones (trans: Frampton J). William Norton, London
Mori S, Boom BM, Carvalho AM, Santos TS (1983) Southern Bahian moist forests. Bot Rev 49:155–232
Myers N, Mittermeier RA, Mittermeier CG, Gustavo da Fonseca AB, Kent J (2000) Biodiversity hotspots for conservation priorities. Nature 403:853–858
Nóbrega P (1886) Cartas do Brasil do Padre Manuel da Nóbrega (1549–1560). In: Materiais e achêgas para a história e geografia do Brasil. No.1. Imprensa Nacional, Rio de Janeiro
Petiver J (1697) Letter sent to Dr. Hans Sloane, a catalogue of some Guinea plants, with their native names and virtues. Philos Trans 19:677–686
Pfeiffer J, Voeks R (2008) Biological invasions and biocultural diversity: linking ecological and cultural systems. Environ Conserv 35(4):281–293
Pieroni A, Vandebroek I (eds) (2008) Traveling cultures and plants: the ethnobiology and ethnopharmacy of human migrations. Berghahn Books, New York
Pires T (1967 [1512–1515]) Suma oriental. In: The Suma oriental of Tomé Pires (trans: Cortesao A). Hakluyt Society, Nendeln, pp 1–289
Pires M, Abreu P, Soares C, Souza B, Mariano D, Costa Silva D, Rocha A (2009) Etnobotânica de terreiros de Candomblé nos municípios de Ilhéus e Itabuna, Bahia, Brasil. Rev Bras Biocien 7:3–8
Piso G (1948 [1648]) História natural do Brasil ilustrada (trans: Taunay A). Companhia Editora Nacional, São Paulo
Poivre P (1770) Travels of a philosopher: or observations on the manners and arts of various nations in Africa and Asia (trans: from French). R. Urie, Glasgow

Purseglove J (1974) Tropical crops: dicotyledons. Longman, London

Rashford J (2012) Candomblé's cosmic tree and Brazil's *Ficus* species. In: Voeks R, Rashford J (eds) African ethnobotany in the Americas. Springer, Dordrecht

Ribeiro A (2008) The transatlantic slave trade to Bahia. In: Eltis D, Richardson D (eds) Extending the frontiers: essays on the new transatlantic slave trade database. Yale University Press, New Haven, pp 130–154

Ribeiro M, Metzger J, Martensen A, Ponzoni F, Hirota M (2009) The Brazilian Atlantic forest: how much is left, and how is the remaining forest distributed? Implications for conservation. Biol Conserv 142:1141–1153

Richards PW (1973) Africa, the "odd man out". In: Meggers BJ, Ayensu ES, Duckworth WD (eds) Tropical forest ecosystems in Africa and South America: a review. Smithsonian Institution, Washington, DC, pp 21–26

Rolander D (2008) Daniel Rolander's Journal (1754–1756). The Suriname Journal: composed during an exotic journey (trans: from Latin by Dobreff J, Dahlman C, Morgan D, Tipton J). In: Hansen L (ed) The Linnaeus Apostles: global science & adventure, vol 3(3). IK Foundation & Company, London, pp 1215–1576

Rutten AMG (2000) Dutch transatlantic medicine trade in the eighteenth century under the cover of the West India Company, 2nd edn. Erasmus, Rotterdam

Scarpa A, Guerci A (1982) Various uses of the castor oil plant (*Ricinus communis* L.): a review. J Ethnopharmacol 5:117–137

Schardong R, Cervi A (2000) Estudos etnobotânicos das plantas de uso medicinal e místico na comunidade de São Benedito, Bairro São Francisco, Campo Grande, MS, Brasil. Acta Bioló Paranaense (Curitiba) 29:187–217

Silva B (1984) Comportamento das chuvas no estudo da Bahia: Uma contribuição cartográfica. Departamento de Geografia, Universidade Federal da Bahia, Salvador

Sloane H (1707, 1725) A voyage to the islands Madera, Barbados, Nieves, S. Christophers and Jamaica with the natural history of the herbs and trees, four-footed beasts, fishes, birds, insects, reptiles, &c. Printed by the author, London

Smith AF (2002) Peanuts: the illustrious history of the goober pea. University of Illinois Press, Urbana

Spary EC (2000) Utopia's garden: French natural history from old regime to revolution. University of Chicago Press, Chicago

Stearn WT (1988) Carl Linnaeus's acquaintance with tropical plants. Taxon 37:776–781

Stedman J (1988 [1790]) In: Price R, Price S (eds) Narrative of a five years expedition against the revolted Negroes of Surinam. The Johns Hopkins University Press, Baltimore

Stepp JR, Moerman DE (2001) The importance of weeds in ethnopharmacology. J Ethnopharmacol 75:19–23

Trapham T (1694) Some observations made upon the Bermudas Berries imported from the Indies shewing their admirable virtues in curing the green-sickness. Np, London

Van Andel TR, Ruysschaert S, Putte K, Groenendijk S (2012) What makes a plant magic? What makes a plant magical? Symbolism and sacred herbs in Afro-Surinamese *Winti* rituals. In: Voeks R, Rashford J (eds) African ethnobotany in the Americas. Springer, Dordrecht

Verger P (1966) The Yoruba high god. Univ Ife J Afr Stud 2:19–40

Verger P (1981) Orixás. Corrupio, São Paulo

Verger P (1987) Fluxo e refluxo: Do tráfico de escravos entre o Golfo do Benin e a Bahia de Todos os Santos (trans: Gadzonis T). Corrupio, São Paulo

Verger P (1995) Ewé: O uso das plantas na sociedade Iorubá. Editora Schwarcz, São Paulo

Voeks RA (1993) African medicine and magic in the Americas. Geogr Rev 83:66–78

Voeks RA (1996) Extraction and tropical rain forest conservation in eastern Brazil. In: Edwards D, Booth W, Choy S (eds) Tropical rainforest research – current issues. Kluwer Academic Publishers, Dordrecht, pp 477–487

Voeks RA (1997) Sacred leaves of Candomblé: African magic, medicine, and religion in Brazil. University of Texas Press, Austin

Voeks RA (2000) Candomblé ethnobotany: African medicinal plant classification in Brazil. In: Minnis P (ed) Ethnobotany: a reader. University of Oklahoma Press, Norman, pp 148–171
Voeks RA (2004) Disturbance pharmacopoeias: medicine and myth from the humid tropics. Ann Assoc Am Geogr 94(4):868–888
Voeks RA (2009) Traditions in transition: African diaspora ethnobotany in lowland South America. In: Alexiades M (ed) Mobility and migration in indigenous Amazonia: contemporary ethnoecological perspectives. Berghahn Books, London, pp 275–294
Voeks RA (2011) Ethnobotany. Chapter 26. In: Millington A, Blumler M, MacDonald G, Schickhoff U (eds) The Sage handbook of biogeography. Sage Publications, London, pp 493–509
Wadstrom CB (1795) An essay on colonization, particularly applied to the west coast of Africa, with some free thoughts on cultivation and commerce. Darton & Harvey, London

Index

R. Voeks and J. Rashford (eds.), *African Ethnobotany in the Americas*,
DOI 10.1007/978-1-4614-0836-9, © Springer Science+Business Media New York 2013

H

I

N

The manufacturer's authorised representative in the EU is Springer Nature Customer Service Centre GmbH, Europaplatz 3, 69115 Heidelberg, Germany. If you have any concerns regarding our products, please contact ProductSafety@springernature.com

Printed and bound by CPI Group (UK) Ltd, Croydon, CR0 4YY
13/10/2026
02232340-0010